STUDY GUIDE AND WORKBOOK:
AN INTERACTIVE APPROACH

for Starr and Taggart's

BIOLOGY

The Unity and Diversity of Life
NINTH EDITION

JANE B. TAYLOR

Northern Virginia Community College

JOHN D. JACKSON

North Hennepin Community College

BROOKS/COLE

TM

THOMSON LEARNING

Australia • Canada • Mexico • Singapore • Spain • United Kingdom • United States

BROOKS/COLE

THOMSON LEARNING

Biology Publisher: Jack Carey
Project Development Editor: Kristin Milotich
Assistant Editor: Daniel Lombardino
Editorial Assistant: Karen Hansten
Marketing Team: Rachel Alvelais, Mandie Houghan, Laura Hubrich, Carla Martin-Falcone
Print Buyer: Micky Lawler
Production Coordinator: Stephanie Andersen
Manuscript Editor: Linda Purrington
Art Coordinator: Roberta Broyer
Permissions Editor: Mary Kay Hancharick
Cover Photo: *Snow Leopard, a currently endangered species,* © Art Wolfe, PhotoResearchers
Cover Design: Gary Head, Gary Head Design
Typesetting: Publishers' Design & Production Services, Inc.
Printing and Binding: Webcom Ltd.

For more information about this or any other Brooks/Cole products: contact:
BROOKS/COLE
511 Forest Lodge Road
Pacific Grove, CA 93950 USA
www.brookscole.com
1-800-423-0563 (Thomson Learning Academic Resource Center)

For permission to use material from this work, contact us by
www.thomsonrights.com
fax: 1-800-730-2215
phone: 1-800-730-2214

Printed in Canada

10   9   8   7   6   5   4   3   2   1

ISBN 0-534-38036-0

CONTENTS

Photo Credits:

Chapter 7
p. 76 (36, 37): David Fisher

Chapter 22
p. 236 (10): George Musil/Visuals Unlimited
p. 236 (12): Tony Brain/SPL/Photo Researchers
p. 236 (14): T.J. Beverige, University of Guelph
p. 236 (15): Tony Brain and David Parker/SPL/Photo Researchers

Chapter 23
p. 224 (20): M. Abbey/Visuals Unlimited
p. 224 (21): T.E. Adams/Visuals Unlimited
p. 224 (22): John Clegg/Ardea, London
p. 248: D.J. Patterson/Seaphot Limited: Planet Earth Pictures
p. 250 (20): Jan Hinsch/SPL/Photo Researchers
p. 250 (22): John Clegg/Ardea, London
p. 251 (23): M. Abbey, Visuals Unlimited

Chapter 24
p. 255 (3): Garry T. Cole, University of Texas, Austin/BPS
p. 256 (3): Ed Reschke
p. 258 (1): Robert C. Simpson/Nature Stock
p. 258 (2): John Hodgin
p. 259 (3): Jane Burton/Bruce Coleman, Ltd.
p. 259 (4): M. Eichelberger/Visuals Unlimited
p. 259 (6): Edward S. Ross
p. 259 (7): Robert C. Simpson/Nature Stock
p. 259 (8): G.L. Barron, University of Guelph

Chapter 25
p. 266: Jane Burton/Bruce Coleman Ltd.
p. 269: A.&E. Boomford, Ardea, London; Lee Casebere
p. 272: Edward S. Ross; Robert & Linda Mitchell Photography; R.J. Erwin/Photo Researchers

Chapter 26
p. 290 (36): Carolina Biological Supply Company
p. 302 (24): Herve Chaumeton/Agence Nature
p. 303 (a): Ian Took/Biofotos
p. 303 (b): John Mason/Ardea, London
p. 303 (c): Chris Huss/The Wildlife Collection
p. 303 (d): Kjell B. Sandved

Chapter 27
p. 313 (a): Bill Wood/Bruce Coleman Ltd.
p. 326 (30): Christopher Crowley
p. 326 (31): Bill Wood/Bruce Coleman Ltd.
p. 326 (33): Reinhard/ZEFA
p. 326 (34): Peter Scoones/Seaphot Ltd.: Planet Earth Pictures
p. 327 (35): Ervin Christian/ZEFA
p. 327 (36): © Norbert Wu/Peter Arnold, Inc.
p. 327 (37) Allan Power/Bruce Coleman, Ltd.
p. 327 (38) Rick M. Harbo
p. 327 (39) Herve Chaumeton/Agence Nature

Chapter 29
p. 341 (1): D.E. Akin and I.L. Rigsby, Richard B. Russel Agricultural Research Center, Agricultural Research Service, U.S. Dept. Agriculture Athens, GA
p. 341 (2): Biophoto Associates
p. 341 (3): Biophoto Associates
p. 341 (4): Kingsley R. Stern
p. 341 (5): Biophoto Associates
p. 341 (6): Jan Robert Factor/Photo Researchers
p. 344 (1-3): Robert & Linda Mitchell Photography
p. 345 (16-18): Carolina Biological Supply Company
p. 345 (20-23): Carolina Biological Supply Company
p. 348 (11): Jeremy Burgess/SPL/Photo Researchers
p. 348 (12): Jeremy Burgess/SPL/Photo Researchers
p. 350 (8-10)(a Inset): Chuck Brown
p. 353 (16-18): H.A. Core, W.A. Cote, and A.C. Day, *Wood Structure and Identification*, 2nd Ed., Syracuse University Press, 1979

Chapter 31
p. 368 (1-5): ©Gary Head
p. 374 (a-b): Patricia Schulz
p. 374 (c-d): Ray F. Evert
p. 374 (e-f): John Limbaugh/Ripon Microslides, Inc.

Chapter 32
p. 382: Herve Chaumeton/Agence Nature
p. 382: Larry L. Runk/Grant Heilman, Inc.; James D. Mauseth

Chapter 35
p. 419 (16-17): Manfred Kage/ Peter Arnold, Inc.
p. 423 (1-10): C. Yokochi and J. Rohen, *Photographic Anatomy of the Human Body*, 2nd Ed., Igaku-Shoin, Ltd., 1979

Chapter 36
p. 434 (24): Ed Reschke

Chapter 38
p. 461 (28,31): Ed Reschke
p. 468: D. Fawcett, *The Cell*, Philadelphia: W.B. Saunders Co., 1966.

Chapter 45
p. 548 (2): Ed Reschke
p. 555 (17-18, 20): Photographs Lennart Nilsson, *A Child Is Born*, ©1966, 1977 Dell Publishing Company, Inc.

PREFACE

Tell me and I will forget, show me and I might remember, involve me and I will understand.
—Chinese Proverb

The proverb outlines three levels of learning, each successively more effective than the method preceding it. The writer of the proverb understood that humans learn most efficiently when they *involve* themselves in the material to be learned. This study guide is like a tutor; when properly used it increases the efficiency of your study periods. The interactive exercises actively involve you in the most important terms and central ideas of your text. Specific tasks ask you to recall key concepts and terms and apply them to life; they test your understanding of the facts and indicate items to reexamine or clarify. Your performance on these tasks provides an estimate of your next test score based on specific material. Most important, though, this biology study guide and text together help you make informed decisions about matters that affect your own well-being and that of your environment. In the years to come, human survival on planet Earth will demand administrative and managerial decisions based on an informed biological background.

HOW TO USE THIS STUDY GUIDE

Following this preface, you will find an outline that will show you how the study guide is organized and will help you use it efficiently. Each chapter begins with a title and an outline list of the 1- and 2-level headings in that chapter. The Interactive Exercises follow, wherein each chapter is divided into sections of one or more of the main (1-level) headings that are labeled 1.1, 1.2, and so on. *For easy reference to an answer or definition, each question and term in this unique study guide is accompanied by the appropriate text page(s), and appears in the form:* [p.352]. The Interactive Exercises begin with a list of Selected Words (other than boldfaced terms) chosen by the authors as those that are most likely to enhance understanding. In the text chapters, the selected words appear in italics, quotation marks, or roman type. This is followed by a list of Boldfaced, Page-Referenced Terms that appear in the text. These terms are essential to understanding each study guide section of a particular chapter. Space is provided by each term for you to formulate a definition in your own words. Next is a series of different types of exercises that may include completion, short answer, true/false, fill-in-the-blanks, matching, choice, dichotomous choice, label and match, problems, labeling, sequencing, multiple choice, and completion of tables.

A Self-Quiz immediately follows the Interactive Exercises. This quiz is composed primarily of multiple-choice questions although sometimes we present another examination device or some combination of devices. Any wrong answers in the Self-Quiz indicate portions of the text you need to reexamine. A series of Chapter Objectives/Review Questions follows each Self-Quiz. These are tasks that you should be able to accomplish if you have understood the assigned reading in the text. Some objectives require you to compose a short answer or long essay while others may require a sketch or supplying correct words.

The final part of each chapter is named Integrating and Applying Key Concepts. It invites you to try your hand at applying major concepts to situations in which there is not necessarily a single pat answer and so none is provided in the chapter answer section. Your text generally will provide enough clues to get you started on an answer, but this part is intended to stimulate your thought and provoke group discussions.

A person's mind, once stretched by a new idea, can never return to its original dimension.
—Oliver Wendell Holmes

STRUCTURE OF THIS STUDY GUIDE

The outline below shows how each chapter in this study guide is organized.

Chapter Number ⟶ **4**

Chapter Title ⟶ **CELL STRUCTURE AND FUNCTION**
Animalcules and Cells Fill'd With Juices

Chapter Outline ⟶

BASIC ASPECTS OF CELL STRUCTURE AND FUNCTION
Structural Organization of Cells
Lipid Bilayer of Cell Membranes
Cell Size and Cell Shape
Focus on Science: **MICROSCOPES: GATEWAYS TO CELLS**

THE DEFINING FEATURES OF EUKARYOTIC CELLS
Major Cellular Components
Which Organelles Are Typical of Plants?
Which Organelles Are Typical of Animals?

THE NUCLEUS
Nuclear Envelope
Nucleolus
Chromosomes
What Happens to the Proteins Specified by DNA?

THE CYTOMEMBRANE SYSTEM
Endoplasmic Reticulum
Golgi Bodies
A Variety of Vesicles

MITOCHONDRIA

SPECIALIZED PLANT ORGANELLES
Chloroplasts and Other Plastids
Central Vacuole

COMPONENTS OF THE CYTOSKELETON
Microtubules
Microfilaments
Myosin and Other Accessory Proteins
Intermediate Filaments

THE STRUCTURAL BASIS OF CELL MOTILITY
Mechanisms of Cell Movement
Flagella and Cilia

CELL SURFACE SPECIALIZATIONS
Eukaryotic Cell Walls
Matrixes Between Animal Cells
Cell-to-Cell Junctions

PROKARYOTIC CELLS—THE BACTERIA

Interactive Exercises ⟶ The interactive exercises are divided into numbered sections by titles of main headings and page references. Each section begins with a list of author-selected words that appear in the text chapter in italics, quotation marks, or roman type. This is followed by a list of important boldfaced, page-referenced terms from each section of the chapter. Each section ends with interactive exercises that vary in type and require constant interaction with the important chapter information.

Self-Quiz ⟶ Usually a set of multiple-choice questions that sample important blocks of text information.

Chapter Objectives/ ⟶ Combinations of relative objectives to be met and questions to be answered.
Review Questions

Integrating and ⟶ Applications of text material to questions for which there may be more than
Applying Key Concepts one correct answer.

Answers to Interactive ⟶ Answers for all interactive exercises can be found at the end of this study
Exercises and Self-Quiz guide by chapter and title, and the main headings with their page references, followed by answers for the Self-Quiz.

1

CONCEPTS AND METHODS IN BIOLOGY

Biology Revisited

DNA, ENERGY, AND LIFE
    Nothing Lives Without DNA
    Nothing Lives Without Energy

ENERGY AND LIFE'S ORGANIZATION
    Levels of Biological Organization
    Interdependencies Among Organisms

IF SO MUCH UNITY, WHY SO MANY SPECIES?

AN EVOLUTIONARY VIEW OF DIVERSITY

    Mutation—Original Source of Variation
    Evolution Defined
    Natural Selection Defined

THE NATURE OF BIOLOGICAL INQUIRY
    Observations, Hypotheses, and Tests
    About the Word "Theory"

Focus on Science: **THE POWER OF EXPERIMENTAL TESTS**

THE LIMITS OF SCIENCE

Interactive Exercises

Note: In the answer sections of this book, a specific molecule is most often indicated by its abbreviation. For example, adenosine triphosphate is ATP.

Biology Revisited [pp.2–3]

1.1. DNA, ENERGY, AND LIFE [pp.4–5]

Selected Words: *DNA to RNA to protein* [p.4], *transfer* of energy [p.5], *ATP molecules* [p.5], *internal environment* [p.5]

In addition to the boldfaced terms, the text features other important terms essential to understanding the assigned material. "Selected Words" is a list of these terms, which appear in the text in italics, in quotation marks, and occasionally in roman type. Latin binomials found in this section are underlined and in roman type to distinguish them from other italicized words.

Boldfaced, Page-Referenced Terms

The page-referenced terms are important; they are in boldface type in the text chapter. Write a definition for each term in your own words without looking at the text. Next, compare your definition with that given in the chapter or in the text glossary. If your definition seems inaccurate, allow some time to pass and repeat this procedure until you can define each term rather quickly (how fast you answer is a gauge of your learning effectiveness).

[p.3] biology ___the scientific study of life_____

[p.4] DNA ___(Deoxyribonucleic Acid) a nucleic acid that controls genetic inheritance._____

[p.4] inheritance _the aquisition of traits from DNA from parent to offspring._

[p.4] reproduction _the production of offspring_

[p.4] development _the transformation of a fertalized egg into a multicelled adult with cells, tissues, and organs specialized for certain tasks._

[p.5] energy _the capacity to do work._

[p.5] metabolism _the capacity to obtain and convert energy from its surroundings, and use energy to maintain itself, grow, and make more cells._

[p.5] receptors _molecules and structures that detect stimuli._

[p.5] stimulus _detected by receptors, they are sun light energy, heat energy, hormone molecules chem. energy, and mechanical energy of a bite._

[p.5] homeostasis _when a body's internal invironment maintains a certain range suitable for cell activity._

Fill-in-the-Blanks

(1) _Cells_ [p.4] are the smallest units of matter having a capacity for life. The signature molecule of cells is a nucleic acid known as (2) _DNA_ [p.4]. Encoded in this molecule's structure are the instructions for assembling a dazzling array of (3) _proteins_ [p.4] from a limited number of smaller building blocks, the (4) _amino_ [p.4] acids. (5) _Enzymes_ [p.4] are a class of worker proteins that swiftly build, split, and rearrange all of the complex molecules of life when they receive an energy boost. (6) _RNA_ [p.4] carry out DNA's protein-building instructions by working with some enzymes. Think about the information encoded in DNA flowing to (7) _RNA_ [p.4] and then to (8) _protein_ [p.4]. One of the key defining features of life is (9) _inheritance_ [p.4], the process by which parents transmit DNA instructions for duplicating their traits to offspring. For frogs and humans and other large organisms, DNA also guides (10) _reproduction_ [p.4], the transformation of a fertilized egg into a multicelled adult with cells, tissues, and organs specialized for particular tasks.

(11) _Energy_ [p.5] is most simply defined as a capacity to do work. Nothing in the universe happens without a complete or partial (12) _use_ [p.5] of energy. (13) _Metabolism_ [p.5] refers to the cell's capacity to obtain and convert energy from its surroundings and use energy to maintain itself, grow, and produce more cells. Leaves contain cells that carry on the process of (14) _photosynthesis_ [p.5] by intercepting energy from the sun and converting it to energy molecules called (15) _ATP_ [p.5]. These molecules transfer energy to metabolic workers—in this case, enzymes that assemble sugar molecules. These ATP energy molecules also form by (16) _aerobic respiration_ [p.5]. This process can release energy that cells have tucked away in sugars and other kinds of molecules.

Organisms sense changes in their surroundings, then they make controlled, compensatory (17) responses [p.5] to them. To accomplish this, each organism has (18) receptors [p.5], which are molecules and structures that detect (19) stimulus (plural) [p.5]. A (20) stimuli (singular) [p.5] is a specific form of energy detected by receptors. Examples are sunlight, heat, chemical, and mechanical energy. Cells adjust metabolic activities in response to signals from receptors. After a snack, simple sugars and other molecules leave the gut and enter the blood, which is part of a(n) (21) internal [p.5] environment. Blood sugar levels then rise and stimulate secretion of the hormone insulin by the pancreas. Most of your cells have receptors for this hormone. Insulin stimulates cells to take up sugar molecules from the internal environment and return blood sugar concentration levels to normal. Organisms respond so exquisitely to energy changes that their internal operating conditions remain within tolerable limits. This is called a state of (22) homeostasis [p.5], one of the key defining features of life.

1.2. ENERGY AND LIFE'S ORGANIZATION [pp.6–7]

Boldfaced, Page-Referenced Terms

[p.6] cell the smallest unit of organization that has the ability to survive, and reproduce on its own.

[p.6] multicelled organisms they consist of specialized, interdependent cells organized as tissues and organs.

[p.6] population a group of the same species occupying an area.

[p.6] community the populations of species occupying an area.

[pp.6–7] ecosystem the community and its physical and chemical environment.

[p.7] biosphere all regions of the Earth's atmosphere.

[p.7] producers plants and organisms that make their own food.

[p.7] consumers organisms that depend on energy that comes from the producers or other organisms that consumed the producers.

[p.7] decomposers break down sugars and other biological molecules to simpler materials - which may be cycled back to the producers.

Matching

Choose the most appropriate answer for each term.

1. _F_ organ system [p.6]
2. _E_ cell [p.6]
3. _H_ community [p.6]
4. _J_ ecosystem [pp.6–7]
5. _G_ molecule [p.6]
6. _C_ organelle [p.6]
7. _L_ population [p.6]
8. _B_ subatomic particle [p.6]
9. _A_ tissue [p.6]
10. _D_ biosphere [pp.6–7]
11. _M_ multicelled organism [p.6]
12. _K_ organ [p.6]
13. _I_ atom [p.6]

A. One or more tissues interacting as a unit
B. A proton, neutron, or electron
C. A well-defined structure within a cell, performing a particular function
D. All the regions of Earth where organisms can live
E. The smallest unit of life capable of surviving and reproducing on its own
F. Two or more organs whose separate functions are integrated to perform a specific task
G. Two or more atoms bonded together
H. All the populations interacting in a given area
I. The smallest unit of a pure substance that has the properties of that substance
J. A community interacting with its nonliving environment
K. An individual composed of cells arranged in tissues, organs, and often organ systems
L. A group of individuals of the same species in a particular place at a particular time
M. A group of cells that work together to carry out a particular function

Sequence

Arrange the following levels of organization in nature in the correct hierarchical order. Write the letter of the most inclusive level next to 14. The letter of the least inclusive level is written next to 26.

14. _8_ A. Tissue [p.6]
15. _3_ B. Community [p.6]
16. _11_ C. Molecule [p.6]
17. _1_ D. Biosphere [p.6]
18. _6_ E. Organ system [p.6]
19. _10_ F. Organelle [p.6]
20. _2_ G. Ecosystem [p.6]
21. _12_ H. Atom [p.6]
22. _9_ I. Cell [p.6]
23. _4_ J. Population [p.6]
24. _13_ K. Subatomic particle [p.6]
25. _5_ L. Multicelled organism [p.6]
26. _7_ M. Organ [p.6]

Fill-in-the-Blanks

Plants and other organisms that make their own food are (27) _producers_ [p.7] and serve to start the great flow of energy into the world of life. Animals feed directly or indirectly on energy stored in tissues of the producers; they are known as (28) _consumers_ [p.7]. (29) _Decomposers_ [p.7] are bacteria and fungi that feed on tissues or remains of other organisms and break down biological molecules to simple materials that may be cycled back to producers. Thus, there is a one-way flow of (30) _energy_ [p.7] through organisms and a (31) _cycling_ [p.7] of materials among them that organizes life in the biosphere. In nearly all cases, energy flow starts with energy from the (32) _sun_ [p.7].

1.3. IF SO MUCH UNITY, WHY SO MANY SPECIES? [pp.8–9]

Selected Words: family [p.8], order [p.8], phylum [p.8], kingdom [p.8], prokaryotic [p.8], eukaryotic [p.9]

Boldfaced, Page-Referenced Terms

[p.8] species _a specific kind of organism_

[p.8] genus _groups all closely related species together in a less specific class._

[p.8] Archaebacteria _-(prokaryotic)- found in extreme environments, much like the forms when life originated. (Anaerobic)_

[p.8] Eubacteria _- (prokaryotic)- found throughout lifes reaches. (Aerobic)_

[p.8] Protista _(eukaryotic) - Different types of multi and single celled producers and consumers._

[p.8] Fungi _(eukaryotic) Decomposing consumers. They secrete digestive enzymes out of their fungal body, and then their cells aborb the digested bits._

[p.8] Plantae _(eukaryotic) Producers which absorb their energy from the sun through photosynthesis._

[p.8] Animalia _(eukaryotic) Consumers that eat producers, other consumers, or both._

Fill-in-the-Blanks

Different "kinds" of organisms are referred to as (1) _species_ [p.8]. A (2) _genus_ [p.8] is designated by the first part of a two-part name of each organism that encompasses all the species that seem closely related by way of their recent descent from a common ancestor. Each genus encompasses all (3) _species_ [p.8] that seem closely related. The second part of the name designates a particular (4) _species_ [p.8]. For example, the pronghorn antelope is known by the two-part name *Antilocapra americana*; *Antilocapra* is the (5) _genus_ [p.8] name and *americana* is the (6) _species_ [p.8] name. In the classification of organisms, genera that share common ancestry are grouped in the same (7) _family_ [p.8]. Related families are grouped in the same (8) _order_ [p.8] and related orders are grouped in the same (9) _class_ [p.8]. Related classes are grouped into a (10) _phylum_ [p.8] that is then assigned to one of the six kingdoms of life. The adjective (11) _prokaryotic_ [p.8] describes single-celled organisms that lack a nucleus, whereas the adjective (12) _eukaryotic_ [p.9] describes single-celled and multicelled organisms whose DNA is located in a nucleus.

Complete the Table

11. Complete the following table by entering the correct name of each kingdom of life described.

Kingdom	Description
[pp.8–9] a. *Plantae*	Eukaryotic, multicelled, photosynthetic producers
[pp.8–9] b. *Euprokaryotic*	Prokaryotic, very successful in distribution, so live nearly everywhere; single cells; producers, consumers, or decomposers
[pp.8–9] c. *Fungi*	Eukaryotic, mostly multicelled, decomposers and consumers; food is digested outside their cells and bodies
[pp.8–9] d. *Archaebacteria*	Prokaryotic, live only in extreme habitats, such as the ones that prevailed when life originated
[pp.8–9] e. *Protista*	Eukaryotic, single-celled species and some multicelled forms, larger than bacteria but internally more complex
[pp.8–9] f. *Animalia*	Eukaryotic, diverse multicelled consumers, actively move at least during some stage of their life

Sequence

Arrange the following categories in correct hierarchical order. Write the letter of the largest, most inclusive category next to 12. The letter of the smallest, most exclusive category is written next to 18. This exercise classifies an animal with the common name of "beaver." Refer to p. 8 and Appendix I in the text.

12. ___   A. Class: Mammalia [p.8]

13. ___   B. Family: Castoridae [p.8]

14. ___   C. Genus: *Castor* [p.8]

15. ___   D. Kingdom: Animalia [p.8]

16. ___   E. Order: Rodentia [p.8]

17. ___   F. Phylum: Chordata [p.8]

18. ___   G. Species: *canadensis* [p.8]

1.4. AN EVOLUTIONARY VIEW OF DIVERSITY [pp.10–11]

Selected Words: *hemophilia A* [p.10], *natural* selection [pp.10–11], *selective agents* [p.10], *antibiotics* [p.11], *Staphylococcus* [p.11]

Boldfaced, Page-Referenced Terms

[p.10] mutation _a molecular change in DNA_

[p.10] adaptive trait _gives the individual an advantage in surviving and reproducing under given environmental conditions_ [p.10] evolution _genetic change over time_

[p.10] artificial selection _change in certain traits which take place in a controled and artificial environment_

[p.11] natural selection _____

Choice

For questions 1–10, choose from the following:

a. evolution through artificial selection   b. evolution through natural selection

1. _a_ Pigeon breeding [p.10]
2. _a_ Antibiotics are powerful agents of this process [p.11]
3. _b_ A favoring of adaptive traits in nature [p.10]
4. _a_ The selection of one form of a trait over another taking place under contrived, manipulated conditions [p.10]
5. _b_ Refers to change that is occurring within a line of descent over time [p.10]
6. _b_ A difference in which individuals of a given generation survive and reproduce, the difference being an outcome of which ones have adaptive forms of traits [p.11]
7. _b_ Darwin viewed this as a simple model for natural selection [p.10]
8. _a_ Breeders are the "selective agents" [p.10]
9. _b_ The mechanism whereby antibiotic resistance evolves [p.11]
10. _b_ Changing of the relative frequencies of different forms of moths through successive generations [p.10]

1.5. THE NATURE OF BIOLOGICAL INQUIRY [pp.12–13]
1.6. *Focus on Science:* THE POWER OF EXPERIMENTAL TESTS [pp.14–15]
1.7. THE LIMITS OF SCIENCE [p.15]

Selected Words: if–then process [p.12], *logic* [p.12], *inductive* logic [p.12], *deductive* logic [p.12], sampling error [p.13], *high or low probability* [p.13], *biological therapy* [p.14], *Escherichia coli* [p.14], *subjective* answers [p.15]

Boldfaced, Page-Referenced Terms
[p.12] hypotheses _an educated guess_

[p.12] prediction _a statement of what you should observe_

[p.12] test _making systematic observations of your predictions, building models, and conducting experiments_

[p.12] models _detailed descriptions or analogies that help us visualize something that has not yet been directly observed._

[p.12] inductive logic _deriving a general statement from specific observations_

[p.12] deductive logic _making an inference about a specific consequence or specific predictions that must follow from a hypothesis._

[p.12] experiments _means of simplifying observations by manipulating and controlling the conditions under which observations are made._

[p.13] control group _a standard of comparison with experimental groups._

[p.13] variables _specific aspects of objects or events that may differ or change over time and among individuals._

[p.13] scientific theory _what a hypothesis becomes once it meets specified criteria_

Sequence

Arrange the following steps of the scientific method in correct chronological sequence. Write the letter of the first step next to 1, the letter of the second step next to 2, and so on.

1. _G_   A. Develop hypotheses about what the solution or answer to a problem might be. [p.12]

2. _A_   B. Devise ways to test the accuracy of predictions drawn from the hypothesis (use of observations, models, and experiments). [p.12]

3. _D_   C. Repeat or devise new tests (different tests might support the same hypothesis). [p.12]

4. _B_   D. Make a prediction, using hypotheses as a guide; the "if–then" process. [p.12]

5. _E_   E. If the tests do not provide the expected results, check to see what might have gone wrong. [p.12]

6. _C_   F. Objectively analyze and report the results from tests and the conclusions drawn. [p.12]

7. _F_   G. Identify a problem or ask a question about nature. [p.12]

Labeling

Assume that you have to identify what object is hidden inside a sealed, opaque box. Your only tools to test the contents are a bar magnet and a triple-beam balance. Label each of the following with an O (for observation) or a C (for conclusion).

8. _O_ The object has two flat surfaces.

9. _O_ The object is composed of nonmagnetic metal.

10. _C_ The object is not a quarter, a half-dollar, or a silver dollar.

11. _O_ The object weighs x grams.

12. _C_ The object is a penny.

Complete the Table

13. Complete the following table of concepts important to understanding the scientific method of problem solving. Choose from scientific experiments, variables, prediction, inductive logic, control group, hypotheses, deductive logic, and scientific theory.

Concept	Definition
[p.12] a. *hypothesis*	Educated guesses about what the possible answers (or solutions) to scientific problems might be
[p.12] b. *prediction*	A statement of what one should be able to observe in nature if one looks; the "if–then" process
[p.13] c. *scientific theorey*	A testable explanation about the cause or causes of a broad range of related phenomena; it remains open to tests, revision, and tentative acceptance or rejection
[p.12] d. *scientific experiments*	Simplify observation in nature or the laboratory by manipulating and controlling the conditions under which observations are made
[p.13] e. *control group*	Identical with an experimental group in all respects *except* for the one variable being studied
[p.13] f. *variable*	Specific aspects of objects or events that may differ or change over time and among individuals
[p.12] g. *inductive logic*	An individual derives a general statement from specific observations
[p.12] h. *deductive logic*	An individual makes inferences about specific consequences or specific predictions that must follow from a hypothesis

Dichotomous Choice

Circle one of two possible answers given between parentheses in each statement; questions 14–18 deal with spontaneous generation.

An Italian physician, Francisco Redi, published a paper in 1688 in which he challenged the doctrine of spontaneous generation, the proposition that living things could arise from dead material. Although many examples of spontaneous generation were described in his day, Redi's work dealt particularly with disproving the notion that decaying meat could be transformed into flies. He tested his ideas in a laboratory.

14. "Two sets of jars are filled with meat or fish. One set is sealed; the other is left open so that flies can enter the jars." This description deals with a(n) (hypothesis/**experiment**). [p.12]
15. The description in the last question also includes a (prediction/**control**). [p.13]
16. Prior to his test, Redi suggested that "Worms are derived directly from the droppings of flies." This statement represents a (theory/**hypothesis**). [p.12]
17. "Worms (maggots) will appear only in the second set of jars" represents a(n) (**prediction**/hypothesis). [p.12]
18. The statement "Mice arise from a dirty shirt and a few grains of wheat placed in a dark corner" is best called a(n) (**belief**/test). [p.15]
19. From a multitude of individual observations he made of the natural world, Charles Darwin proposed the theory of organic evolution. This was an example of (deductive logic/**inductive logic**). [p.12]
20. Since the time that Darwin proposed the theory of organic evolution, countless numbers of biologists have discovered evidence of various kinds that conform to the general theory. This is an example of (**deductive logic**/inductive logic). [p.12]

21. The control group is identical to the experimental group except for the (hypothesis /(variable) being studied. [p.13]
22. Systematic observations, model development, and conducting experiments are all methods employed to (make predictions /(test predictions). [p.12]
23. Science is distinguished from faith in the supernatural by ((cause and effect/ experimental design). [p.12]
24. Through their failure to use large-enough samples in their experiments, scientists encounter (bias in reporting results /(sampling error). [p.13]
25. Science emphasizes reporting test results in ((quantitative/ qualitative terms). [p.15]

Completion

26. Questions whose answers are _subjective_ in nature do not readily lend themselves to scientific analysis and experimentation. [p.15]
27. Scientists often stir up controversy when they explain a part of the world that was considered beyond natural explanation—that is, belonging to the "_supernatural_." [p.15]
28. The external world, not internal _conviction_, must be the testing ground for scientific beliefs. [p.15]

Self-Quiz

_d_ 1. About 12 to 24 hours after a meal, a person's blood sugar level normally varies from about 60 to 90 mg per 100 ml of blood, though its blood sugar level is maintained within a fairly narrow range despite uneven intake of sugar is caused by the body's ability to carry out _____. [p.5]
   a. predictions
   b. inheritance
   c. metabolism
   d. homeostasis

_a_ 2. Different species of Galápagos Island finches have different beak types to obtain different kinds of food. One species removes tree bark with a sharp beak to forage for insect larvae and pupae, whereas another species has a large, powerful beak capable of crushing and eating large, heavy, coated seeds. These statements illustrate _____. [p.10]
   a. adaptation
   b. metabolism
   c. puberty
   d. homeostasis

_b_ 3. A boy is color-blind just as his grandfather was, even though his mother had normal vision. This situation is the result of _____. [p.4]
   a. adaptation
   b. inheritance
   c. metabolism
   d. homeostasis

_c_ 4. The digestion of food, the production of ATP by photosynthesis and respiration, the construction of the body's proteins, reproduction of cells, and the contraction of a muscle are all activities associated with _____. [p.5]
   a. adaptation
   b. inheritance
   c. metabolism
   d. homeostasis

_d_ 5. Which of the following involves using energy to do work? [p.5]
   a. atoms bonding together to form molecules
   b. the division of one cell into two cells
   c. the digestion of food
   d. all of these

_b_ 6. The experimental group and control group are identical except for _____. [p.13]
   a. the number of variables studied
   b. the variable under study
   c. the two variables under study
   d. the number of experiments performed on each group

_d_ 7. A hypothesis should *not* be accepted as valid if _____. [pp.12–13]
   a. the sample studied is determined to be representative of the entire group
   b. a variety of different tools and experimental designs yield similar observations and results
   c. other investigators can obtain similar results when they conduct the experiment under similar conditions
   d. several different experiments, each without a control group, systematically eliminate each of the variables except one

_a_ 8. The principal point of evolution by natural selection is that _____. [p.11]
   a. it measures the difference in survival and reproduction that has occurred among individuals who differ from one another in one or more heritable traits
   b. even bad mutations can improve survival and reproduction of organisms in a population
   c. evolution does not occur when some forms of traits increase in frequency and others decrease or disappear with time
   d. individuals lacking adaptive traits make up more of the reproductive base for each new generation

_c_ 9. Which match is incorrect? [pp.8–9]
   a. Kingdom Animalia—multicelled consumers, most move about
   b. Kingdom Plantae—mostly multicelled producers
   c. Kingdom Monera—relatively simple, multicelled organisms
   d. Kingdom Fungi—mostly multicelled decomposers
   e. Kingdom Protista—many complex single cells, some multicellular

_e_ 10. The least inclusive of the taxonomic categories listed is _____. [p.8]
   a. family
   b. phylum
   c. class
   d. order
   e. genus

Chapter Objectives/Review Questions

This section lists general and detailed chapter objectives that can be used as review questions. You can make maximum use of these items by writing answers on a separate sheet of paper. Fill in answers where blanks are provided. To check for accuracy, compare your answers with information given in the chapter or glossary.

1. At the _____ level, differences between living and nonliving things begin to emerge. [p.4]
2. The signature molecule of cells is a nucleic acid known as _____. [p.4]
3. Describe the nature of the instructions that are encoded in DNA's structure. [p.4]
4. Cells arise only from cells that already exist, through the process of _____. [p.4]
5. Briefly describe the role of DNA in the development of organisms. [pp.4–5]
6. Relate the concept of energy to the capacity of cells to metabolize. [p.5]
7. _____ is an energy carrier that helps drive hundreds of metabolic activities. [p.5]
8. By the process of aerobic _____, cells can release stored energy in sugars and other kinds of molecules. [p.5]
9. _____ are certain molecules and structures that can detect stimuli. [p.5]
10. List some typical stimuli that are detected by receptors. [p.5]
11. _____ refers to the internal operating conditions of organisms remaining within tolerable limits. [p.5]
12. Arrange in order, from least inclusive to most inclusive, the levels of organization that occur in nature. Define each level as you list it. [pp.6–7]
13. Explain how the actions of producers, consumers, and decomposers create an interdependency among organisms. [p.7]
14. Describe the general pattern of energy flow through Earth's life forms, and explain how Earth's resources are used again and again (cycled). [p.7]

15. Explain the use of genus and species names by considering your Latin name, *Homo sapiens*. [p.8]
16. Arrange in order, from greater to fewer organisms included, the following categories of classification: class, family, genus, kingdom, order, phylum, and species. [p.8]
17. List the six kingdoms of life; briefly describe the general characteristics of the organisms placed in each. [pp.8–9]
18. Distinguish these terms: *prokaryotic, eukaryotic.* [pp.8–9]
19. Explain the origin of trait variations that function in inheritance. [p.10]
20. An _____ trait is any form of a trait that helps an organism survive and reproduce under a given set of environmental conditions. [p.10]
21. _____ means genetically based changes in a line of descent over time. [p.10]
22. Darwin used _____ selection as a model for natural selection. [p.10]
23. Define *natural selection*, and briefly describe what is occurring when a population is said to evolve. [pp.10–11]
24. Explain what is meant by the term *diversity* and speculate about what caused the great diversity of life forms on Earth. [p.11]
25. List and explain the general steps used in scientific research. [p.12]
26. Distinguish between inductive logic and deductive logic. [p.12]
27. _____ are tests that simplify observation in nature or the laboratory by manipulating and controlling the conditions under which observations are made. [p.12]
28. Generally, members of a control group should be identical to those of the experimental group except for the one _____ being studied. [p.13]
29. Define what is meant by scientific theory; cite an actual example. [p.13]
30. Because of possible bias on the part of experimenters, science emphasizes presenting test results in _____ terms. [p.15]
31. Explain how the methods of science differ from answering questions by using subjective thinking and systems of belief. [p.15]

Interpreting and Applying Key Concepts

1. Humans have the ability to maintain body temperature very close to 37°C.
   a. What conditions would tend to make body temperature drop?
   b. What measures do you think your body takes to raise body temperature when it drops?
   c. What conditions would cause body temperature to rise?
   d. What measures do you think your body takes to lower body temperature when it rises?
2. Do you think that all humans on Earth today should be grouped in the same species?
3. What sorts of topics do scientists usually regard as untestable by the kinds of methods that scientists generally use?

2

CHEMICAL FOUNDATIONS FOR CELLS

Checking Out Leafy Clean-Up Crews

REGARDING THE ATOMS
    Structure of Atoms
    Isotopes—Variant Forms of Atoms

Focus on Science: **USING RADIOISOTOPES TO TRACK CHEMICALS AND SAVE LIVES**

WHAT HAPPENS WHEN ATOM BONDS WITH ATOM?
    Electrons and Energy Levels
    From Atoms to Molecules

IMPORTANT BONDS IN BIOLOGICAL MOLECULES
    Ion Formation and Ionic Bonding

    Covalent Bonding
    Hydrogen Bonding

PROPERTIES OF WATER
    Polarity of the Water Molecule
    Water's Temperature-Stabilizing Effects
    Water's Cohesion
    Water's Solvent Properties

ACIDS, BASES, AND BUFFERS
    The pH Scale
    How Do Acids Differ From Bases?
    Buffers Against Shifts in pH
    Salts

Interactive Exercises

Checking Out Leafy Clean-Up Crews [pp.20–21]

2.1. REGARDING THE ATOMS [p.22]

2.2. *Focus on Science:* USING RADIOISOTOPES TO TRACK CHEMICALS AND SAVE LIVES [p.23]

Selected Words: trace element [p.20], *atomic* number [p.22], *mass* number [p.22], *PET* [p.23], *radiation therapy* [p.23]

Boldfaced, Page-Referenced Terms

[p.20] bioremediation <ins>the use of living organisms to withdraw harmful substances from the environmental.</ins>

[p.20] elements <ins>fundamental forms of matter that have mass and take up space.</ins>

[p.22] atoms <ins>the smallest particles that retain the properties of an element.</ins>

[p.22] protons <ins>positively charged subatomic particles ⊕</ins>

[p.22] electrons _negatively charged subatomic particles ⊖_

[p.22] neutrons _neutral or no charged subatomic particles 0_

[p.22] isotopes _even though all atoms of an element have the same number of_
protons, they don't all have the same number of neutrons.

[p.22] radioisotope _an isotope that has an unstable nucleus and that stabilizes_
itself by spontaneously emitting energy and particles.

[p.23] tracer _any substance with a radioisotope attached to it._

Matching

Choose the most appropriate answer for each.

1. _M_ atoms [p.22]
2. _C_ protons [p.22]
3. _H_ trace element [p.20]
4. _D_ PET [p.23]
5. _N_ neutrons [p.22]
6. _B_ electrons [p.22]
7. _F_ atomic number [p.22]
8. _K_ mass number [p.22]
9. _J_ bioremediation [p.20]
10. _L_ elements [p.20]
11. _E_ isotope [p.22]
12. _G_ radioisotopes [p.22]
13. _A_ tracer [p.23]
14. _I_ radiation therapy [p.23]

A. Radioisotope used with scintillation counters to reveal the pathway or destination of a substance
B. Subatomic particles with a negative charge
C. Positively charged subatomic particles within the nucleus
D. Positron emission tomography (PET); obtains images of particular body tissues
E. Atoms of a given element that differ in the number of neutrons
F. The number of protons in an atom
G. Radioactive isotopes
H. Chemical elements representing less than 0.01 percent of body weight
I. Destroys or impairs living cancer cells
J. The use of living organisms to withdraw harmful substances from the environment
K. The number of protons and neutrons in the nucleus of one atom
L. Fundamental forms of matter that occupy space, have mass, and cannot be broken down into something else
M. Smallest units that retain the properties of a given element
N. Subatomic particles within the nucleus carrying no charge

2.3. WHAT HAPPENS WHEN ATOM BONDS WITH ATOM? [pp.24–25]

Selected Words: *lowest available energy level* [p.24], *higher energy levels* [p.24], *inert* atoms [p.24]; *formulas* [p.25], *chemical equations* [p.25], *mole* [p.25], molar weight [p.25], atomic weight [p.25]

Boldfaced, Page-Referenced Terms

[p.24] shell model _a way to look at how electrons are distributed in atoms._

[p.24] chemical bond _a union between the electron structure of atoms_

[p.25] molecule *when two or more atoms bond*

[p.25] compounds *consists of two or more different elements*

[p.25] mixture *two or more elements are simply intermingling in proportions that can vary*

Matching

Choose the most appropriate answer for each.

1. __C__ mixture [p.25]
2. __E__ shell [p.24]
3. __I__ lowest available energy level [p.24]
4. __H__ inert atoms [p.24]
5. __A__ orbitals [p.24]
6. __D__ chemical bond [p.24]
7. __F__ compounds [p.25]
8. __B__ molecule [p.25]
9. __G__ higher energy levels [p.24]

A. Regions of space around an atom's nucleus where electrons are likely to be at any one instant
B. Results when two or more atoms bond together
C. Two or more elements are simply intermingling in proportions that usually vary
D. A union between the electron structures of atoms
E. A series of these enclose all orbitals available to an atom's electrons
F. Types of molecules composed of two or more different elements in proportions that never vary
G. Energy of electrons farther from the nucleus than the first orbital
H. Refers to those with no vacancies in their shells and hence show little tendency to enter chemical reactions
I. Energy of electrons in the orbital closest to the nucleus

Complete the Table

10. Complete the table (text Figure 2.8, p. 25) by entering the name of the element and its symbol in the appropriate spaces.

Element	Symbol	Atomic Number	Electron Distribution			
			First Shell	Second Shell	Third Shell	Fourth Shell
Calcium [p.25] a.	Ca	20	2	8	8	2
Carbon [p.25] b.	C	6	2	4		
Chlorine [p.25] c.	Cl	17	2	8	7	
Hydrogen [p.25] d.	H	1	1			
Sodium [p.25] e.	Na	11	2	8	1	
Nitrogen [p.25] f.	N	7	2	5		
Oxygen [p.25] g.	O	8	2	6		

Fill-in-the-Blanks

The expression $12H_2O$ plus $6\ CO_2 \rightarrow 6O_2$ plus $C_6H_{12}O_6$ plus H_2O is known as the chemical (11) _equation_ [p.25] for photosynthesis. H_2O is the (12) _formula_ [p.25] for water. The arrow in the expression above means (13) _yields_ [p.25]. The (14) _reactants_ [p.25] are to the left of the reaction arrow and the (15) _products_ [p.25] are to the right of the reaction arrow. In the expression one can count 12 hydrogen atoms on the left side of the arrow, so one should be able to count (16) _12_ [p.25] hydrogen atoms on the right side of the arrow. By reference to text Table 2.1, p. 22, one can determine the weight of one mole of H_2SO_4 to be (17) _98 grams_ [p.25].

Identification

18. Following the model (number of protons and neutrons shown in the nucleus), identify the indicated atoms of the elements illustrated by entering appropriate electrons in this form: (2e⁻). [pp.24–25]

MODEL:

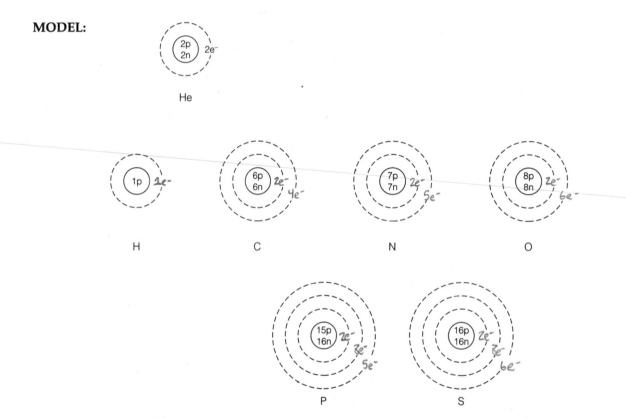

2.4. IMPORTANT BONDS IN BIOLOGICAL MOLECULES [pp.26–27]

Selected Words: *sharing* [p.26], *single* covalent bond [p.26], *double* covalent bond [p.26], *triple* covalent bond [p.26], *nonpolar* covalent bond [p.26], *polar* covalent bond [p.27], no *net* charge [p.27]

Boldfaced, Page-Referenced Terms

[p.26] ion _any atom that atom that has gained or lost an electron_

[p.26] ionic bond _when two ions join and do not seperate from the attraction of_
their opposite charges. ex: NaCl aka Salt.

[p.26] covalent bond _a sharing of a pair of electrons_

[p.27] hydrogen bond _a weak attraction between an electronegative atom and a_
hydrogen atom taking part in a second polar covalent bond.

Identification

1. Following the model, complete the diagram by adding arrows to identify the transfer of electron(s), showing how positive magnesium and negative chlorine ions form ionic bonds to create a molecule of $MgCl_2$ (magnesium chloride). [p.26]

MODEL:

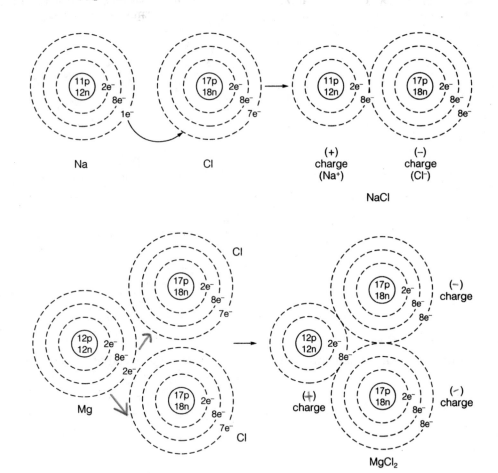

Identification

2. Following the model of hydrogen gas shown, complete the diagram by placing electrons (as dots) in the outer shells to identify the nonpolar covalent bonding that forms oxygen gas; similarly identify polar covalent bonds by completing electron structures to form a water molecule. [pp.26–27]

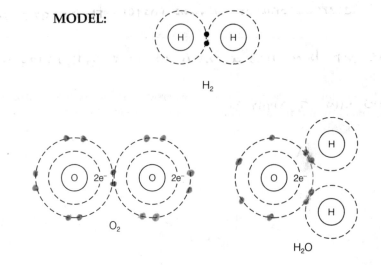

MODEL:

H_2

O_2

H_2O

Short Answer

3. Distinguish between a nonpolar covalent bond and a polar covalent bond. Cite an example of each. [pp.26–27] _____

_____

_____

_____

_____

_____

_____

4. Describe one example of a large biological molecule within which hydrogen bonds exist. [p.27] _____

_____

_____

_____

2.5. PROPERTIES OF WATER [pp.28–29]

Selected Words: dissolved [p.29]

Boldfaced, Page-Referenced Terms

[p.28] hydrophilic substances _(water loving) they readily hydrogen-bond with water_

_____

[p.28] hydrophobic substances _(water dreading) water polarity repels nonpolar molecules, including oils._

[p.29] temperature _the molecular motion of a given substance._

[p.29] evaporation _heat energy converts liquid water to the gaseous state._

[p.29] cohesion _means something has a capacity to resist rupturing when placed under tension_

[p.29] solute _any dissolved substance._

Fill-in-the-Blanks

The (1) _polarity_ [p.28] of water molecules allows them to hydrogen-bond with each other. Water molecules hydrogen-bond with polar molecules, which are (2) _hydrophilic_ (water-loving) substances [p.28]. Polarity causes water to repel oil and other nonpolar substances, which are (3) _hydrophobic_ (water-dreading) [p.28]. (4) _Temperature_ [p.29] is a measure of the molecular motion of a given substance. Liquid water changes its temperature more slowly than air because of the great amount of heat required to break the high number of (5) _hydrogen_ [p.29] bonds between water molecules; this property helps stabilize temperature in aquatic habitats and cells. When large energy inputs increase molecular motion to the point where hydrogen bonds stay broken, releasing individual molecules from the water surface, the process is known as (6) _evaporation_ [p.29]. Below 0ºC, water molecules become locked in the less-dense, lattlicelike bonding pattern of (7) _ice_ [p.29], which is less dense than water. Collective hydrogen bonding creates a high tension on surface water molecules, resulting in (8) _cohesion_ [p.29], the property of water that explains how long, narrow water columns rise to the tops of tall trees. Water is an excellent (9) _solvent_ [p.29] in which ions and polar molecules readily dissolve. Substances dissolved in water are known as (10) _solute_ [p.29]. A substance is (11) _dissolved_ in water when spheres of (12) _hydration_ [p.29] form around its individual ions or molecules.

2.6. ACIDS, BASES, AND BUFFERS [pp.30–31]

Selected Words: *donate* H+ [p.30], *accept* H+ [p.30], *acidic* solutions [p.30], *basic* solutions [p.30], *acid stomach* [p.30], *chemical burns* [p.31], *acid rain* [p.31], *coma* [p.31], *tetany* [p.31], *acidosis* [p.31], *alkalosis* [p.31]

Boldfaced, Page-Referenced Terms

[p.30] hydrogen ions _H+_

[p.30] hydroxide ions _OH-_

[p.30] pH scale the ionization of water

[p.30] acids donate protons (H^+) to other solutes or water molecules

[p.30] bases accept protons H^+ when dissolved in water and OH^- indirectly or directly when they do this.

[p.31] buffer system a partnership between a weak acid and the base that forms when the acid dissolves in water

[p.31] salts are compounds that release ions other than H^+ and OH^- in solutions

Matching

Choose the most appropriate answer for each.

1. _N_ acid stomach [p.30]
2. _K_ acids [p.30]
3. _E_ pH scale [p.30]
4. _J_ chemical burns [p.31]
5. _H_ H^+ [p.30]
6. _D_ bases [p.30]
7. _I_ examples of basic solutions [p.30]
8. _A_ coma [p.31]
9. _B_ acidosis [p.31]
10. _C_ OH^- [p.30]
11. _O_ tetany [p.31]
12. _L_ examples of acid solutions [p.30]
13. _E_ alkalosis [p.31]
14. _G_ buffer system [p.31]
15. _M_ acid rain [p.31]

A. A sometimes irreversible state of unconsciousness
B. CO_2 builds up in the blood, too much H_2CO_3 forms, and blood pH severely decreases
C. Hydroxide ion
D. Substances that accept H^+ when dissolved in water
E. An uncorrected increase in blood pH
F. Used to measure H^+ concentration in various fluids
G. A partnership between a weak acid and the base that forms when it dissolves in water; counters slight pH shifts
H. Hydrogen ion or proton
I. Baking soda, seawater, egg white
J. Can be caused by ammonia, drain cleaner, and sulfuric acid in car batteries
K. Substances that donate H^+ when dissolved in water
L. Lemon juice, gastric fluid, coffee
M. Sulfur dioxide is a major component; some regions are very sensitive to its pH
N. Can be caused by eating too much fried chicken or certain other foods
O. A potentially lethal pH stage in which the body's skeletal muscles enter a state of uncontrollable contraction

Complete the Table

16. Complete the following table by consulting text Figure 2.17, page 30.

Fluid	pH Value	Acid/Base
[p.30] a. Blood	7.3-7.5	basic
[p.30] b. Saliva	6.2-7.4	acidic
[p.30] c. Urine	5.0-7.0	acidic
[p.30] d. Stomach acid	1.0-3.0	acidic
[p.30] e. Seawater	7.8-8.3	basic
[p.30] f. Some acid rain	3.0	acidic

Self-Quiz

_c_ 1. Each element has a unique _____, which refers to the number of protons present in its atoms. [p.22]
   a. isotope
   b. mass number
   c. atomic number
   d. radioisotope

_a_ 2. A molecule is _____. [p.25]
   a. a bonding together of two or more atoms
   b. less stable than its constituent atoms separated
   c electrically charged
   d. a carrier of one or more extra neutrons

_d_ 3. If lithium has an atomic number of 3 and an atomic mass of 7, it has _____ neutron(s) in its nucleus. [p.22]
   a. one
   b. two
   c. three
   d. four
   e. seven

_d_ 4. Substances that are nonpolar and repelled by water are _____. [p.28]
   a. hydrolyzed
   b. nonpolar
   c. hydrophilic
   d. hydrophobic

_c_ 5. A hydrogen bond is _____. [p.27]
   a. a sharing of a pair of electrons between a hydrogen nucleus and an oxygen nucleus
   b. a sharing of a pair of electrons between a hydrogen nucleus and either an oxygen or a nitrogen nucleus
   c. formed when an electronegative atom of a molecule weakly interacts with a hydrogen atom that is already participating in a polar covalent bond
   d. none of the above

_c_ 6. An ionic bond is one in which _____. [p.26]
   a. electrons are shared equally
   b. electrically neutral atoms have a mutual attraction
   c. two charged atoms have a mutual attraction due to electron transfer
   d. electrons are shared unequally

_a_ 7. A covalent bond is one in which _____. [p.26]
   a. electrons are shared
   b. electrically neutral atoms have a mutual attraction
   c. two charged atoms have a mutual attraction due to electron transfer
   d. electrons are lost

_C_ 8. A nonpolar covalent bond implies that _____. [pp.26–27]
  a. one negative atom bonds with a hydrogen atom
  b. the bond is double
  c. there is no difference in charge at the ends (the two poles) of the bond
  d. atoms of different elements do not exert the same pull on shared electrons

_A_ 9. The shapes of large molecules are controlled by _____ bonds. [p.27]
  a. hydrogen
  b. ionic
  c. covalent
  d. inert
  e. single

_d_ 10. A solution with a pH of 10 is _____ times as basic as one with a pH of 8. [p.30]
  a. 2
  b. 3
  c. 10
  d. 100
  e. 1,000

_e_ 11. A control that minimizes unsuitable pH shifts is a(n) _____. [p.31]
  a. neutral molecule
  b. salt
  c. base
  d. acid
  e. buffer

Chapter Objectives/Review Questions

1. The use of living organisms to withdraw harmful substances from the environment is known as _____. [p.20]
2. Explain what an element is. [p.20]
3. Define *trace* element. [p.20]
4. List and describe the three types of subatomic particles, and explain why hydrogen is an exception. [p.22]
5. The number of protons in an atom is referred to as the _____ number of that element; the combined number of protons and neutrons in the atomic nucleus is referred to as the _____ number of that element. [p.22]
6. Distinguish between isotopes and radioisotopes. [p.22]
7. Describe what tracers are and how they are used. [p.23]
8. _____ uses radioactively labeled compounds to yield images of metabolically active and inactive tissues in a patient. [p.23]
9. Describe the use of radiation therapy. [p.23]
10. Explain the relationship of orbitals to shells. [p.24]
11. Sketch shell models of the atoms described in the text. [text Figure 2.8, p.25]
12. Read a chemical equation. [text Figure 2.9, p. 25]
13. Explain why helium and neon are known as inert elements. [p.25]
14. A _____ is formed when two or more atoms bond together. [p.25]
15. A _____ contains atoms of two or more elements whose proportions never vary. [p.25]
16. How does a mixture differ from a compound? [p.25]
17. An _____ is an atom that becomes positively or negatively charged. [p.26]
18. An association of two oppositely charged ions is a(n) _____ bond. [p.26]
19. In a(n) _____ bond, two atoms share electrons. [p.26]
20. Explain why H_2 is an example of a nonpolar covalent bond and H_2O has two polar covalent bonds. [pp.26–27]
21. In a(n) _____ bond, a small, highly electronegative atom of a molecule weakly interacts with a hydrogen atom that is already participating in a polar covalent bond. [p.27]
22. Polar molecules attracted to water are _____; all nonpolar molecules are _____ and are repelled by water. [p.28]
23. _____ is a measure of molecular motion; during _____, an input of heat energy converts liquid water to the gaseous state. [p.29]
24. Describe the formation of ice in terms of hydrogen bonding. [p.29]
25. Completely describe the properties of water that move water from the roots to the tops of the tallest trees. [p.29]

26. Distinguish a solvent from a solute. [p.29]
27. Describe what happens when a substance is dissolved in water. [p.29]
28. The ionization of water is the basis of the _____ scale. [p.30]
29. _____ are substances that donate H$^+$ ions when they dissolve in water, and _____ are substances that accept H$^+$ when dissolved in water. [p.30]
30. Black coffee with a pH of 5 is a(n) _____ solution, whereas baking soda with a pH of 9 is a(n) _____ solution. [p.30]
31. Define *buffer system*; cite an example and describe how buffers operate. [p.31]
32. Define the following terms: *acid stomach, chemical burns, coma, tetany, acidosis,* and *alkalosis.* [pp.30–31]
33. A _____ and water is produced by a chemical reaction between an acid and a base. [p.31]

Integrating and Applying Key Concepts

1. Explain what would happen if water were a nonpolar molecule instead of a polar molecule. Would water be a good solvent for the same kinds of substances? Would the nonpolar molecule's specific heat likely be higher or lower than that of water? Would surface tension be affected? cohesive nature? ability to form hydrogen bonds? Is it likely that the nonpolar molecules could form unbroken columns of liquid? What implications would that hold for trees?

3

CARBON COMPOUNDS IN CELLS

Interactive Exercises

Carbon, Carbon, in the Sky—Are You Swinging Low and High? [pp.34–35]

3.1. PROPERTIES OF ORGANIC COMPOUNDS [pp.36–37]

Selected Words: *global warming* [p.35], "organic" substances [p.36], "inorganic" substances [p.36], *functional-group transfer* [p.37], *electron transfer* [p.37], *rearrangement* [p.37], *condensation* [p.37], *cleavage* [p.37]

Boldfaced, Page-Referenced Terms

[p.36] organic compounds _the molecules of life, with hydrogen and often other elements covalently bonded to carbon atoms._

[p.36] functional groups _Various kinds of atoms or clusters of them covalently bonded to the backbone_

[p.36] alcohols _dissolve quickly in water, because water molecules easily form hydrogen bonds with -OH groups._

[p.37] enzymes _proteins which make specific metabolic reactions proceed faster than they would on their own._

[p.37] condensation reactions _enzymes remove a hydroxl group from one molecule and an H atom from another, then speed the formation of a covalent bond btw. the two molecules at their exposed sites._

[p.37] hydrolysis _like condensation in reverse_

Labeling

Study the structural formulas of the following organic compounds. Note that each carbon atom can share pairs of electrons with as many as four other atoms. Reference text Figure 3.2, page 36, to identify the circled functional groups (sometimes repeated) by entering the correct name in the blanks with matching numbers.

1. _methyl_ [p.36]
2. _hydroxl_ [p.36]
3. _ketone_ [p.36]
4. _amino_ [p.36]
5. _phosphate_ [p.36]
6. _carboxl_ [p.36]
7. _aldehyde_ [p.36]

Short Answer

8. State the general role of enzymes as they relate to the chemistry of life. [p.37] _____

_____

_____

_____

Complete the Table

9. Complete the following table summarizing five categories of reactions that are mediated by enzymes.

Reaction Category	Reaction Description
[p.37] a. Rearrangement	
[p.37] b. Cleavage	
[p.37] c. Condensation	
[p.37] d. Electron transfer	
[p.37] e. Functional-group transfer	

Identification

10. Study the structural formulas of the two adjacent amino acids. Identify the enzyme action causing formation of a covalent bond and a water molecule (through a condensation reaction) by circling an H atom from one amino acid and an -OH group from the other amino acid. Also circle the covalent bond that formed the dipeptide [pp.37,43].

amino acid   amino acid   dipeptide

Short Answer

11. Describe hydrolysis through enzyme action for the molecules in exercise 10. [p.37] _____

_____

_____

_____

_____

_____

3.2. CARBOHYDRATES [pp.38–39]

*Selected Words: mono*saccharide [p.38], *oligo*saccharide [p.38], *di*saccharides [p.38], *poly*saccharides [p.38]

Boldfaced, Page-Referenced Terms

[p.38] carbohydrates _consist of carbon, hydrogen, and oxygen in a 1:2:1 ratio._

[p.38] monosaccharides _simplest form, one monomer of sugar_

[p.38] oligosaccharides _a short chain of two or more sugar monomers that are bonded covalently_

[p.38] polysaccharides _hundreds or thousands of straight or branched sugar monomers_

Identification

1. In the diagram, identify condensation reaction sites between the two glucose molecules by circling the components of the water removed that allow a covalent bond to form between the glucose molecules (text Figure 3.5, p. 38). Note that the reverse reaction is hydrolysis and that both condensation and hydrolysis reactions require enzymes in order to proceed. [p.38]

glucose
(a monosaccharide)   glucose
(a monosaccharide)   maltose
(a disaccharide)   water

Complete the Table

2. Complete the following table by entering the name of the carbohydrate described by its carbohydrate class and functions.

Carbohydrate	Carbohydrate Class	Function
[p.38] a.	Sucrose, Oligosaccharide	Most plentiful sugar in nature; transport form of carbohydrates in plants; table sugar; formed from glucose and fructose
[p.38] b.	Deoxyribose, Monosaccharide	Five-carbon sugar occurring in DNA
[p.38] c.	Glucose, Monosaccharide	Main energy source for most organisms; precursor of many organic molecules; serve as building blocks for larger carbohydrates
[p.38] d.	Cellulose Polysaccharide	Structural material of plant cell walls; formed from glucose chains
[p.38] e.	Ribose, Monosaccharide	Five-carbon sugar occurring in RNA
[p.38] f.	Lactose, Oligosaccharide	Sugar present in milk; formed from glucose and galactose
[p.39] g.	Glycogen Polysaccharide	Muscle cells tap their stores of this compound for a rapid burst of energy
[p.39] h.	Chitin Polysaccharide	Main structural material in some external skeletons and other hard body parts of some animals and fungi
[p.38] i.	Glycogen, Polysaccharide	Animal starch; stored especially in liver and muscle tissue; formed from glucose chains
[p.38] j.	Starch, Polysaccharide	Storage form for photosynthetically produced sugars

3.3. LIPIDS [pp.40–41]

Selected Words: *unsaturated* tails [p.40], *saturated* tails [p.40], eicosanoids [p.41]

Boldfaced, Page-Referenced Terms

[p.40] lipids _show little tendency to dissolve in water, although they readily dissolve in nonpolar substances._

[p.40] fats _have one, two, or three fatty acids attached to glycerol_

[p.40] fatty acid _has a backbone of as many as thirty-six carbon atoms, a carboxl group at one end, and hydrogen occupying most or all of the remaining bonding sites_

[p.40] triglycerides _neutral fats having three fatty acid tails attached to glycerol_

[p.41] phospholipid _has a glycerol backbone, two fatty acid tails, and a hydrophilic head with a phosphate group and another polar group_

[p.41] sterols _are among the many lipids that have no fatty acids._

[p.41] waxes _have long-chain fatty acids tightly packed and linked to long-chain alcohols or to carbon rings._

Labeling

1. In the appropriate blanks, label the molecules shown as saturated or unsaturated.

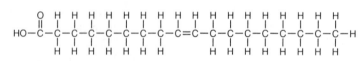

oleic acid

a.

stearic acid

b.

a. _____ [p.40]   b. _____ [p.40]

Identification

2. Combine glycerol with three fatty acids to form a triglyceride by circling the participating atoms that will identify three covalent bonds. Also circle the covalent bonds in the triglyceride. [p.40]

$$H-\overset{\overset{H}{|}}{\underset{|}{C}}-OH \ + \ HO-\overset{\overset{O}{\|}}{C}-R$$

$$H-\overset{|}{\underset{|}{C}}-OH \ + \ HO-\overset{\overset{O}{\|}}{C}-R$$

$$H-\overset{|}{\underset{\underset{H}{|}}{C}}-OH \ + \ HO-\overset{\overset{O}{\|}}{C}-R$$

yields ⟶

$$H-\overset{\overset{H}{|}}{\underset{|}{C}}-O-\overset{\overset{O}{\|}}{C}-R$$

$$H-\overset{|}{\underset{|}{C}}-O-\overset{\overset{O}{\|}}{C}-R \ + \ 3H_2O$$

$$H-\overset{|}{\underset{\underset{H}{|}}{C}}-O-\overset{\overset{O}{\|}}{C}-R$$

glycerol   three fatty
                acids

triglyceride
(a complete fat
molecule)

Short Answer

3. Describe the structure and biological functions of phospholipid molecules. [p.41] _____

_____

_____

_____

_____

Choice

For questions 4–18, choose from the following:

        a. fatty acids   b. triglycerides   c. phospholipids   d. waxes   e. sterols

4. ___ richest source of body energy [p.40]

5. ___ honeycomb material [p.41]

6. ___ cholesterol [p.41]

7. ___ saturated tails [p.40]

8. ___ butter and lard [p.40]

9. ___ lack fatty acid tails [p.41]

10. ___ occur in eukaryotic cell membranes [p.41]

11. ___ all possess a rigid backbone of four fused carbon rings [p.41]

12. ___ plant cuticles [p.41]

13. ___ precursors of vitamin D, steroids, and bile salts [p.41]

14. ___ unsaturated tails [p.40]

15. ___ vegetable oil [p.40]

16. ___ vertebrate insulation [p.40]

17. ___ furnishes protection and lubrication for hair, skin, and feathers [p.41]

18. ___ neutral fats [p.40]

3.4. AMINO ACIDS AND THE PRIMARY STRUCTURE OF PROTEINS [pp.42–43]

3.5. HOW DOES A THREE-DIMENSIONAL PROTEIN EMERGE? [pp.44–45]

3.6. *Focus on the Environment:* FOOD PRODUCTION AND A CHEMICAL ARMS RACE [p.46]

Selected Words: primary structure [p.43], secondary structure [p.44], tertiary structure [p.44], quaternary structure [p.45], lipoproteins [p.45], glycoproteins [p.45], *herbicides* [p.47], *insecticides* [p.47], *fungicides* [p.47]

Boldfaced, Page-Referenced Terms

[p.42] proteins _____

_____

[p.42] amino acid _____

_____

[p.42] polypeptide chain _____

_____

[p.44] hemoglobin _____

_____

[p.45] globular proteins _____

_____

[p.45] fibrous proteins _____

_____

[p.45] denaturation _____

_____

[p.46] toxin _____

_____

Matching

For exercises 1, 2, and 3, match the major parts of *every* amino acid by entering the letter of the part in the blank corresponding to the number on the molecule.

1. ___[p.42]
2. ___[p.42]
3. ___[p.42]

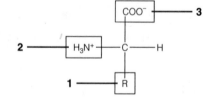

A. R group (a symbol for a characteristic group of atoms that differ in number and arrangement from one amino acid to another)
B. Carboxyl group (ionized)
C. Amino group (ionized)

Identification

4. Four ionized amino acids (in cellular solution) are shown in the upper illustration below; form the proper covalent bonds by circling the acids and ions involved in polypeptide formation. Circle the newly formed peptide bonds of the completed polypeptide on the lower illustration. [p.43]

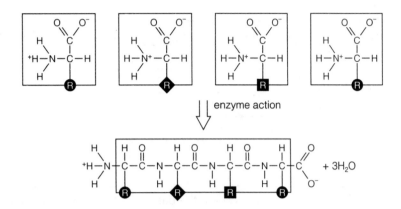

Matching

Choose the most appropriate answer for each term.

5. ___amino acid [p.42]

6. ___peptide bond [p.42]

7. ___polypeptide chain [pp.42–43]

8. ___primary structure [p.43]

9. ___proteins [p.42]

10. ___secondary structure [p.44]

11. ___tertiary structure [p.44]

12. ___dipeptide [p.42]

13. ___quaternary structure [p.45]

14. ___lipoproteins [p.45]

15. ___glycoproteins [p.45]

16. ___denaturation [p.45]

A. A coiled or extended pattern of protein structure caused by regular intervals of H bonds
B. Three or more amino acids joined in a linear chain
C. Proteins with linear or branched oligosaccharides covalently bonded to them; found on animal cell surfaces, in cell secretion, or on blood proteins
D. Folding of a protein through interactions among R groups of a polypeptide chain
E. Form when freely circulating blood proteins encounter and combine with cholesterol, or phospholipids
F. The type of covalent bond linking one amino acid to another
G. Hemoglobin, a globular protein of four chains, is an example
H. Breaking weak bonds in large molecules (such as protein) to change its shape so it no longer functions
I. Formed when two amino acids join together
J. Lowest level of protein structure; has a linear, unique sequence of amino acids
K. A small organic compound having an amino group, an acid group, a hydrogen atom, and an R group
L. The most diverse of all the large biological molecules; constructed from pools of only twenty kinds of amino acids

3.7. NUCLEOTIDES [p.47]

3.8. NUCLEIC ACIDS [pp.48–49]

Selected Words: NAD+ [p.47], FAD [p.47], cAMP [p.47]

Boldfaced, Page-Referenced Terms

[p.47] nucleotides _____

[p.47] ATP _____

_____

[p.47] coenzymes _____

_____

[p.48] nucleic acids _____

_____

[p.48] DNA _____

_____

[p.48] RNAs _____

_____

Matching

For exercises 1, 2, and 3, match the following answers to the parts of a nucleotide shown in the diagram.

1. ___[p.47]
2. ___[p.47]
3. ___[p.47]

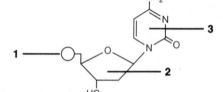

A. A five-carbon sugar (ribose or deoxyribose)
B. Phosphate group
C. A nitrogen-containing base that has either a single-ring or double-ring structure

Identification

4. In the diagram of a single-stranded nucleic acid molecule, encircle as many complete nucleotides as possible. How many complete nucleotides are present [p.47]?

Matching

Choose the most appropriate answer for each term.

5. ___adenosine triphosphate [p.47]

6. ___NAD+, FAD, cAMP [p.47]

7. ___RNAs [p.47]

8. ___DNA [p.47]

A. Single nucleotide strands; function in processes by which genetic instructions are used to build proteins
B. ATP, a cellular energy carrier
C. Coenzymes that serve as enzyme helpers
D. Double nucleotide strand; encodes genetic instructions with nucleotide sequences

---

Self-Quiz

Complete the Table

1. Complete the following table by entering the correct name of the major cellular organic compounds suggested in the "types" column (choose from carbohydrates, lipids, proteins, and nucleic acids).

Cellular Organic Compounds	Types
[p.41] a. lipid	Phospholipids
[p.48] b. nucleic acid	Nucleotides
[p.42] c. Proteins	Weapons against bacteria-causing bacteria
[pp.37,42] d. proteins	Enzymes
[p.47] e. nucleic acid	Genetic information, inherited instructions
[pp.38–39] f. carbohydrate	Glycogen, starch, cellulose, and chitin
[p.48] g. nucleic acid	DNA and RNAs
[p.40] h. Lipid	Triglycerides
[p.47] i. nucleic acid	ATP
[p.40] j. lipid	Saturated and unsaturated fats
[pp.40–41] k. lipid	Sterols, oils, and waxes
[p.38] l. carbohydrate	Glucose and sucrose

_d_ 2. Amino acids are linked by _____ bonds to form the primary structure of a protein. [p.42]
   a. disulfide
   b. hydrogen
   c. ionic
   d. peptide

_a_ 3. Proteins _____. [p.42]
   a. are weapons against disease-causing bacteria and other invaders
   b. are composed of nucleotide subunits
   c. translate protein-building instructions into actual protein structures
   d. lack diversity of structure and function

_C_ 4. Lipids _____. [p.40]
   a. include fats that are broken down into one fatty acid molecule and three glycerol molecules
   b. are composed of monosaccharides
   c. include triglycerides that serve as energy sources
   d. include cartilage and chitin

_C_ 5. DNA _____. [p.48]
   a. is one of the adenosine phosphates
   b. is one of the nucleotide coenzymes
   c. contains protein-building instructions
   d. is composed of monosaccharides

_C_ 6. Most of the chemical reactions in cells must have _____ present before they proceed. [pp.37,42]
   a. RNA
   b. salt
   c. enzymes
   d. fats

_b_ 7. Carbon is part of so many different substances because _____. [p.36]
   a. carbon generally forms two covalent bonds with a variety of other atoms
   b. a carbon atom generally forms four covalent bonds with a variety of atoms
   c. carbon ionizes easily
   d. carbon is a polar compound

_d_ 8. The information built into a protein's amino acid sequence plus a coiled pattern of that chain and the addition of more folding yields the _____ level of protein structure. [p.44]
   a. quaternary
   b. primary
   c. secondary
   d. tertiary

_C_ 9. _____ are molecules used by cells as structural materials and transportable or storage forms of energy. [p.38]
   a. Lipids
   b. Nucleic acids
   c. Carbohydrates
   d. Proteins

_bde_10. Hydrolysis could be correctly described as the _____. [p.37]
   a. heating of a compound in order to drive off its excess water and concentrate its volume
   b. breaking of a long-chain compound into its subunits by adding water molecules to its structure between the subunits
   c. linking of two or more molecules by the removal of one or more water molecules
   d. constant removal of hydrogen atoms from the surface of a carbohydrate
   e. condensation in reverse

_b_11. Genetic instructions are encoded in the base sequence of _____; molecules of _____ function in processes using genetic instructions to construct proteins. [pp.48–49]
   a. DNA; DNA
   b. DNA; RNA
   c. RNA; DNA
   d. RNA; RNA

Chapter Objectives/Review Questions

1. The molecules of life are _____ compounds. [p.36]
2. Each carbon atom can share pairs of electrons with as many as _____ other atoms. [p.36]
3. A _____ has only hydrogen atoms attached to a carbon backbone. [p.36]
4. Hydroxyl, amino, and carboxyl are examples of _____ groups. [p.36]
5. Describe the role of enzymes in the metabolism of life. [pp.36–37]
6. _____ use some assortment of the simple sugars, fatty acids, amino acids, and nucleotides for building all the organic compounds they require for their structure and functioning. [p.36]
7. _____ make specific metabolic reactions proceed faster than they would on their own and different _____ mediate different kinds of reactions. [p.37]

8. Define the following: *functional-group transfer, electron transfer, rearrangement, condensation,* and *cleavage.* [p.37]
9. Define and distinguish between *condensation reactions* and *hydrolysis;* cite a general example. [p.37]
10. Define *carbohydrates;* list their two general functions. [p.38]
11. Most carbohydrates consist of _____, _____, and _____ in a 1:2:1 ratio. [p.38]
12. Name and generally define the three classes of carbohydrates. [p.38]
13. A _____ means "one monomer of sugar"; give common examples and their functions. [p.38]
14. An _____ is a short chain of two or more covalently bonded sugar monomers; give examples of well-known disaccharides and their functions. [p.38]
15. A _____ is a straight or branched chain of hundreds of thousands of sugar monomers, of the same or different kinds; be able to give common examples and their functions. [pp.38–39]
16. Define *lipids;* list their general functions. [p.40]
17. Describe a "fatty acid"; a _____ molecule has three fatty acid tails attached to a backbone of glycerol; distinguish a saturated fatty acid from an unsaturated fatty acid. [p.40]
18. Describe the structure of triglycerides; list examples and their functions. [p.40]
19. A _____ has two fatty acid tails and a hydrophilic head attached to a glycerol backbone; list examples and their cellular functions. [p.41]
20. Define *sterols,* and describe their chemical structure; cite the importance of the sterols known as cholesterol and the steroids. [p.41]
21. Lipids called _____ have long-chain fatty acids, tightly packed and linked to long-chain alcohols or to carbon rings; list examples and their functions. [p.41]
22. Describe protein structure and cite their general functions; sketch the three general parts of every amino acid. [p.42]
23. Describe how the primary, secondary, tertiary, and quaternary structure of proteins results in complex three-dimensional structures. [pp.42–45]
24. Distinguish lipoproteins from glycoproteins. [p.45]
25. _____ refers to the loss of a molecule's three-dimensional shape through disruption of the weak bonds responsible for it. [p.45]
26. A _____ is an organic compound, a normal metabolic product of one species, but its chemical effects can harm or kill individuals of a different species that come in contact with it. [p.46]
27. Distinguish between *herbicides, insecticides,* and *fungicides.* [p.47]
28. List the three parts of every nucleotide. [p.47]
29. The nucleic acids _____ and _____, built of nucleotides, are the basis of inheritance and reproduction. [p.48]

Integrating and Applying Key Concepts

1. Humans can obtain energy from many different food sources. Do you think this ability is an advantage or a disadvantage in terms of long-term survival? Why?
2. If the ways that atoms bond affect molecular shapes, do the ways that molecules behave toward one another influence the shapes of organelles? Do the ways that organelles behave toward one another influence the structure and function of the cells?

4

CELL STRUCTURE AND FUNCTION

Interactive Exercises

Animalcules and Cells Fill'd With Juices [pp.52–53]

4.1. BASIC ASPECTS OF CELL STRUCTURE AND FUNCTION [pp.54–55]

4.2. *Focus on Science:* MICROSCOPES—GATEWAYS TO THE CELL [pp.56–57]

Selected Words: *cellulae* [p.52], *compound light microscope* [p.56], *transmission electron microscope* [p.56], *scanning electron microscope* [p.56], *scanning tunneling microscope* [p.57], *phase-contrast process* [p.57], Nomarski process [p.57]

Boldfaced, Page-Referenced Terms

[p.53] cell theory Every organism is composed of one or more cells, the cell is the smallest unit having properties of life, and the continuity of life arises directly from the growth and division of single cells.

[p.54] cell _is the smallest unit that retains the properties of life_

[p.54] plasma membrane _a thin outermost membrane that maintains the cell as a distinct entity._

[p.54] nucleus _(eukaryotic) the DNA occupies a membraneous sac in the cell interior_

[p.54] nucleoid _(prokaryotic) the area in the cell where the DNA is foun_

[p.54] cytoplasm _everything between the plasma membrane and region of DNA_

[p.54] ribosomes _structures on which proteins are built._

[p.54] eukaryotic cells _includes organelles, which are other compartments bounded by membranes. the nucleus is the defining feature._

[p.54] prokaryotic cells _does not have a nucleus. No membranes intervene between the nucleoid and the cytoplasm surrounding it._

[p.54] lipid bilayer _the continuous boundary that bars free passage of water-soluble substances into and out of the cell._

[p.55] surface-to-volume ratio _if a cell expands in diameter during growth, its volume will increase more rapidly than its surface area will._

[p.56] micrograph _a photograph of an image that came into view with the help of a microscope._

[p.56] wavelength _is the distance from one wave's peak to the peak of the wave behind it._

Matching

Choose the most appropriate answer for each term.

1. ___prokaryotic cells [p.54}
2. ___plasma membrane [p.54]
3. ___cytoplasm [p.54]
4. ___ribosomes [p.54]
5. ___nucleus [p.54]
6. ___eukaryotic cells [p.54]
7. ___surface-to-volume ratio [p.55]
8. ___lipid bilayer [p.54]
9. ___nucleoid [p.54]
10. ___phospholipid molecules [p.54]
11. ___cell [p.54]

A. An interior region of bacterial cells where DNA is found
B. Structural arrangement of all cell membranes
C. Bacteria cells lacking a nucleus
D. A physical relationship that constrains increases in cell size
E. The smallest unit that retains the properties of life
F. Cytoplasmic structures on which proteins are built
G. The most abundant components of cell membranes
H. The thin outermost membrane of cells; maintains the cell as a distinct entity; substances and signals move across it
I. A membranous sac containing DNA; found in eukaryotic cells
J. The area of a cell between the plasma membrane and the region of DNA
K. Their cytoplasm includes organelles, which are tiny sacs and other compartments bounded by membranes

12. Explain why most cells are, of necessity, very small. [p.55] _____

_____

_____

_____

Matching

Choose the most appropriate answer for each term.

13. ___electron wavelength [p.56]

14. ___micrograph [p.56]

15. ___transmission electron microscope [p.56]

16. ___scanning tunneling microscope [p.57]

17. ___wavelength [p.56]

18. ___compound light microscope [p.56]

19. ___phase-contrast and Nomarski processes [p.57]

20. ___scanning electron microscope [p.56]

A. Glass lenses bend incoming light rays to form an enlarged image of a cell or some other specimen

B. The distance from one wave's peak to the peak of the wave behind it

C. A computer analyzes a tunnel formed in electron orbitals, produced between the tip of a needlelike probe and an atom

D. A narrow beam of electrons moves back and forth across the surface of a specimen coated with a thin metal layer

E. A photograph of an image formed with a microscope

F. About 100,000 times shorter than those of visible light

G. Electrons pass through a thin section of cells to form an image

H. Create optical contrasts without staining the cells; enhance the usefulness of light micrographs

4.3. DEFINING FEATURES OF EUKARYOTIC CELLS [pp.58–61]

Selected Words: compartmentalization [p.58]

Boldfaced, Page-Referenced Terms

[p.58] organelle _is an internal membrane-bound sac or compartment that serves one or more specialized functions in eukaryotic cells._

Complete the Table

1. Complete the following table to identify eukaryotic organelles and their functions.

Organelle or Structure	Main Function
[p.58] a.	Localizing the cell's DNA
[p.58] b.	Synthesis of polypeptide chains
[p.58] c.	Route and modify newly formed polypeptide chains; lipid synthesis
[p.58] d.	Modifying polypeptide chains into mature proteins; sort and ship proteins and lipids for secretion or use inside the cell
[p.58] e.	Transport or store a variety of substances; digest substances and structures in the cell; other functions
[p.58] f.	Producing many ATP molecules in highly efficient fashion
[p.58] g.	Overall cell shape and internal organization; moving the cell and its internal structures

Short Answer

2. Compare the illustrations of plant and animal cells shown in text Figure 4.8, p. 59; Figure 4.9, p. 60; and Figure 4.10, p. 61. List the organelles and structures that do not seem to be common in both plant and animal cells.

_____

_____

_____

4.4. THE NUCLEUS [pp.62–63]

Selected Words: chromatin [p.63], chromosomes [p.63]

Boldfaced, Page-Referenced Terms

[p.62] nucleus (eukaryotic) It protects the DNA and is the boundary for what passes to and from the cytoplasm

[p.62] nuclear envelope the double-membrane system that wraps around the nucleus.

[p.62] nucleolus (plural, nucleoli) inside the nucleus, is where a large number of proteins and RNA molecules are constructed. These materials build ribosomes.

[p.63] chromatin – the cells collection of DNA, together with all proteins associated with it,

[p.63] chromosome is one DNA molecule and its associated proteins, reguardless of whether it is in threadlike or condensed form.

Short Answer

1. List the two major functions of the nucleus. [p.62] _____

_____

_____

_____

_____

Complete the Table

2. Complete this table about the eukaryotic nucleus by entering the name of each nuclear component described.

Nuclear Component	Description
[pp.62–63] a.	Dense cluster of RNA and proteins that will be assembled into subunits of ribosomes
[p.62] b.	Consists of two lipid bilayers in which numerous protein molecules are embedded; surrounds the fluid portion of the nucleus
[p.63] c.	The cell's collection of DNA, together with all proteins associated with it
[p.62] d.	Fluid interior portion of the nucleus
[p.63] e.	One DNA molecule and its associated proteins, regardless if it is in threadlike or condensed form

Fill-in-the-Blanks

Outside the nucleus, polypeptide chains for proteins are assembled on (3) _____ [p.63]. Many new chains become stockpiled in the (4) _____ [p.63] or get used at once. Many others enter a (5) _____ [p.63] system. This system consists of different organelles, including (6) _____ [p.63] reticulum, (7) _____ [p.63] bodies, and vesicles.

Thanks to (8) _____ [p.63] instructions, many (9) _____ [p.63] take on particular final forms in the cytomembrane system. (10) _____ [p.63] also are packaged and assembled in the system by (11) _____ [p.63] and other proteins constructed according to DNA's instructions. (12) _____ [p.63] deliver the proteins and lipids to specific sites within the cell or to the plasma membrane, for export.

4.5. THE CYTOMEMBRANE SYSTEM [pp.64–65]

Selected Words: rough ER [p.64], *smooth* ER [p.64], lysosomes [p.65], peroxisomes [p.65]

Boldfaced, Page-Referenced Terms

[p.64] cytomembrane system _is a series of organelles in which lipids are assembled and new polypeptide chains are modified into final proteins_ [p.64] endoplasmic reticulum, or ER _____

[p.64] Golgi bodies *is where enzymes put the finishing touches on proteins and lipids, sort them out, and package them inside vescles for shipment to specific locations.*

[p.65] vesicles *are tiny, membranous sacs that move through the cytoplasm or take up position in it.*

Matching

Study the following illustration and match each component of the cytomembrane system with the most correct description of function. Some components may be used more than once.

1. ___Assembly of polypeptide chains [p.64]

2. ___Lipid assembly [p.64]

3. ___DNA instructions for building polypeptide chains [p.64]

4. ___Initiate protein modification following assembly [p.64]

5. ___Proteins and lipids take on final form [p.64]

6. ___Sort and package lipids and proteins for transport to proper destinations following modification [p.64]

7. ___Vesicles formed at plasma membrane transport substances into cytoplasm [p.64]

8. ___Sacs of enzymes that break down fatty acids and amino acids, forming hydrogen peroxide [p.65]

9. ___Special vesicles budding from Golgi bodies that become organelles of intracellular digestion [p.65]

10. ___Transport unfinished proteins to a Golgi body [p.64]

11. ___Transport finished Golgi products to the plasma membrane [p.64]

12. ___Release Golgi products at the plasma membrane [p.64]

13. ___Transport unfinished lipids to a Golgi body [p.64]

A. spaces within smooth membranes of ER
B. nucleus
C. Golgi body
D. vesicles from Golgi
E. vesicles budding from rough ER
F. endocytosis with vesicles
G. exocytosis with vesicles
H. spaces within rough ER
I. ribosomes in the cytoplasm
J. vesicles budding from smooth ER
K. lysosomes
L. peroxisomes

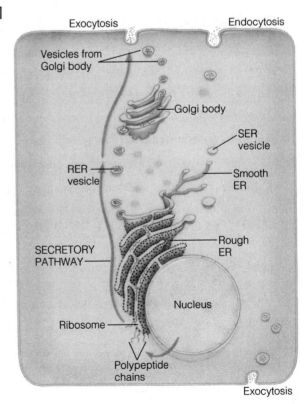

4.6. MITOCHONDRIA [p.66]

4.7. SPECIALIZED PLANT ORGANELLES [p.67]

Selected Words: stroma [p.67], thylakoid membrane [p.67], granum [p.67], chlorophylls [p.67], carotenoids [p.67], chromoplasts [p.67], amyloplasts [p.67]

Boldfaced, Page-Referenced Terms

[p.66] mitochondrion _specializes in the protection of large quantaties of ATP, the main energy-carrying molecule between different reaction sites in cells._

[p.67] chloroplasts _____

[p.67] central vacuole _____

Choice

For questions 1–20, choose from the following:

a. mitochondria   b. chloroplasts   c. amyloplasts   d. central vacuole   e. chromoplasts

1. ___ Occur only in photosynthetic eukaryotic cells [p.67]
2. ___ ATP molecules form when organic compounds are completely broken down to carbon dioxide and water [p.66]
3. ___ Plastids that lack pigments [p.67]
4. ___ A muscle cell might have a thousand or more [p.66]
5. ___ Plastids that have an abundance of carotenoids but no chlorophylls [p.67]
6. ___ ATP-forming reactions require oxygen [p.66]
7. ___ Causes fluid pressure to build up inside a living plant cell [p.67]
8. ___ Organelles that convert sunlight energy into the chemical energy of ATP [p.67]
9. ___ The source of the red-to-yellow colors of many flowers, autumn leaves, ripening fruits, and carrots or other roots [p.67]
10. ___ Inner folds are called *cristae* [p.66]
11. ___ May increase so much in volume that it takes up 50 to 90 percent of the cell's interior [p.67]
12. ___ Plastid with internal areas known as *grana* and *stroma* [p.67]
13. ___ Plastids that resemble photosynthetic bacteria [p.67]
14. ___ Two distinct compartments are created by a double-membrane system [p.66]
15. ___ Store starch grains and are abundant in cells of stems, potato tubers, and seeds [p.67]
16. ___ Resemble bacteria in terms of size and biochemistry [p.66]
17. ___ The site of photosynthesis in plant cells [p.67]
18. ___ Fluid-filled, stores amino acids, sugars, ions, and toxic wastes [p.67]
19. ___ All eukaryotic cells have one or more [p.66]
20. ___ The thylakoid membrane [p.67]

4.8. COMPONENTS OF THE CYTOSKELETON [pp.68–69]
4.9. THE STRUCTURAL BASIS OF CELL MOBILITY [pp.70–71]
4.10. CELL SURFACE SPECIALIZATIONS [pp.72–73]

Selected Words: tubulin [p.68], centrosome [p.68], MTOCs [p.68], GTP [p.68], GDP [p.68], actin [p.69], myosin [p.69], dynein [p.69], crosslinking proteins [p.69], intermediate filaments [p.69], *Amoeba proteus* [p.70], "9 + 2 array" [p.70], pectin [p.72], lignin [p.72], plasmodesmata [p.73], *tight* junctions [p.73], *adhering* junctions [p.73], *gap* junctions [p.73]

Boldfaced, Page-Referenced Terms

[p.68] cytoskeleton gives cells their internal organization, shape, and capacity to move.

[p.68] microtubules is a long, hollow cylinder, 25 nm wide, made of tubulin monomers. Responsible for movement in the cell

[pp.68–69] microfilaments thinnest of the cytoskeletal elements, 5-7 nm wide. Responsible for movement in the cell.

[p.69] intermediate filaments ropelike cytoskeletal elements that impart mechanical strength to cells and tissues

[p.69] motor protein attach to the surface of microtubules and microfilaments in ways that cause cell movement.

[p.69] cell cortex is an extensive, three-dimensional network of microfilaments and other proteins just beneath the plasma membrane.

[p.70] pseudopods "false feet" temporary, lobelike protrusions from the body.

[p.70] flagellum (plural, flagella) usually just one tube that helps move the whole cell "9+2 array"

[p.70] cilium (plural, cilia) multiple tubes that help move objects over the cell. "9+2 array"

[p.70] centriole a barrel-shaped structure that is one type of microtubule-producing center

[p.70] basal body

[p.72] cell wall

[p.72] primary wall

[p.72] secondary wall

Dichotomous Choice

Circle one of two possible answers given between parentheses in each statement.

1. (Protein/Carbohydrate) subunits form the basic components of microtubules. [p.68]
2. Microtubules, microfilaments, or both take part in most aspects of eukaryotic cell (movement/chemistry). [p.68]
3. Hollow microtubule cylinders consist of (tubulin/actin) protein subunits or monomers [p.68].
4. A microtubule grows in length by adding monomers on its (minus/plus) end. [p.68]
5. Sites giving rise to microtubules are known as (MTOCs/GTP-GDP). [p.68]
6. Doctors have used (taxol/colchicine) to decrease the uncontrolled cell divisions underlying the growth of some benign and malignant tumors. [pp.68–69]
7. Microfilaments consist of two polypeptide chains of (tubulin/actin) monomers helically twisted together. [p.69]
8. Myosin and dynein serve as (construction/motor) proteins. [p.69]
9. The extensive, three-dimensional network of microfilaments and other proteins just beneath the plasma membrane is known as the *cell* (*cortex*/MTOC). [p.69]
10. (Intermediate filaments/Microfilaments) mechanically strengthen cells or cell parts and help maintain their shape. [p.69]
11. The pseudopods of *Amoeba proteus* grow by controlled (assembly/disassembly) of microfilaments inside each lobe. [p.70]
12. Within muscle cells, ATP-activated myosin repeatedly bind and release microfilament neighbors; this action causes the microfilaments to slide over the myosin strands, toward the center of the contractile unit, which (lengthens/shortens) as a result. [p.70]
13. In response to the sun's position overhead, chloroplasts flowing in plant cells due to myosin monomers "walking" over bundles of microfilaments is known as (amoeboid motion/cytoplasmic streaming). [p.70]
14. Sperm and many other free-living cells use (flagella/cilia) as whiplike tails for swimming. [p.70]
15. The human respiratory tract is lined with beating (flagella/cilia). [p.70]
16. 9 + 2 arrays of microtubules arise from a (centriole/pseudopod), one type of microtubule-producing center. [p.70]
17. The cross-sectional array of 9 + 2 microtubules is found in (cilia/centrioles). [pp.70–71]
18. Flagella and cilia beat by a sliding mechanism involving motor proteins known as (crosslinking proteins/dynein). [p.71]

Matching

Choose the most appropriate answer for each term.

19. ___primary wall [p.72]
20. ___secondary wall [p.72]
21. ___plasmodesmata [p.72]
22. ___tight junctions [p.73]
23. ___adhering junctions [p.73]
24. ___gap junctions [p.73]

A. Link the cells of epithelial tissues lining the body's outer surface, inner cavities, and organs
B. Numerous tiny channels crossing the adjacent primary walls of living plant cells and connect their cytoplasm
C. Quite sticky, and they cement adjacent cells together
D. Link the cytoplasm of neighboring animal cells and are open channels for the rapid flow of signals and substances
E. Formed of rigid cellulose, lignin, and additional deposits; reinforces plant cell shape
F. Join cells in tissues of the skin, heart, and other organs subject to stretching

Short Answer

25. What particular quality of secondary cell walls that is not found in primary walls allows woody plants such as shrubs and trees to stand erect? [p.72]_____

_____

_____

4.11. PROKARYOTIC CELLS—THE BACTERIA [pp.70–75]

Selected Words: prokaryotic [p.74], pathogenic bacteria [p.74], bacterial flagellum [p.74], pili [p.74], *Escherichia coli* [p.74], *Nostoc* [p.74], *Pseudomonas marginalis* [p.74]

Fill-in-the-Blanks

With one rare exception, (1) _____ [p.74] are the smallest and most structurally simple cells. The word *prokaryotic* means "before the (2) _____" [p.74], which implies that bacteria evolved before cells possessing nuclei existed. Most bacteria have a semirigid or rigid cell (3) _____ [p.74] that wraps around the plasma membrane, structurally supports the cell, and imparts shape to it [p.74]. Dissolved substances can move freely to and from the plasma membrane because the wall is (4) _____ [p.74]. Often, sticky (5) _____ [p.74] surround the cell wall and form a protective, thick, jellylike capsule around it. Bacteria have a plasma membrane that helps to (6) _____ [p.74] the movement of substances to and from the cytoplasm. A bacterial plasma membrane, too, has (7) _____ [p.74] that serve as channels, transporters, and receptors for signals and substances. Bacterial cells also have many cytoplasmic (8) _____ [p.74] on which polypeptide chains are assembled. These small and internally simple cells do not require a (9) _____ [p.74]. The cytoplasm of a bacterium is continuous with an irregularly shaped region of (10) _____ [p.74] named the (11) _____ [p.74]. A circular molecule of DNA, also called the *bacterial* (12) _____ [p.74], occupies this region. Extending from the surface of many bacterial cells are one or more threadlike motile structures known as *bacterial* (13) _____ [p.74]. These structures help a cell move rapidly in its fluid surroundings. Other surface projections include (14) _____ [p.74]. These are the main (15) _____ [p.74] filaments that help many kinds of bacteria attach to various surfaces, even to each other. There are two kingdoms of metabolically diverse organisms, the (16) _____ [p.74] and (17) _____ [p.74]. Ancient (18) _____ [p.75] cells gave rise to all the protistans, plants, fungi, and animals ever to appear on Earth.

Self-Quiz

Labeling and Matching

Identify each indicated part of the accompanying illustrations. Complete the exercise by matching and entering the letter of the proper function description in the parentheses following each label. Some letter choices must be used more than once.

1. _____ _____ (   ) [p.64]

2. _____ (   ) [p.65]

3. _____ (cytoskeletal component) (   ) [p.69]

4. _____ (   ) [p.66]

5. _____ (   ) [p.67]

6. _____ (cytoskeletal component) (   ) [p.68]

7. _____ (   ) [p.67]

8. _____ _____ _____ (   ) [p.64]

9. _____ (   ) [p.64]

10. _____ _____ _____ (   ) [p.64]

11. _____ + _____ (   ) [pp.62–63]

12. _____ (   ) [pp.62–63]

13. _____ (   ) [p.62]

14. _____ (   ) [p.62]

15. _____ _____ (   ) [p.54]

16. _____ _____ (   ) [p.72]

17. microfilaments (cytoskeletal component) (L) [p.69]

18. micro tubules (cytoskeletal component) (H) [p.68]

19. plasma membrane (N) [p.54]

20. mitochondrion (O) [p.66]

21. nuclear envelope (A) [p.62]

22. nucleolus (K) [pp.62–63]

23. DNA + nucleoplasm (Q) [pp.62–63]

24. nucleus (I) [p.62]

25. vesicle (G) [p.65]

26. lysosome (E) [p.65]

27. rough ER (P) [p.64]

28. ribosomes (R) [p.64]

29. smooth ER (D) [p.64]

30. vesicle (G) [p.65]

31. Golgi Body (J) [p.64]

32. centrioles (F) [p.70]

A. Pore-riddled two-membrane structure; a profusion of ribosomes is found on its outer surface

B. Porous, but provides protection and structural support for some cells

C. Present in mature, living plant cells; a fluid-filled organelle that stores amino acids, sugars, ions, and toxic wastes

D. Free of ribosomes; the main site of lipid synthesis in many cells

E. Formed as buds from Golgi membranes of animal cells and some fungal cells; an organelle of intracellular digestion

F. Barrel-shaped structures that serve as a type of microtubule-producing center

G. Tiny membranous sacs that move through the cytoplasm or take up positions in it; a common type functions as an organelle of intracellular digestion

H. Takes part in diverse cell movements; consist of protein subunits, called *tubulins*, that form a hollow cylinder

I. A membrane-bound compartment that houses DNA in eukaryotic cells

J. Enzymes put the finishing touches on proteins and lipids, sort them out, and package them inside vesicles for shipment to specific locations

K. Site of assembly of protein and RNA molecules into ribosomal subunits

L. Takes part in diverse cell movements; consist of two chains of actin subunits, twisted together

M. Organelles that convert sunlight energy to the chemical energy of ATP, which is used to make sugars and other organic compounds

N. Thin, outermost membrane that maintains the cell as a distinct entity; substances and signals continually move across it in highly controlled ways

O. Site of aerobic respiration; ATP-producing powerhouse of all eukaryotic cells

P. Stacks of flattened sacs with many ribosomes attached; accomplishes initial modification of protein structure after formation on ribosomes

Q. Genetic material and the fluid interior portion of the nucleus

R. Site of the synthesis of every new polypeptide chain

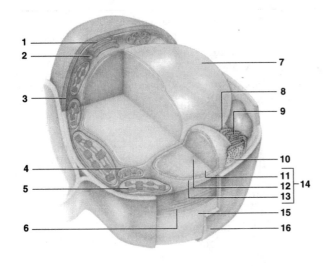

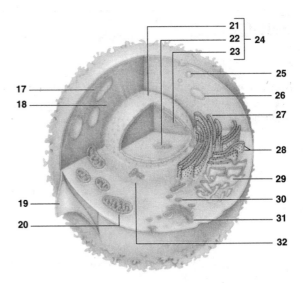

Multiple Choice

_D_ 33. Membranes consist of _____. [pp.54–55]
   a. a lipid bilayer
   b. a protein bilayer
   c. phospholipids and proteins
   d. both a and c are correct

_c_ 34. Which of the following is *not* found as a part
   of prokaryotic cells? [p.58]
   a. Ribosomes
   b. DNA
   c. Nucleus
   d. Cytoplasm
   e. Cell wall

_d_ 35. Which of the following statements most
   correctly describes the relationship between
   cell surface area and cell volume? [p.55]
   a. As a cell expands in volume, its diameter
      increases at a rate faster than its surface
      area does.
   b. Volume increases with the square of the
      diameter, but surface area increases only
      with the cube.
   c. If a cell were to grow four times in diam-
      eter, its volume of cytoplasm increases
      sixteen times and its surface area in-
      creases sixty-four times.
   d. Volume increases with the cube of the di-
      ameter, but surface area increases only
      with the square.

_d_ 36. Most cell membrane functions are carried
   out by _____. [pp.54–55]
   a. carbohydrates
   b. nucleic acids
   c. lipids
   d. proteins

_b_ 37. Animal cells dismantle and dispose of waste
   materials by _____. [p.65]
   a. using centrally located vacuoles
   b. several lysosomes fusing with a vesicle
      formed at the plasma membrane that en-
      closes the wastes
   c. microvilli packaging and exporting the
      wastes
   d. mitochondrial breakdown of the wastes

_a_ 38. The nucleolus is the site where _____.
   [pp.62–63]
   a. the protein and RNA subunits of ribo-
      somes are assembled
   b. the chromatin is formed
   c. chromosomes are bound to the inside of
      the nuclear envelope
   d. chromosomes duplicate themselves

_C_ 39. The _____ is free of ribosomes and curves through the cytoplasm like connecting pipes; the main site of lipid synthesis. [p.64]
   a. lysosome
   b. Golgi body
   c. smooth ER
   d. rough ER

_a_ 40. Which of the following is *not* present in all cells? [pp.58–59]
   a. Cell wall
   b. Plasma membrane
   c. Ribosomes
   d. DNA molecules

_b_ 41. As a part of the cytomembrane system, the _____ put the finishing touches on lipids and proteins to permit sorting and packaging for specific locations. [p.64]
   a. endoplasmic reticulum
   b. Golgi bodies
   c. peroxisomes
   d. lysosomes

_C_ 42. Chloroplasts _____. [p.67]
   a. are specialists in oxygen-requiring reactions
   b. function as part of the cytoskeleton
   c. trap sunlight energy and produce organic compounds
   d. assist in carrying out cell membrane functions

_b_ 43. Mitochondria convert energy stored in _____ to forms that the cell can use, principally ATP. [p.66]
   a. water
   b. carbon compounds
   c. NADPH$_2$
   d. carbon dioxide

_d_ 44. _____ are sacs of enzymes that bud from ER; they produce potentially harmful hydrogen peroxide while breaking down fatty acids and amino acids. [p.65]
   a. Lysosomes
   b. Glyoxysomes
   c. Golgi bodies
   d. Peroxisomes

_C_ 45. Two classes of cytoskeletal elements underlie nearly all movements of eukaryotic cells; they are _____. [p.68]
   a. desmins and vimentins
   b. actin and microfilaments
   c. microtubules and microfilaments
   d. microtubules and myosin

Choice

Cells of the organisms in the six kingdoms of life share the following characteristics: plasma membrane, DNA, RNA, and ribosomes. For questions 46–57, choose the kingdom(s) possessing the characteristics listed below. Some questions will require more than one kingdom as the answer.

   a. Archaebacteria, Eubacteria   b. Protistans   c. Fungi   d. Plants   e. Animals

46. _bcde_ cytoskeleton [p.76]
47. _abcd_ cell wall [p.76]
48. _bcde_ nucleus [p.76]
49. _bcde_ nucleolus [p.76]
50. _cd_ central vacuole [p.76]
51. _a_ simple flagellum [p.76]
52. _abd_ photosynthetic pigment [p.76]
53. _bd_ chloroplast [p.76]
54. _bcde_ endoplasmic reticulum [p.76]

55. _bcde_ Golgi body [p.76]
56. _bcde_ lysosome [p.76]
57. _bcde_ complex flagella and cilia [p.76]

Chapter Objectives/Review Questions

1. List the three generalizations that together constitute the cell theory. [p.52]
2. List and describe the three major regions that all cells have in common. [p.54]
3. The cytoplasm of _____ cells includes tiny sacs called *organelles*; one sac houses the DNA. [p.54]
4. _____ cells, by contrast, have no nuclei. [p.54]
5. Describe the lipid bilayer arrangement for the plasma membrane. [pp.54–55]
6. Cell size is necessarily limited because its volume increases with the _____ but surface area increases only with the _____. [p.55]
7. Briefly describe the operating principles of light microscopes, phase-contrast microscopes, scanning tunneling microscopes, transmission electron microscopes, and scanning electron microscopes. [pp.56–57]
8. Briefly describe the cellular location and function of the organelles typical of most eukaryotic cells: nucleus, ribosomes, endoplasmic reticulum, Golgi body, various vesicles, mitochondria, and the cytoskeleton. [p.58]
9. In eukaryotic cells, _____ separate different incompatible chemical reactions in space and time. [p.62]
10. Eukaryotic cells protect their _____ inside a nucleus. [p.62]
11. Describe the nature of the nuclear envelope and relate its function to its structure. [p.62]
12. _____ are sites where the protein and RNA subunits of ribosomes are assembled. [pp.62–63]
13. _____ is the cell's collection of DNA molecules and associated proteins; a _____ is an individual DNA molecule and associated proteins. [p.63]
14. Explain how the endoplasmic reticulum (rough and smooth types), peroxisomes, Golgi bodies, lysosomes, and a variety of vesicles function together as the cytomembrane system. [pp.64–65]
15. _____ are organelles of intracellular digestion that bud from Golgi membranes of animal cells and some fungal cells. [p.65]
16. Define and describe the function of peroxisomes. [p.65]
17. Within _____, energy stored in organic molecules is released by enzymes and used to form many ATP molecules in the presence of oxygen. [p.66]
18. Describe the detailed structure of the chloroplast, the site of photosynthesis (include grana and stroma). [p.67]
19. Give the general function of the following plant organelles: chloroplasts, chromoplasts, amyloplasts, and the central vacuole. [p.67]
20. Elements of the _____ give eukaryotic cells their internal organization, overall shape, and capacity to move. [p.68]
21. List the three major structural elements of the cytoskeleton and give the general function of each. [p.68]
22. All three classes of cytoskeletal elements grow through _____, by which many monomers are joined together. [p.68]
23. _____ proteins such as myosin and dynein play roles in cell movement. [p.69]
24. _____ filaments are the most stable elements of the cytoskeleton and are present only in animal cells or specific tissues. [p.69]
25. Explain the three mechanisms of cell movements. [p.70]
26. *Amoeba proteus*, a soft-bodied protistan, crawls on _____. [p.70]
27. Both cilia and flagella have an internal microtubule arrangement called the "_____ _____" array. [p.70]
28. A centriole remains at the base of a completed microtubule-producing center, where it is often called a _____ _____. [p.70]
29. Explain the sliding mechanism by which flagella and cilia beat. [p.71]
30. Distinguish a primary cell wall from a secondary cell wall in leafy plants. [p.72]
31. Describe the location and function of plasmodesmata. [p.73]

32. _____ junctions link the cells of epithelial tissues; _____ junctions join cells in tissues of the skin, heart, and other organs subjected to stretching; _____ junctions link the cytoplasm of neighboring cells. [p.73]

33. Describe the structure of a generalized prokaryotic cell. Include the bacterial flagellum, nucleoid, pili, capsule, cell wall, plasma membrane, cytoplasm, and ribosomes. [p.74]

Integrating and Applying Key Concepts

1. Which parts of a cell constitute the minimum necessary for keeping the simplest of living cells alive?

2. How did the existence of a nucleus, compartments, and extensive internal membranes confer selective advantages on cells that developed these features?

5

A CLOSER LOOK AT CELL MEMBRANES

Interactive Exercises

It Isn't Easy Being Single [pp.78–79]

5.1. MEMBRANE STRUCTURE AND FUNCTION [pp.80–81]

5.2. *Focus on Science:* TESTING IDEAS ABOUT CELL MEMBRANES [pp.82–83]

Selected Words: mosaic [p.80], *fluid* [p.80], *transport* proteins [p.81], *receptor* proteins [p.81], *recognition* proteins [p.81], *adhesion* proteins [p.81], *observed* ratio of proteins to lipids [p.82], ratio *predicted* [p.82], *freeze-fracturing, freeze-etching* [p.82]

Boldfaced, Page-Referenced Terms

[p.80] phospholipid *has a phosphate-containing head and two fatty acid tails attached to a glycerol backbone*

[p.80] lipid bilayer *is the structural basis of cell membranes*

[p.80] fluid mosaic model *cell membranes are a mixed composition - a mosaic - of phospholipids, glysolipids, sterols, and proteins.*

[p.82] centrifuge *is a motor-driven rotary device that can spin test tubes at very high speed.*

Labeling

Identify each numbered membrane protein in the following illustration. [p.81]

1. <u>Adhesion</u> protein   2. through 5. are types of <u>Transport</u> proteins.

6. <u>Receptor</u> protein   7. <u>Recognition Protein</u>

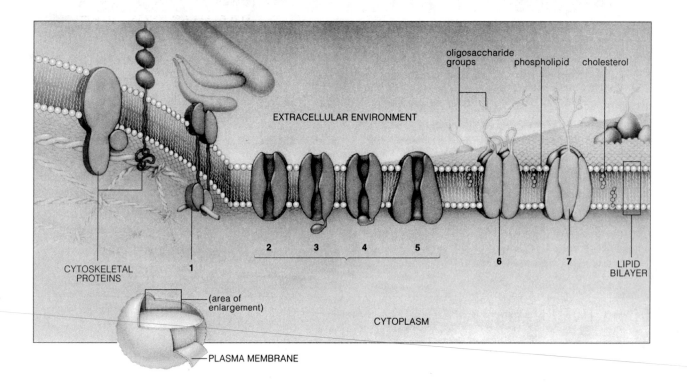

Matching

Choose the most appropriate answer for each term.

8. ___ "fluid" [p.80]

9. ___ phospholipid [p.80]

10. ___ adhesion proteins [p.81]

11. ___ transport proteins [p.81]

12. ___ "mosaic" [p.80]

13. ___ centrifuge [p.82]

14. ___ recognition proteins [p.81]

15. ___ freeze-fracturing and freeze-etching [p.82]

16. ___ fluid mosaic model [p.80]

17. ___ receptor proteins [p.81]

18. ___ lipid bilayer arrangement [p.80]

19. ___ protein coat model [p.82]

A. Allow water-soluble substances to move through their interior, thus crossing the bilayer

B. The mixed composition of diverse phospholipids, glycolipids, sterols, and proteins

C. A motor-driven rotary device that can spin test tubes at very high speeds; materials of lesser mass remain near the top of the tube

D. Of multicelled organisms; helps cells of the same type locate and stick to one another and stay positioned in the proper tissues

E. By this model, membranes have a mixed composition of lipids and proteins

F. Bind extracellular substances, such as hormones, that trigger changes in cell activities

G. Comparing the observed ratio of proteins to lipids against the ratio predicted on the basis of the model

H. Determination of the nature of cell membranes by observation

I. Like molecular fingerprints, their oligosaccharide chains identify a cell as being of a specific type

J. Consists of a hydrophilic head and two hydrophobic tails

K. Refers to the motions and interactions of component parts of the membrane

L. The structural basis of cell membranes

5.3. CROSSING SELECTIVELY PERMEABLE MEMBRANES [pp.84–85]

5.4. PROTEIN-MEDIATED TRANSPORT [pp.86–87]

Selected Words: concentration [p.84], *gradient* [p.84], *passive transport* [p.85], *active transport* [p.85], *exocytosis* [p.85], *endocytosis* [p.85], net direction of movement [p.86], *net* movement [p.87]

Boldfaced, Page-Referenced Terms

[p.84] selective permeability _the membrane permits some some substances but not others to to cross it in certain ways, at certain times._

[p.84] concentration gradient _is a difference in the number of molecules or ions of a given substance between adjoining regions._

[p.84] diffusion _is the name for the net movement of like molecules or ions down a concentration gradient._

[p.85] electric gradient _is a difference in electric charges of adjoining regions_

[p.85] pressure gradient _is a difference in the pressure being exerted in adjoining regions._

[p.86] passive transport _is the name for a flow of solutes through the interior of transport proteins, down their concentration gradients._

[p.87] active transport _energy-driven mechanisms called "membrane pumps" make solutes cross membranes against concentration gradients._

[p.87] calcium pump _helps keep the calcium concentration in a cell at least a thousand times lower than outside._

[p.87] sodium–potassium pump _moves potassium ions (k+) across the plasma membrane._

_____

Fill-in-the-Blanks

If a membrane has selective (1) _____ [p.84], it possesses a molecular structure that permits some substances but not others to cross it in certain ways, at certain times. (2) _____ [p.84] refers to the number of molecules or ions of a substance in a specified region, as in a volume of fluid or air. (3) _____ _____ [p.84] is a difference in the number of molecules or ions of a given substance between adjoining regions. (4) _____ [p.84] is the name for the net movement of like molecules or ions down a concentration gradient; it is a key factor in the movement of substances across cell membranes and through cytoplasmic fluid. Diffusion is faster when a gradient is (5) _____ [p.84]. A net distribution of molecules that is nearly uniform through two adjoining regions is called "dynamic (6) _____" [p.85]. The rates of diffusion are faster at (7) _____ temperatures [p.85]. Molecular (8) _____ [p.85] also affects diffusion rates. The rate and direction of diffusion may also fall under the influence of an (9) _____ [p.85] gradient, a difference in electric charges of adjoining regions. The presence of a (10) _____ [p.85] gradient may also affect the rate and direction of diffusion.

Choice

For questions 11–24, choose from the following mechanisms of protein-mediated transport:

    a. passive transport   b. active transport   c. applies to both active and passive transport

11. ___ The calcium pump [p.87]

12. ___ Part of the transport protein closes in behind the bound solute—and part opens up to the opposite side of the membrane [pp.86–87]

13. ___ Involves a carrier protein that is not energized [p.86]

14. ___ A carrier protein has a specific site that weakly binds a substance [pp.86–87]

15. ___ During a given interval, the *net* direction of movement depends on how many solute molecules make random contact with vacant binding sites in the interior of proteins [p.86]

16. ___ The transport protein must receive an energy boost, usually from ATP [p.87]

17. ___ Solute binding to a carrier protein leads to changes in protein shape [pp.86–87]

18. ___ The sodium–potassium pump [p.87]

19. ___ Transport proteins span the bilayer, and their interior is able to open on both sides of it [pp.86–87]

20. ___ A solute is pumped across the cell membrane *against* its concentration gradient [p.87]

21. ___ The solute binding site improves when ATP donates energy to the carrier protein to allow a better chemical fit [p.87]

22. ___ Passive two-way transport will continue until solute concentrations become equal on both sides of the membrane [p.86]

23. ___ Net movement will be down the solute's concentration gradient [p.86]

24. ___ Think of the solute as hopping onto the transport protein on one side of the membrane, then hopping off on the other side [p.86]

Complete the Table

25. Complete the table below listing substances by entering the name of the correct membrane transport mechanism that will move that substance across a membrane. Choose from diffusion, osmosis, facilitated diffusion, (passive transport), and active transport.

Substance	Transport Mechanism
a. H_2O [p.84]	
b. CO_2 [p.84]	
c. Na^+ [p.87]	
d. Glucose [pp.84,86]	
e. O_2 [p.84]	
f. K^+ [p.87]	
g. Substances pumped through interior of a transport protein; energy input required [p.87]	
h. Substances move through interior of a transport protein; no energy input required [pp.86–87]	

5.5. MOVEMENT OF WATER ACROSS MEMBRANES [pp.88–89]

Selected Words: *tonicity* [p.89], *osmotic pressure* [p.89], *plasmolysis* [p.89]

Boldfaced, Page-Referenced Terms

[p.88] bulk flow _is the mass movement of one or more substances in response to pressure, gravity, or some other external force._

[p.88] osmosis _is the diffusion of water molecules in response to a water concentration gradient between two regions seperated by a selective permeable membrane._

[p.89] hypotonic solution _solution that has fewer solutes on one side of a membrane._

[p.89] hypertonic solution _solution that has more solutes one one side of a membrane._

[p.89] isotonic solution _have the same solute consentrations, so water shows no net osmotic movement from one to the other._

[p.89] hydrostatic pressure _a force directed against a wall, membrane, or some other structure that encloses the fluid._

Matching

Choose the most appropriate answer for each.

1. ___bulk flow [p.88]
2. ___osmosis [p.88]
3. ___tonicity [p.89]
4. ___hypotonic solution [p.89]
5. ___hypertonic solution [p.89]
6. ___isotonic solutions [p.89]
7. ___hydrostatic pressure [p.89]
8. ___osmotic pressure [p.89]
9. ___plasmolysis [p.89]

A. Refers to the relative solute concentrations of two fluids
B. Having the same solute concentrations
C. Mass movement of one or more substances in response to pressure, gravity, or other external force
D. The amount of force that prevents further increase in a solution's volume
E. The fluid on one side of a membrane that contains more solutes than the fluid on the other side of the membrane
F. The diffusion of water in response to a water concentration gradient between two regions separated by a selectively permeable membrane
G. Osmotically induced shrinkage of cytoplasm
H. The fluid on one side of a membrane that contains fewer solutes than the fluid on the other side of the membrane
I. A fluid force exerted against a cell wall and/or membrane enclosing the fluid

True–False

If the statement is true, write a T in the blank. If the statement is false, make it correct by changing the underlined word(s) and writing the correct word(s) in the answer blank.

_____10. Because membranes exhibit selective permeability, concentrations of dissolved substances can <u>increase</u> on one side of the membrane or the other. [p.88]

_____11. A water concentration gradient is influenced by the number of <u>solute</u> molecules present on both sides of the membrane. [p.88]

_____12. The relative concentrations of solutes in two fluids are referred to as <u>isotonic</u>. [p.89]

_____13. An animal cell placed in a <u>hypertonic</u> solution would swell and perhaps burst. [p.89]

_____14. Water tends to move from <u>hypotonic</u> solutions to areas with more solutes. [p.89]

_____15. Physiological saline is 0.9 percent NaCl; red blood cells placed in such a solution will not gain or lose water; therefore, one could state that the fluid in red blood cells is <u>hypertonic</u>. [p.89]

_____16. A solution of 80 percent water, 20 percent solute is <u>more</u> concentrated than a solution of 70 percent water, 30 percent solute. [p.89]

_____17. The mass movement of one or more substances in response to pressure, gravity, or some other external force is called <u>osmosis</u>. [p.88]

_____18. Plant cells placed in a <u>hypotonic</u> solution will swell. [p.89]

5.6. BULK TRANSPORT ACROSS MEMBRANES [pp.90–91]

Selected Words: *receptor-mediated* endocytosis [p.90], *bulk-phase* endocytosis [p.90], phagocytic vesicle [p.91]

Boldfaced, Page-Referenced Terms

[p.90] exocytosis _when a vesicle moves to the cell surface, and the protein-studded lipid bilayer of its membrane fuses with the plasma membrane._

[p.90] endocytosis _when a cell takes in substances next to its surface._

[p.90] phagocytosis _(a form of endocytosis) when a cell engulfs microorganisms, large edible particles, and cellular debris. "cell eating"_

Matching

Choose the most appropriate answer for each.

1. ___exocytosis [p.90]
2. ___receptor-mediated endocytosis [p.90]
3. ___bulk-phase endocytosis [p.90]
4. ___phagocytosis [p.90]
5. ___membrane cycling [p.91]

A. A cell engulfs microorganisms, large edible particles, and cellular debris
B. Membrane initially used for endocytic vesicles returns receptor proteins and lipids back to the plasma membrane
C. Vesicles form around small volumes of extracellular fluid of various content
D. A cytoplasmic vesicle moves to the cell surface, its own membrane fuses with the plasma membrane while its contents are released to the environment
E. Chemical recognition and binding of specific substances; pits of clathrin baskets sink into the cytoplasm and close on themselves

Self-Quiz

___ 1. White blood cells use _____ to devour disease agents invading your body. [p.90]
   a. diffusion
   b. bulk flow
   c. osmosis
   d. phagocytosis

___ 2. _____ cells depend on the calcium-pump mechanism. [p.87]
   a. Intestine
   b. Nerve
   c. Muscle
   d. Amoeba

___ 3. _____ proteins bind extracellular substances, such as hormones, that trigger changes in cell activities. [p.81]
   a. Receptor
   b. Adhesion
   c. Transport
   d Recognition

___ 4. In a lipid bilayer, tails point inward and form a(n) _____ region that excludes water. [p.80]
   a. acidic
   b. basic
   c. hydrophilic
   d. hydrophobic

___ 5. A protistan adapted to life in a freshwater pond is collected in a bottle and transferred to a saltwater bay. Which of the following is likely to happen? [p.89]
   a. The cell bursts.
   b. Salts flow out of the protistan cell.
   c. The cell shrinks.
   d. Enzymes flow out of the protistan cell.

___ 6. Which of the following is *not* a form of active transport? [p.88]
   a. Sodium–potassium pump
   b. Endocytosis
   c. Exocytosis
   d. Bulk flow

___ 7. Which of the following is *not* a form of passive transport? [p.90]
   a. Osmosis
   b. Passive transport
   c. Bulk flow
   d. Exocytosis

___ 8. O_2, CO_2, H_2O, and other small, electrically neutral molecules move across the cell membrane by _____. [p.85]
   a. electric gradients
   b. receptor-mediated endocytosis
   c. passive transport
   d. active transport

___ 9. Ions such as H^+, Na^+, K^+, and Ca^{++} move across cell membranes by _____. [p.87]
   a. receptor-mediated endocytosis
   b. pressure gradients
   c. passive transport
   d. active transport

___10. Receptors, pits, and clathrin baskets participate in _____. [p.90]
   a. passive transport
   b. receptor-mediated endocytosis
   c. bulk-phase endocytosis
   d. active transport

Chapter Objectives/Review Questions

1. _____ molecules are the most abundant component of cell membranes. [p.80]
2. Being a composite of phospholipids, glycolipids, sterols, and proteins, the membrane is said to have a _____ quality. [p.80]
3. A lipid bilayer shows a _____ behavior as a result of motions and interactions of its component parts. [p.80]
4. Most membrane functions are carried out by _____ embedded in or attached to one of the surfaces of the membrane. [p.81]
5. State the general functions of transport proteins, receptor proteins, recognition proteins, and adhesion proteins. [p.81]
6. What is the value of a centrifuge to cellular studies? Freeze-fracturing and freeze-etching? [p.82]
7. Explain the concept of selective permeability as it applies to cell membrane function. [p.84]
8. The difference in the number of molecules or ions of a given substance in two adjoining regions is defined as _____ _____. [p.84]
9. The net movement of like molecules or ions down a concentration gradient is called _____. [p.84]
10. In addition to concentration gradients, diffusion rates can be influenced by temperature, _____ gradients, and _____ gradients. [pp.84–85]
11. Generally distinguish passive transport from active transport. [p.85]
12. _____ involves fusion between the plasma membrane and a membrane-bound vesicle that formed inside the cytoplasm; _____ involves an inward sinking of a small patch of plasma membrane, which then seals back on itself to form a cytoplasmic vesicle. [p.85]
13. Describe the mechanism and result of solute binding to a transport protein. [p.86]
14. _____ transport is the name for a flow of solutes through the interior of transport proteins, down their concentration gradients. [p.86]
15. In passive transport, _____ movement will be down the concentration gradient (from higher to lower) until the concentrations are the _____ on both sides of the membrane. [p.86]
16. Describe how ATP is used in active transport and how it improves the passage of solutes. [p.87]
17. The calcium and sodium–potassium pump are examples of _____ transport, wherein net solute movement is against the concentration gradient. [p.87]
18. Define and cite an example of *bulk flow*. [p.88]

19. The diffusion of water molecules in response to water concentration gradients between two regions separated by a selectively permeable membrane is known as _____. [p.88]
20. Define *tonicity*. [pp.88–89]
21. Water tends to move from a _____ solution (less solutes) to a _____ solution (more solutes). [p.89]
22. Water shows no net osmotic movement in an _____ solution. [p.89]
23. A fluid force directed against a wall or membrane is _____ pressure. [p.89]
24. Define *osmotic pressure*. [p.89]
25. _____ is responsible for wilting lettuce. [p.89]
26. Describe mechanisms involved in receptor-mediated endocytosis. [p.90]
27. How does bulk-phase endocytosis differ from phagocytosis? [p.90]
28. Explain the role of endocytosis and exocytosis in membrane recycling; cite an example. [p.91]

Integrating and Applying Key Concepts

If there were no such thing as active transport, how would the lives of organisms be affected?

6

GROUND RULES OF METABOLISM

Interactive Exercises

You Light Up My Life [pp.94–95]

6.1. ENERGY AND THE UNDERLYING ORGANIZATION OF LIFE [pp.96–97]

Selected Words: fluorescent light [p.94], *bioluminescent gene transfers* [p.94], kittyboo [p.94], *thermal* energy [p.96], *chemical* work [p.96], *mechanical* work [p.96], *electrochemical* work [p.96]

Boldfaced, Page-Referenced Terms

[p.94] bioluminescence _when organisms flash with flouresent light._

[p.95] metabolism _a cell's capacity to accuire energy and use it to build, break apart, store, and release substances in controlled ways._

[p.96] potential energy _____

[p.96] kinetic energy _____

[p.96] heat _____

[p.96] chemical energy _____

_____

[p.96] kilocalorie _____

_____

[p.96] first law of thermodynamics _____

_____

[p.97] entropy _____

_____

[p.97] second law of thermodynamics _____

_____

Fill-in-the-Blanks

Living cells must get (1) _____ [p.95] from their environment so that they can do various kinds of tasks (referred to as (2) _____ [p.95] by scientists). Cells need to build specific kinds of molecules and store them for a time when they will need to be used, at which time the cells may need to rearrange the molecules or break them apart; these activities are a kind of chemical (2) known as (3) _____ [p.95]. Energy released when specific kinds of molecules are broken apart is used to power thousands of energy-requiring chemical (4) _____ [p.95] that sustain life. When you rise from sitting in a chair, some of the potential energy stored in food molecules is transformed into (5) _____ _____ [p.96]: the energy of motion involved in the doing of work. 100% of the potential energy cannot be transformed into (5) because some will be released as (6) _____ [p.96] during the conversion.

Choice

In the blank preceding each item, indicate if the first law of thermodynamics (I) or the second law of thermo-dynamics (II) is best described.

7. ___ Apple trees absorbing energy from the sun and storing the energy in the chemical bonds of starch and sugar. [p.96]

8. ___ A hydroelectric plant at a waterfall, producing electricity. [p.96]

9. ___ Egyptian pyramids crumbling slowly to dust over long periods of time. [p.97]

10. ___ The glow of an incandescent bulb following the flow of electrons through a wire. [p.96]

11. ___ Earth's sun is continuously losing energy to its surroundings. [p.97]

12. ___ The movement of a gasoline-powered automobile. [p.96, logic]

13. ___ The glow of a firefly. [p.94]

14. ___ Humans running the 100-meter dash following usual food intake. [p.96]

15. ___ The death and decay of an organism. [p.97]

Choice

For questions 16–20, choose from these possibilities:

    a. chemical energy   b. entropy   c. heat   d. kilocalorie   e. metabolism

16. ___ is a measure of the amount of disorder in a system. [p.97]

17. ___ includes all of the activities by which a cell acquires energy and materials and uses them to build, break apart, store, and release substances in controlled processes that are typical for that cell. [p.95]

18. ___ The measure of energy that can heat 1,000 grams of water from 14.5°C to 15.5°C at standard pressure is one _____. [p.96]

19. ___ is the potential energy stored in the attractive forces (bonds) that cause atoms to group together into molecules. [p.96]

20. ___ that results from collisions among molecules and their surroundings is a kind of kinetic energy also called *thermal energy*. [p.96]

True–False

If the statement is true, write a T in the blank. If the statement is false, make it correct by changing the underlined word(s) and writing the correct word(s) in the answer blank.

_____21. The <u>first</u> law of thermodynamics states that entropy is constantly increasing in the universe. [p.97]

_____22. At rest your body steadily gives off heat equal to that from a <u>100-watt</u> light bulb. [p.96]

_____23. When you eat a potato, some of the stored chemical energy of the food is converted into <u>kinetic</u> energy that moves your muscles. [p.96]

_____24. <u>Energy</u> is the capacity to accomplish work. [p.96]

_____25. The amount of low-quality energy in the universe is <u>decreasing</u>. [p.97]

_____26. <u>No</u> energy conversion can ever be 100 percent efficient. [p.97]

Fill-in-the-Blanks

Some organisms, such as fireflies and kittyboos, emit light, a phenomenon known as (27) _____ [p.94].

(27) occurs as (28) _____ [p.94] first receives a phosphate group from (29) _____ [p.94] and then is

oxidized by (30) _____ [p.94] acting together with the enzyme (31) _____ [p.94]. Light is emitted

as excited (32) _____ [p.94] return to their lower energy levels.

6.2. ENERGY CHANGES AND CELLULAR WORK [pp.98–99]
6.3. A LOOK AT TYPICAL PHOSPHATE-GROUP TRANSFERS [p.100]
6.4. A LOOK AT TYPICAL ELECTRON TRANSFERS [p.101]
6.5. THE DIRECTIONAL NATURE OF METABOLISM [pp.102–103]

Selected Words: *energy inputs* [p.98], "oxidation" [p.101], "reduction" [p.101], reversible [p.102], reactant [p.102], product [p.102], *biosynthetic pathway* [p.103], *degradative pathway* [p.103]

Boldfaced, Page-Referenced Terms

[p.98] endergonic reaction _____

[p.98] exergonic reaction _____

[p.100] ATP, adenosine triphosphate _____

[p.100] ADP, adenosine diphosphate _____

[p.100] phosphorylation _____

[p.100] ATP/ADP cycle _____

[p.101] oxidation–reduction reaction _____

[p.101] electron transport system _____

[p.102] anabolism _____

[p.102] catabolism _____

[p.102] chemical reaction _____

[p.102] chemical equilibrium _____

[p.103] law of conservation of mass _____

[p.103] metabolic pathways _____

[p.103] substrates _____

[p.103] intermediates _____

[p.103] end products _____

[p.103] energy carriers _____

[p.103] enzymes _____

_____

[p.103] cofactors _____

_____

[p.103] transport proteins _____

_____

Labeling

Classify each of the following reactions as *endergonic* or *exergonic*. [p.98]

_____ 1. Burning wood at a campfire.

_____ 2. The products of a chemical reaction have more energy than the reactants.

_____ 3. Glucose + oxygen → carbon dioxide + water plus energy

_____ 4. The reactants of a chemical reaction have more energy than the product.

_____ 5. The reaction releases energy.

Fill-in-the-Blanks

Photosynthetic cells possess the chemical mechanisms to convert sunlight energy to the chemical energy of molecules known as (6) _____ [p.99], a type of nucleotide. It is important to recognize that all cells lack the ability to extract energy *directly* from food molecules such as glucose. The energy in the food molecules must first be harnessed and converted to the energy of (7) _____ _____ [p.99]. Cells must obtain their energy from these molecules. An ATP molecule consists of (8) _____ [p.100] (a nitrogen-containing compound), the five-carbon sugar (9) _____ [p.100], and a string of three (10) _____ [p.100] groups. These components are held together by (11) _____ [p.100] bonds, some of which are rather unstable. Many different enzymes carry out (12) _____ [p.100] reactions to release the terminal phosphate group. This releases (13) _____ [p.100] that can be used by cells to carry out many types of biological tasks. ATP molecules serve as the cell's renewable (14) _____ [p.100] carrier. In the ATP/ADP cycle, an energy input drives the binding of ADP to a phosphate group or to unbound phosphate (P$_i$), forming (15) _____ [p.100]. Attaching a phosphate group to a molecule is called (16) _____ [p.100]; this energizes that molecule and prepares it to enter a specific reaction in a metabolic pathway.

Labeling

Identify the following molecule and label its parts.

17. _____ [p.100]

18. _____ [p.100]

19. _____ [p.100]

20. The name of this molecule is _____ _____. [p.100]

Fill-in-the-Blanks

Oxidation–reduction refers to (21) _____ [p.101] transfers. In terms of oxidation–reduction reactions, a molecule in the sequence that donates electrons is said to be (22) _____ [p.101], while molecules accepting electrons are said to be (23) _____ [p.101]. Electron transport systems "intercept" excited electrons and make use of the (24) _____ [p.101] they release. If we think of the electron transport system as a staircase, excited electrons at the top of the staircase have the (25) [choose one] ❏ most ❏ least energy [p.101]. As the electrons are transferred from one electron carrier to another, some (26) _____ [p.101] can be harnessed to do biological (27) _____ [p.101]. One type of biological work occurs when energy released during electron transfers (down the steps) is used to move ions in ways that set up ion concentration and electric gradients across a membrane. These gradients are essential for the formation of (28) _____ [p.101].

Matching

Match the lettered statements to the numbered items on the sketch. [All from p.101]

29. ___
30. ___
31. ___
32. ___
33. ___

A. Represent the cytochrome molecules in an electron transport system
B. Electrons at their highest energy level
C. Released energy harnessed and used to produce ATP
D. Electrons at their lowest level
E. The separation of hydrogen atoms into protons and electrons

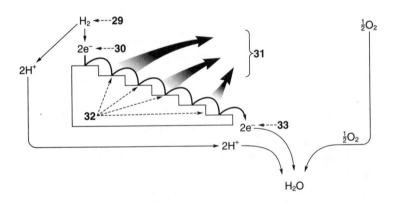

Fill-in-the-Blanks

The substances present at the end of a chemical reaction, the (34) _____ [p.102], may have less or more energy than did the starting substances, the (35) _____ [p.102]. Most reactions are (36) _____ [p.102] in that they can proceed in forward and reverse directions. Such reactions tend to approach chemical (37) _____ [p.102], a state in which the reactions are proceeding at about the same (38) _____ [p.102] in both directions.

Matching

Study the sequence of reactions below. Identify the components of the reactions by selecting items from the following list and entering the correct letter in the appropriate blank.

39. ___ [p.103]

40. ___ [p.103]

41. ___ [p.102]

42. ___ [p.103]

43. ___ [p.102]

44. ___ [p.103]

A. cofactor
B. intermediates
C. reactants
D. end product
E. reversible reaction
F. enzymes

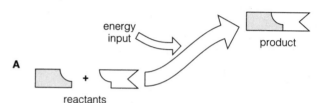

Labeling

Study the diagram at the right; choose the most appropriate answer for each: A or B.

45. _____ a degradative reaction [p.103]

46. _____ an endergonic reaction [p.98]

47. _____ a biosynthetic reaction [p.103]

48. _____ an exergonic reaction [p.98]

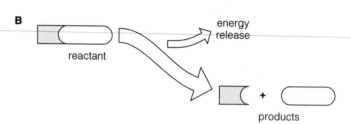

Matching

Match the most appropriate letter to its number.

49. ___ intermediates [p.103]

50. ___ degradative pathways [p.103]

51. ___ chemical equilibrium [p.102]

52. ___ cofactors [p.103]

53. ___ transport proteins [p.103]

54. ___ metabolic pathway [p.103]

55. ___ reactants (substrates) [p.103]

56. ___ energy carriers [p.103]

57. ___ biosynthetic pathways [p.103]

58. ___ enzymes [p.103]

A. An orderly series of reactions catalyzed by enzymes
B. Small organic molecules are assembled into larger organic molecules
C. Mainly ATP; donate(s) energy to reactions
D. Small molecules and metal ions that assist enzymes or serve as carriers
E. Substances able to enter into a reaction
F. Compounds formed between the beginning and end of a metabolic pathway
G. Organic compounds are broken down in stepwise reactions
H. Proteins (usually) that catalyze reactions
I. Rate of forward reaction = rate of reverse reaction
J. Membrane-bound substances that adjust concentration gradients in ways that influence the direction of metabolic reactions

Fill-in-the-Blanks

In cells, the release of energy from glucose proceeds in controlled steps of a degradative pathway, so that (59) _____ [p.103] molecules form along the route from glucose to carbon dioxide and water. Each step of the pathway is controlled and helped by a(n) (60) _____ [p.103], which is usually a protein that speeds up a specific reaction. At each step in the pathway, only some of the bond energy is released. In the internal membrane systems of chloroplasts and mitochondria, the liberated electrons released from the breaking of chemical bonds are sent through (61) _____ _____ [p.101] systems; these organized systems consist of enzymes and (62) _____ [p.101], bound in a cell membrane, that transfer electrons in a highly organized sequence.

Identification

Name the types of pathways in the following diagram of chemical reaction sequences: [p.103]

63. _____ pathway

64. _____ pathway

65. _____ pathway

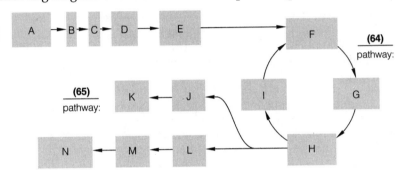

6.6. ENZYME STRUCTURE AND FUNCTION [pp.104–105]
6.7. FACTORS INFLUENCING ENZYME ACTIVITY [pp.106–107]
6.8. *Focus on Health:* GROWING OLD WITH MOLECULAR MAYHEM [pp.108–109]

Selected Words: transition state [p.105], hormone [p.107], superoxide dismutase [p.108], catalase [p.108], "age spots" [p.109]

Boldfaced, Page-Referenced Terms

[p.104] enzymes _____

_____

[p.104] activation energy _____

_____

[p.105] active sites _____

_____

[p.105] induced-fit model _____

_____

[p.106] allosteric control _____

_____

[p.106] feedback inhibition _____

_____

[p.107] coenzymes _____

_____

[p.108] free radical _____

_____

[p.108] antioxidants _____

_____

Fill-in-the-Blanks

At each step in a metabolic pathway, a specific (1) _____ [p.104] lowers the activation energy for the formation of an intermediate compound. (2) _____ [p.104] are highly selective proteins that act as catalysts, which means they greatly enhance the rate at which specific reactions approach (3) _____ [p.104]. The specific substance upon which a particular enzyme acts is called its (4) _____ [p.104]; this substance fits into the enzyme's crevice, which is called its (5) _____ _____ [p.105]. The (6) _____ - _____ [p.105] model describes how a substrate contacts the site without a perfect fit. Enzymes increase reaction rates by lowering the required (7) _____ _____ [p.105]. Sometimes (7) is lowered because the reactant molecule(s) are oriented by the enzyme into positions that put their mutually attractive chemical groups on precise (8) _____ [p.105] courses much more frequently than if the enzyme weren't there. In other situations, acidic or basic side groups of amino acids occupy active sites and easily donate or accept (9) _____ [p.105] atoms to or from substrate molecules; (10) _____ [p.105] and stepwise transfers of electrons from one molecule to another work this way. Enzymes change the (11) _____ [p.105], not the outcome, of a chemical reaction.

Identification

Below is an "energy hill" reaction diagram. Identify the different amounts of energy in #s 12, 13, and 16. Identify the two reaction pathways in #s 14 and 15. [p.104]

12. _____ _____

_____ _____

13. _____ _____

_____ _____

14. _____

15. _____

16. _____ _____ _____

_____ _____

Matching

Match the items on the following sketch with the list of descriptions. Some answers may require more than one letter. [p.105]

17. ___

18. ___

19. ___

20. ___

21. ___

22. ___

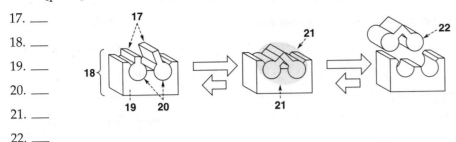

A. Transition state, the time of the most precise fit between enzyme and substrate
B. Complementary active site of the enzyme
C. Enzyme, a protein with catalytic power
D. Product or reactant molecules that an enzyme can specifically recognize
E. Product or reactant molecule
F. Bound enzyme–substrate complex

Fill-in-the-Blanks

In the graph (below left) maximum enzyme activity occurs at (23) _____ [p.106] °C, and minimum activity occurs at (24) _____ [p.106] °C. In the graph (below right), enzyme (25) _____ [p.106] functions best in basic solutions, enzyme (26) _____ [p.106] works best in neutral solutions and enzyme (27) _____ [p.106] functions best in acidic solutions.

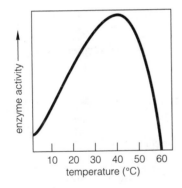

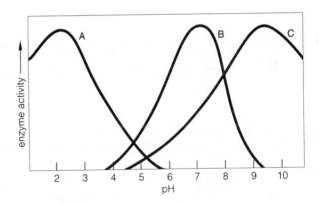

    (28) _____ [p.106] and (29) _____ [p.106] are two factors that influence the rates of enzyme activity. Extremely high fevers can destroy the three-dimensional shape of an enzyme, which may adversely affect (30) _____ [p.106] and cause death. The increased heat energy disrupts weak (31) _____ [p.106] holding the enzyme in its three-dimensional shape. Most enzymes function best at about pH (32) _____ [p.106].

    One example of a control governing enzyme activity is the (33) _____ [p.106] pathway in bacteria. The bacterium synthesizes tryptophan and other (34) _____ _____ [p.106] used to construct its proteins. When no more tryptophan is needed, the rate of protein synthesis slows and the cellular concentration of tryptophan (the end product) (35) [choose one] ❏ rises ❏ falls [pp.106–107], as a control mechanism called (36) _____ _____ [p.106] begins to operate. In this case, excess tryptophan molecules shut down their own production when the unused molecules of tryptophan act to inhibit a key (37) _____ [p.107] in the biochemical pathway. The inhibited enzyme is governed by (38) _____ [p.107] control. Such enzymes have an active site and an (39) _____ _____ [p.107] site where specific substances may bind and alter enzyme activity. When the pathway producing tryptophan is blocked, fewer tryptophan molecules are present to inhibit the key enzyme and production (40) [choose one] ❏ rises ❏ falls [p.107]. Another type of biochemical control in humans and other multicelled organisms operates through signaling agents called (41) _____ [p.107].

Matching

Choose the one most appropriate answer for each. [p.107]

42. ___metal ions

43. ___FAD

44. ___coenzymes

45. ___NADP⁺

46. ___NAD⁺

A. Large organic molecules; derived in part from vitamins; transfer hydrogens and electrons

B. A coenzyme, nicotinamide adenine dinucleotide phosphate; in reduced form carries H^+ and electrons to other reaction sites

C. A coenzyme, nicotinamide adenine dinucleotide, transfers hydrogens and electrons to other reaction sites

D. Example: Fe^{++} serves as a cofactor and is a component of enzymatic cytochromes

E. A coenzyme, flavin adenine dinucleotide; transfers hydrogens and electrons to other reaction sites

Self-Quiz

_d_ 1. An important principle of the second law of thermodynamics states that _____. [p.97]
   a. energy can be transformed into matter, and because of this, we can get something for nothing
   b. energy can only be destroyed during nuclear reactions, such as those that occur inside the sun
   c. if energy is gained by one region of the universe, another place in the universe also must gain energy in order to maintain the balance of nature
   d. matter tends to become increasingly more disorganized

_c_ 2. Essentially, the first law of thermodynamics states that _____. [p.96]
   a. one form of energy cannot be converted into another
   b. entropy is increasing in the universe
   c. energy cannot be created or destroyed
   d. energy cannot be converted into matter or matter into energy

_d_ 3. An enzyme is best described as _____. [p.104]
   a. an acid
   b. protein
   c. a catalyst
   d. a fat
   e. both b and c

_e_ 4. Which is *not* true of enzyme behavior? [p.104]
   a. Enzyme shape may change during catalysis.
   b. The active site of an enzyme orients its substrate molecules, thereby promoting interaction of their reactive parts.
   c. All enzymes have an active site where substrates are temporarily bound.
   d. An individual enzyme can catalyze a wide variety of different reactions.

a 5. When NAD^+ combines with hydrogen, the NAD^+ is _____. [p.107]
   a. reduced
   b. oxidized
   c. phosphorylated
   d. denatured

c 6. A substance that gains electrons is _____. [p.101]
   a. oxidized
   b. a catalyst
   c. reduced
   d. a substrate

d 7. In _____ pathways, carbohydrates, lipids, and proteins are broken down in stepwise reactions that lead to products of lower energy. [p.103]
   a. intermediate
   b. biosynthetic
   c. induced
   d. degradative

c 8. As to major function, NAD^+, FAD, and $NADP^+$ are classified as _____. [p.107]
   a. enzymes
   b. phosphate carriers
   c. cofactors that function as coenzymes
   d. end products of metabolic pathways

d 9. When a phosphate bond is linked to ADP, the bond _____. [p.100]
   a. absorbs a large amount of free energy when the phosphate group is attached during hydrolysis
   b. is formed when ATP is hydrolyzed to ADP and one phosphate group
   c. is usually found in each glucose molecule; that is why glucose is chosen as the starting point for glycolysis
   d. releases a large amount of usable energy when the phosphate group is split off during hydrolysis

a 10. An allosteric enzyme _____. [pp.106–107]
   a. has an active site where substrate molecules bind and another site that binds with intermediate or end-product molecules
   b. is an important energy-carrying nucleotide
   c. carries out either oxidation reactions or reduction reactions but not both
   d. raises the activation energy of the chemical reaction it catalyzes

Chapter Objectives/Review Questions

1. Fireflies, various beetles, and some other organisms display _____ when luciferases excite the electrons of luciferins. [p.94]
2. _____ is the controlled capacity to acquire and use energy for stockpiling, breaking apart, building, and eliminating substances in ways that contribute to survival and reproduction. [p.95]
3. Explain what is meant by a "system" as related to the laws of thermodynamics. [p.96]
4. Define *energy*; be able to state the first and second laws of thermodynamics. [pp.96–97]
5. _____ is a measure of the degree of randomness or disorder of systems. [p.97]
6. Explain how the world of life maintains a high degree of organization. [p.97]
7. Explain the functioning of the ATP/ADP cycle. [pp.98–100]
8. Adding a phosphate to a molecule is called _____. [p.100]
9. _____ molecules are the cell's main, renewable energy carriers between sites of metabolic reactions in cells. [p.100]
10. _____-_____ reactions refer to the electron transfers occurring in an electron transport system. [p.101]
11. Describe the condition known as *chemical equilibrium*. [p.102]
12. What is meant by a reversible reaction? [p.102]
13. A _____ pathway is an orderly sequence of reactions with specific enzymes acting at each step; the pathways are always linear or circular. [p.103]
14. Give the function of each of the following participants in metabolic pathways: substrates, intermediates, enzymes, cofactors, energy carriers, and end products. [p.103]
15. What are enzymes? Explain their importance. [p.104]
16. The location on the enzyme where specific reactions are catalyzed is the active _____. [p.105]

17. According to Koshland's _____-_____ model, each substrate has a surface region that almost but not quite matches chemical groups in an active site. [p.105]
18. When the "energy hill" is made smaller by enzymes in order that particular reactions may proceed, it may be said that the enzyme has lowered the _____ energy. [pp.104–105]
19. Explain what happens to enzymes in the presence of extreme temperatures and pH. [p.106]
20. An enzyme control called _____ inhibition operates in the tryptophan pathway when excess tryptophan molecules inhibit a key allosteric enzyme in the pathway. [pp.106–107]
21. Cofactors called _____ are organic molecules that may be vitamins or are derived in part from vitamins; _____ ions may also serve as cofactors. [p.107]
22. _____ are small molecules or metal ions that assist enzymes or carry atoms or electrons from one reaction site to another. [p.107]
23. Explain the differences between NAD^+ and $NADH$; $NADP^+$ and $NADPH$; FAD and $FADH_2$. [p.107]

Integrating and Applying Key Concepts

A piece of dry ice left sitting on a table at room temperature vaporizes. As the dry ice vaporizes into CO_2 gas, does its entropy increase or decrease? Tell why you answered as you did.

7

HOW CELLS ACQUIRE ENERGY

Sunlight and Survival

PHOTOSYNTHESIS—AN OVERVIEW
　　Where the Reactions Take Place
　　Energy and Materials for the Reactions

SUNLIGHT AS AN ENERGY SOURCE
　　Properties of Light
　　Pigments—Molecular Bridge from Sunlight to
　　　　Photosynthesis

THE RAINBOW CATCHERS
　　The Chemical Basis of Color
　　On the Variety of Photosynthetic Pigments
　　Where Are Photosynthetic Pigments Located?
　　About Those Roving Pigments

THE LIGHT-DEPENDENT REACTIONS
　　What Happens to the Absorbed Energy?
　　Cyclic and Noncyclic Electron Flow
　　The Legacy—A New Atmosphere

**A CLOSER LOOK AT ATP FORMATION IN
CHLOROPLASTS**

THE LIGHT-INDEPENDENT REACTIONS
　　How Do Plants Capture Carbon?
　　How Do Plants Build Glucose?

FIXING CARBON—SO NEAR, YET SO FAR
　　C4 Plants
　　CAM Plants

Focus on Science: LIGHT IN THE DEEP DARK SEA?

Focus on the Environment: AUTOTROPHS, HUMANS,
　　AND THE BIOSPHERE

Interactive Exercises

Sunlight and Survival [pp.112–113]

7.1. PHOTOSYNTHESIS—AN OVERVIEW [pp.114–115]

Selected Words: photoautotroph [p.112], Englemann [p.112], *Spirogyra* [p.112], aerobic respiration [p.112], thylakoid membrane system [p.114], granum (pl., grana) [p.114], *light-dependent* reactions [p.114], *light-independent* reactions [p.114], $NADP^+$ [p.114], NADPH [p.114], radioisotopes [p.114], glucose [p.114]

Boldfaced, Page-Referenced Terms

[p.112] autotrophs _self-nourishing organisms._

[p.112] photosynthesis _a process where sunlight energy is captured to create ATP. This is performed by plants, some bacteria, and many protistans._

[p.112] heterotrophs _organisms that feed on autotrophs, other heterotrophs and organic wastes._

[p.114] chloroplasts _the organelles of photosynthesis in plants and protistans called algae. It stores food._

[p.114] stroma _one of two outermost membranes of the chloroplasts which surround a largely fluid interior._

[p.114] thylakoid _a membrane that weaves through the stroma. Parts of the system are often folded into disk-shaped sacs, stacked on top of one another_

Fill-in-the-Blanks

(1) _Autotrophs_ [p.112] obtain carbon and energy from the physical environment; their carbon source is
(2) _carbon   dioxide_ [p.112]. (3) _Phosphosynthetic_ [p.112] autotrophs obtain energy from sunlight.
T. (4) _Engleman_ [p.112] demonstrated that aerobic bacteria moved to the regions of a strand of *Spirogyra*
where red and (5) _violet_ [p.112] light was falling on the strand; O_2 was produced here. (6) _Heterotrophs_
[p.112] feed on autotrophs, each other, and organic wastes; representatives include (7) _animals_ [p.112],
fungi, many protistans, and most bacteria. Although energy stored in organic compounds such as glucose
may be released by several pathways, the pathway known as (8) _aerobic   respiration_ [p.112] releases the
most energy.

9. In the space below, supply the missing information to complete the summary equation for photosynthesis:
   [All from p.113]

$$12 \; \underline{H_2O} \; + \; \underline{6} \; CO_2 \rightarrow \underline{6} \; O_2 + C_6H_{12}O_6 + 6 \; \underline{H_2O}$$

10. Supply the appropriate information to state the equation (above) for photosynthesis in words:

(a) _Twelve_ [p.113] molecules of water plus six molecules of (b) _carbon   dioxide_ [p.113] (in the presence
of pigments, enzymes, and sunlight) yield six molecules of (c) _oxygen_ [p.113] plus one molecule of
(d) _glucose_ [p.113] plus (e) _six_ [p.113] molecules of water.

The two major sets of reactions of photosynthesis are the (11) _light_ - _dependent_ [p.114] reactions and
the (12) _light_ - _independent_ [p.114] reactions. (13) _Carbon   Dioxide_ [p.114] and (14) _water_ [p.114] are
the reactants of photosynthesis, and the end product is usually given as (15) _glucose_ [p.114]. The internal
membranes and channels of the chloroplast are the (16) _thylakoids_ [p.114] membrane system and are
organized into stacks, called (17) _grana_ [p.114]. Spaces inside the thylakoid disks and channels form a
continuous compartment where (18) _hydrogen_ [p.114] ions accumulate to be used to produce ATP. The
semifluid interior area surrounding the grana is known as the (19) _stroma_ [p.114] and is the area where
the products of photosynthesis are produced.

Choice

For questions 20–31, choose the area of the chloroplast that correctly relates to the listed structures and events.

a. thylakoid membrane system   b. stroma

20. _b_ light-independent reactions [p.115]
21. _a_ the coenzyme NADP⁺ picks up liberated hydrogen and electrons [p.114]
22. _a_ granum [p.115]
23. _b_ sugars are assembled [p.114]
24. _a_ light-dependent reactions [p.115]
25. _a_ ATP production [p.114]
26. _b_ carbon dioxide provides the carbon [p.115]
27. _a_ sunlight energy is absorbed [p.114]
28. _a_ where the first stage of photosynthesis proceeds [p.114]
29. _a_ water molecules are split [p.114]
30. _b_ NADPH delivers the hydrogen received from water [p.114]
31. _a_ oxygen is formed [p.115]

Labeling

Identify the following structures.

32. _oxygen_ [p.115]

33. _ATP_ [p.115]

34. _NADPH_ [p.115]

35. _CO₂_ [p.115]

36. _chloroplasts_ [p.114]

37. _thylakoid_ _membrane_ _system_ [p.115]

38. _stroma_ [p.115]

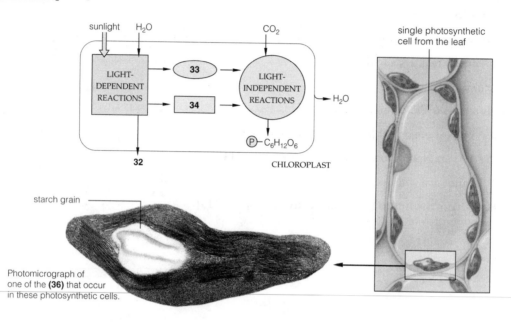

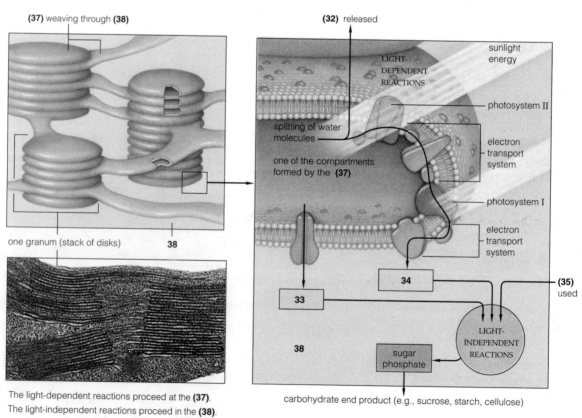

Overview of the sites where the key steps of both stages of
reactions proceed. The sections to follow will fill in the details.

7.2. SUNLIGHT AS AN ENERGY SOURCE [pp.116–117]

7.3. THE RAINBOW CATCHERS [pp.118–119]

Selected Words: radiant energy [p.116], *visible* light [p.116], ultraviolet radiation [p.117], ozone (O_3) [p.117], *accessory* pigments [p.117], chlorophyll *a* [p.118], chlorophyll *b* [p.118], beta-carotene [p.118], melanin [p.119], photosystem I [p.119], photosystem II [p.119]

Boldfaced, Page-Referenced Terms

[p.116] wavelength   *the distance between crests of every two successive waves.*

[p.116] electromagnetic spectrum   *the entire range of all the wavelengths.*

[p.116] photons   *when absorbed by matter, it is the energy of visible light that can be measured as if it were organized in packets.*

[p.117] pigments   *are different molecules that absorb wavelengths of light.*

[p.117] absorption spectrum   *is a diagram that shows how effectively a pigment molecule absorbs different wavelengths in the spectrum of visible light.*

[p.118] fluorescence   *when any destabilized molecule emits light as it reverts to its more stable configuration.*

[p.118] chlorophylls   *are the main pigments in all but one marginal group of photoautotrophs.*

[p.118] carotenoids   *these accessory pigments absorb blue-violet and blue-green wavelengths that chlorophylls miss.*

[p.118] anthocyanins   *pigments include red + blue*

[p.118] phycobilins   *pigments include red + blue.*

[p.119] photosystems   *the thylakoid membrane system of chloroplasts. has pigments organized in clusters called (photosystems)*

Fill-in-the-Blanks

The light-capturing phase of photosynthesis takes place on a system of (1) _____ [p.119] membranes.
A(n) (2) _____ [p.117] is a packet of light energy. Thylakoid membranes contain (3) _____ [p.117],
which absorb photons of light. The principal pigments are the (4) _____ [p.118], which reflect green
wavelengths but absorb (5) _____ [p.118] and (6) _____ [p.118] wavelengths. (7) _____
[p.118] are pigments that absorb violet and blue wavelengths but reflect yellow, orange, and red.

A cluster of 200 to 300 of these pigment proteins is a(n) (8) _____ [p.119]. When pigments absorb (9) _____ [p.117] energy, an (10) _____ [p.118] is transferred from a photosystem to a(n) (11) _____ [p.119] molecule.

Matching

Choose the most appropriate answer.

12. __H__ chlorophylls [p.118]
13. __G__ chlorophyll *b* and carotenoids [p.118]
14. __E__ carotenoids [p.118]
15. __F__ violet-blue-green-yellow-red [p.116]
16. __B__ photons [p.117]
17. __A__ chlorophyll *a* [p.118]
18. __C__ chloroplast [p.119]
19. __I__ phycobilins [p.118]
20. __J__ granum [recall p.114]
21. __D__ pigments [p.117]

A. The main pigment of photosynthesis
B. Packets of energy that have an undulating motion through space
C. The two stages of photosynthesis occur here
D. Molecules that can absorb light
E. Absorb violet and blue wavelengths but transmit red, orange, and yellow
F. Visible light portion of the electromagnetic spectrum
G. Pigments that transfer energy to chlorophyll *a*
H. Absorb violet-to-blue and red wavelengths; the reason leaves appear green
I. Red and blue pigments
J. The site of the first stage of photosynthesis

7.4. THE LIGHT-DEPENDENT REACTIONS [pp.120–121]

7.5. A CLOSER LOOK AT ATP FORMATION IN CHLOROPLASTS [p.122]

Selected Words: electron acceptor molecule [p.120], NADPH [p.120], *cyclic* pathway of ATP formation [p.120], *type I* photosystem [p.120], P700 [p.120], *noncyclic* pathway of ATP and NADPH formation [p.120], *type II* photosystem [p.120], P680 [p.120], ATP synthases [p.121], stroma [p.122], chemiosmotic model of ATP formation [p.122]

Boldfaced, Page-Referenced Terms

[p.120] light-dependent reactions _____

_____

[p.120] reaction center _the chlorophyll takes place here for the photosystem_

_____

[p.120] electron transport systems _are organized arrays of enzymes, coenzymes, and other proteins associated with a cell membrane._

[p.120] photolysis _water molecules split into oxygen, hidrogen ions (H+), and electrons._

_____

Complete the Table

1. Complete the following table on elements of the cyclic pathway of the light-dependent reactions.

Component	Function
a. Type I photosystem [p.120]	
b. Electrons [pp.120–121]	
c. P700 [p.120]	
d. Electron acceptor [p.120]	
e. Electron transport system [p.120]	
f. ADP [p.122]	

Labeling

The following diagram illustrates noncyclic photophosphorylation. Identify each numbered part of the illustration.

2. *electron   acceptor* [p.121]

3. *first   transport   system* [p.121]

4. *photosystem   II* [p.121]

5. *photosystem   I* [p.121]

6. *photolysis* [p.121]

7. *NADPH* [p.121]

8. *ATP* [p.122]

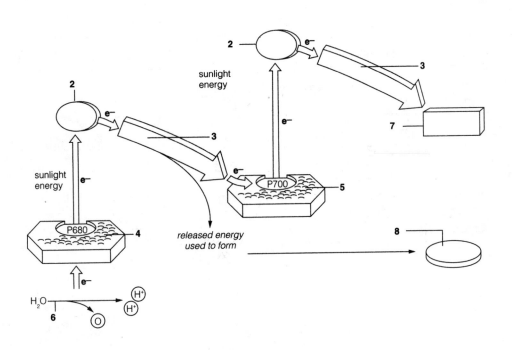

Fill-in-the-Blanks

ATP forms in both the cyclic and noncyclic pathways. When (9) electrons [p.122] flow through the membrane-bound transport systems, they pick up hydrogen ions (H^+) outside the membrane and shunt them into the (10) thylakoid [p.122] compartment. This sets up H^+ concentration and electric (11) gradients [p.122] across the membrane. Hydrogen ions that were split away from (12) water [p.122] molecules increase the gradients. The ions respond by flowing out through the interior of (13) ATP synthases [p.122] proteins that span the membrane. Energy associated with the flow drives the binding of unbound phosphate (P_i) to ADP, the result being (14) ATP [p.122]. The above description is known as the (15) chemiosmotic [p.122] theory of ATP formation.

The noncyclic pathway also produces (16) NADPH [p.121] by using (17) electrons [p.121] from water and H^+ ions from the thylakoid compartment to reduce $NADP^+$.

Complete the Table

With a check mark (√), indicate for each phase of the light-dependent reactions all items from the left-hand column that are applicable.

Light-Dependent Reactions :	Cyclic Pathway	Noncyclic Pathway	Photolysis Alone
Uses H_2O as a reactant [p.121]	(18)	(30)	(42)
Produces a net gain of H_2O as a product [recall p.113]	(19)	(31)	(43)
Photosystem I involved (P700) [pp.120–121]	(20)	(32)	(44)
Photosystem II involved (P680) [pp.120–122]	(21)	(33)	(45)
ATP produced [pp.121–122]	(22)	(34)	(46)
NADPH produced [pp.120–121]	(23)	(35)	(47)
Uses CO_2 as a reactant [logic; not mentioned on pp.120–122]	(24)	(36)	(48)
Causes H^+ to be shunted into the thylakoid compartments from the stroma [p.122]	(25)	(37)	(49)
Produces O_2 as a product [p.121]	(26)	(38)	(50)
Produces H^+ by breaking apart H_2O [p.121]	(27)	(39)	(51)
Uses ADP and P_i as reactants [pp.121–122]	(28)	(40)	(52)
Uses $NADP^+$ as a reactant [p.121]	(29)	(41)	(53)

7.6. THE LIGHT-INDEPENDENT REACTIONS [p.123]

7.7. FIXING CARBON—SO NEAR, YET SO FAR [pp.124–125]

7.8. *Focus on Science:* LIGHT IN THE DEEP DARK SEA? [p.125]

7.9. *Focus on the Environment:* AUTOTROPHS, HUMANS, AND THE BIOSPHERE [p.126]

Selected Words: "synthesis" [p.123], stroma [p.123], *photorespiration* [p.124], mesophyll cells [p.124], bundle-sheath cells [p.124], glycolate [pp.124,125], succulents [p.125], *chemoautotrophs* [p.126]

Boldfaced, Page-Referenced Terms

[p.123] light-independent reactions _are the synthesis part of photo synthesis._

[p.123] Calvin–Benson cycle _Uses ATP and NADPH from light-dependent reactions. RuBP or some compound to which carbon has been affixed is rearranged and regenerated, and a sugar phosphate forms._

[p.123] RuBP (ribulose bisphosphate) _is a compound with a backbone of five carbon atoms._

[p.123] Rubisco (RuBP carboxylase) _is the enzyme that attaches the carbon atom of CO_2 to RuBP_

[p.123] PGA (phosphoglycerate) _a stable molecule with a three-carbon backbone_

[p.123] carbon fixation _incorporating a carbon from CO_2 into a stable organic compound._

[p.123] PGAL (phosphoglyceraldehyde) _is the resultant when PGA accepts a phosphate group from ATP, plus hydrogen and electrons from NADPH_

[p.124] stomata (singular, stoma) _microscopic openings across the leaf surface to release oxygen (O_2)_

[p.124] C3 plants _refers to the three-carbon PGA the first intermediate of its carbon-fixing pathway._

[p.124] C4 plants _____

[p.125] CAM plants _____

[p.125] hydrothermal vent _fissures in the sea floor where molten rock mixes with cold sea water._

Labeling and Matching

Identify each part of the following illustration. Complete the exercise by matching and entering the letter of the proper function description in the parentheses following each label. [All from p.123]

1. _____ _____ ( )

2. _____ _____ ( )

3. _____ ( )

4. _____ _____ ( )

5. _____ ( )

6. _____ ( )

7. _____ _____ ( )

8. _____-_____ _____ ( )

9. _____ _____ ( )

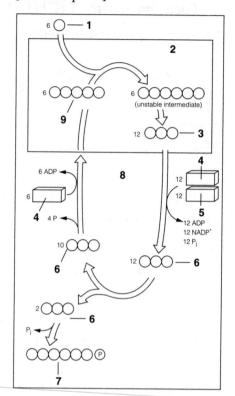

A. A three-carbon sugar, the first sugar produced; goes on to form sugar phosphate and RuBP

B. Typically used at once to form carbohydrate end products of photosynthesis

C. A five-carbon compound produced from PGALs; attaches to incoming CO_2

D. A compound that diffuses into leaves; attached to RuBP by enzymes in photosynthetic cells

E. Includes all the chemical reactions that "fix" carbon into an organic compound

F. Three-carbon compounds formed from the splitting of the six-carbon intermediate compound

G. A molecule that was reduced in the noncyclic pathway; furnishes hydrogen atoms to construct sugar molecules

H. A product of the light-dependent reactions; necessary in the light-independent reactions to energize molecules in metabolic pathways.

I. Includes all the chemistry that fixes CO_2; converts PGA to PGAL and PGAL to RuBP and sugar phosphates

Fill-in-the-Blanks

The light-independent reactions can proceed without sunlight as long as (10) _____ [p.123] and (11) _____ [p.123] are available. The reactions begin when an enzyme links (12) _____ _____ [p.123] to (13) _____ _____ [p.123], a five-carbon compound. The resulting six-carbon compound is highly unstable and breaks apart at once into two molecules of a three-carbon compound, (14) _____ [p.123]. This entire reaction sequence is called carbon (15) _____ [p.123]. Each ATP gives a phosphate group to each (16) _____ [p.123]. This intermediate compound takes on H^+ and electrons from NADPH to form (17) _____ [p.123]. It takes (18) _____ [p.123] carbon dioxide molecules to produce twelve PGAL. Most of the PGAL becomes rearranged into new (19) _____ [p.123] molecules—which can be used to fix more (20) _____ _____ [p.123]. Two (21) _____ [p.123] are joined together to form a (22) _____ _____ [p.123], primed for further reactions. The Calvin–Benson cycle yields enough RuBP to replace those used in carbon (23) _____ [p.123]. ADP, $NADP^+$, and phosphate leftovers are sent back to the (24) _____-_____ [p.123] reaction sites, where they are again converted to (25) _____ [p.123] and (26) _____ [p.123]. (27) _____ _____ [p.123] formed in the cycle serves as a building block for the plant's main carbohydrates. When RuBP attaches to oxygen instead of carbon dioxide, (28) _____ [p.124] results; this is typical of (29) _____ [p.124] plants in hot, dry conditions. If less PGA is available, leaves produce a reduced amount of (30) _____ [p.124]. C4 plants can still construct carbohydrates when the ratio of carbon dioxide to (31) _____ [p.124] is unfavorable because of the attachment of carbon dioxide to form (32) _____ [p.124] in certain leaf cells. CAM plants generally live in hot climates; they often have the fleshy, water-storing tissues and thick water-restricting surface structure characteristic of (33) _____ [p.125] plants. CAM plants open their stomata and fix CO_2 at (34) _____ [p.125], thereby minimizing water loss.

Nisbet and Van Dover believe that the first cells living on Earth arose approximately 3.8 billion years ago at (35) _____ _____ [p.125] habitats, where they could have used (36) _____ _____ [p.125] (as a source of hydrogen and electrons) and (37) _____ _____ [p.125] as a source of carbon atoms. Absorption spectra for the (38) _____ [p.125] in evolutionarily ancient photosynthetic bacteria correspond to light measured at (35) and photosynthetic coenzymes contain iron, sulfur, manganese and other minerals, which are abundant at (35).

Organisms that obtain energy from oxidation of inorganic substances such as ammonium compounds, and iron or sulfur compounds, are known as (39) _____–autotrophs [p.126]. Such organisms use this energy to build (40) _____ [p.126] compounds. Organisms that obtain energy from sunlight are known as (41) _____–autotrophs [p.126].

Complete the Table

With a check mark (√), indicate for each phase of the light-dependent reaction all items from the left-hand column that are applicable.

Light-Independent Reactions	CO₂ Fixation Alone	Conversion of PGA to PGAL	Regeneration of RuBP	Formation of Glucose and Other Organic Compounds
Requires RuBP as a reactant [p.123]	(42)	(53)	(64)	(75)
Requires ATP as a reactant [p.123]	(43)	(54)	(65)	(76)
Produces ADP as a product [p.123]	(44)	(55)	(66)	(77)
Requires NADPH as a reactant [p.123]	(45)	(56)	(67)	(78)
Produces NADP⁺ as a reactant [p.123]	(46)	(57)	(68)	(79)
Produces PGA [p.123]	(47)	(58)	(69)	(80)
Produces PGAL [p.123]	(48)	(59)	(70)	(81)
Requires PGAL as a reactant [p.123]	(49)	(60)	(71)	(82)
Produces P$_i$ as a product [p.123]	(50)	(61)	(72)	(83)
Produces H₂O as a product [p.123]	(51)	(62)	(73)	(84)
Requires CO₂ as a reactant [p.123]	(52)	(63)	(74)	(85)

Self-Quiz

_a_ 1. The electrons that are passed to NADPH during the noncyclic pathway of photosynthesis were obtained from _____. [p.121]
a. water
b. CO₂
c. glucose
d. sunlight

_b_ 2. The cyclic pathway of the light-dependent reactions functions mainly to _____. [p.120]
a. fix CO₂
b. make ATP
c. produce PGAL
d. regenerate ribulose bisphosphate

_c_ 3. Chemosynthetic autotrophs obtain energy by oxidizing such inorganic substances as _____. [p.126]
a. PGA
b. PGAL
c. sulfur
d. water

_a_ 4. The ultimate electron and hydrogen acceptor in noncyclic photophosphorylation is _____. [p.121]
a. NADP⁺
b. ADP
c. O₂
d. H₂O

_c_ 5. C4 plants have an advantage in hot, dry conditions because _____. [p.124]
   a. their leaves are covered with thicker wax layers than those of C3 plants
   b. their stomates open wider than those of C3 plants, thus cooling their surfaces
   c. special leaf cells possess a means of capturing CO_2 even in stress conditions
   d. they also are capable of carrying on photorespiration

_d_ 6. Chlorophyll is _____. [p.119]
   a. on the outer chloroplast membrane
   b. inside the mitochondria
   c. in the stroma
   d. in the thylakoid membrane system

_a_ 7. Which of the following is applicable to C3 plants? [p.123]
   a. At the end of carbon fixation, the intermediate compound is PGA
   b. At the end of carbon fixation, the intermediate compound is oxaloacetate
   c. They are more sensitive to cold temperatures than are C4 plants
   d. Corn, crabgrass, and sugarcane are examples of C3 plants

_b_ 8. Plant cells produce O_2 during photosynthesis by _____. [p.119]
   a. splitting CO_2
   b. splitting water
   c. degradation of the stroma
   d. breaking up sugar molecules

_c_ 9. Plants need _____ and _____ to carry on photosynthesis. [p.114]
   a. oxygen; water
   b. oxygen; CO_2
   c. CO_2; H_2O
   d. sugar; water

_d_ 10. The two products of the light-dependent reactions that are required for the light-independent chemistry are _____ and _____. [p.121]
   a. CO_2; H_2O
   b. O_2; NADPH
   c. O_2; ATP
   d. ATP; NADPH

Chapter Objectives/Review Questions

1. Distinguish between organisms known as *autotrophs* and those known as *heterotrophs*. [p.112]
2. State what T. Englemann's 1882 experiment with *Spirogyra* revealed. [pp.112,117]
3. Study the general equation for photosynthesis until you can remember the reactants and products. Reproduce the equation from memory on another piece of paper. [p.113]
4. List the two major stages of photosynthesis as well as where in the cell they occur and what reactions occur there. [pp.114–115]
5. Describe the details of a familiar site of photosynthesis, the green leaf. Begin with the layers of a leaf cross-section and complete your description with the minute structural sites within the chloroplast where the major sets of photosynthetic reactions occur. [pp.114–115]
6. The flattened channels and disklike compartments inside the chloroplast are organized into stacks, the _____, which are surrounded by a semifluid interior, the _____; this is the _____ membrane system. [pp.114–115]
7. The energy-poor molecules that act as raw materials in the photosynthetic equation are _____ and _____. [p.115]
8. _____ are packets of light energy. [p.117]
9. _____ absorbs violet-to-blue as well as red wavelengths. [p.118]
10. The main pigments of photosynthesis are _____. [p.118]
11. What pigments are responsible for the red and blue colors of red algae and cyanobacteria? [p.118]
12. Name the wavelengths absorbed and transmitted by the carotenoids. [p.118]
13. Describe how the pigments found on thylakoid membranes are organized into photosystems and how they relate to photon light energy. [pp.119–120]
14. Contrast the components and functioning of the cyclic and noncyclic pathways of the light-dependent reactions. [pp.120–121]

15. Explain what the water split during photolysis contributes to the noncyclic pathway of the light-dependent reactions. [pp.120–121]
16. Two energy-carrying molecules produced in the noncyclic pathways are _____ and _____; explain why these molecules are necessary for the light-independent reactions. [pp.120–121]
17. _____ _____ is the metabolic pathway most efficient in releasing the energy stored in organic compounds. [p.121]
18. Following evolution of the noncyclic pathway, _____ accumulated in the atmosphere and made _____ respiration possible. [p.121]
19. Explain how the chemiosmotic theory is related to thylakoid compartments and the production of ATP. [p.122]
20. Explain why the light-independent reactions are called by that name. [p.123]
21. Describe the process of carbon dioxide fixation by stating which reactants are necessary to initiate the process and what stable products result from this process. [p.123]
22. Describe the Calvin–Benson cycle in terms of its reactants and products. [p.123]
23. State the fate of all the phosphorylated glucose produced by photosynthetic reactions in photoautotrophs. [p.123]
24. When carbon fixation occurs, sunlight energy, which excited electrons during the light-dependent reactions, becomes stored as _____ energy in an organic compound. [p.123]
25. Describe the mechanism by which C4 plants thrive under hot, dry conditions; distinguish this CO_2 capturing mechanism from that of C3 plants. [pp.124–125]
26. Describe the carbon-fixing adaptation of the CAM plants living in arid environments. [p.125]
27. Explain why some scientists are hypothesizing that the first autotrophic cells may have evolved near hydrothermal vents. [p.125]
28. Bacteria that are able to obtain energy from ammonium ions, iron, or sulfur compounds are known as _____. [p.126]

Integrating and Applying Key Concepts

Suppose that humans acquired all the enzymes needed to carry out photosynthesis. Speculate about the attendant changes in human anatomy, physiology, and behavior that would be necessary for those enzymes actually to carry out photosynthetic reactions.

8

HOW CELLS RELEASE STORED ENERGY

Interactive Exercises

The Killers Are Coming! The Killers Are Coming! [pp.130–131]

8.1. HOW DO CELLS MAKE ATP? [pp.132–133]

Selected Words: anaerobic [p.132], *coenzymes* [p.133]

Boldfaced, Page-Referenced Terms

[p.132] aerobic respiration *an oxygen-dependent pathway of ATP formation.*

[p.132] glycolysis *enzymes cleave and rearrange a glucose molecule into two molecules of pyruvate*

[p.132] pyruvate *has a backbone of three carbon atoms.*

[p.133] the Krebs cycle *when enzymes break down pyruvate to carbon dioxide and water.*

    → FAD (flavin adenine dinucleotide)
[p.133] NAD⁺ (nicotinamide adenine dinucleotide) *take part in the kreb cycle. It is derived from vitamins, and are coenzymes. It helps enzymes by accepting electrons (e-) and hydrogen removed from intermediates.*

[p.133] FAD (flavin adenine dinucleotide) *Same as NAD+*

[p.133] electron transport phosphorylation *sets up hydrogen concentration and electric gradients, which drive ATP formation at nearby membrane proteins.*

Short Answer

1. Although various organisms use different energy sources, what is the usual form of chemical energy that will drive metabolic reactions? [p.131] _____

2. Describe the function of oxygen in the main degradative pathway, aerobic respiration. [p.131] _____

3. List the most common anaerobic pathways, and describe the conditions in which they function. [pp.131–132] _____

Fill-in-the-Blanks

Virtually all forms of life depend on a molecule known as (4) _____ [p.132] as their primary energy carrier. Plants produce adenosine triphosphate during (5) _____ [p.132], but plants and all other organisms also can produce ATP through chemical pathways that degrade (take apart) food molecules. The main degradative pathway requires free oxygen and is called (6) _____ _____ [p.132].

There are three stages of aerobic respiration. In the first stage, (7) _____ [p.132], glucose is partially degraded to (8) _____ [p.132]. By the end of the second stage, which includes the (9) _____ [p.133] cycle, glucose has been completely degraded to carbon dioxide and (10) _____ [p.133]. Neither of the first two stages produces much (11) _____ [p.133]. During both stages, protons and (12) _____ [p.133] are stripped from intermediate compounds and delivered to a (13) _____ [p.133] system. That system is used in the third stage of reactions, electron transport (14) _____ [p.133]; passage of electrons along the transport system drives the enzymatic "machinery" that phosphorylates ADP to produce a high yield of (15) _____ [p.133]. (16) _____ [p.133] accepts "spent" electrons from the transport system and keeps the pathway clear for repeated ATP production.

Other degradative pathways are (17) _____ [p.132], in that something other than oxygen serves as the final electron acceptor in energy-releasing reactions. (18) _____ [pp.131,132] and anaerobic (19) _____ _____ [pp.131,132] are the most common anaerobic pathways.

Completion

20. Complete the following equation, which summarizes the degradative pathway known as *aerobic respiration:* [p.132]

_____ + _____ $O_2 \rightarrow 6$ _____ $+ 6$ _____

21. Supply the appropriate information to state the equation (see #20) for aerobic respiration in words:

One molecule of glucose plus six molecules of _____ [p.132] (in the presence of appropriate enzymes)

yield _____ [p.132] molecules of carbon dioxide plus _____ [p.132] molecules of water.

Matching

Choose the one most appropriate answer for each.

22. ___Krebs cycle [p.133]

23. ___oxygen [p.132]

24. ___mitochondrion [p.132]

25. ___electron transport phosphorylation [p.133]

26. ___enzymes [p.132]

27. ___ATP [p.133]

28. ___glycolysis [p.132]

29. ___aerobic respiration [p.132]

30. ___cytoplasm [p.132]

31. ___fermentation pathways and anaerobic electron transport [p.132]

A. Starting point for three energy-releasing pathways
B. Main energy-releasing pathway for ATP formation
C. Site of glycolysis
D. Third and final stage of aerobic respiration; high ATP yield
E. Oxygen is not the final electron acceptor
F. Catalyze each reaction step in the energy-releasing pathways
G. Second stage of aerobic respiration; pyruvate is broken down to CO_2 and H_2O
H. Site of the aerobic pathway
I. The final electron acceptor in aerobic pathways
J. The energy form that drives metabolic reactions

8.2. GLYCOLYSIS: FIRST STAGE OF ENERGY-RELEASING PATHWAYS [pp.134–135]

Selected Words: pyruvate [p.134], *energy-requiring* steps [p.134], *energy-releasing* steps [p.134], NAD⁺ [p.134], NADH [p.134], *net* energy yield [p.134]

Boldfaced, Page-Referenced Terms

[p.134] substrate-level phosphorylation *the direct transfer of a phosphate group from a substrate of a reaction to some other molecule – in this case ADP.*

Fill-in-the-Blanks

(1) _____ [recall Ch.7] organisms can synthesize and stockpile energy-rich carbohydrates and other food molecules from inorganic raw materials. (2) _____ [p.134] is partially broken down by the glycolytic pathway; at the end of this process some of its stored energy remains in two (3) _____ [p.134] molecules. Some of the energy of glucose is released during the breakdown reactions and used in forming the energy carrier (4) _____ [p.134] and the reduced coenzyme (5) _____ [p.134]. These reactions take place in the cytoplasm. Glycolysis begins with two phosphate groups being transferred to (6) _____ [p.134] from two (7) _____ [p.134] molecules. The addition of two phosphate groups to (6) energizes it and causes it to become unstable and split apart, forming two molecules of (8) _____ [p.134]. Each (8) gains one (9) _____ [p.134] group from the cytoplasm, then (10) _____ [p.134] atoms and electrons from each PGAL are transferred to NAD⁺, changing this coenzyme to NADH. Two (11) _____ [p.135] molecules form by substrate-level phosphorylation; the cell's energy investment is

paid off. One (12) _____ [p.135] molecule is released from each 2-PGA as a waste product. The resulting intermediates are rather unstable; each gives up a(n) (13) _____ [p.135] group to ADP. Once again, two (14) _____ [p.134] molecules have formed by (15) _____-_____ [p.135] phosphorylation. For each (16) _____ [p.135] molecule entering glycolysis, the net energy yield is two ATP molecules that the cell can use anytime to do work. The end products of glycolysis are two molecules of (17) _____ [p.135], each with a (18) _____-carbon [p.135] backbone.

(number)

Sequence

Arrange the following events of the glycolysis pathway in correct chronological sequence. Write the letter of the first step next to 19, the letter of the second step next to 20, and so on. [All from p.135]

19. ___
20. ___
21. ___
22. ___
23. ___
24. ___
25. ___
26. ___

A. Two ATPs form by substrate-level phosphorylation; the cell's energy debt is paid off
B. Diphosphorylated glucose (fructose 1,6-bisphosphate) molecules split to form 2 PGALs; this is the first energy-releasing step
C. Two 3-carbon pyruvate molecules form as the end products of glycolysis
D. Glucose is present in the cytoplasm
E. Two more ATPs form by substrate-level phosphorylation, the cell gains ATP; net yield of ATP from glycolysis is 2 ATPs
F. The cell invests two ATPs; one phosphate group is attached to each end of the glucose molecule (fructose 1,6-bisphosphate)
G. Two PGALs gain two phosphate groups from the cytoplasm
H. Hydrogen atoms and electrons from each PGAL are transferred to NAD$^+$, reducing this carrier to NADH

8.3. SECOND STAGE OF THE AEROBIC PATHWAY [pp. 136–137]

8.4. THIRD STAGE OF THE AEROBIC PATHWAY [pp. 138–139]

Selected Words: coenzymes [p.136], coenzyme A [p.136], ATP synthases [pp.136,138], NADH [p.136], FAD [p.136], FADH$_2$ [p.136], electron transport *phosphorylation* [p.138], chemiosmotic model [p.138], kilocalorie [p.139]

Boldfaced, Page-Referenced Terms

[p.136] mitochondrion _in this organelle, the second and third stages of the aerobic pathway_ _run to completion._

[p.136] acetyl-CoA _is formed when an enzyme helper, coenzyme A combines with the_ _remaining two-carbon fragment._

[p.136] oxaloacetate _the entry point for the kreb cycle_

Fill-in-the-Blanks

If sufficient oxygen is present, the end product of glycolysis enters a preparatory step, (1) _____ _____ [p.136] formation. This step converts pyruvate into acetyl CoA, the molecule that enters the (2) _____ [p.136] cycle, which is followed by (3) _____ _____ [p.138] phosphorylation.

During these three processes, a total of (4) _____ [p.138] (number) (5) _____ [p.138] (energy-carrier

molecules) are generated. In the preparatory conversions prior to the Krebs cycle and within the Krebs

cycle, the food molecule fragments are further broken down into (6) _____ _____ [p.137]. During

these reactions, hydrogen atoms (with their (7) _____ [p.137]) are stripped from the fragments and

transferred to the coenzymes (8) _____ [p.137] and (9) _____ [p.137].

Labeling

In exercises 10–14, identify the structure or location; in exercises 15–18, identify the chemical substance involved. In exercise 19, name the metabolic pathway. [All from p.136]

10. _____ _____ of mitochondrion   15. _____

11. _____ _____ of mitochondrion   16. _____

12. _____ _____ of mitochondrion   17. _____

13. _____ _____ of mitochondrion   18. _____

14. _____   19. _____ _____ _____

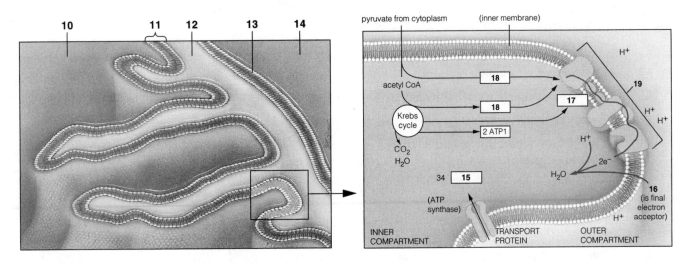

Fill-in-the-Blanks

NADH delivers its electrons to the highest possible point of entry into a transport system; from each NADH

enough H+ is pumped to produce (20) _____ [p.139] (number) ATP molecules. FADH$_2$ delivers its

electrons at a lower point of entry into the transport system; fewer H+ are pumped, and (21) _____

[p.139] (number) ATPs are produced. The electrons are then sent down highly organized (22) _____

[p.138] systems located in the inner membrane of the mitochondrion; hydrogen ions are pumped into the

outer mitochondrial compartment. According to the (23) _____ [p.138] model, the hydrogen ions

accumulate and then follow a gradient to flow through channel proteins, called ATP (24) _____ [p.138],

that lead into the inner compartment. The energy of the hydrogen ion flow across the membrane is used to

phosphorylate ADP to produce (25) _____ [p.138]. Electrons leaving the electron transport system

combine with hydrogen ions and (26) _____ [p.138] to form water. These reactions occur only in

(27) _____ [p.138]. From glycolysis (in the cytoplasm) to the final reactions occurring in the mitochondria,

the aerobic pathway commonly yields (28) _____ [p.138] (number) ATP or (29) _____ [p.139] (number) ATP for every glucose molecule degraded.

Choice

For questions 30–51, choose from the following. Some correct answers may require more than one letter.

a. preparatory steps to Krebs cycle   b. Krebs cycle   c. electron transport phosphorylation

30. ___ Chemiosmosis occurs to form ATP molecules [p.138]

31. ___ Three carbon atoms in pyruvate leave as three CO_2 molecules [p.137]

32. ___ Chemical reactions occur at transport systems [pp.138–139]

33. ___ Coenzyme A picks up a two-carbon acetyl group [p.137]

34. ___ Makes two turns for each glucose molecule entering glycolysis [p.137]

35. ___ Two NADH molecules form for each glucose entering glycolysis [p.137]

36. ___ Oxaloacetate forms from intermediate molecules [p.137]

37. ___ Named for a scientist who worked out its chemical details [p.136]

38. ___ Occurs within the mitochondrion [p.138]

39. ___ Two $FADH_2$ and six NADH form from one glucose molecule entering glycolysis [p.137]

40. ___ Hydrogens collect in the mitochondrion's outer compartment [p.138]

41. ___ Hydrogens and electrons are transferred to NAD^+ and FAD [p.137]

42. ___ Two ATP molecules form by substrate-level phosphorylation [p.137]

43. ___ Free oxygen withdraws electrons from the system and then combines with H^+ to form water molecules [p.138]

44. ___ No ATP is produced [p.137]

45. ___ Thirty-two ATPs are produced [p.138]

46. ___ Delivery point of NADH and $FADH_2$ [p.138]

47. ___ Two pyruvates enter for each glucose molecule entering glycolysis [p.137]

48. ___ The carbons in the acetyl group leave as CO_2 [p.137]

49. ___ One carbon in pyruvate leaves as CO_2 [p.137]

50. ___ An electron transport system and channel proteins are involved [p.138]

51. ___ Two $FADH_2$ and ten NADH are sent to this stage [p.138]

8.5. ANAEROBIC ROUTES OF ATP FORMATION [pp. 140–141]

Selected Words: *pyruvate* [p.140], *Lactobacillus* [p.140], *Saccharomyces cerevisiae* [p.140], *Saccharomyces ellipsoideus* [p.140].

Boldfaced, Page-Referenced Terms

[p.140] lactate fermentation – the pyruvate molecules from the first stage of reactions accept the hydrogen and electrons from NADH. This regenerates NAD^+ and converts pyruvate into lactate.

[p.140] alcoholic fermentation _enzymes convert pyruvate molecules that are formed during glycolysis to an intermediate form: acetaldehyde._

[p.141] anaerobic electron transport _ATP-forming pathway in which electrons stripped from an organic compound move through transport systems in the bacterial plasma membrane._

Fill-in-the-Blanks

If (1) _____ [p.140] is not present in sufficient amounts, the end product of glycolysis enters (2) _____ [p.140] pathways; in some bacteria and muscle cells, pyruvate is converted into such products as (3) _____ [p.140], or in yeast cells it is converted into (4) _____ [p.140] and (5) _____ _____ [p.140].

(6) _____ [p.140] pathways do not use oxygen as the final (7) _____ [p.140] acceptor that ultimately drives the ATP-forming machinery. Anaerobic routes must be used by many bacteria and protistans that live in an oxygen-free environment. (8) _____ [p.140] precedes any of the fermentation pathways. During (8), a glucose molecule is split into two (9) _____ [p.140] molecules, two energy-rich (10) _____ [p.140] intermediate molecules form, and the net energy yield from one glucose molecule is two ATP.

In one kind of fermentation pathway, (11) _____ [p.140] itself accepts hydrogen and electrons from NADH. (11) is then converted to a three-carbon compound, (12) _____ [p.140], during this process in a few bacteria species and some animal cells. Human muscle cells can carry on (12) fermentation in times of oxygen depletion; this provides a low yield of ATP.

In yeast cells, each pyruvate molecule from glycolysis forms an intermediate called (13) _____ [p.140] as a gas (14) _____ _____ [p.140] is detached from pyruvate with the help of an enzyme. This intermediate accepts hydrogen and electrons from NADH and is then converted to (15) _____ [p.140], the end product of alcoholic fermentation.

In both types of fermentation pathways, the net energy yield of two ATPs is formed during (16) _____ [p.140]. The reactions of the fermentation chemistry regenerate the (17) _____ [p.140] needed for glycolysis to occur.

Anaerobic electron transport is an energy-releasing pathway occurring among the (18) _____ [p.141]. For example, sulfate-reducing bacteria living in soil or water produce (19) _____ [p.141] by stripping electrons from a variety of compounds and sending them through membrane transport systems. The inorganic compound (20) _____ [p.141] ($SO_4^=$) serves as the final electron acceptor and is converted into foul-smelling hydrogen sulfide gas (H_2S).

Complete the Table

Include a check mark (√) in each box that correctly links an occurrence (left-hand column) with a process (or processes).

	Glycolysis	Lactate Fermentation	Alcoholic Fermentation	Anaerobic Electron Transport
6-C → 3C / 3C [p.140]	(21)	(33)	(45)	(57)
NADH → NAD⁺ [p.140]	(22)	(34)	(46)	(58)
NAD⁺ → NADH [p.140]	(23)	(35)	(47)	(59)
$SO_4^= \rightarrow H_2S$ [p.141]	(24)	(36)	(48)	(60)
CO_2 is a waste product [p.141]	(25)	(37)	(49)	(61)
3-C → 3-C [p.140]	(26)	(38)	(50)	(62)
3-C → 2-C [p.141]	(27)	(39)	(51)	(63)
ATP is used as a reactant [p.140]	(28)	(40)	(52)	(64)
ATP is produced [pp.140–141]	(29)	(41)	(53)	(65)
pyruvate → ethanol / CO_2 [p.141]	(30)	(42)	(54)	(66)
Occurs in animal cells [p.140]	(31)	(43)	(55)	(67)
Occurs in yeast cells [p.140]	(32)	(44)	(56)	(68)

8.6. ALTERNATIVE ENERGY SOURCES IN THE HUMAN BODY [pp.142–143]

8.7. Commentary: PERSPECTIVE ON LIFE [p.144]

True–False

If the statement is true, write a T in the blank. If the statement is false, explain why in the answer blank.

_____1. Glucose is the only carbon-containing molecule that can be fed into the glycolytic pathway. [p.143]

_____2. Simple sugars, fatty acids, and glycerol that remain after a cell's biosynthetic and storage needs have been met are generally sent to the cell's respiratory pathways for energy extraction. [p.142]

_____3. Carbon dioxide and water, the products of aerobic respiration, generally get into the blood and are carried to gills or lungs, kidneys, and skin, where they are expelled from the animal's body. [p.144; common knowledge]

_____4. Energy is recycled along with materials. [p.144]

_____5. The first forms of life on Earth were most probably photosynthetic eukaryotes. [p.144]

Labeling

Identify the process or substance indicated in the illustration below. [All from p.143]

6. _____ _____

7. _____

8. _____

9. _____ _____

10. _____ _____

11. _____ _____

12. _____

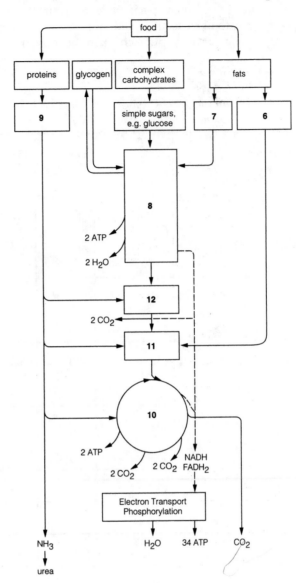

Complete the Table

Across the top of each column are the principal phases of degradative pathways into which food molecules (in various stages of breakdown) enter or in which specific events occur. Put a check (√) in each box that indicates the phase into which a specific food molecule is fed, or in which a specific event occurs. For example, if simple sugars can enter the glycolytic pathway, put a check in the top left-hand box; if not, let the box remain blank.

	Glycolysis (includes pyruvate)	Acetyl-CoA Formation	Krebs Cycle	Electron Transport Phosphorylation	Fermentation	
					Alcoholic	Lactate
Complex Carbohydrates → Simple Sugars → which enter [p.142]	(13)	(28)	(43)	(58)	(73)	(88)
Fats — Fatty acids, which enter	(14)	(29)	(44)	(59)	(74)	(89)
Glycerol, which enters [pp.137,142–143]	(15)	(30)	(45)	(60)	(75)	(90)
Proteins → Amino acids, which enter [pp.142–143]	(16)	(31)	(46)	(61)	(76)	(91)
Intermediate energy carriers (NADH, $FADH_2$) are produced [pp.135,137,143]	(17)	(32)	(47)	(62)	(77)	(92)
	(18)	(33)	(48)	(63)	(78)	(93)
ATPs produced directly as a result of this process alone [p.135]	(19)	(34)	(49)	(64)	(79)	(94)
NAD^+ produced [pp.138,140–141]	(20)	(35)	(50)	(65)	(80)	(95)
FAD produced [p.137]	(21)	(36)	(51)	(66)	(81)	(96)
ADP produced [p.135]	(22)	(37)	(52)	(67)	(82)	(97)
Unbound phosphate (P_i) required [pp.135,137]	(23)	(38)	(53)	(68)	(83)	(98)
CO_2 produced (waste product) [pp.137,141,143]	(24)	(39)	(54)	(69)	(84)	(99)
H_2O produced (waste product) [pp.135,138]	(25)	(40)	(55)	(70)	(85)	(100)
One-half O_2 reacts with H^+ [p.138]	(26)	(41)	(56)	(71)	(86)	(101)
NADH required to drive this process [pp.138,140,141]	(27)	(42)	(57)	(72)	(87)	(102)

Choice

For questions 103–117, refer to the text and Figure 8.12; choose from the following: [All pp.142,143]

a. glucose	b. glucose-6-phosphate	c. glycogen	d. fatty acids	e. triglycerides
f. PGAL	g. acetyl-CoA		h. amino acids	i. glycerol	j. proteins

103. _e_ Fats that are broken down between meals or during exercise as alternatives to glucose

104. _b_ Used between meals when free glucose supply dwindles; enters glycolysis after conversion

105. _d_ Its breakdown yields much more ATP than glucose

106. _a_ Absorbed in large amounts immediately following a meal

107. _c_ Represents only 1 percent or so of the total stored energy in the body

108. _h_ Following removal of amino groups, the carbon backbones may be converted to fats or carbohydrates or they may enter the Krebs cycle

109. _e_ On the average, represents 78 percent of the body's stored food

110. _b_ Between meals liver cells can convert it back to free glucose and release it

111. _b_ Can be stored in cells but not transported across plasma membranes

112. _h_ Amino groups undergo conversions that produce urea, a nitrogen-containing waste product excreted in urine

113. _i_ Converted to PGAL in the liver, a key intermediate of glycolysis

114. _e_ Accumulate inside the fat cells of adipose tissues, at strategic points under the skin

115. _c_ A storage polysaccharide produced from glucose-6-phosphate following food intake that exceeds cellular energy demand (and increases ATP production to inhibit glycolysis)

116. _h_ Building blocks of the compounds that represent 21 percent of the body's stored food

117. _g_ A product resulting from enzymes cleaving circulating fatty acids; enters the Krebs cycle

---

Self-Quiz

_c_ 1. Glycolysis would quickly halt if the process ran out of _____, which serves as the hydrogen and electron acceptor. [p.134]
   a. NADP$^+$
   b. ADP
   c. NAD$^+$
   d. H$_2$O

_c_ 2. The ultimate electron acceptor in aerobic respiration is _____. [p.138]
   a. NADH
   b. carbon dioxide (CO$_2$)
   c. oxygen (1/2 O$_2$)
   d. ATP

_d_ 3. When glucose is used as an energy source, the largest amount of ATP is generated by the _____ portion of the entire respiratory process. [p.138]
   a. glycolytic pathway
   b. acetyl-CoA formation
   c. Krebs cycle
   d. electron transport phosphorylation

_b_ 4. The process by which about 10 percent of the energy stored in a sugar molecule is released as it is converted into two small organic-acid molecules is _____. [pp.134,139]
   a. photolysis
   b. glycolysis
   c. fermentation
   d. the dark reactions

_d_ 5. During which of the following phases of respiration is ATP produced directly by substrate-level phosphorylation? [p.135]
   a. glucose formation
   b. ethanol production
   c. acetyl-CoA formation
   d. glycolysis

_d_ 6. What is the name of the process by which reduced NADH transfers electrons along a chain of acceptors to oxygen so as to form water and in which the energy released along the way is used to generate ATP? [p.138]
   a. glycolysis
   b. acetyl-CoA formation
   c. the Krebs cycle
   d. electron transport phosphorylation

_a_ 7. Pyruvic acid can be regarded as the end product of _____. [p.135]
   a. glycolysis
   b. acetyl-CoA formation
   c. fermentation
   d. the Krebs cycle

_d_ 8. Which of the following is *not* ordinarily capable of being reduced at any time? [p.138]
   a. NAD^+
   b. FAD
   c. oxygen, O_2
   d. water

_d_ 9. ATP production by chemiosmosis involves _____. [p.138]
   a. H^+ concentration and electric gradients across a membrane
   b. ATP synthases
   c. formation of ATP in the inner mitochondrial compartment
   d. all of the above

_d_ 10. During the fermentation pathways, a net yield of two ATP is produced from _____; the NAD^+ necessary for _____ is regenerated during the fermentation reactions. [p.140]
   a. the Krebs cycle; glycolysis
   b. glycolysis; electron transport phosphorylation
   c. the Krebs cycle; electron transport phosphorylation
   d. glycolysis; glycolysis

Matching

Match the following components of respiration to the list of words below. Some components may have more than one answer.

11. _C_ lactic acid, lactate [p.140]

12. _ABD_ $NAD^+ \rightarrow$ NADH [pp.135,137]

13. _BCD_ carbon dioxide is a product [p.137]

14. _CE_ NADH $\rightarrow NAD^+$ [p.138]

15. _BC_ pyruvate used as a reactant [p.137]

16. _AD_ ATP produced by substrate-level phosphorylation [pp.135,137]

17. _A_ glucose [p.135]

18. _BD_ acetyl-CoA is both a reactant and a product [p.137]

19. _E_ oxygen [p.138]

20. _AE_ water is a product [p.138]

A. Glycolysis
B. Preparatory conversions prior to the Krebs cycle
C. Fermentation
D. Krebs cycle
E. Electron transport phosphorylation

Chapter Objectives/Review Questions

1. No matter what the source of energy may be, organisms must convert it to _____, a form of chemical energy that can drive metabolic reactions. [p.132]
2. The main energy-releasing pathway is _____ respiration. [p.132]
3. Give the overall equation for the aerobic respiratory route. [p.132]
4. In the first of the three stages of aerobic respiration, one _____ is partially degraded to two pyruvate molecules. [p.133]
5. By the end of the second stage of aerobic respiration, which includes the _____ cycle, _____ has been completely degraded to carbon dioxide and water. [p.133]
6. Do the first two stages of aerobic respiration yield a high or low quantity of ATP? [p.133]
7. The third stage of aerobic respiration is called *electron transport* _____, which yields many ATP molecules. [p.133]
8. Explain, in general terms, the role of oxygen in aerobic respiration. [p.133]
9. Glycolysis occurs in the _____ of the cell. [p.134]
10. Explain the purpose served by the cell investing two ATP molecules into the chemistry of glycolysis. [p.134]
11. State the kinds of factors responsible for pyruvic acid entering the acetyl-CoA formation pathway. [p.134]
12. Four ATP molecules are produced by _____-_____ phosphorylation for every two used during glycolysis. (Consult Figure 8.4 in the main text.) [pp.134–135]
13. Glycolysis produces _____ (number) NADH, _____ (number) ATP (net) and _____ (number) pyruvate molecules for each glucose molecule entering the reactions. [p.135]
14. Consult Figure 8.6 in the main text. Relate the events that happen during acetyl-coenzyme A formation and explain how the process of acetyl-CoA formation relates glycolysis to the Krebs cycle. [p.137]
15. What happens to the CO_2 produced during acetyl-CoA formation and the Krebs cycle? [p.137]
16. Calculate the number of ATP molecules produced during the Krebs cycle for each glucose molecule that enters glycolysis. [p.137]
17. Explain how chemiosmosis theory operates in the mitochondrion to account for the production of ATP molecules. [p.138]
18. Consult Figure 8.8 in the main text and predict what will happen to the NADH produced during acetyl-CoA formation and the Krebs cycle. [p.139]
19. Briefly describe the process of electron transport phosphorylation by stating what reactants are needed and what the products are. State how many ATP molecules are produced through operation of the transport system. [p.139]
20. Account for the total *net yield* of thirty-six ATP molecules produced through aerobic respiration; that is, state how many ATPs are produced in glycolysis, the Krebs cycle, and electron transport system. [p.139]
21. List some environments where there is very little oxygen present and where anaerobic organisms might be found. [p.140]
22. In fermentation chemistry _____ molecules from glycolysis are accepted to construct either lactate or ethyl alcohol; thus, a low yield of _____ molecules continues in the absence of oxygen. [pp.140–141]
23. List the main anaerobic energy-releasing pathways and the examples of organisms that use them. [pp.140–141]
24. Describe what happens to pyruvate in anaerobic organisms. Then explain the necessity for pyruvate to be converted to a fermentation product. [pp.140–141]
25. State which factors determine whether the pyruvate (pyruvic acid) produced at the end of glycolysis will enter into the alcoholic fermentation pathway, the lactate fermentation pathway, or the acetyl–coenzyme A formation pathway. [pp.140–141]
26. List some sources of energy (other than glucose) that can be fed into the respiratory pathways. [pp.142–143]
27. Predict what your body would do to synthesize its needed carbohydrates and fats if you switched to a diet of 100 percent protein. [pp.142–143]
28. After reading "Perspective on Life" in the main text, outline the supposed evolutionary sequence of energy-extraction processes. [p.144]
29. Scrutinize the sketch in the Commentary in the main text closely; then reproduce the carbon cycle from memory. [p.144]

Integrating and Applying Key Concepts

How is the "oxygen debt" experienced by runners and sprinters related to aerobic and anaerobic respiration in humans?

9

CELL DIVISION AND MITOSIS

Interactive Exercises

Silver in the Stream of Time [pp.148–149]

9.1. DIVIDING CELLS: THE BRIDGE BETWEEN GENERATIONS [p.150]

9.2. THE CELL CYCLE [p.151]

Selected Words: *cell division* [p.148], *nuclear* division [p.150], *2n* [p.150], *haploid* chromosome number [p.150], *n* [p.150], G1–S–G2–M [p.151]

Boldfaced, Page-Referenced Terms

[p.148] reproduction _____

_____

[p.150] mitosis _____

_____

[p.150] meiosis _____

_____

[p.150] somatic cells _____

_____

[p.150] germ cells _____

_____

[p.150] chromosome _____

_____

[p.150] sister chromatids _____

_____

[p.150] centromere _____

_____

[p.150] chromosome number _____

_____

[p.150] diploid _____

_____

[p.151] cell cycle _____

_____

[p.151] interphase _____

_____

Matching

Choose the most appropriate answer for each term.

1. ___centromere [p.150]
2. ___haploid or *n* cell [p.150]
3. ___diploid or *2n* cell [p.150]
4. ___chromosome [p.150]
5. ___nuclear division [p.150]
6. ___germ cells [p.150]
7. ___somatic cells [p.150]
8. ___sister chromatids [p.150]
9. ___mitosis and meiosis [p.150]
10. ___chromosome number [p.150]

A. Cell lineage set-aside for forming gametes and sexual reproduction
B. Necessary for the reproduction of eukaryotic cells
C. Any cell having two of each type of chromosome characteristic of the species
D. The number of each type of chromosome present in a cell (*n* or *2n*)
E. Each DNA molecule with attached proteins
F. Sort out and package parent DNA molecules into new cell nuclei
G. A small chromosome region with attachment sites for microtubules
H. The two attached DNA molecules of a duplicated chromosome
I. Body cells that reproduce by mitosis and cytoplasmic division
J. Possess only one of each type of chromosome characteristic of the species

Labeling

Identify the stage in the cell cycle indicated by each number.

11. _____ [p.151]

12. _____ [p.151]

13. _____ [p.151]

14. _____ [p.151]

15. _____ [p.151]

16. _____ [p.151]

17. _____ [p.151]

18. _____ [p.151]

19. _____ [p.151]

20. _____ [p.151]

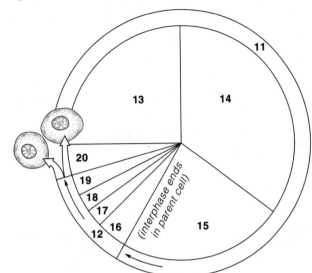

Matching

Link each time span identified below with the most appropriate number in the preceding labeling section.

21. ___Period after duplication of DNA during which the cell prepares for division [p.151]

22. ___The complete period of nuclear division that is followed by cytoplasmic division (a separate event) [p.151]

23. ___DNA duplication occurs now; a time of "synthesis" of DNA and proteins [p.151]

24. ___Period of cell growth before DNA duplication; a "gap" of interphase [p.151]

25. ___Usually the longest part of a cell cycle [p.151]

26. ___Period of cytoplasmic division [p.151]

27. ___Period that includes G1, S, G2 [p.151]

9.3. THE STAGES OF MITOSIS—AN OVERVIEW [pp.152–153]

Selected Words: *mitos* [p.152], *new* microtubules [p.152], prometaphase [p.153], *meta-* [p.153]

Boldfaced, Page-Referenced Terms

[p.152] prophase _____

_____

[p.152] metaphase _____

_____

[p.152] anaphase _____

_____

[p.152] telophase _____

_____

[p.152] spindle apparatus _____

_____

[p.152] centrioles _____

_____

Labeling and Matching

Identify each of the mitotic stages shown by entering the correct stage in the blank beneath the sketch. Select from late prophase, transition to metaphase (prometaphase), cell at interphase, metaphase, early prophase, telophase, interphase-daughter cells, and anaphase. Complete the exercise by matching and entering the letter of the correct phase description in the parentheses following each label.

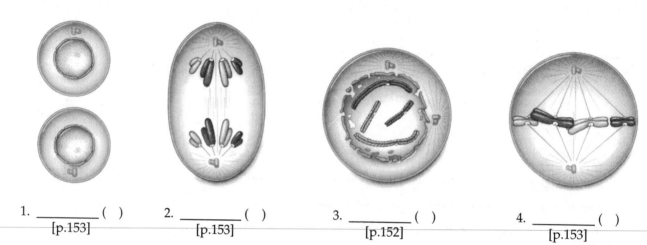

1. _____ ( )
   [p.153]

2. _____ ( )
   [p.153]

3. _____ ( )
   [p.152]

4. _____ ( )
   [p.153]

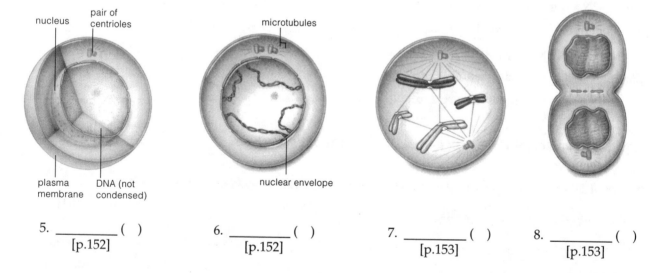

5. _____ ( )
   [p.152]

6. _____ ( )
   [p.152]

7. _____ ( )
   [p.153]

8. _____ ( )
   [p.153]

A. Attachments between two sister chromatids of each chromosome break; the two are now separate chromosomes that move to opposite spindle poles

B. Microtubules penetrate the nuclear region and collectively form the spindle apparatus; microtubules become attached to the two sister chromatids of each chromosome

C. The DNA and its associated proteins have started to condense

D. All the chromosomes are now fully condensed and lined up at the equator of the spindle

E. DNA is duplicated and the cell prepares for nuclear division

F. Two daughter cells have formed, each diploid with two of each type of chromosome, just like the parent cell's nucleus

G. Chromosomes continue to condense. New microtubules are assembled, and they move one of two centriole pairs toward the opposite end of the cell. The nuclear envelope begins to break up

H. Patches of new membrane fuse to form a new nuclear envelope around the decondensing chromosomes

9.4. DIVISION OF THE CYTOPLASM [pp.154–155]

Selected Words: cytokinesis [p.154]

Boldfaced, Page-Referenced Terms

[p.154] cytoplasmic division _____

_____

[p.154] cell plate formation _____

_____

[p.155] cleavage furrow _____

_____

Choice

For questions 1–10 on cytoplasmic division, choose from the following:

a. plant cells   b. animal cells

1. ___ Formation of a cell plate [p.154]

2. ___ Contractions continue and in time will divide the cell in two [pp.154–155]

3. ___ Cellulose deposits form a crosswall between the two daughter cells [p.154]

4. ___ Possess rigid walls that cannot be pinched in two [p.154]

5. ___ A cleavage furrow [p.155]

6. ___ Deposits form a cementing middle lamella [p.154]

7. ___ A band of microfilaments beneath the cell's plasma membrane generates the force for cytoplasmic division [p.155]

8. ___ Two daughter nuclei are cut off in separate cells, each with its own cytoplasm and plasma membrane [p.155]

9. ___ At the location of a disklike structure, deposits of cellulose accumulate [p.154]

10. ___ The structure grows at its margins until it fuses with the parent cell's plasma membrane [p.154]

9.5. A CLOSER LOOK AT THE CELL CYCLE [pp.156–157]

9.6. *Focus on Science:* HENRIETTA'S IMMORTAL CELLS [p.158]

Selected Words: histone-DNA spool [p.156], *Colchicum* [p.157], Henrietta Lacks [p.158]

Boldfaced, Page-Referenced Terms

[p.156] histones _____

_____

[p.156] nucleosome _____

_____

[p.156] kinetochores _____

_____

[p.156] motor proteins _____

_____

[p.158] HeLa cells _____

_____

Matching

Choose the most appropriate answer for each term.

1. ___G2 [p.156]
2. ___histones [p.156]
3. ___nucleosome [p.156]
4. ___late prophase [p.156]
5. ___kinetochore [p.156]
6. ___centromere [p.156]
7. ___G1 [p.156]
8. ___motor proteins [p.156]
9. ___*Colchicum* [p.157]
10. ___DNA topoisomerase [p.156]

A. The time in the cell cycle when a chromosome acquires its distinct shape and size
B. A long period when cells assemble most of the carbohydrates, lipids, and proteins they use or export
C. An enzyme that recognizes the tangling occurring in condensing DNA molecules and then prevents the tangling
D. Small disk-shaped structures at the surface of centromeres that serve as docking sites for spindle microtubules
E. Produces a poison that blocks assembly and promotes disassembly of microtubules
F. Dividing cells produce proteins that will drive mitosis to completion
G. Accessory for cytoskeletal elements; attached to and project from microtubules and microfilaments that take part in cell movements
H. Term for a histone-DNA spool
I. Many act like spools for winding up small stretches of DNA
J. The most prominent constriction of a metaphase chromosome

Short Answer

11. Describe the origin and the significance of HeLa cells. [p.158] _____

_____

_____

_____

_____

_____

Self-Quiz

____ 1. The replication of DNA occurs _____. [p.151]
- a between the growth phases of interphase
- b. immediately before prophase of mitosis
- c. during prophase of mitosis
- d. during prophase of meiosis

____ 2. In the cell life cycle of a particular cell, _____. [p.151]
- a. mitosis occurs immediately prior to G1
- b. G2 precedes S
- c. G1 precedes S
- d. mitosis and S precede G1

____ 3. In eukaryotic cells, which of the following can occur during mitosis? [pp.152–153]
- a. Two mitotic divisions to maintain the parental chromosome number
- b. The replication of DNA
- c. A long growth period
- d. The disappearance of the nuclear envelope

____ 4. Diploid refers to _____. [p.150]
- a. having two chromosomes of each type in somatic cells
- b. twice the parental chromosome number
- c. half the parental chromosome number
- d. having one chromosome of each type in somatic cells

____ 5. Somatic cells are _____ cells; germ cells are _____ cells. [p.150]
- a. meiotic; body
- b. body; body
- c. meiotic; meiotic
- d. body; meiotic

____ 6. If a parent cell has sixteen chromosomes and undergoes mitosis, the resulting cells will have _____ chromosomes. [p.150]
- a. sixty-four
- b. thirty-two
- c. sixteen
- d. eight
- e. four

____ 7. The correct order of the stages of mitosis is _____. [pp.152–153]
- a. prophase, metaphase, telophase, anaphase
- b. telophase, anaphase, metaphase, prophase
- c. telophase, prophase, metaphase, anaphase
- d. anaphase, prophase, telophase, metaphase
- e. prophase, metaphase, anaphase, telophase

____ 8. "The nuclear envelope breaks completely into numerous tiny, flattened vesicles. Now the chromosomes are free to interact with microtubules that are extending toward them, from the poles of the forming spindle." These sentences describe the _____ of mitosis. [pp.152–153]
- a. prophase
- b. metaphase
- c. prometaphase
- d. anaphase
- e. telophase

____ 9. During _____, sister chromatids of each chromosome are separated from each other, and those former partners, now chromosomes, move to opposite poles. [p.153]
- a. prophase
- b. metaphase
- c. anaphase
- d. telophase

____ 10. Each histone-DNA spool is a single structural unit called a _____. [p.156]
- a. kinetochore
- b. motor protein
- c. centromere
- d. nucleosome

____ 11. In the process of cytokinesis, cleavage furrows are associated with _____ cell division, and cell plate formation is associated with _____ cell division. [pp.154–155]
- a. animal; animal
- b. plant; animal
- c. plant; plant
- d. animal; plant

Chapter Objectives/Review Questions

1. Define the word *reproduction*. [p.148]
2. The terms *mitosis* and *meiosis* refer to the division of the cell's _____. [p.150]
3. Distinguish between somatic cells and germ cells as to their location and function. [p.150]
4. The eukaryotic chromosome is composed of _____ and _____. [p.150]
5. The two attached threads of a duplicated chromosome are known as sister _____. [p.150]
6. The _____ is a small region with attachment sites for the microtubules that move the chromosome during nuclear division. [p.150]
7. Any cell having two of each type of chromosome is a _____ cell; a _____ cell is one set aside for the formation of gametes. [p.150]
8. List and describe, in order, the various activities occurring in the eukaryotic cell life cycle. [p.151]
9. Interphase of the cell cycle consists of G1, _____, and G2. [p.151]
10. S is the time in the cell cycle when _____ replication occurs. [p.151]
11. Describe the structure and function of the spindle apparatus. [p.152]
12. Describe the number and movements of centrioles in the cell division of some cells. [p.152]
13. The _____ is a time of transition when the nuclear envelope breaks up into tiny, flattened vesicles prior to metaphase. [p.153]
14. Give a detailed description of the cellular events occurring in the prophase, metaphase, anaphase, and telophase of mitosis. [pp.152–153]
15. Compare and contrast cytokinesis as it occurs in plant and animal cell division; use the following concepts: cleavage furrow, microfilaments at the cell's midsection, and cell plate formation. [pp.154–155]
16. Characterize the organization of metaphase chromosomes using the following terms: *histones, nucleosome, kinetochore,* and *motor proteins.* [p.156]
17. Describe particular events occurring in G1, S, and G2 of interphase. [p.156]
18. Explain the value of the plant *Colchicum* to cellular scientists. [p.157]
19. Explain how cells from Henrietta Lacks continue to benefit humans everywhere more than forty years after her death. [p.158]

Integrating and Applying Key Concepts

Runaway cell division is characteristic of cancer. Imagine the various points of the mitotic process that might be sabotaged in cancerous cells in order to halt their multiplication. Then try to imagine how one might discriminate between cancerous and normal cells in order to guide those methods of sabotage most effective in combating cancer.

10

MEIOSIS

Interactive Exercises

Octopus Sex and Other Stories [pp.160–161]

10.1. COMPARING SEXUAL WITH ASEXUAL REPRODUCTION [p.162]

10.2. HOW MEIOSIS HALVES THE CHROMOSOME NUMBER [pp.162–163]

Selected Words: *unfertilized* egg cells [p.161], *sexual* reproduction [p.161], *pairs of genes* [p.162], *hom-* [p.162], *two consecutive divisions* [p.163], *homologue to homologue* [p.163]

Boldfaced, Page-Referenced Terms

[p.161] germ cells _____

_____

[p.161] gametes _____

_____

[p.162] asexual reproduction _____

_____

[p.162] genes _____

_____

[p.162] sexual reproduction _____

_____

[p.162] allele _____

_____

[p.162] meiosis _____

_____

[p.162] chromosome number _____

_____

[p.162] diploid number (2*n*) _____

_____

[p.162] homologous chromosomes _____

_____

[p.163] haploid number (*n*) _____

_____

[p.163] sister chromatids _____

_____

Choice

For questions 1–10, choose from the following:

a. asexual reproduction   b. sexual reproduction

1. ___ Offspring inherit new combinations of alleles [p.162]
2. ___ One parent alone produces offspring [p.162]
3. ___ Each offspring inherits the same number and kinds of genes as its parent [p.162]
4. ___ Commonly involves two parents [p.162]
5. ___ The production of "clones" [p.162]
6. ___ Involves meiosis, formation of gametes, and fertilization [p.162]
7. ___ Offspring are genetically identical copies of the parent [p.162]
8. ___ Produces the variation in traits that is the foundation of evolutionary change [p.162]
9. ___ Change can only occur by mutations [p.162]
10. ___ The first cell of a new individual has pairs of genes on pairs of homologous chromosomes that are of maternal and paternal origin [p.162]

Dichotomous Choice

Circle one of two possible answers given between parentheses in each statement.

11. (Meiosis/Mitosis) divides chromosomes into separate parcels not once but twice prior to cell division. [p.162]
12. Sperm and eggs are sex cells known as (germ/gamete) cells. [p.162]
13. (Haploid/Diploid) cells possess pairs of homologous chromosomes. [p.162]
14. (Haploid/Diploid) germ cells produce haploid gametes. [p.163]
15. (Meiosis/Mitosis) produces cells that have one member of each pair of homologous chromosomes possessed by the species. [p.163]
16. Identical alleles are found on (homologous chromosomes/sister chromatids). [p.163]
17. Two attached DNA molecules are known as (sister chromatids/homologous chromosomes). [p.163]
18. Two attached sister chromatids represent (two/one) chromosome(s). [p.163]

19. One pair of duplicated chromosomes would be composed of (two/four) chromatids. [p.163]
20. With meiosis, chromosomes proceed through (one/two) divisions to yield four haploid nuclei. [p.163]
21. DNA is replicated during (prophase I of meiosis I/interphase preceding meiosis I). [p.163]
22. During meiosis I, each duplicated (chromosome/chromatid) lines up with its partner, homologue to homologue, and then the partners are moved apart from one another. [p.163]
23. Cytoplasmic division following meiosis I results in two (diploid/haploid) daughter cells. [p.163]
24. During (meiosis I/meiosis II), the two sister chromatids of each chromosome are separated from each other, and each sister chromatid is now a chromosome in its own right. [p.163]
25. If human body cell nuclei contain twenty-three pairs of homologous chromosomes, each resulting gamete will contain (twenty-three/forty-six) chromosomes. [p.163]

Matching

To review the major stages of meiosis, match the following written descriptions with the appropriate sketch. Assume that the cell in this model initially has only one pair of homologous chromosomes (one from a paternal source and one from a maternal source) and crossing over does not occur. Complete the exercise by indicating the diploid (2*n* = 2) or haploid (*n* = 1) chromosome number of the cell chosen in the parentheses following each blank.

26. ___ ( ) A pair of homologous chromosomes prior to S of interphase in a diploid germ cell [p.163]

27. ___ ( ) While the germ cell is in S of interphase, chromosomes are duplicated through DNA replication; the two sister chromatids are attached at the centromere [p.163]

28. ___ ( ) During meiosis I, each duplicated chromosome lines up with its partner, homologue to homologue [p.163]

29. ___ ( ) Also during meiosis I, the chromosome partners separate from each other in anaphase I; cytokinesis occurs, and each chromosome goes to a different cell [p.163]

30. ___ ( ) During meiosis II (in two cells), the sister chromatids of each chromosome are separated from each other; four haploid nuclei form; cytokinesis results in four cells (potential gametes) [p.163]

10.3. A VISUAL TOUR OF THE STAGES OF MEIOSIS [pp.164–165]

10.4. A CLOSER LOOK AT KEY EVENTS OF MEIOSIS I [pp.166–167]

*Selected Words: non*sister chromatids [p.166], *maternal* chromosomes [p.167], *paternal* chromosomes [p.167]

Boldfaced, Page-Referenced Terms

[p.166] crossing over _____

_____

Labeling and Matching

Identify each of the meiotic stages shown below by entering the correct stage of either meiosis I or meiosis II in the blank beneath the sketch. Choose from prophase I, metaphase I, anaphase I, telophase I, prophase II, metaphase II, anaphase II, and telophase II. Complete the exercise by matching and entering the letter of the correct stage description in the parentheses following each label.

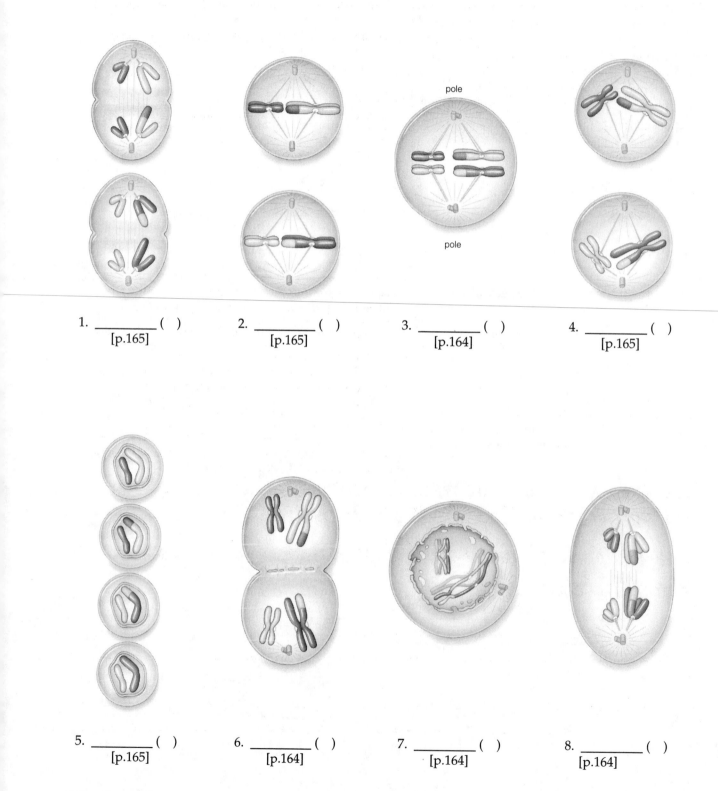

1. _____ ( )
   [p.165]

2. _____ ( )
   [p.165]

3. _____ ( )
   [p.164]

4. _____ ( )
   [p.165]

5. _____ ( )
   [p.165]

6. _____ ( )
   [p.164]

7. _____ ( )
   [p.164]

8. _____ ( )
   [p.164]

A. The spindle is now fully formed; all chromosomes are positioned midway between the poles of one cell
B. In each of two daughter cells, microtubules attach to the kinetochores of chromosomes, and motor proteins drive the movement of chromosomes toward the spindle's equator
C. Four daughter nuclei form; when the cytoplasm divides, each new cell has a haploid chromosome number, all in the unduplicated state; the cells may develop into gametes in animals or spores in plants
D. In one cell, each duplicated chromosome is pulled away from its homologous partner; the partners are moved to opposite spindle poles
E. Duplicated chromosomes condense; each chromosome pairs with its homologous partner; crossing over and genetic recombination (swapping of gene segments) occurs; each chromosome becomes attached to some microtubules of a newly forming spindle
F. Motor proteins and spindle microtubule interactions have moved all the duplicated chromosomes so that they are positioned at the spindle equator, midway between the poles
G. Two haploid cells form, each having one of each type of chromosome that was present in the parent cell; the chromosomes are still in the duplicated state
H. Attachment between the two chromatids of each chromosome breaks; former "sister chromatids" are now chromosomes in their own right and are moved to opposite poles by interactions between motor proteins and kinetochore microtubules

Matching

Choose the most appropriate answer for each.

9. ___2^{23} [p.167]

10. ___paternal chromosomes [p.167]

11. ___early prophase I [p.166]

12. ___nonsister chromatids [pp.166–167]

13. ___late prophase I [p.166]

14. ___crossing over [pp.166–167]

15. ___metaphase I [p.167]

16. ___each chromosome zippers to its homologue [p.166]

17. ___function of meiosis [p.166]

18. ___maternal chromosomes [p.167]

A. Random positioning of maternal and paternal chromosomes at the spindle equator
B. Chromosomes continue to condense and become thicker rodlike forms
C. Reduction of the chromosome number by half for forthcoming gametes
D. Break at the same places along their length and then exchange corresponding segments
E. Twenty-three chromosomes inherited from your mother
F. Combinations of maternal and paternal chromosomes possible in gametes from one germ cell
G. Each duplicated chromosome is in thin, threadlike form
H. An intimate parallel array of homologues that favors crossing over
I. Twenty-three chromosomes inherited from your father
J. Breaks up old combinations of alleles and puts new ones together in pairs of homologous chromosomes

10.5. FROM GAMETES TO OFFSPRING [pp.168–169]

Selected Words: *gamete*-producing bodies [p.168], *spore*-producing bodies [p.168]

Boldfaced, Page-Referenced Terms

[p.168] spores _____

_____

[p.168] sperm _____

_____

[p.168] oocyte _____

_____

[p.168] egg _____

_____

[p.168] fertilization _____

_____

Choice

For questions 1–14, choose from the following:

      a. animal life cycle   b. plant life cycle   c. both animal and plant life cycles

1. ___ Meiosis results in the production of haploid spores [p.168]
2. ___ A zygote divides by mitosis [p.168]
3. ___ Meiosis results in the production of haploid gametes [p.168]
4. ___ Haploid gametes fuse in fertilization to form a diploid zygote [p.168]
5. ___ A zygote divides by mitosis to form a diploid sporophyte [p.168]
6. ___ During meiosis, one egg and three polar bodies form [p.168]
7. ___ A spore divides by mitosis to produce a haploid gametophyte [p.168]
8. ___ Gametes form by oogenesis and spermatogenesis [p.168]
9. ___ A haploid gametophyte divides by mitosis to produce haploid gametes [p.168]
10. ___ A secondary oocyte gets nearly all the cytoplasm [p.168]
11. ___ A haploid spore divides by mitosis to produce a gametophyte [p.168]
12. ___ A diploid body forms from mitosis of a zygote [p.168]
13. ___ A gamete-producing body and a spore-producing body develop during the life cycle [p.168]
14. ___ One daughter cell of the secondary oocyte develops into a second polar body [p.168]

Sequence

Arrange the following entities in correct order of development, entering a 1 by the stage that appears first and a 5 by the stage that completes the process of spermatogenesis. Complete the exercise by indicating, in the parentheses following each blank, if each cell is *n* or *2n*.

15. ___ ( ) primary spermatocyte [p.169]
16. ___ ( ) sperm [p.169]
17. ___ ( ) spermatid [p.169]
18. ___ ( ) spermatogonium [p.169]
19. ___ ( ) secondary spermatocyte [p.169]

Matching

Choose the most appropriate answer to match with each oogenesis concept.

20. ___primary oocyte [p.169]

21. ___oogonium [p.169]

22. ___secondary oocyte [p.169]

23. ___ovum and three polar bodies [p.169]

24. ___first polar body [p.169]

A. The cell in which synapsis, crossing over, and recombination occur
B. A cell that is equivalent to a diploid germ cell
C. A haploid cell formed after division of the primary oocyte that does not form an ovum at second division
D. Haploid cells, but only one of which functions as an egg
E. A haploid cell formed after division of the primary oocyte, the division of which forms a functional ovum

Short Answer

25. List the various mechanisms that contribute to the huge number of new gene combinations that may result from fertilization. [pp.168–169] _____

_____

_____

_____

10.6. MEIOSIS AND MITOSIS COMPARED [pp.170–171]

Selected Words: clones [p.170], *variation in traits* [p.170], *germ cell* [p.170], somatic cell [p.170]

Complete the Table

1. Complete the following table by entering the word *mitosis* or the word *meiosis* in the blank adjacent to the statement describing one of these processes.

Description	Mitosis/Meiosis
[p.171] a. Involves one division cycle	
[p.170] b. Functions in growth and tissue repair	
[p.171] c. Daughter cells are haploid	
[p.170] d. Occurs only in germ cells	
[pp.170–171] e. Involves two division cycles	
[p.171] f. Daughter cells have one chromosome from each homologous pair	
[p.170] g. Asexual reproduction by single-celled eukaryotic species	
[p.171] h. Daughter cells have the diploid chromosome number	
[p.171] i. Completed when four daughter cells are formed	

Matching

The cell model used in this exercise has two pairs of homologous chromosomes, one long pair and one short pair. Match the descriptions to the numbers of chromosomes shown in the following sketches.

2. ___one cell at the beginning of meiosis II [p.170]

3. ___a daughter cell at the end of meiosis II [p.171]

4. ___metaphase I of meiosis [p.170]

5. ___metaphase of mitosis [p.171]

6. ___G1 in a daughter cell after mitosis [p.171]

7. ___prophase of mitosis [p.171]

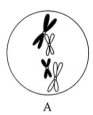

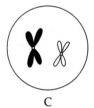

| A | B | C | D | E | F |

The following thought questions refer to the sketches above; enter answers in the blanks following each question.

8. How many chromosomes are present in cell E? ___
9. How many chromatids are present in cell E? ___
10. How many chromatids are present in cell C? ___
11. How many chromatids are present in cell D? ___
12. How many chromosomes are present in cell F? ___

Self-Quiz

___ 1. Which of the following does not occur in prophase I of meiosis? [p.166]
    a. a cytoplasmic division
    b. a cluster of four chromatids
    c. homologues pairing tightly
    d. crossing over

___ 2. Crossing over is one of the most important events in meiosis because _____. [pp.166–167]
    a. it leads to genetic recombination
    b. homologous chromosomes must be separated into different daughter cells
    c. the number of chromosomes allotted to each daughter cell must be halved
    d. homologous chromatids must be separated into different daughter cells

___ 3. Crossing over _____. [pp.166–167]
    a. generally results in pairing of homologues and binary fission
    b. is accompanied by gene-copying events
    c. involves breakages and exchanges between sister chromatids
    d. is a molecular interaction between two of the nonsister chromatids of a pair of homologous chromosomes

___ 4. The appearance of chromosome ends lapped over each other in meiotic prophase I provides evidence of _____. [p.166]
    a. meiosis
    b. crossing over
    c. chromosomal aberration
    d. fertilization
    e. spindle-fiber formation

___ 5. Which of the following does not increase genetic variation? [p.170]
   a. crossing over
   b. random fertilization
   c. prophase of mitosis
   d. random homologue alignments at metaphase I

___ 6. Which of the following is the most correct sequence of events in animal life cycles? [p.168]
   a. meiosis ⟶ fertilization ⟶ gametes ⟶ diploid organism
   b. diploid organism ⟶ meiosis ⟶ gametes ⟶ fertilization
   c. fertilization ⟶ gametes ⟶ diploid organism ⟶ meiosis
   d. diploid organism ⟶ fertilization ⟶ meiosis ⟶ gametes

___ 7. In sexually reproducing organisms, the zygote is _____. [p.168]
   a. an exact genetic copy of the female parent
   b. an exact genetic copy of the male parent
   c. completely unlike either parent genetically
   d. a genetic mixture of male parent and female parents

___ 8. Which of the following is the most correct sequence of events in plant life cycles? [p.168]
   a. fertilization ⟶ zygote ⟶ sporophyte ⟶ meiosis ⟶ spores ⟶ gametophytes ⟶ gametes
   b. fertilization ⟶ sporophyte ⟶ zygote ⟶ meiosis ⟶ spores ⟶ gametophytes ⟶ gametes

   c. fertilization ⟶ zygote ⟶ sporophyte ⟶ meiosis ⟶ gametes ⟶ gametophyte ⟶ spores
   d. fertilization ⟶ zygote ⟶ gametophyte ⟶ meiosis ⟶ gametes ⟶ sporophyte ⟶ spores

___ 9. The cell in the diagram is a diploid that has three pairs of chromosomes. From the number and pattern of chromosomes, the cell _____. [p.171]
   a. could be in the first division of meiosis
   b. could be in the second division of meiosis
   c. could be in mitosis
   d. could be in neither mitosis nor meiosis, because this stage is not possible in a cell with three pairs of chromosomes

___ 10. You are looking at a cell from the same organism as in the previous question. Now the cell _____. [p.171]
   a. could be in the first division of meiosis
   b. could be in the second division of meiosis
   c. could be in mitosis
   d. could be in neither mitosis nor meiosis, because this stage is not possible in this organism

Chapter Objectives/Review Questions

1. Distinguish between germ cells and gametes. [p.161]
2. "One parent alone produces offspring, and each offspring inherits the same number and kinds of genes as its parent" describes _____ reproduction. [p.162]
3. _____ reproduction involves meiosis, gamete formation, and fertilization. [p.162]
4. _____ divides chromosomes into separate parcels not once but twice prior to cell division. [p.162]
5. Describe the relationships among the following: homologous chromosomes, diploid number, and haploid number. [pp.162–163]
6. If the diploid chromosome number for a particular plant species is 18, the haploid gamete number is _____. [p.163]
7. During interphase a germ cell duplicates its DNA; a duplicated chromosome consists of two DNA molecules that remain attached to a constriction called the _____. [p.163]
8. As long as the two DNA molecules remain attached, they are referred to as _____ _____ of the chromosome. [p.163]

9. During meiosis _____, each duplicated chromosome lines up with its partner, homologue to homologue. [p.163]

10. During meiosis II, the two sister _____ of each _____ are separated from each other. [p.163]

11. Sketch cell models of the various stages of meiosis I and meiosis II. Using color pencils or crayons to distinguish two pairs of homologous chromosomes (a long pair and a short pair) is helpful. [pp.164–165]

12. _____ _____ breaks up old combinations of alleles and puts new ones together during prophase I of meiosis. [p.167]

13. The _____ attachment and subsequent positioning of each pair of maternal and paternal chromosomes at metaphase I leads to different _____ of maternal and paternal traits in each generation of offspring. [p.167]

14. Meiosis in the animal life cycle results in haploid _____; meiosis in the plant life cycle results in haploid _____. [p.168]

15. Using the special terms for the cells at the various stages, describe spermatogenesis in male animals and oogenesis in female animals. [p.169]

16. Crossing over, the distribution of random mixes of homologous chromosomes into gametes, and fertilization all contribute to _____ in the traits of offspring. [p.169]

17. Mitotic cell division produces only _____; meiotic cell division, in conjunction with subsequent fertilization, promotes _____ in traits among offspring. [p.170]

18. List three ways that meiosis promotes variation in offspring. [p.170]

Integrating and Applying Key Concepts

In 1997, the first mammal was cloned. It was a sheep named Dolly, which resulted from a cloning experiment in Scotland. More recently, the successful cloning of pigs was accomplished. If at some time in the future cloning of humans becomes possible, what might be the effects of reproduction without sex on human populations? (Speculate.)

11

OBSERVABLE PATTERNS OF INHERITANCE

Interactive Exercises

A Smorgasbord of Ears and Other Traits [pp.174–175]

11.1. MENDEL'S INSIGHT INTO PATTERNS OF INHERITANCE [pp.176–177]

11.2. MENDEL'S THEORY OF SEGREGATION [pp.178–179]

11.3. INDEPENDENT ASSORTMENT [pp.180–181]

Selected Words: *observable* evidence [p.176], *Pisum sativum* [p.176], identical alleles [p.177], *nonidentical* alleles [p.177], *homozygous* condition [p.177], *heterozygous* condition [p.177], *dominant* allele [p.177], *recessive* allele [p.177], *P*, *F*$_1$, *F*$_2$ [p.177]

Boldfaced, Page-Referenced Terms

[p.177] genes _____

_____

[p.177] alleles _____

_____

[p.177] true-breeding lineage _____

_____

[p.177] hybrid offspring _____

_____

[p.177] homozygous dominant _____

_____

[p.177] homozygous recessive _____

_____

[p.177] heterozygous _____

_____

[p.177] genotype _____

_____

[p.177] phenotype _____

_____

[p.178] monohybrid crosses _____

_____

[p.179] probability _____

_____

[p.179] Punnett-square method _____

_____

[p.179] testcrosses _____

_____

[p.179] segregation _____

_____

[p.180] dihybrid crosses _____

_____

[p.181] independent assortment _____

_____

Matching

Choose the most appropriate answer for each.

1. ___genotype [p.177]
2. ___alleles [p.177]
3. ___heterozygous [p.177]
4. ___dominant allele [p.177]
5. ___phenotype [p.177]
6. ___genes [p.177]
7. ___true-breeding lineage [p.177]
8. ___homozygous recessive [p.177]
9. ___recessive allele [p.177]
10. ___homozygous [p.177]
11. ___P, F_1, F_2 [p.177]
12. ___hybrid offspring [p.177]
13. ___diploid organism [p.177]
14. ___gene locus [p.177]
15. ___homozygous dominant [p.177]

A. Parental, first-generation, and second-generation offspring
B. All the different molecular forms of the same gene
C. Particular location of a gene on a chromosome
D. Describes an individual having a pair of nonidentical alleles
E. An individual with a pair of recessive alleles, such as *aa*
F. Allele whose effect is masked by the effect of the dominant allele paired with it
G. Offspring of a genetic cross that inherit a pair of nonidentical alleles for a trait
H. Refers to an individual's observable traits
I. Refers to the particular genes an individual carries
J. When the effect of an allele on a trait masks that of any recessive allele paired with it
K. When both alleles of a pair are identical
L. An individual with a pair of dominant alleles, such as *AA*
M. Units of information about specific traits; passed from parents to offspring
N. Has a pair of genes for each trait, one on each of two homologous chromosomes
O. When offspring of genetic crosses inherit a pair of identical alleles for a trait, generation after generation

Fill-in-the-Blanks

Offspring of (16) _____ [p.178] crosses are heterozygous for the one trait being studied. (17) _____ [p.179] means that the chance that each outcome of a given event will occur is proportional to the number of ways it can be reached. Because (18) _____ [p.179] is a chance event, the rules of probability apply to genetics crosses. The separation of *A* and *a* as members of a pair of homologous chromosomes move to different gametes during meiosis is known as Mendel's theory of (19) _____ [p.179]. Crossing F_1 hybrids (possibly of unknown genotype) back to a plant known to be a true-breeding recessive plant is known as a (20) _____ [p.179]. From this cross, a ratio of (21) _____ [p.179] is expected. When F_1 offspring inherit two gene pairs, each consisting of two nonidentical alleles, the cross is known as a (22) _____ [p.180] cross. "Gene pairs assorting into gametes independently of other gene pairs located on nonhomologous chromosomes" describes Mendel's theory of (23) _____ _____ [p.181].

Problems

24. In garden pea plants, tall (*T*) is dominant over dwarf (*t*). In the cross *Tt* × *tt*, the *Tt* parent would produce a gamete carrying *T* (tall) and a gamete carrying *t* (dwarf) through segregation; the *tt* parent could only produce gametes carrying the *t* (dwarf) gene. Use the Punnett-square method (refer to text Figures 11.4, 11.6, and 11.7) to determine the genotype and phenotype probabilities of offspring from the above cross, *Tt* × *tt*: [pp.178–179]

Although the Punnett-square (checkerboard) method is a favored method for solving single-factor genetics problems, there is a quicker way. Only six different outcomes are possible from single-factor crosses. Studying the following relationships allows one to obtain the result of any such cross by inspection.

1. *AA* × *AA* = all *AA*
   (Each of the four blocks of the Punnett square would be *AA*.)
2. *aa* × *aa* = all *aa*
3. *AA* × *aa* = all *Aa*
4. *AA* × *Aa* = 1/2 *AA*; 1/2 *Aa*
   or *Aa* × *AA*
   (Two blocks of the Punnett square are *AA*, and two blocks are *Aa*.)
5. *aa* × *Aa* = 1/2 *aa*; 1/2 *Aa*
   or *Aa* × *aa*
6. *Aa* × *Aa* = 1/4 *AA*; 1/2 *Aa*; 1/4 *aa*
   (One block in the Punnett square is *AA*, two blocks are *Aa*, and one block is *aa*.)

Complete the Table

25. Using the gene symbols (tall and dwarf pea plants) in exercise 24, apply the six Mendelian ratios listed above to complete the following table of single-factor crosses by inspection. State results as phenotype and genotype ratios. [pp.178–179]

Cross	Phenotype Ratio	Genotype Ratio
a. *Tt* × *tt*		
b. *TT* × *Tt*		
c. *tt* × *tt*		
d. *Tt* × *Tt*		
e. *tt* × *Tt*		
f. *TT* × *tt*		
g. *TT* × *TT*		
h. *Tt* × *TT*		

When working genetics problems dealing with two gene pairs, you can visualize the independent assortment of gene pairs located on nonhomologous chromosomes into gametes by using a fork-line device. Assume that in humans, pigmented eyes (*B*) (an eye color other than blue) are dominant over blue (*b*), and right-handedness (*R*) is dominant over left-handedness (*r*). To learn to solve a problem, cross the parents *BbRr* × *BbRr*. A sixteen-block Punnett square is required with gametes from each parent arrayed on two sides of the Punnett square (refer to text Figures 11.8 and 11.9). The gametes receive genes through independent assortment using a fork-line method, as follows. [pp.180–181]

26. Array the gametes at the right on two sides of the Punnett square; combine these haploid gametes to form diploid zygotes within the squares. In the blank spaces below, enter the probability ratios derived within the Punnett square for the phenotypes listed: [pp.180–181]

$$B\ b\ R\ r \quad\times\quad B\ b\ R\ r$$
$$BR,\ Br,\ bR,\ br \qquad\qquad BR,\ Br,\ bR,\ br$$

a. _____ pigmented eyes, right-handed
b. _____ pigmented eyes, left-handed
c. _____ blue-eyed, right-handed
d. _____ blue-eyed, left-handed

27. Albinos cannot form the pigments that normally produce skin, hair, and eye color, so albinos exhibit white hair and pink eyes and skin (because the blood shows through). To be an albino, one must be homozygous recessive (aa) for the pair of genes that code for the key enzyme in pigment production. Suppose a woman of normal pigmentation (A __) with an albino mother marries an albino man. State the possible kinds of pigmentation possible for this couple's children, and specify the ratio of each kind of child the couple is likely to have. Show the genotype(s) and state the phenotype(s): [pp.178–179]

_____

_____

_____

28. In horses, black coat color is influenced by the dominant allele (B), and chestnut coat color is influenced by the recessive allele (b). Trotting gait is due to a dominant gene (T), pacing gait to the recessive allele (t). A homozygous black trotter is crossed to a chestnut pacer. [pp.180–181]
   a. What will be the appearance and gait of the F_1 and F_2 generations?

   _____

   b. Which phenotype will be most common?_____

   c. Which genotype will be most common? _____

   d. Which of the potential offspring will be certain to breed true?_____

11.4. DOMINANCE RELATIONS [p.182]

11.5. MULTIPLE EFFECTS OF SINGLE GENES [p.183]

11.6. INTERACTIONS BETWEEN GENE PAIRS [pp.184–185]

Selected Words: ABO blood typing [p.182], transfusions [p.182], sickle-cell anemia [p.183], albinism [p.184]

Boldfaced, Page-Referenced Terms

[p.182] incomplete dominance _____

_____

[p.182] codominance _____

_____

[p.182] multiple allele system _____

_____

[p.183] pleiotropy _____

_____

[p.184] epistasis _____

_____

Complete the Table

1. Complete the following table by supplying the type of inheritance illustrated by each example. Choose from these gene interactions: pleiotropy, multiple allele system, incomplete dominance, codominance, and epistasis.

Type of Inheritance	Example
[p.182] a.	Pink-flowered snapdragons produced from red-and white–flowered parents
[p.182] b.	AB type blood from a gene system of three alleles, *A*, *B*, and *O*
[p.182] c.	A gene with three or more alleles such as the *ABO* blood-typing alleles
[p.184] d.	Black, brown, or yellow fur of Labrador retrievers and comb shape in poultry
[p.183] e.	The multiple phenotypic effects of the gene causing human sickle-cell anemia

Problems

2. Genes that are not always dominant or recessive may blend to produce a phenotype of a different appearance. This is termed *incomplete dominance*. In four o'clock plants, red flower color is determined by gene R and white flower color by R', while the heterozygous condition, RR', is pink. Complete the table below by determining the phenotypes and genotypes of the offspring of the following crosses: [p.182]

Cross	Phenotype	Genotype
a. $RR \times R'R' =$		
b. $R'R' \times R'R' =$		
c. $RR \times RR' =$		
d. $RR \times RR =$		

Sickle-cell anemia is a genetic disease in which children that are homozygous for a defective gene produce defective hemoglobin (Hb^S/Hb^S). The genotypes of normal persons are Hb^A/Hb^A. If the level of blood oxygen drops below a certain level, in a person with the Hb^S/Hb^S genotype, the hemoglobin chains stiffen and cause the red blood cells to form sickle, or crescent, shapes. These cells clog and rupture capillaries, which result in oxygen-deficient tissues where metabolic wastes collect. Several body functions are badly damaged. Severe anemia and other symptoms develop, and death nearly always occurs before adulthood. The sickle-cell gene is considered *pleiotropic*. Persons that are heterozygous (Hb^A/Hb^S) are said to possess *sickle-cell trait*. They are able to produce enough normal hemoglobin molecules to appear normal, but their red blood cells will sickle if they encounter oxygen tension (such as at high altitudes).

3. A man whose sister died of sickle-cell anemia married a woman whose blood is found to be normal. What advice would you give this couple about the inheritance of this disease as they plan their family? [p.183]

_____

_____

_____

4. If a man and a woman, each with sickle-cell trait, planned to marry, what information could you provide for them regarding the genotypes and phenotypes of their future children? [p.183] _____

_____

_____

In one example of a *multiple allele system* with *codominance*, the three genes I^Ai, I^Bi, and i produce proteins found on the surfaces of red blood cells that determine the four blood types in the ABO system, A, B, AB, and O. Genes I^A and I^B are both dominant over i but not over each other. They are codominant. Recognize that blood types A and B may be heterozygous or homozygous (I^AI^A, I^Ai or I^BI^B, I^Bi), whereas blood type O is homozygous (*ii*). Indicate the genotypes and phenotypes of the offspring and their probabilities from the parental combinations in exercises 5–9. [p.182]

5. $I^Ai \times I^AI^B$ = _____

6. $I^Bi \times I^Ai$ = _____

7. $I^AI^A \times ii$ = _____

8. $ii \times ii$ = _____

9. $I^AI^B \times I^AI^B$ = _____

In one type of gene interaction, two alleles of a gene mask the expression of alleles of another gene, and some expected phenotypes never appear. *Epistasis* is the term given such interactions. Work the following problems on scratch paper to understand epistatic interactions. In sweet peas, genes C and P are necessary for colored flowers. In the absence of either (_ _ *pp* or *cc*_ _), or both (*ccpp*), the flowers are white. What will be the color of the offspring of the following crosses, and in what proportions will they appear? [p.184]

10. $CcPp \times ccpp$ = _____

11. $CcPP \times Ccpp$ = _____

12. $Ccpp \times ccPp$ = _____

In the inheritance of the coat (fur) color of Labrador retrievers, allele B specifies black that is dominant over brown (chocolate), b. Allele E permits full deposition of color pigment, but two recessive alleles, *ee*, reduce deposition, and a yellow coat results.

13. Predict the phenotypes of the coat color and their proportions resulting from the following cross: $BbEe \times Bbee$ = [p.184] _____

_____

In poultry, an epistatic interaction occurs in which two genes produce a phenotype that neither gene can produce alone. The two interacting genes (R and P) produce comb shape in chickens. The possible genotypes and phenotypes are as follows:

Genotypes	Phenotypes
R_ P_	walnut comb
R_ pp	rose comb
rrP_	pea comb
rrpp	single comb

Hint: Where a blank appears in the preceding genotypes, either the dominant or recessive symbol in that blank yields the same phenotype.

14. What are the genotype and phenotype ratios of the offspring of a heterozygous walnut-combed male and a single-combed female? [p.185]

_____

_____

15. Cross a homozygous rose-combed rooster with a homozygous single-combed hen, and list the genotype and phenotype ratios of their offspring. [p.185]

_____

_____

_____

11.7. HOW CAN WE EXPLAIN LESS PREDICTABLE VARIATIONS? [pp.186–187]
11.8. EXAMPLES OF ENVIRONMENTAL EFFECTS ON PHENOTYPE [p.188]

Selected Words: camptodactyly [p.186], "bell-shaped" curve [p.187], *Hydrangea macrophylla* [p.188]

Boldfaced, Page-Referenced Terms

[p.186] continuous variation _____

_____

Choice

For questions 1–5, choose from the following primary contributing factors:

a. environment   b. a number of genes affect a trait

1. ___ Height of human beings [p.187]

2. ___ Continuous variation in a trait [p.186]

3. ___ Flower color in *Hydrangea macrophylla* [p.188]

4. ___ The range of eye colors in the human population [p.186]

5. ___ Heat-sensitive version of one of the enzymes required for melanin production in Himalayan rabbits [p.188]

Self-Quiz

___ 1. The best statement of Mendel's principle of independent assortment is that _____. [pp.180–181]
   a. one allele is always dominant to another
   b. hereditary units from the male and female parents are blended in the offspring
   c. the two hereditary units that influence a certain trait separate during gamete formation
   d. each hereditary unit is inherited separately from other hereditary units

___ 2. All the different molecular forms of the same gene are called _____. [p.177]
   a. chiasma
   b. alleles
   c. autosome
   d. locus

___ 3. In the F_2 generation of a monohybrid cross involving complete dominance, the expected phenotypic ratio is _____. [pp.178–179]
   a. 3:1
   b. 1:1:1:1
   c. 1:2:1
   d. 1:1

___ 4. In the F_2 generation of a cross between a red-flowered snapdragon (homozygous) and a white-flowered snapdragon, the expected phenotypic ratio of the offspring is _____. [p.182]
   a. 3/4 red, 1/4 white
   b. 100 percent red
   c. 1/4 red, 1/2 pink, 1/4 white
   d. 100 percent pink

___ 5. In a testcross, F_1 hybrids are crossed to an individual known to be _____ for the trait. [p.179]
   a. heterozygous
   b. homozygous dominant
   c. homozygous
   d. homozygous recessive

___ 6. The tendency for dogs to bark while trailing is determined by a dominant gene, S, whereas silent trailing is due to the recessive gene, s. In addition, erect ears, D, is dominant over drooping ears, d. What combination of offspring would be expected from a cross between two erect-eared barkers who are heterozygous for both genes? [pp.180–181]
   a. 1/4 erect barkers, 1/4 drooping barkers, 1/4 erect silent, 1/4 drooping silent
   b. 9/16 erect barkers, 3/16 drooping barkers, 3/16 erect silent, 1/16 drooping silent
   c. 1/2 erect barkers, 1/2 drooping barkers
   d. 9/16 drooping barkers, 3/16 erect barkers, 3/16 drooping silent, 1/16 erect silent

___ 7. A man with type A blood could be the father of _____. [p.182]
   a. a child with type A blood
   b. a child with type B blood
   c. a child with type O blood
   d. a child with type AB blood
   e. all of the above

___ 8. A single gene that affects several seemingly unrelated aspects of an individual's phenotype is said to be _____. [p.183]
   a. pleiotropic
   b. epistatic
   c. mosaic
   d. continuous

___ 9. Suppose two individuals, each heterozygous for the same characteristic, are crossed. The characteristic involves complete dominance. The expected genotype ratio of their progeny is _____. [pp.178–179]
   a. 1:2:1
   b. 1:1
   c. 100 percent of one genotype
   d. 3:1

___ 10. If the two homozygous classes in the F_1 generation of the cross in exercise 9 are allowed to mate, the observed genotype ratio of the offspring will be _____. [pp.178–179]
   a. 1:1
   b. 1:2:1
   c. 100 percent of one genotype
   d. 3:1

___ 11. Applying the types of inheritance learned in this chapter in the text, the skin color trait in humans exhibits _____. [pp.186–187]
   a. pleiotropy
   b. epistasis
   c. environmental effects
   d. continuous variation

Chapter Objectives/Review Questions

1. What was the prevailing method of explaining the inheritance of traits before Mendel's work with pea plants? [p.176]
2. Garden pea plants are naturally _____-fertilizing, but Mendel took steps to _____-fertilize them for his experiments. [p.177]
3. _____ are units of information about specific traits; they are passed from parents to offspring. [p.177]
4. What is the general term applied to the location of a gene on a chromosome? [p.177]
5. Define *allele*; how many types of alleles are present in the genotypes *Tt*? *tt*? *TT*? [p.177]
6. Explain the meaning of a true-breeding lineage. [p.177]
7. When two alleles of a pair are identical, it is a _____ condition; if the two alleles are different, this is a _____ condition. [p.177]
8. Distinguish a dominant allele from a recessive allele. [p.177]
9. _____ refers to the genes present in an individual; _____ refers to an individual's observable traits. [p.177]
10. Offspring of _____ crosses are heterozygous for the one trait being studied. [p.178]
11. Explain why probability is useful to genetics. [p.179]
12. Use the Punnett-square method of solving genetics problems. [p.179]
13. Define the *testcross* and cite an example. [p.179]
14. Mendel's theory of _____ states that during meiosis, the two genes of each pair separate from each other and end up in different gametes. [p.179]
15. Solve dihybrid genetic crosses. [pp.180–181]
16. Mendel's theory of _____ _____ states that gene pairs on homologous chromosomes tend to be sorted into one gamete or another independently of how gene pairs on other chromosomes are sorted out. [p.181]
17. Distinguish among complete dominance, incomplete dominance, and codominance. [p.182]
18. Define *multiple allele system* and cite an example. [p.182]
19. Explain why sickle-cell anemia is a good example of pleiotropy. [p.183]
20. Gene interaction involving two alleles of a gene that mask expression of another gene's alleles is called _____. [p.184]
21. List possible explanations for less predictable trait variations that are observed. [pp.186–187]
22. List two human traits that are explained by continuous variation. [pp.186–187]
23. Himalayan rabbits and garden hydrangeas are good examples of environmental effects on _____ expression. [p.188]

Integrating and Applying Key Concepts

Solve the following genetics problem: In garden peas, one pair of alleles controls the height of the plant and a second pair of alleles controls flower color. The allele for tall (*D*) is dominant to the allele for dwarf (*d*), and the allele for purple (*P*) is dominant to the allele for white (*p*). A tall plant with purple flowers crossed with a tall plant with white flowers produces offspring that are 3/8 tall purple, 3/8 tall white, 1/8 dwarf purple, and 1/8 dwarf white. What are the genotypes of the parents?

12

CHROMOSOMES AND HUMAN GENETICS

Interactive Exercises

The Philadelphia Story [pp.192–193]

12.1. THE CHROMOSOMAL BASIS OF INHERITANCE—AN OVERVIEW [p.194]
12.2. *Focus on Science:* KARYOTYPING MADE EASY [p.195]

Selected Words: *Philadelphia chromosome* [p.192], *leukemias* [p.192], *spectral* karyotyping [p.192], Walther
Flemming [p.193], August Weismann [p.193], *wild-type* allele [p.194], *Colchicum* [p.195]

Boldfaced, Page-Referenced Terms

[p.192] karyotype _____

_____

[p.194] genes _____

_____

[p.194] homologous chromosomes _____

_____

[p.194] alleles _____

_____

[p.194] crossing over _____

_____

[p.194] genetic recombination _____

_____

[p.194] X chromosome _____

_____

[p.194] Y chromosome _____

_____

[p.194] sex chromosomes _____

_____

[p.194] autosomes _____

_____

[p.195] in vitro _____

_____

Fill-in-the-Blanks

The first abnormal chromosome to be associated with cancer was named the (1) _____ [p.192] chromosome. (2) _____ [p.192] arise when stem cells in bone marrow overproduce the white blood cells necessary for the body's housekeeping and defense. A preparation of metaphase chromosomes based on their defining features is a (3) _____ [p.192]. A Philadelphia chromosome is not easy to identify in standard karyotypes but should be easier using the newer (4) _____ [p.192] karyotyping, which artificially colors chromosomes in an unambiguous way.

Heritable traits arise from units of molecular information (located on chromosomes) that are known as (5) _____ [p.194]. Paired chromosomes that have the same length, shape, gene sequence, and interact during meiosis, are known as (6) _____ [p.194] chromosomes. (7) _____ [p.194] are different molecular forms of the same gene that arose through mutation. The most common form of a gene is called a (8) _____-type [p.194] allele. An event known as (9) _____ _____ [p.194] functions to exchange gene segments between the members of a homologous pair and thus provide genetic (10) _____ [p.194]. Human X and Y chromosomes that determine gender are examples of (11) _____ [p.194] chromosomes. Chromosomes that are identical in both sexes of a species are called (12) _____ [p.194].

The (13) _____ [p.195] of an individual (or species) is a preparation of sorted metaphase chromosomes. To prepare a karyotype, technicians culture cells (14) _____ _____ [p.195] (in glass). During culture, a chemical compound named (15) _____ [p.195] is added to the culture medium. The purpose of this chemical is to block spindle formation and arrest nuclear divisions at the (16) _____ [p.195] stage of mitosis, when the chromosomes are the most condensed and easiest to identify in cells. Cells are then moved to the bottoms of culture tubes by a spinning force called (17) _____ [p.195]. Following saline treatment to separate the chromosomes, cells are fixed, stained, and photographed for study. The photograph is cut apart, one chromosome at a time, and the individual cutouts are arranged according to size, shape, and length of arms. Then all the pairs of homologous chromosomes are horizontally aligned by their (18) _____ [p.195].

12.3. SEX DETERMINATION IN HUMANS [pp.196–197]
12.4. EARLY QUESTIONS ABOUT GENE LOCATIONS [pp.198–199]

Selected Words: nonsexual traits [p.196], *SRY* gene [p.196], *Drosophila melanogaster* [p.198], *Zea mays* [p.198], Thomas Hunt Morgan [p.198]

Boldfaced, Page-Referenced Terms

[p.198] X-linked genes _____

_____

[p.198] Y-linked genes _____

_____

[p.198] linkage group _____

_____

[p.198] reciprocal crosses _____

_____

[p.199] cytological markers _____

_____

Matching

Choose the most appropriate answer for each term.

1. ___X-linked genes [p.198]
2. ___linkage group [p.198]
3. ___*SRY* [p.196]
4. ___Y-linked genes [p.198]
5. ___reciprocal crosses [p.198]

A. Male-determining gene found on the Y chromosome; expression leads to testes formation
B. In the first cross one parent displays the trait of interest, whereas in the second cross, the other parent displays it
C. The block of genes located on each type of chromosome; these genes tend to travel together in inheritance
D. Found only on the Y chromosome
E. Found only on the X chromosome

Complete the Table

6. Complete the Punnett-square table at the right, which brings Y-bearing and X-bearing sperm together randomly in fertilization. [p.196]

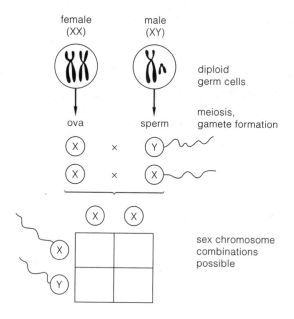

Dichotomous Choice

Answer the following questions related to the Punnett-square table just completed.

7. Male humans transmit their Y chromosome only to their (sons/daughters). [p.196]
8. Male humans receive their X chromosome only from their (mothers/fathers). [p.196]
9. Human mothers and fathers each provide an X chromosome for their (sons/daughters). [p.196]

Problems

10. In Morgan's experiments with *Drosophila*, white eyes are determined by a recessive X-linked gene and the wild-type or normal brick-red eyes are due to its dominant allele. Use symbols of the following types: X^WY = a white-eyed male; X^WX^W = a homozygous normal red female. [p.198]
    a. What offspring can be expected from a cross of a white-eyed male and a homozygous normal female? [p.198]
    b. In addition, show the genotypes and phenotypes of the F_2 offspring. [p.198]
11. Which of the following represents a chromosome that has undergone crossover and recombination? It is assumed that the organism involved is heterozygous with the following genotype:

$$\begin{array}{c|c} A| & |a \\ B| & |b \end{array}$$

$$\begin{array}{cccc} |A & a| & B| & |a \\ |B & b| & A| & |B \end{array}$$

a. [p.199]   b. [p.199]   c. [p.199]   d. [p.199]

12.5. RECOMBINATION PATTERNS AND CHROMOSOME MAPPING [pp.200–201]

Selected Words: actual physical differences [p.201], *map distance* [p.201], *physical distance* [p.201], *color blindness* [p.201], *hemophilia* [p.201]

Boldfaced, Page-Referenced Terms

[p.201] linkage mapping _____

_____

Early investigators realized that the frequency of crossing over (data from actual genetic crosses) could be used as a tool to map genes on chromosomes. Genetic maps are graphic representations of the relative positions and distances between genes on each chromosome (linkage group). Geneticists arbitrarily equate one percent of crossover (recombination) with 1 genetic map unit. In the following questions, a line is used to represent a chromosome with its linked genes.

Problems

1. If genes A and B are twice as far apart on a chromosome as genes C and D, how often would you expect that crossing over occurs between genes A and B as compared to between genes C and D? [pp.200–201]

_____

2. Following breeding experiments and a study of crossovers, the following map distances were recorded: gene C to gene B = 35 genetic map units; B to A = 10 map units, and A to C = 45 map units. Which of the following maps is correct? [p.201]

a. $\underset{A\ B\ C}{\llcorner\!\sqcup\!\lrcorner}$ b. $\underset{B\ A\ C}{\llcorner\!\sqcup\!\lrcorner}$ c. $\underset{B\ \ C\ A}{\llcorner\!\sqcup\!\lrcorner}$ d. $\underset{B\ A\ \ C}{\llcorner\!\sqcup\!\lrcorner}$

Individuals with dominant gene *N* have the nail-patella syndrome and exhibit abnormally developed fingernails and absence of kneecaps; the recessive gene *n* is normal. Geneticists have learned that the genes determining the ABO blood groups and the nail-patella gene are linked, both found on chromosome 9. A woman with blood type A and nail-patella syndrome marries a man who has blood type O and has normal fingernails and kneecaps. This couple has three children. One daughter has blood type A like her mother, but has normal fingernails and kneecaps. A second daughter has blood type O with nail-patella syndrome. A son has blood type O and normal fingernails and kneecaps. The chromosomes of the parents are diagrammed below.

3. Complete the exercise by showing the linked genes of each of the children below. [pp.200–201]

♀	♂
A n	*O n*
_____	_____

×   (or *An/ON × On/On*)

_____	_____
O N	*O n*

Daughter 1	Daughter 2	Son
_____	_____	_____
_____	_____	_____

4. Explain how the son can have the genotype *On/On* when only one parent has a chromosome with genes *O* and *n* linked. _____ [pp.200–201]

12.6. HUMAN GENETIC ANALYSIS [pp.202–203]

Selected Words: polydactyly [p.202], Huntington disorder [p.202], genetic disease [p.203]

Boldfaced, Page-Referenced Terms

[p.202] pedigree _____

_____

[p.203] genetic abnormality _____

_____

[p.203] genetic disorder _____

_____

[p.203] syndrome _____

_____

[p.203] disease _____

_____

Matching

The following standardized symbols are used in constructing pedigree charts. Choose the appropriate description for each.

1. ___[p.202] ■ ●
2. ___[p.202] ●
3. ___[p.202] ◆
4. ___[p.202] ■
5. ___[p.202] **I, II, III, IV...**
6. ___[p.202] ■—●
7. ___[p.202] ●■■●

A. Individual showing the trait being tracked
B. Male
C. Successive generations
D. Offspring with birth order left to right
E. Female
F. Sex unspecified; numerals present indicate number of children
G. Marriage/mating

Short Answer

8. Distinguish among these terms: *genetic abnormality, genetic disorder, syndrome,* and *genetic disease.* [p.203] ___

_____

_____

_____

Complete the Table

9. Complete the following table by indicating whether the genetic disorder listed is caused by inheritance that is autosomal recessive, autosomal dominant, X-linked recessive, or involves changes in chromosome number or in chromosome structure.

Genetic Disorder	Inheritance Pattern
[p.203] a. Galactosemia	
[p.203] b. Achondroplasia	
[p.203] c. Hemophilia	
[p.203] d. Anhidrotic dysplasia	
[p.203] e. Huntington disorder	
[p.203] f. Turner and Down syndrome	
[p.203] g. Achoo syndrome	
[p.203] h. Cri-du-chat syndrome	
[p.203] i. XYY condition	
[p.203] j. Color blindness	
[p.203] k. Fragile X syndrome	
[p.203] l. Progeria	
[p.203] m. Klinefelter syndrome	
[p.203] n. Testicular feminization syndrome	
[p.203] o. Phenylketonuria	

12.7. INHERITANCE PATTERNS [pp.204–205]

12.8. *Focus on Health:* TOO YOUNG TO BE OLD [p.206]

Selected Words: *galactosemia* [p.204], *Huntington disorder* [p.204], *achondroplasia* [p.204], *color blindness* [p.205], *hemophilia A* [p.205], fragile X syndrome [p.205], *expansion* mutations [p.205], *Hutchinson-Gilford progeria syndrome* [p.206]

Choice

For questions 1–17, choose from the following patterns of inheritance; some items may require more than one letter.

a. autosomal recessive   b. autosomal dominant   c. X-linked recessive

1. ___ The trait is expressed in heterozygous females. [p.204]

2. ___ Heterozygotes can remain undetected. [pp.204–205]

3. ___ The trait appears in each generation. [p.204]

4. ___ The recessive phenotype shows up far more often in males than in females. [p.205]

5. ___ Both parents may be heterozygous normal. [p.204]

6. ___ If one parent is heterozygous and the other homozygous recessive, there is a 50 percent chance any child of theirs will be heterozygous. [p.204]

7. ___ The allele is usually expressed, even in heterozygotes. [p.204]

8. ___ The trait is expressed in heterozygous females. [p.204]

9. ___ Heterozygous normal parents can expect that one-fourth of their children will be affected by the disorder. [p.204]

10. ___ A son cannot inherit the recessive allele from his father, but his daughter can. [p.205]

11. ___ Females can mask this gene, males cannot. [p.205]

12. ___ The trait is expressed in heterozygotes of either sex. [p.204]

13. ___ Heterozygous women transmit the allele to half their offspring, regardless of sex. [p.204]

14. ___ Individuals displaying this type of disorder will always be homozygous for the trait. [p.204]

15. ___ The trait is expressed in both the homozygote and the heterozygote. [p.204]

16. ___ Heterozygous females will transmit the recessive gene to half their sons and half their daughters. [p.205]

17. ___ If both parents are heterozygous, there is a 50 percent chance that each child will be heterozygous. [p.204]

Problems

18. The autosomal allele that causes galactosemia (g) is recessive to the allele for normal lactose metabolism (G). A normal woman whose father had galactosemia marries a man with galactosemia who had normal parents. They have three children, two normal and one with galactosemia. List the genotypes for each person involved. [p.204] _____

19. Huntington disorder is a rare form of autosomal dominant inheritance, H; the normal gene is h. The disease causes progressive degeneration of the nervous system with onset exhibited near middle age. An apparently normal man in his early twenties learns that his father has recently been diagnosed as having Huntington disorder. What are the chances that the son will develop this disorder? [p.204] _____

20. A color-blind man and a woman with normal vision whose father was color blind have a son. Color blindness, in this case, is caused by an X-linked recessive gene. If only the male offspring are considered, what is the probability that their son is color blind? [p.205] _____

21. Hemophilia A is caused by an X-linked recessive gene. A woman who is seemingly normal but whose father was a hemophiliac marries a normal man. What proportion of their sons will have hemophilia? What proportion of their daughters will have hemophilia? What proportion of their daughters will be carriers? [p.205] _____

22. The accompanying pedigree shows the pattern of inheritance of color blindness in a family (people with the trait are indicated by black circles). What is the chance that the third-generation female indicated by the arrow (below) will have a color-blind son if she marries a normal male? a color-blind male? [p.205]

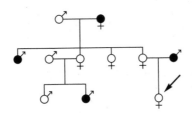

12.9. CHANGES IN CHROMOSOME STRUCTURE [pp.206–207]

12.10. CHANGES IN CHROMOSOME NUMBER [pp.208–209]

*Selected Words: non*homologous chromosome [p.206], *cri-du-chat* [p.207], *miscarriages* [p.208], *tetraploid* [p.208], trisomic [p.208], monosomic [p.208], *Down syndrome* [p.208], *Turner syndrome* [p.209], *Klinefelter syndrome* [p.209]; *XYY males* [p.209]

Boldfaced, Page-Referenced Terms

[p.206] duplications _____

_____

[p.206] inversion _____

_____

[p.206] translocation _____

_____

[p.207] deletion _____

_____

[p.208] aneuploidy _____

_____

[p.208] polyploidy _____

_____

[p.208] nondisjunction _____

_____

[p.209] double-blind studies _____

_____

Labeling and Matching

On rare occasions, chromosome structure becomes abnormally rearranged. Such changes may have profound effects on the phenotype of an organism. Label the following diagrams of abnormal chromosome structure as a deletion, a duplication, an inversion, or a translocation. Complete the exercise by matching and entering the letter of the proper description in the parentheses following each label.

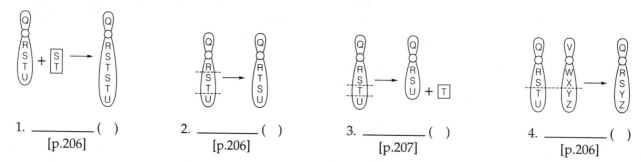

1. _____ ( )
   [p.206]

2. _____ ( )
   [p.206]

3. _____ ( )
   [p.207]

4. _____ ( )
   [p.206]

A. The loss of a chromosome segment; an example is the cri-du-chat disorder
B. A gene sequence in excess of its normal amount in a chromosome; an example is the fragile X syndrome
C. A chromosome segment that separated from the chromosome and then was inserted at the same place, but in reverse; this alters the position and order of the chromosome's genes; possibly promoted human evolution
D. The transfer of part of one chromosome to a nonhomologous chromosome; an example is when chromosome 14 ends up with a segment of chromosome 8; the Philadelphia chromosome is also an example

Complete the Table

5. Complete the following table to summarize the major categories and mechanisms of chromosome number change in organisms.

Category of Change	Description
[p.208] a. Aneuploidy	
[p.208] b. Polyploidy	
[p.208] c. Nondisjunction	

Short Answer

6. If a nondisjunction occurs at anaphase I of the first meiotic division, what will be the proportion of abnormal gametes (for the chromosomes involved in the nondisjunction)? [p.208] _____

7. If a nondisjunction occurs at anaphase II of the second meiotic division, what will be the proportion of abnormal gametes (for the chromosomes involved in the nondisjunction)? [p.208] _____

8. Contrast the effects of polyploidy in plants and humans. [p.208] _____

_____

_____

9. Define the following terms: *tetraploid, trisomic,* and *monosomic.* [p.208] _____

_____

_____

_____

_____

Choice

For questions 10–19, choose from the following:

a. Down syndrome   b. Turner syndrome   c. Klinefelter syndrome   d. XYY condition

10. ___ XXY male [p.209]

11. ___ Ovaries nonfunctional and secondary sexual traits fail to develop at puberty [p.209]

12. ___ Testes smaller than normal, sparse body hair, and some breast enlargement [p.209]

13. ___ Could only be caused by a nondisjunction in males [p.209]

14. ___ As a group, they tend to be cheerful and affectionate; about 40 percent develop heart defects [p.208]

15. ___ X0 female; often abort early; distorted female phenotype [p.209]

16. ___ Males that tend to be taller than average; some mildly retarded but most are phenotypically normal [p.209]

17. ___ Injections of testosterone reverse feminized traits but not the fertility [p.209]

18. ___ Trisomy 21; muscles and muscle reflexes weaker than normal [p.208]

19. ___ At one time these males were thought to be genetically predisposed to become criminals [p.209]

12.11. *Focus on Bioethics:* PROSPECTS IN HUMAN GENETICS [pp.210–211]

Selected Words: *phenylketonuria* [p.210], *cleft lip* [p.210], *genetic counseling* [p.210], *prenatal diagnosis* [p.210], *embryo* [p.210], *fetus* [p.210], *amniocentesis* [p.210], *chorionic villi sampling* [p.211], *fetoscopy* [p.211], *preimplantation diagnosis* [p.211], *pre*-pregnancy stage [p.211], "test-tube babies" [p.211]

Boldfaced, Page-Referenced Terms

[p.210] abortion _____

_____

[p.211] in vitro fertilization _____

_____

Complete the Table

1. Complete the following table which summarizes methods of dealing with the problems of human genetics. Choose from phenotypic treatments, abortion, genetic screening, preimplantation diagnosis, genetic counseling, and prenatal diagnosis.

Method	Description
[pp.210–211] a.	Detects embryo or fetal conditions before birth; may use amniocentesis, CVS, in vitro fertilization and fetoscopy; an example is a pregnancy at risk in a mother forty-five years old
[p.211] b.	A controversial option if prenatal diagnosis reveals a serious heritable problem; "pro-life" and "pro-choice" factions
[p.210] c.	If a severe heritable problem exists, it includes diagnosis of parental genotypes, detailed pedigrees, and genetic testing for known metabolic disorders; geneticists may be contacted for predictions
[p.211] d.	Relies on in vitro fertilization; a fertilized egg mitotically divides into a ball of eight cells that provides a cell to be analyzed for genetic defects
[p.210] e.	Suppressing or minimizing symptoms of genetic disorders by surgical intervention, controlling diet or environment; PKU and cleft lip are examples of dietary control and surgery
[p. 210] f.	Large-scale programs to detect affected persons or carriers in a population; early detection may allow introduction of preventive measures before symptoms develop; PKU screening for newborns is an example

Self-Quiz

____ 1. All the genes located on a given chromosome compose a _____. [p.198]
   a. karyotype
   b. bridging cross
   c. wild-type allele
   d. linkage group

____ 2. Chromosomes other than those involved in sex determination are known as _____. [p.194]
   a. nucleosomes
   b. heterosomes
   c. alleles
   d. autosomes

____ 3. The farther apart two genes are on a chromosome, _____. [pp.200–201]
   a. the less likely that crossing over and recombination will occur between them
   b. the greater will be the frequency of crossing over and recombination between them

   c. the more likely they are to be in two different linkage groups
   d. the more likely they are to be segregated into different gametes when meiosis occurs

____ 4. Karyotype analysis is _____. [p.195]
   a. a means of detecting and reducing mutagenic agents
   b. a surgical technique that separates chromosomes that have failed to segregate properly during meiosis II
   c. used in prenatal diagnosis to detect chromosomal mutations and metabolic disorders in embryos
   d. a process that substitutes defective alleles with normal ones

___ 5. Which of the following did Morgan and his research group not do? [pp.198–199]
   a. They isolated and kept under culture fruit flies with the sex-linked recessive white-eyed trait.
   b. They developed the technique of amniocentesis.
   c. They discovered X-linked genes.
   d. Their work reinforced the concept that each gene is located on a specific chromosome.

___ 6. Red–green color blindness is a sex-linked recessive trait in humans. A color-blind woman and a man with normal vision have a son. What are the chances that the son is color blind? If the parents ever have a daughter, what is the chance for each birth that the daughter will be color blind? (Consider only the female offspring.) [p.205]
   a. 100 percent, 0 percent
   b. 50 percent, 0 percent
   c. 100 percent, 100 percent
   d. 50 percent, 100 percent
   e. none of the above

___ 7. Suppose that a hemophilic male (X-linked recessive allele) and a female carrier for the hemophilic trait have a nonhemophilic daughter with Turner syndrome. Nondisjunction could have occurred in _____. [pp.205, 208]
   a. both parents
   b. neither parent
   c. the father only
   d. the mother only
   e. the nonhemophilic daughter

___ 8. Nondisjunction involving the X chromosome occurs during oogenesis and produces two kinds of eggs, XX and O (no X chromosome). If normal Y sperm fertilize the two types, which genotypes are possible? [p.208]
   a. XX and XY
   b. XXY and YO
   c. XYY and XO
   d. XYY and YO
   e. YY and XO

___ 9. Of all phenotypically normal males in prisons, the type once thought to be genetically predisposed to becoming criminals was the group with _____. [p.209]
   a. XXY disorder
   b. XYY disorder
   c. Turner syndrome
   d. Down syndrome
   e. Klinefelter syndrome

___10. Amniocentesis is _____. [pp.210–211]
   a. a surgical means of repairing deformities
   b. a form of chemotherapy that modifies or inhibits gene expression or the function of gene products
   c. used in prenatal diagnosis; a small sample of amniotic fluid is drawn to detect chromosomal mutations and metabolic disorders in embryos
   d. a form of gene-replacement therapy
   e. a diagnostic procedure; cells for analysis are withdrawn from the chorion

Chapter Objectives/Review Questions

1. A _____ is a preparation of metaphase chromosomes based on their defining features. [p.192]
2. The units of information about heritable traits are known as _____. [p.194]
3. Diploid (2n) cells have pairs of _____ chromosomes. [p.194]
4. _____ are different molecular forms of the same gene that arise through mutation; a _____-type allele is the most common form of a gene. [p.194]
5. State the circumstances required for crossing over, and describe the results. [p.194]
6. Name and describe the sex chromosomes in human males and females. [p.194]
7. Human X and Y chromosomes fall in the general category of _____ chromosomes; all other chromosomes in an individual's cells are the same in both sexes and are called _____. [p.194]
8. Define *karyotype*; briefly describe its preparation and value. [pp.194–195]
9. Explain meiotic segregation of sex chromosomes to gametes and the subsequent random fertilization that determines sex in many organisms. [p.196]
10. A newly identified region of the Y chromosome called _____ appears to be the master gene for male sex determination. [p.196]

11. All the genes on a specific chromosome are called a _____ group. [p.198]
12. Define the terms *X-linked* and *Y-linked genes*. [p.198]
13. In whose laboratory was sex linkage in fruit flies discovered? When? [p.198]
14. State the relationship between crossover frequency and the location of genes on a chromosome. [p.199]
15. The principle of recombination frequencies between _____ on the same chromosome is based on the distances between them. [p.200]
16. The probability that a crossover will disrupt gene linkage is proportional to the _____ that separates them. [pp.200–201]
17. The percentage of recombinants in gametes is a quantitative measure of the relative positions of genes along a chromosome; this investigative approach is now called _____ _____. [p.201]
18. State the difference between actual physical distances between genes and what is termed *map distance*. [p.201].
19. The farther apart two genes are on a chromosome, the _____ will be the frequency of crossing over and therefore of genetic recombination between them. [p.201]
20. A _____ chart or diagram is used to study genetic connections between individuals. [p.202]
21. A genetic _____ is a rare, uncommon version of a trait, whereas an inherited genetic _____ causes mild to severe medical problems. [p.203]
22. A _____ is a recognized set of symptoms that characterize a given disorder. [p.203]
23. Describe what is meant by a genetic disease. [p.203]
24. Carefully characterize patterns of autosomal recessive inheritance, autosomal dominant inheritance, and X-linked recessive inheritance. [pp.204–205]
25. Describe the Hutchinson–Gilford progeria syndrome and its probable mode of inheritance. [p.206]
26. A(n) _____ is a loss of a chromosome segment; a(n) _____ is a gene sequence separated from a chromosome but then was inserted at the same place, but in reverse; a(n) _____ is a repeat of several gene sequences on the same chromosome; a(n) _____ is the transfer of part of one chromosome to a nonhomologous chromosome. [pp.206–207]
27. When gametes or cells of an affected individual end up with one extra or one less than the parental number of chromosomes, the condition is known as _____; relate this concept to monosomy and trisomy. [pp.208–209]
28. Having three or more complete sets of chromosomes is called _____. [p.208]
29. _____ is the failure of the chromosomes to separate in either meiosis or mitosis. [p.208]
30. Trisomy 21 is known as _____ syndrome; Turner syndrome has the chromosome constitution, _____; XXY chromosome constitution is _____ syndrome; taller-than-average males with sometimes slightly lowered IQs have the _____ condition. [p.209]
31. Explain what is meant by double-blind studies. [p.209]
32. Define *phenotypic treatment* and describe one example. [p.210]
33. List some benefits of genetic screening and genetic counseling to society. [p.210]
34. Explain the procedures and purpose of three types of prenatal diagnosis: amniocentesis, chorionic villi analysis, and fetoscopy; compare the risks. [pp.210–211]
35. Discuss some of the ethical considerations that might be associated with a decision to induce abortion. [p.211]
36. A procedure known as *preimplantation diagnosis* relies on _____ - _____ fertilization. [p.211]

Integrating and Applying Key Concepts

1. The parents of a young boy bring him to their doctor. They explain that the boy does not seem to be going through the same vocal developmental stages as his older brother. The doctor orders a common cytogenetics test to be done, and it reveals that the young boy's cells contain two X chromosomes and one Y chromosome. Describe the test that the doctor ordered and explain how and when such a genetic result, XXY, most logically occurred.
2. Solve the following genetics problem. Show rationale, genotypes, and phenotypes. A husband sues his wife for divorce, arguing that she has been unfaithful. His wife gave birth to a girl with a fissure in the iris of her eye, an X-linked recessive trait. Both parents have normal eye structure. Can the genetic facts be used to argue for the husband's suit? Explain your answer.

13

DNA STRUCTURE AND FUNCTION

Cardboard Atoms and Bent-Wire Bonds

DISCOVERY OF DNA FUNCTION
    Early and Puzzling Clues
    Confirmation of DNA Function

DNA STRUCTURE
    What Are the Components of DNA?
    Patterns of Base Pairing

Focus on Bioethics: ROSALIND'S STORY

DNA REPLICATION AND REPAIR
    How Is a DNA Molecule Duplicated?
    Monitoring and Fixing the DNA

Focus on Science: DOLLY, DAISIES, AND DNA

Interactive Exercises

Cardboard Atoms and Bent-Wire Bonds [pp.214–215]

13.1. DISCOVERY OF DNA FUNCTION [pp.216–217]

Selected Words: J. F. Miescher [p.214], L. Pauling [p.214], J. Watson and F. Crick [p.214], F. Griffith [p.216], *Streptococcus pneumoniae* [p.216], O. Avery [p.216], A. Hershey and M. Chase [pp.216–217], *Escherichia coli* [pp.216–217], ^{35}S and ^{32}P [p.217]

Boldfaced, Page-Referenced Terms

[p.214] deoxyribonucleic acid, DNA _____

_____

[p.216] bacteriophages _____

_____

Complete the Table

1. Complete the following table, which traces the discovery of DNA function.

Investigators	Year	Contribution
[p.124] a. Miescher	1868	
[p.216] b. Griffith	1928	
[p.216] c. Avery (also MacLeod and McCarty)	1944	
[pp.216–217] d. Hershey and Chase	1952	

Fill-in-the-Blanks

A bacteriophage is a kind of (2) _____ [p.216] that can infect (3) _____ [p.216] cells. Enzymes from the (2) take over enough of the host cell's metabolic processes to make substances that are necessary to construct new (4) _____ [p.216]. Some of these substances are (5) _____ [p.217] that can be labeled with a radioisotope of sulfur, ^{35}S. The genetic material of the bacteriophage used in the experiments by Hershey and Chase was labeled with the radioisotope of phosphorus known as (6) _____ [p.217]. When (7) _____ [p.217] particles were allowed to infect *Escherichia coli* cells, the (8) _____ [p.217] radioisotope remained outside the bacterial cells; the (9) _____ [p.217] became part of the (10) _____ _____ [p.217] injected *into* the bacterial cells. Through many experiments, researchers accumulated strong evidence that (11) _____ [p.217], not (12) _____ [p.217], serves as the molecule of inheritance in all living cells.

13.2. DNA STRUCTURE [pp.218–219]
13.3. *Focus on Bioethics:* ROSALIND'S STORY [p.220]

Selected Words: E. Chargaff [p.218], R. Franklin and M. Wilkins [p.218], DNA double helix [p.219], x-ray crystallography [p.220]

Boldfaced, Page-Referenced Terms

[p.218] nucleotide _____

_____

[p.218] adenine, A _____

_____

[p.218] guanine, C _____

_____

[p.218] thymine, T _____

_____

[p.218] cytosine, C _____

_____

[p.218] x-ray diffraction images _____

_____

Short Answer

1. List the three parts of a nucleotide. [p.218] _____

_____

_____

Labeling

Four nucleotides are illustrated below. [All are from text p.218.] In the blank, label each nitrogen-containing base correctly as guanine, thymine, cytosine, or adenine. In the parentheses following each blank, indicate whether that nucleotide base is a purine (pu) or a pyrimidine (py). (See next section for definitions.)

2. _____ ( )   3. _____ ( )   4. _____ ( )   5. _____ ( )

Labeling and Matching

Identify each indicated part of the following DNA illustration. Choose from these answers: phosphate group, double-ring nitrogen base, single-ring nitrogen base, nucleotide, and deoxyribose. Complete the exercise by matching and entering the letter of the proper structure description in the parentheses following each label.

The following memory devices may be helpful: use *pyrCUT* to remember that the single-ring nucleotides (pyrimidines) are cytosine, uracil (in RNA), and thymine. Use *purAG* to remember that the double-ring nucleotides (purines) are adenine and guanine; to help recall the number of hydrogen bonds between the DNA bases, remember that AT = 2 and CG = 3.

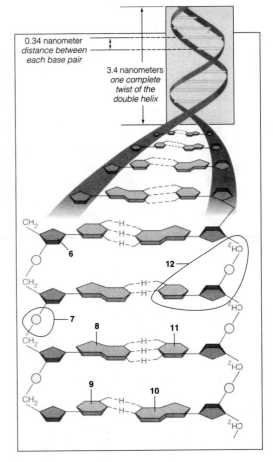

0.34 nanometer distance between each base pair

3.4 nanometers one complete twist of the double helix

6. _____ ( ) [p.219]

7. _____ _____ ( ) [p.219]

8. _____-ring nitrogen base ( ) [p.219]

9. _____-ring nitrogen base ( ) [p.219]

10. _____-ring nitrogen base ( ) [p.219]

11. _____-ring nitrogen base ( ) [p.219]

12. A complete _____ ( ) [p.219]

A. The single-ring nitrogen base is thymine, because it has two hydrogen bonds.
B. A five-carbon sugar joined to two phosphate groups in the upright portion of the DNA ladder.
C. The double-ring nitrogen base is guanine, because it has three hydrogen bonds.
D. The single-ring nitrogen base is cytosine, because it has three hydrogen bonds.
E. The double-ring nitrogen base is adenine, because it has two hydrogen bonds.
F. Composed of three smaller molecules: a phosphate group, five-carbon deoxyribose sugar, and a nitrogenous base (in this case, a single-ring nitrogen base).
G. A chemical group that joins two sugars in the upright portion of the DNA ladder.

True–False

_____13. DNA is composed of <u>four</u> different types of nucleotides. [p.218]

_____14. In the DNA of every species, the amount of adenine present always equals the amount of <u>thymine</u>, and the amount of cytosine always equals the amount of <u>guanine</u> (A = T and C = G). [p.218]

_____15. In a nucleotide, the phosphate group is attached to the <u>nitrogen</u>-containing base. [p.218]

_____16. Watson and Crick built their model of DNA in the early <u>1950s</u>. [p.214, recall]

_____17. Guanine pairs with cytosine and adenine pairs with thymine by forming hydrogen bonds between them. [p.219]

Fill-in-the-Blanks

The pattern for base (18) _____ [p.219] between the two nucleotide strands in DNA is (19) _____

[p.219] for all species (A–T; G–C). The base (20) _____ [p.219] (determining which base follows the next

in a nucleotide strand) is (21) _____ [p.219] from species to species.

Short Answer

22. Explain why understanding the structure of DNA helps scientists understand how living organisms can have so much in common at the molecular level and yet be so diverse at the whole organism level. [p.219]

_____

_____

23. The term *antiparallel* means that parallel strands of a material run in opposite directions. Study text Figure 13.7, read the descriptive messages, and identify the features of the DNA molecule that make biochemists describe its structure as antiparallel. [p.219]_____

_____

_____

_____

_____

_____

13.4. DNA REPLICATION AND REPAIR [pp.220–221]

13.5. *Focus on Science:* DOLLY, DAISIES, AND DNA [p.222]

Selected Words: x-ray crystallography [p.220], *semiconservative* replication [p.221], R. Okazaki [p.221], *continuous* assembly [p.221], *discontinuous* assembly [p.221], complementary strand [p.221], clone [p.222], differentiated [p.222], Ian Wilmut [p.222], uncommitted [p.222], enucleated [p.222], R. Yanagimachi [p.222]

Boldfaced, Page-Referenced Terms

[p.220] DNA replication _____

_____

[p.221] DNA polymerases _____

_____

[p.221] DNA ligases _____

_____

[p.221] DNA repair _____

_____

Labeling

1. The term *semiconservative replication* refers to the fact that each new DNA molecule resulting from the replication process is "half old, half new." In the following illustration, complete the replication required in the middle of the molecule by adding the required letters representing the missing nucleotide bases. Recall that ATP energy and the appropriate enzymes are actually required in order to complete this process. [p.220]

T – ___	___ – A
G – ___	___ – C
A – ___	___ – T
C – ___	___ – G
C – ___	___ – G
C – ___	___ – G
old new	new old

True–False

If a statement below is true, write T in the blank. If false, explain why by changing one or more of the underlined words.

_____2. The hydrogen bonding of adenine to <u>guanine</u> is an example of complementary base pairing. [p.220]

_____3. The replication of DNA is considered a <u>conservative</u> process because each new molecule is really half new and half old. [pp.220–221]

_____4. Each parent strand remains intact during replication, and a new companion strand is assembled on <u>each</u> <u>of</u> <u>those</u> <u>parent</u> strands. [p.220]

_____5. DNA <u>ligases</u> govern the assembly of nucleotides on a parent strand. [p.221]

_____6. DNA polymerases, DNA ligases, and other enzymes also engage in DNA <u>replication</u>. [p.221]

Self-Quiz

___ 1. Each DNA strand has a backbone that consists of alternating _____. [p.218]
   a. purines and pyrimidines
   b. nitrogen-containing bases
   c. hydrogen bonds
   d. sugar and phosphate molecules

___ 2. In DNA, complementary base-pairing occurs between _____. [p.218]
   a. cytosine and uracil
   b. adenine and guanine
   c. adenine and uracil
   d. adenine and thymine

___ 3. Adenine and guanine are _____. [p.218]
   a. double-ringed purines
   b. single-ringed purines
   c. double-ringed pyrimidines
   d. single-ringed pyrimidines

___ 4. Franklin used the technique known as _____ to determine many of the physical characteristics of DNA. [p.220]
   a. transformation
   b. cloning
   c. density-gradient centrifugation
   d. x-ray diffraction

___ 5. The significance of Griffith's experiment in which he used two strains of pneumonia-causing bacteria is that _____. [p.216]
a. the conserving nature of DNA replication was finally demonstrated
b. it demonstrated that harmless cells had become permanently transformed into pathogens through a change in the bacterial hereditary material
c. it established that pure DNA extracted from disease-causing bacteria and injected into harmless strains transformed them into "pathogenic strains"
d. it demonstrated that radioactively labeled bacteriophages transfer their DNA but not their protein coats to their host bacteria

___ 6. The significance of the experiments in which ^{32}P and ^{35}S were used is that _____. [p.217]
a. the semiconservative nature of DNA replication was finally demonstrated
b. it demonstrated that harmless cells had become permanently transformed through a change in the bacterial hereditary system
c. it established that pure DNA extracted from disease-causing bacteria transformed harmless strains into "killer strains"
d. it demonstrated that radioactively labeled bacteriophages transfer their DNA but not their protein coats to their host bacteria

___ 7. Franklin's research contribution was essential in _____. [pp.218–220]
a. establishing the double-stranded nature of DNA
b. establishing the principle of base pairing
c. establishing most of the principal structural features of DNA
d. all of the above

___ 8. Chargaff's requirement that A = T and G = C suggested that _____. [p.218]
a. cytosine molecules pair up with guanine molecules, and thymine molecules pair up with adenine molecules
b. the two strands in DNA run in opposite directions (are antiparallel)
c. the number of adenine molecules in DNA relative to the number of guanine molecules differs from one species to the next
d. the replication process must necessarily be semiconservative

___ 9. A single strand of DNA with the base-pairing sequence C–G–A–T–T–G is compatible only with the sequence _____. [p.218]
a. C–G–A–T–T–G
b. G–C–T–A–A–G
c. T–A–G–C–C–T
d. G–C–T–A–A–C

___10. Rosalind Franklin's data indicated that the DNA molecule had to be long and thin with a width (diameter) that is 2 nanometers along its length. Watson and Crick declared that _____ ensured that the width of the DNA molecule must be uniform. [p.219]
a. the antiparallel nature of DNA
b. semiconservative replication processes
c. hydrogen bonding of the sugar–phosphate backbones
d. complementary base-pairing processes that match purine with pyrimidine

Chapter Objectives/Review Questions

1. Before 1952, _____ molecules and _____ _____ molecules were suspected of housing the genetic code. [p.214]
2. The two scientists who assembled the clues to DNA structure and produced the first model were _____ and _____. [p.214]
3. Summarize the research carried out by Miescher, Griffith, Avery, and colleagues, and Hershey and Chase; state the specific advances made by each in the understanding of genetics. [pp.214, 216–217]
4. Viruses called _____ were used in early research efforts to discover the genetic material. [p.216]
5. Summarize the specific research that demonstrated that DNA, not protein, governed inheritance. [pp.216–217]
6. Draw the basic shape of a deoxyribose molecule, and show how a phosphate group is joined to it when forming a nucleotide. [p.219]
7. Show how a nucleotide base would be joined to the sugar–phosphate combination drawn in objective 6. [p.219]
8. DNA is composed of double-ring nucleotides known as _____ and single-ring nucleotides known as _____; the two purines are _____ and _____, whereas the two pyrimidines are _____ and _____. [p.218]
9. Assume that the two parent strands of DNA have been separated and that the base sequence on one parent strand is A–T–T–C–G–C; the base sequence that will complement that parent strand is _____. [p.219]
10. List the pieces of information about DNA structure that Rosalind Franklin discovered through her x-ray diffraction research. [pp.218–220]
11. Explain what is meant by the pairing of nitrogen-containing bases (base pairing), and explain the mechanism that causes bases of one DNA strand to join with bases of the other strand. [pp.218–219]
12. Generally describe how double-stranded DNA replicates from stockpiles of nucleotides. [p.220]
13. Explain what is meant by "each parent strand is conserved in each new DNA molecule." [pp.220–221]
14. During DNA replication, enzymes called DNA _____ assemble new DNA strands. [p.221]
15. Distinguish between continuous strand assembly and discontinuous strand assembly. [p.221]
16. Describe the process of making a genetically identical copy of yourself. [p.222]

Integrating and Applying Key Concepts

Review the stages of mitosis and meiosis, as well as the process of fertilization. Relate what was learned in the chapter about DNA replication and the relationship of DNA to a chromosome. As you pass through fertilization and the stages of both types of cell division, use a diploid number of $2n = 4$. Show the proper number of DNA threads in each cell at each of the stages of mitosis and meiosis. Include the cells that represent the end products of mitosis and meiosis.

14

FROM DNA TO PROTEINS

Interactive Exercises

Beyond Byssus [pp.224–225]

14.1. *Focus on Science:* CONNECTING GENES WITH PROTEINS [pp.226–227]
14.2. HOW IS DNA TRANSCRIBED INTO RNA? [pp.228–229]

Selected Words: byssus [p.224], Garrod [p.226], Beadle and Tatum [p.226], *Neurospora crassa* [p.226], sickle-cell anemia [p.226], Pauling and Itano [p.226], Ingram [p.227], *protein-building* instructions [p.228], poly-A tail [p.229]

Boldfaced, Page-Referenced Terms

[p.225] base sequence _____

_____

[p.225] transcription _____

_____

[p.225] translation _____

_____

[p.225] ribonucleic acid, RNA _____

_____

[p.226] gel electrophoresis _____

_____

[p.228] messenger RNA, mRNA _____

_____

[p.228] ribosomal RNA, rRNA _____

_____

[p.228] transfer RNA, tRNA _____

_____

[p.228] uracil _____

_____

[p.228] RNA polymerase _____

_____

[p.228] promoter _____

_____

[p.229] introns _____

_____

[p.229] exons _____

_____

Fill-in-the-Blanks

The pattern of which base follows the next in a strand of DNA is referred to as the base (1) _____
[p.225]. A region of DNA that calls for the assembly of specific amino acids into a polypeptide chain is a(n)
(2) _____ [p.225]. The two steps from genes to proteins are called (3) _____ [p.225] and
(4) _____ [p.225]. In (5) _____ [p.225], single-stranded molecules of RNA are assembled on DNA
templates in the nucleus. In (6) _____ [p.225], the RNA molecules are shipped from the nucleus into
the cytoplasm, where they are used as templates for assembling (7) _____ [p.225] chains. Following
translation, one or more chains become (8) _____ [p.225] into the three-dimensional shape of protein
molecules. Proteins have (9) _____ [p.225] and (10) _____ [p.225] roles in cells, including control
of DNA. Garrod hypothesized that an enzyme that operated at the next step of a (11) _____ [p.226]
pathway was (12) _____ [p.226] and could explain why the molecules of a particular substance were
accumulating in excess amounts in the body fluids of affected individuals. Experimenting with red bread
mold, Beadle and Tatum found that each inherited (13) _____ [p.226] corresponded to a defective
(14) _____ [p.226]. A more precise hypothesis regarding the relationship of genes with proteins
emerged from Ingram's work with hemoglobin: the amino acids of polypeptide chains—the structural units
of proteins—are encoded in (15) _____ [p.227].

Complete the Table

16. Three types of RNA are transcribed from DNA in the nucleus (from genes that code only for RNA). Complete the following table, which summarizes information about these molecules.

RNA Molecule	Abbreviation	Description/Function
[p.228] a. Ribosomal RNA		
[p.228] b. Messenger RNA		
[p.228] c. Transfer RNA		

Short Answer

17. List three ways in which a molecule of RNA differs structurally from a molecule of DNA. [p.228] _____

_____

_____

18. Cite two similarities in DNA replication and transcription. [p.228] _____

_____

_____

19. What are the three key ways in which transcription differs from DNA replication? [p.228] _____

_____

_____

Sequence

Arrange the steps of transcription in correct chronological sequence. Write the letter of the first step next to 20, the letter of the second step next to 21, and so on.

20. ___ A. The RNA strand grows along exposed bases until RNA polymerase meets a DNA base sequence that signals "stop." [p.230]

21. ___ B. RNA polymerase binds with the DNA promoter region to open up a local region of the DNA double helix. [p.229]

22. ___ C. An RNA polymerase enzyme locates the DNA bases of the promoter region of one DNA strand by recognizing DNA-associated proteins near a promoter. [p.229]

23. ___ D. RNA is released from the DNA template as a free, single-stranded transcript. [p.229]

24. ___ E. RNA polymerase moves stepwise along exposed nucleotides of one DNA strand; as it moves, the DNA double helix keeps unwinding. [p.229]

Completion

25. Suppose the following line represents the DNA strand that will act as a template for the production of mRNA through the process of transcription. Fill in the blanks below the DNA strand with the sequence of complementary bases that will represent the message carried from DNA by mRNA to the ribosome in the cytoplasm [see text Figure 14.8 on pp.228–229].

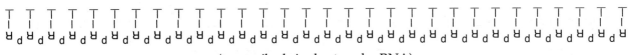

T A C A A G A T A A C A T T A T T T C C T A C C G T C A T C

(transcribed single-strand mRNA)

Labeling and Matching

Newly transcribed mRNA contains more genetic information than is necessary to code for a chain of amino acids. Before the mRNA leaves the nucleus for its ribosome destination, an editing process occurs as certain portions of nonessential information are snipped out. Identify each indicated part of the illustration below; use abbreviations for the nucleic acids. Complete the exercise by matching and entering the letter of the description in the parentheses following each label.

26. _____ ( ) [p.229]

27. _____ ( ) [p.229]

28. _____ ( ) [p.229]

29. _____ ( ) [p.229]

30. _____ ( ) [p.229]

31. _____ _____

_____ ( ) [p.229]

A. The actual coding portions of mRNA
B. Noncoding portions of the newly transcribed mRNA
C. Presence of cap and tail, introns snipped out, and exons spliced together
D. Acquiring of a poly-A tail by the modified mRNA transcript
E. The region of the DNA template strand to be copied
F. Reception of a nucleotide cap by the 5′ end of mRNA (the first synthesized)

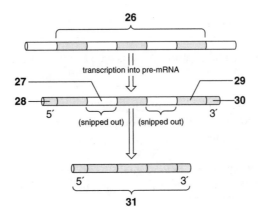

14.3. DECIPHERING THE mRNA TRANSCRIPTS [pp.230–231]

14.4. HOW IS mRNA TRANSLATED? [pp.232–233]

Selected Words: Khorana and Nirenberg [p.230], bases *three at a time* [p.230], START signal [p.230], tRNA "hook" [p.230], "wobble effect" [p.231], *initiation* [p.232], *elongation* [p.232], *termination* [p.232], polysome [p.233], polypeptide chains [p.233]

Boldfaced, Page-Referenced Terms

[p.230] codons _____

_____

[p.230] genetic code _____

_____

[p.230] anticodon _____

_____

Matching

Choose the most appropriate answer for each term.

1. ___codon [p.230]
2. ___three bases at a time [p.230]
3. ___sixty-one [p.230]
4. ___the genetic code [p.230]
5. ___molecular "hook" [p.230]
6. ___ribosome [p.231]
7. ___anticodon [p.230]
8. ___the "stop" codons [p.230]

A. Composed of two subunits, the small subunit with P and A amino acid binding sites as well as a binding site for mRNA
B. Reading frame of the nucleotide bases in mRNA
C. on tRNA, an attachment site for an amino acid
D. UAA, UAG, UGA
E. A sequence of three nucleotide bases that can pair with a specific mRNA codon
F. Name for each base triplet in mRNA
G. The number of codons that actually specify amino acids
H. Term for how the nucleotide sequences of DNA and then mRNA correspond to the amino acid sequence of a polypeptide chain

Complete the Table

9. Complete the following table, which distinguishes the stages of translation.

Translation Stage	Description
[p.232] a.	Special initiator tRNA loads onto small ribosomal subunit and recognizes AUG; small subunit binds with mRNA, and large ribosomal subunit joins small one.
[p.232] b.	Amino acids are strung together in sequence dictated by mRNA codons as the mRNA strand passes through the two ribosomal subunits; two tRNAs interact at P and A sites.
[p.232] c.	MRNA "stop" codon signals the end of the polypeptide chain; release factors detach the ribosome and polypeptide chain from the mRNA.

Completion

10. Given the following DNA sequence, deduce the composition of the mRNA transcript: [p.230]

TAC AAG ATA ACA TTA TTT CCT ACC GTC ATC

___ ___ ___ ___ ___ ___ ___ ___ ___ ___

(mRNA transcript)

11. Deduce the composition of the tRNA anticodons that would pair with the specific mRNA codons, from question 10, as these tRNAs deliver the amino acids (identified below) to the P and A binding sites of the small ribosomal subunit. [p.230]

___ ___ ___ ___ ___ ___ ___ ___ ___ ___
(tRNA anticodons)

12. From the mRNA transcript in question 10, use Figure 14.11 of the text to deduce the composition of the amino acids of the polypeptide sequence. [p.230]

___ ___ ___ ___ ___ ___ ___ ___ ___ ___
(amino acids)

Fill-in-the-Blanks

The order of (13) _____ _____ [p.230] in a protein is specified by a sequence of nucleotide bases. The genetic code is read in units of (14) _____ [p.230] nucleotides; each unit of three codes for (15) _____ [p.230] amino acid(s). In the table that showed which triplet specified a particular amino acid, the triplet code was incorporated in (16) _____ [p.230] molecules. Each of these triplets is referred to as a(n) (17) _____ [p.230]. (18) _____ [p.230] alone carries the instructions for assembling a particular sequence of amino acids from the DNA to the ribosomes in the cytoplasm, where (19) _____ [p.231] of the polypeptide occurs. (20) _____ [pp.230–231] RNA acts as a shuttle molecule as each type brings its particular (21) _____ _____ [pp.230–231] to the ribosome where it is to be incorporated into the growing (22) _____ [p.231]. A(n) (23) _____ [p.230] is a triplet on mRNA that forms hydrogen bonds with a(n) (24) _____ [p.230], which is a triplet on tRNA. During the stage of translation called (25) _____ [p.232], a particular tRNA that can start transcription and an mRNA transcript are both loaded onto a ribosome. In the (26) _____ [p.232] stage of translation, a polypeptide chain is assembled as the mRNA passes between two ribosomal subunits, like a thread being moved through the eye of a needle. During the last stage of translation, (27) _____ [p.232], a STOP codon in the mRNA moves onto the platform, and no tRNA has a corresponding anticodon. Now proteins called (28) _____ [p.232] factors bind to the ribosome. They trigger (29) _____ [p.232] activity that detaches the mRNA and the chain from the ribosome.

14.5. DO MUTATIONS AFFECT PROTEIN SYNTHESIS? [pp.234–235]
14.6. *Focus on Health:* MUTAGENIC RADIATION [p.236]

Selected Words: frameshift mutation [p.234], mutation *frequency* [p.235], *ionizing* radiation [p.236], *nonionizing* radiation [p.236]

Boldfaced, Page-Referenced Terms

[p.234] gene mutations _____

_____

[p.234] base-pair substitution _____

_____

[p.234] insertions _____

_____

[p.234] deletions _____

_____

[p.235] transposons _____

_____

[p.235] mutation rate _____

_____

[p.235] ionizing radiation _____

_____

[p.235] alkylating agents _____

_____

Fill-in-the-Blanks

In addition to changes in chromosomes (crossing over, recombination, deletion, addition, translocation, and inversion) changes can also occur in the structure of DNA; these modifications are referred to as *gene mutations*. Complete the following exercise on types of spontaneous gene mutations.

Ultraviolet radiation, gamma rays, x-rays, and certain natural and synthetic chemicals are examples of environmental agents called (1) _____ [p.235] that may enter cells and damage strands of DNA. If A becomes paired with C instead of T during DNA replication, this spontaneous mutation is a base-pair (2) _____ [p.234]. Sickle-cell anemia is a genetic disease whose cause has been traced to a single DNA base pair; the result is that one (3) _____ _____ [p.234] is substituted for another during protein synthesis. When an extra base becomes inserted into a gene region, it shifts the "three-bases-at-a-time" reading frame by (4) _____ [p.234] base; hence the name (5) _____ [p.234] mutation. Some DNA regions "jump" to new DNA locations and often inactivate the genes in their new environment; such (6) _____ [pp.234–235] elements may give rise to observable changes in the phenotype of an organism.

Each gene has a characteristic (7) _____ [p.235] rate, which is the probability that it will change spontaneously during a specified interval. Not all mutations are spontaneous. Many result after exposure to (8) _____ [p.235] or mutation-causing agents in the environment. The wavelength DNA absorbs most strongly is (9) _____ [p.235] radiation. It may induce crosslinks to form between two (10) _____ [p.235] neighbors on the same DNA strand. (11) _____ [p.235] cancers are one outcome. Other mutagens are gamma rays and x-rays. They can ionize water and other molecules around the DNA, so (12) _____ [p.235] radicals form that may attack the structure of DNA. Substances called (13) _____ [p.235] agents transfer methyl or ethyl groups to reactive sites on the bases or phosphate groups of DNA. At an alkylated site, DNA becomes more susceptible to base-pair disruptions that invite (14) _____. [p.235]. Many (15) _____ [p.235], which are cancer-causing agents, operate by alkylating the DNA. A (16) _____ [p.235] that is specified by a heritable mutation may have harmful, neutral, or beneficial effects on an individual's ability to function in the prevailing environment. Thus, the outcomes of gene mutations can have powerful (17) _____ [p.235] consequences.

Labeling and Matching

A summary of the flow of genetic information in protein synthesis is useful as an overview. Identify the indicated parts of the illustration on the next page by filling in the blanks with the names of the appropriate structures or functions. Choose from the following: DNA, mRNA, tRNA, polypeptide, rRNA subunits, intron, exon, mature mRNA transcript, new mRNA transcript, anticodon, amino acids, ribosome-mRNA complex. Complete the exercise by matching and entering the letter of the description in the parentheses following each label.

18. _____ (   ) [pp.236–237]

19. _____ (   ) [pp.236–237]

    (process)

20. _____ (   ) [pp.236–237]

21. _____ (   ) [pp.236–237]

22. _____ _____ [pp.236–237]

    _____

23. _____ (   ) [pp.236–237]

24. _____ _____ (   ) [pp.236–237]

25. _____ (   ) [pp.236–237]

26. _____ (   ) [pp.236–237]

27. _____ _____ (   ) [pp.236–237]

28. _____ (   ) [pp.236–237]

29. _____ - _____

    _____ (   ) [pp.236–237]

30. _____ (   ) [pp.236–237]

A. Coding portion of mRNA that will translate into proteins
B. Carries a modified form of the genetic code from DNA in the nucleus to the cytoplasm
C. Transports amino acids to the ribosome and mRNA
D. The building blocks of polypeptides
E. Noncoding portions of newly transcribed mRNA
F. tRNA after delivering its amino acid to the ribosome-mRNA complex
G. Join when translation is initiated
H. Holds the genetic code for protein production
I. Place where translation occurs
J. DNA template creates new RNA transcript
K. A sequence of three bases that can pair with a specific mRNA codon
L. Snipping out of introns, only exons remaining
M. May serve as a functional protein (enzyme) or a structural protein

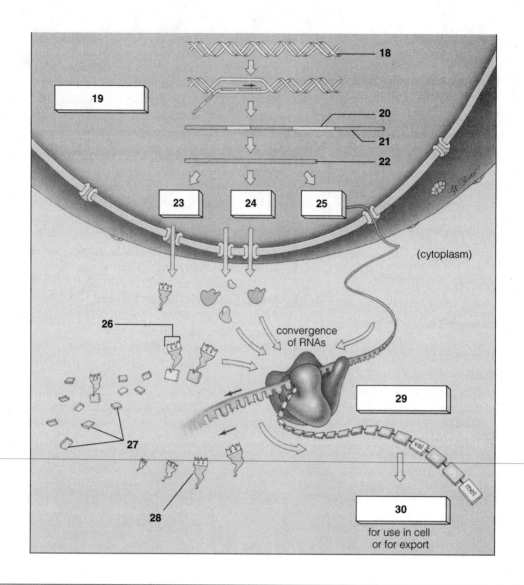

(cytoplasm)

convergence
of RNAs

for use in cell
or for export

Self-Quiz

___ 1. Transcription _____. [p.225]
   a. occurs on the surface of a ribosome
   b. is the final process in the assembly of protein DNA template
   c. occurs during the synthesis of any type of RNA by use of a DNA template
   d. is catalyzed by DNA polymerase

___ 2. _____ experimented with red bread mold to learn that each inherited mutation corresponded to a defective enzyme in a biochemical pathway. [p.226]
   a. Pauling and Itano
   b. Garrod
   c. Beadle and Tatum
   d. Ingram

___ 3. _____ carry(ies) amino acids to ribosomes, where amino acids are linked into the primary structure of a polypeptide. [p.228]
   a. mRNA
   b. tRNA
   c. Introns
   d. rRNA

___ 4. Transfer RNA differs from other types of RNA because it _____. [p.231]
   a. transfers genetic instructions from cell nucleus to cytoplasm
   b. specifies the amino acid sequence of a particular protein
   c. carries an amino acid at one end
   d. contains codons

___ 5. _____ dominates the process of transcription. [p.228]
   a. RNA polymerase
   b. DNA polymerase
   c. Phenylketonuria
   d. Transfer RNA

___ 6. _____ and _____ are found in RNA but not in DNA. [p.228]
   a. Deoxyribose; uracil
   b. Uracil; ribose
   c. Deoxyribose; thymine
   d. Thymine; ribose

___ 7. Each "word" in the mRNA language consists of _____ letters. [p.230]
   a. three
   b. four
   c. five
   d. more than five

___ 8. If each kind of nucleotide is coded for only one kind of amino acid, how many different types of amino acids could be selected? [p.230]
   a. four
   b. sixteen
   c. twenty
   d. sixty-four

___ 9. The genetic code is composed of _____ codons. [p.230]
   a. three
   b. twenty
   c. sixteen
   d. sixty-four

___10. The cause of sickle-cell anemia has been traced to _____. [p.234]
   a. a mosquito-transmitted virus
   b. two DNA mutations that result in two incorrect amino acids in a hemoglobin chain
   c. three DNA mutations that result in three incorrect amino acids in a hemoglobin chain
   d. one DNA mutation that results in one incorrect amino acid in a hemoglobin chain

Chapter Objectives/Review Questions

1. In the rather long road scientists followed in connecting genes to proteins, several pieces of research stood out. Summarize the contributions of the following: Garrod, Beadle and Tatum, Pauling and Itano, and Ingram. [pp.226–227]
2. State how RNA differs from DNA in structure and function, and indicate what features RNA has in common with DNA. [p.228]
3. _____ RNA combines with certain proteins to form the ribosome; _____ RNA carries genetic information for protein construction from the nucleus to the cytoplasm; _____ RNA picks up specific amino acids and moves them to the area of mRNA and the ribosome. [p.228]
4. Describe the process of transcription, and indicate three ways in which it differs from replication. [p.228]
5. Transcription starts at a(n) _____, a specific sequence of bases on one of the two DNA strands that signals the start of a gene. [p.228]
6. What RNA code would be formed from the following DNA code: TAC–CTC–GTT–CCC–GAA? [p.229]
7. Describe how the three types of RNA participate in the process of translation. [pp.228–229]
8. The first end of the mRNA to be synthesized is the _____ end; at the opposite end, the most mature transcripts acquire a(n) _____ tail. [pp.228–229]
9. Distinguish introns from exons. [p.230]
10. Each base triplet in mRNA is called a(n) _____. [p.230]
11. State the relationship between the DNA genetic code and the order of amino acids in a protein chain. [pp.230–231]
12. Scrutinize Figure 14.11 in the text and decide whether the genetic code in this instance applies to DNA, mRNA, or tRNA. [p.230]

13. Explain how the DNA message TAC–CTC–GTT–CCC–GAA would be used to code for a segment of protein, and state what its amino acid sequence would be. [p.230]
14. Describe the "wobble effect." [p.231]
15. Describe events occurring in the following stages of translation: initiation, elongation, and termination. [p.232]
16. What is the fate of the new polypeptides produced by protein synthesis? [p.233]
17. Cite an example of a change in one DNA base pair that has profound effects on the human phenotype. [p.234]
18. Briefly describe the spontaneous DNA mutations known as *base-pair substitution*, *frameshift mutation*, and *transposable element*. [pp.234–235]
19. List some of the environmental agents, or _____, that can cause mutations. [p.235]
20. Using a diagram, summarize the steps involved in the transformation of genetic messages into proteins. [Use text Figure 14.19 on p.237.]

Integrating and Applying Key Concepts

Genes code for specific polypeptide sequences. Not every substance in living cells is a polypeptide. Explain how genes might be involved in the production of a storage starch (such as glycogen) that is constructed from simple sugars.

15

CONTROLS OVER GENES

When DNA Can't Be Fixed

OVERVIEW OF GENE CONTROL

CONTROL IN BACTERIAL CELLS
    Negative Control of Transcription
    Positive Control of Transcription

CONTROL IN EUKARYOTIC CELLS
    Cell Differentiation and Selective Gene
      Expression

EVIDENCE OF GENE CONTROL
    Transcription in Lampbrush Chromosomes
    X Chromosome Inactivation

EXAMPLES OF SIGNALING MECHANISMS
    Hormonal Signals
    Sunlight as a Signal

Focus on Science: LOST CONTROLS AND CANCER

Interactive Exercises

When DNA Can't Be Fixed [pp.240–241]

15.1. OVERVIEW OF GENE CONTROL [p.242]

15.2. CONTROL IN BACTERIAL CELLS [pp.242–243]

Selected Words: *malignant melanoma* [p.240], *xeroderma pigmentosum* [p.240], *basal cell carcinoma* [p.240], *squamous cell carcinoma* [p.240], *benign* tumor [p.241], *malignant* tumor [p.241], *E. coli* [p.242], CAP [p.243], cAMP [p.243]

Boldfaced, Page-Referenced Terms

[p.241] cancers _____

_____

[p.241] metastasis _____

_____

[p.242] regulatory proteins _____

_____

[p.242] negative control systems _____

_____

[p.242] positive control systems _____

_____

[p.242] promoter _____

_____

[p.242] operator _____

_____

[p.242] repressor _____

_____

[p.242] operon _____

_____

[p.243] activator proteins _____

_____

Complete the Table

1. All the diploid cells in an organism possess the same genes, and every cell uses most of the same genes; yet specialized cells must activate only certain genes. Some agents of gene control have been discovered. Transcriptional controls are the most common. Complete the following table to summarize the agents of gene control.

Agents of Gene Control	Method of Gene Control
a. Repressor protein [p.242]	
b. [p.243]	Encourage binding of RNA polymerases to DNA; this is positive control of transcription
c. Promoter [p.242]	
d. [p.242]	short DNA base sequences between promoter and the start of a gene; a binding site for control agents that turn on transcription of mRNA

2. When do gene controls come into play? [p.242]_____

_____

Fill-in-the-Blanks

A promoter and operator provide (3) _____ _____ [p.242]. The (4) _____ [p.243] codes for the

formation of mRNA, which assembles a repressor protein. The affinity of the (5) _____ [p.243] for RNA

polymerase dictates the rate at which a particular operon will be transcribed. Repressor protein allows

(6) _____ _____ [p.243] over the lactose operon. Repressor binds with operator and overlaps

promoter when lactose concentrations are (7) _____ [p.243]. This blocks (8) _____ _____

[p.243] from the genes that will process lactose. This (9) [choose one] ❏ blocks, ❏ promotes [p.243]

production of lactose-processing enzymes. When lactose is present, lactose molecules bind with the

(10) _____ _____ [p.243]. Thus, repressor cannot bind to (11) _____ [p.243], and RNA

polymerase has access to the lactose-processing genes. This gene control works well because lactose-

degrading enzymes are not produced unless they are (12) _____ [p.243].

Labeling and Matching

Escherichia coli, a bacterial cell living in mammalian digestive tracts, is able to exert a negative type of gene control over lactose metabolism. Use the numbered blanks to identify each part of the illustration below. Use abbreviations for nucleic acids. Choose from the following:

lactose   regulator gene   repressor–operator complex   promoter   mRNA transcript
gene with instructions to synthesize permease   lactose operon   lactose-metabolizing enzymes
repressor–lactose complex   repressor protein   RNA polymerase   operator

Complete the exercise by matching and entering the letter of the proper function description in the parentheses following each label. [p.243]

13. _____ _____ ( )
14. _____ _____ ( )
15. _____ _____ ( )
16. _____ ( )
17. _____ ( )
18. _____ _____ ( )
19. _____ –
    _____ ( )
20. _____ _____ ( )
21. _____ –
    _____
    _____ ( )
22. _____ ( )
23. _____ _____ ( )
24. _____ _____ ( )

A. Includes promoter, operator, and the lactose-metabolizing enzymes
B. Short DNA base sequence between promoter and the beginning of a gene
C. The nutrient molecule in the lactose operon
D. Major enzyme that catalyzes transcription
E. Capable of preventing RNA polymerases from binding with DNA
F. Prevents repressor from binding to the operator
G. Genes that produce lactose-metabolizing enzymes
H. Catalyze the digestion of lactose
I. Binds to operator and overlaps promoter; this prevents RNA polymerase from binding to DNA and initiating transcription
J. Specific base sequence that signals the beginning of a gene
K. Gene that contains coding for production of repressor protein
L. Carries genetic instructions to ribosomes for production of lactose enzymes

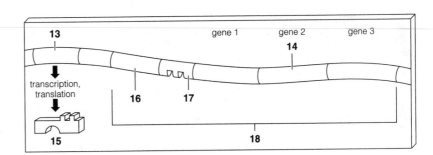

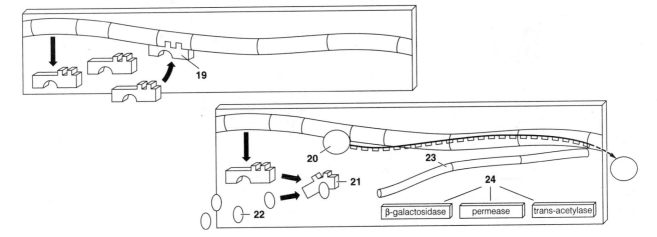

15.3. CONTROL IN EUKARYOTIC CELLS [pp.244–245]
15.4. EVIDENCE OF GENE CONTROL [pp.246–247]

Selected Words: gene amplification [p.244], *DNA rearrangements* [p.244], *chemical modification* [p.244], *alternative splicing* [p.245], *"masked messengers"* [p.245], *chromomeres* [p.246], *"mosaic" female* [p.247], *"calico" cat* [p.247], *"spotting gene"* [p.247], *anhidrotic ectodermal dysplasia* [p.247]

Boldfaced, Page-Referenced Terms

[p.244] cell differentiation _____

_____

[p.246] lampbrush chromosomes _____

_____

[p.247] Barr body _____

_____

[p.247] mosaic tissue effect _____

_____

Short Answer

1. Although a complex organism such as a human being arises from a single cell, the zygote, differentiation occurs in development. Define *differentiation* and relate it to a definition of *selective gene expression*. [p.244]

_____

_____

Labeling and Matching

Identify each numbered part of the illustration showing eukaryotic gene control. Choose from translational control, replicational controls, transcriptional controls, transcript processing controls, and post-translational controls. Complete the exercise by correctly matching and entering the letter of the corresponding gene control description in the parentheses following each label.

2. _____ _____ (   ) [p.244]
3. _____ _____ (   ) [pp.244–245]
4. _____ _____ _____ (   ) [p.245]
5. _____ _____ (   ) [p.245]
6. _____ - _____ _____ (   ) [p.245]

A. Govern the rates at which mRNA transcripts that reach the cytoplasm will be translated into polypeptide chains at ribosomes
B. Govern modification of the initial mRNA transcripts in the nucleus
C. Govern how the polypeptide chains become modified into functional proteins
D. Gene amplification can produce many copies of specific genes needed to produce a lot of a specific protein
E. Influence when and to what degree a particular gene will be transcribed (if at all)

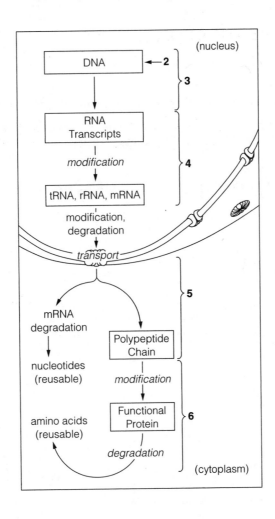

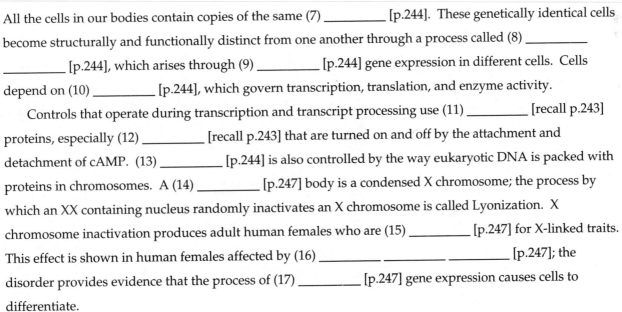

Fill-in-the-Blanks

All the cells in our bodies contain copies of the same (7) _____ [p.244]. These genetically identical cells become structurally and functionally distinct from one another through a process called (8) _____ _____ [p.244], which arises through (9) _____ [p.244] gene expression in different cells. Cells depend on (10) _____ [p.244], which govern transcription, translation, and enzyme activity.

Controls that operate during transcription and transcript processing use (11) _____ [recall p.243] proteins, especially (12) _____ [recall p.243] that are turned on and off by the attachment and detachment of cAMP. (13) _____ [p.244] is also controlled by the way eukaryotic DNA is packed with proteins in chromosomes. A (14) _____ [p.247] body is a condensed X chromosome; the process by which an XX containing nucleus randomly inactivates an X chromosome is called Lyonization. X chromosome inactivation produces adult human females who are (15) _____ [p.247] for X-linked traits. This effect is shown in human females affected by (16) _____ _____ _____ [p.247]; the disorder provides evidence that the process of (17) _____ [p.247] gene expression causes cells to differentiate.

True–False

If false, explain why.

_____18. "Lampbrush chromosomes" of meiotic prophase I seen in electron micrographs of amphibian eggs provide evidence of translational control. [p.246]

_____19. When the same primary mRNA transcript is spliced in alternative ways, the result may be different mRNAs, each coding for a slightly different protein. [p.244]

15.5. EXAMPLES OF SIGNALING MECHANISMS [pp.248–249]

15.6. *Focus on Science*: LOST CONTROLS AND CANCER [pp.250–251]

Selected Words: amplification [p.248], somatotropin [p.248], prolactin [p.248], *erbB* [p.250], *myc* [p.250], *anti*-oncogenes [p.250], *p53* [p.250], *bcl-21* [p.250], carcinogen [p.251]

Boldfaced, Page-Referenced Terms

[p.248] hormones _____

_____

[p.248] enhancers _____

_____

[p.248] ecdysone _____

_____

[p.248] polytene chromosome _____

_____

[p.249] phytochrome _____

_____

[p.250] protein kinases _____

_____

[p.250] growth factors _____

_____

[p.250] oncogene _____

_____

[p.251] apoptosis _____

_____

[p.251] ICE-like proteases _____

_____

Fill-in-the-Blanks

Several kinds of (1) _____ [p.248] influence gene activity. In both animals and plants, molecules called (2) _____ [p.248] stimulate or inhibit gene activity in their target cells. Any cell with (3) _____ [p.248] for a specific hormone is a target. Sometimes a hormone molecule must first bind with a(n) (4) _____ [p.248] (generally, a base sequence on the same DNA molecule that contains the promoter to which RNA polymerase will attach and begin transcription) before transcription can be either turned on or turned off. Many insect larvae produce a hormone known as (5) _____ [p.248], which is an (4) that binds to a receptor on salivary gland cells and triggers very rapid transcription of the genes that contain the instructions for assembling the protein components of saliva. Thus insect larvae can usually digest leafy vegetation as fast as they consume it by producing large amounts of salivary proteins. The genes responsible for salivary protein production have been copied many times by (6) _____ [p.248] and lie parallel to each other, causing that portion of the (7) _____ [p.248] chromosome to appear puffy during transcription.

In humans and other mammals, mammary gland cells do not begin producing milk until the hormone (8) _____ [p.248] attaches to receptors on the surfaces of the gland cells and activates the genes responsible for producing milk proteins. Other cells in mammalian bodies have these genes, but lack the (9) _____ [p.248] to which the hormone attaches.

Red wavelengths of sunlight activate (10) _____ [p.249], a blue-green pigment whose activity influences transcription of certain genes at certain times of the day and season as enzymes and other proteins play key roles in germination, growth, and the formation of flowers, fruits and seeds.

Most cells respond to signals that tell them it's time to die, but (11) _____ [p.251] cells do not. Many (11) cells form a tissue mass called a (12) _____ [recall p.241], that may be either benign or malignant. Malignant cells can break loose from their original growth area, travel through the body and invade other tissues—a process called (13) _____ [recall p.241].

True–False

If the statement is true, write a T in the blank. If false, make it true by changing the underlined word.

_____ 14. Polytene chromosomes of many insect larvae are unusual in that their DNA has been repeatedly replicated; thus, these chromosomes have multiple copies of the same genes. When these genes are undergoing transcription, they "puff." [p.248]

_____ 15. When cells become cancerous, cell populations decrease to very low densities and stop dividing. [p.250]

_____ 16. All abnormal growths and massings of new tissue in any region of the body are called tumors. [recall p.241]

_____ 17. Malignant tumors have cells that migrate and divide in other organs. [recall p.241]

_____ 18. Oncogenes are genes that combat cancerous transformations. [p.250]

_____ 19. Proto-oncogenes rarely trigger cancer. [p.250]

_____ 20. The normal expression of proto-oncogenes is vital, even though their normal expression may be lethal. [p.250]

Self-Quiz

_____ 1. _____ refers to the processes by which cells with identical genotypes become structurally and functionally distinct from one another according to the genetically controlled developmental program of the species. [p.244]
a. Metamorphosis
b. Metastasis
c. Cleavage
d. Differentiation

_____ 2. _____ binds to operator whenever lactose concentrations are low. [p.243]
a. Operon
b. Repressor
c. Promoter
d. Operator

_____ 3. Any gene or group of genes together with its promoter and operator sequence is a(n) _____. [p.242]
a. repressor
b. operator
c. promoter
d. operon

_____ 4. The operon model explains the regulation of _____ in prokaryotes. [p.242]
a. replication
b. transcription
c. induction
d. Lyonization

_____ 5. In multicelled eukaryotes, cell differentiation occurs as a result of _____. [p.244]
a. growth
b. selective gene expression
c. repressor molecules
d. the death of certain cells

_____ 6. One type of gene control discovered in female mammals is _____. [p.247]
a. a conflict in maternal and paternal alleles
b. slow embryo development
c. X chromosome inactivation
d. operon

_____ 7. Due to X inactivation of either the paternal or maternal X chromosome, human females with anhidrotic ectodermal dysplasia _____. [p.247]
a. completely lack sweat glands
b. develop benign growths
c. have mosaic patches of skin that lack sweat glands
d. develop malignant growths

_____ 8. Which of the following characteristics seems to be most uniquely correlated with metastasis? [p.241]
a. loss of nuclear-cytoplasm controls governing cell growth and division
b. changes in recognition proteins on membrane surfaces
c. "puffing" in the chromosomes
d. the massive production of benign tumors

_____ 9. Genes with the potential to induce cancerous formations are known as _____. [p.250]
a. proto-oncogenes
b. oncogenes
c. carcinogens
d. malignant genes

_____ 10. _____ controls govern the rates at which mRNA transcripts that reach the cytoplasm will be translated into polypeptide chains at the ribosomes. [p.245]
a. Transport
b. Transcript processing
c. Translational
d. Transcriptional

Chapter Objectives/Review Questions

1. Define *tumor* and distinguish between benign and malignant tumors. [p.241]
2. _____ is a process in which a cancer cell leaves its proper place and invades other tissues to form new growths. [p.241]
3. A _____ is a specific base sequence that signals the beginning of a gene in a DNA strand. [p.242]
4. Some control agents bind to _____, which are short base sequences between promoters and the beginning of a gene. [p.242]
5. A gene-control arrangement in which the same promoter–operator sequence services more than one gene is called an _____. [p.242]
6. Gene expression is controlled through regulatory _____, _____ that speed up reaction rates, and _____ proteins that bind to enhancer sites on DNA. [pp.242–243]
7. The negative control of _____ protein prevents the enzymes of transcription from binding to DNA; the positive control of _____ protein enhances the binding of RNA polymerases to DNA. [pp.242–243]
8. Describe the sequence of events that occurs on the chromosome of *E. coli* after you drink a glass of milk. [p.243]
9. The cells of *E. coli* manage to produce enzymes to degrade lactose when those molecules are _____ and to stop production of lactose-degrading enzymes when lactose is _____. [p.243]
10. Define *selective gene expression* and explain how this concept relates to cell differentiation in multicelled eukaryotes. [p.244]
11. Cell _____ occurs in multicelled eukaryotes as a result of _____ gene expression. [p.244]
12. List and define the levels of gene control in eukaryotes. [pp.244–245]
13. The "lampbrush" chromosomes seen in amphibian eggs and larvae represent a change in chromosome structure that has been correlated directly with _____ of the genes necessary for growth. [p.246]
14. The condensed X chromosome seen on the edge of the nuclei of female mammals is known as the _____ body. [p.247]
15. Explain how X chromosome inactivation provides evidence for selective gene expression; use the example of anhidrotic ectodermal dysplasia. [p.247]
16. Explain how hormones act as a major agent of gene control. [p.248]
17. Describe the gene control mechanism afforded many insect larvae by polytene chromosomes. [p.248]
18. Describe the relationship of proto-oncogenes, environmental irritants, and oncogenes. [pp.250–251]
19. Cigarette smoke, X-rays, gamma rays, and ultraviolet radiation are examples of _____. [p.251]

Integrating and Applying Key Concepts

Suppose you have been restricting yourself to a completely vegetarian diet for the past six months. Quite unexpectedly, you find yourself in a social situation that requires you to eat a half-pound sirloin steak. Would you expect to digest the steak as easily as you digest soybean burgers? Explain your yes or no answer in terms of transcriptional controls or feedback inhibition.

16

RECOMBINANT DNA AND GENETIC ENGINEERING

Interactive Exercises

Mom, Dad, and Clogged Arteries [pp.254–255]

16.1. A TOOLKIT FOR MAKING RECOMBINANT DNA [pp.256–257]

Selected Words: HDLs [p.254], *LDLs* [p.254], *familial cholesterolemia* [p.254], *Haemophilus influenzae* [p.256], *staggered* cuts [p.256], *Taq I* [p.256], *Eco*RI [p.256], *Not*I [p.256], *sticky* ends [p.256], "cloning factory" [p.257], introns [p.257]

Boldfaced, Page-Referenced Terms

[p.254] gene therapy _____

_____

[p.255] recombinant DNA technology _____

_____

[p.255] genetic engineering _____

_____

[p.256] restriction enzyme _____

_____

[p.256] genome _____

_____

[p.256] DNA ligase _____

_____

[p.256] plasmid _____

_____

[p.257] DNA clone _____

_____

[p.257] cloning vector _____

_____

[p.257] cDNA _____

_____

[p.257] reverse transcriptase _____

_____

Short Answer

1. Describe and distinguish between the bacterial chromosome and plasmids present in a bacterial cell. [pp.256–257] _____

_____

_____

_____

True–False

If the statement is true, write T in the blank. If the statement is false, make it correct by changing the underlined words and writing the correct word(s) in the answer blank.

_____ 2. Plasmids are underline{organelles on the surfaces of which amino acids are assembled into polypeptides}. [p.256]

_____ 3. Gene mutations and recombination are common in nature. [p.255]

Fill-in-the-Blanks

Genetic experiments have been occurring in nature for billions of years as a result of gene (4) _____ [p.255], crossing over and recombination, and other events. Humans now are causing genetic change by using (5) _____ _____ [p.255] technology in which researchers cut and splice together gene regions from different (6) _____ [p.255], then greatly (7) _____ [p.255] the number of copies of the genes that interest them. The genes, and in some cases their (8) _____ [p.255] products, are produced in quantities that are large enough for (9) _____ [p.255] and for practical applications. (10) _____ _____ [p.255] involves isolating, modifying, and inserting particular genes back into the same organism or into a different one.

True–False

A genetic engineer used restriction enzymes to prepare fragments of DNA from two different species that were then mixed. Four of these fragments are illustrated below. Fragments (a) and (c) are from one species, (b) and (d) from the other species. Answer exercises 11–15. If false, explain why.

————————   ————————   ————————   ————————
————————TACA   ————————TTCA   ————————CGTA   ————————ATGT
a.   b.   c.   d.

————————11. Some of the fragments represent sticky ends. [p.256]

————————12. The same restriction enzyme was used to cut fragments (b), (c), and (d). [p.256]

————————13. Different restriction enzymes were used to cut fragments (a) and (d). [p.256]

————————14. Fragment (a) will base-pair with fragment (d) but not with fragment (c). [p.256]

————————15. The same restriction enzyme was used to cut the different locations in the DNA of the two species shown. [p.256]

Matching

Match the steps in the formation of a DNA library with the parts of the illustration below. [All from p.257]

16. ____
17. ____
18. ____
19. ____
20. ____
21. ____

A. Joining of chromosomal and plasmid DNA using DNA ligase
B. Restriction enzyme cuts chromosomal DNA at specific recognition sites
C. Cut plasmid DNA
D. Recombinant plasmids containing cloned library
E. Fragments of chromosomal DNA
F. Same restriction enzyme is used to cut plasmids

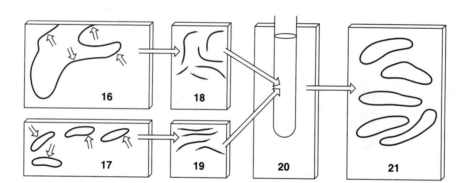

Fill-in-the-Blanks

Synthesizing the appropriate (22) _____ [p.257] does not automatically follow the successful taking in of a modified gene by a host cell; genes must be suitably (23) _____ [p.257] first, which involves getting rid of the (24) _____ [p.257] from the mRNA transcripts, then using the enzyme reverse (25) _____ [p.257] to produce cDNA. The process goes like this:

a. An mRNA transcript of a desired gene is used as a (26) _____ [p.257] for assembling a DNA strand. An enzyme (25) _____ does the assembling.

b. An mRNA- (27) _____ [p.257] hybrid molecule results.

c. (28) _____ [p.257] action removes the mRNA and assembles a second strand of (29) _____ [p.257] on the first strand.

d. The result is double-stranded (30) _____ [p.257], "copied" from an mRNA (31) _____ [p.257].

a.

b.

c.

d.

16.2. PCR—A FASTER WAY TO AMPLIFY DNA [p.258]

16.3. *Focus on Science:* DNA FINGERPRINTS [p.259]

16.4. HOW IS DNA SEQUENCED? [p.260]

16.5. FROM HAYSTACKS TO NEEDLES—ISOLATING GENES OF INTEREST [p.261]

Selected Words: DNA polymerase [p.258], restriction fragment length polymorphisms (RFLPs) [p.259], *genomic* library [p.261], *cDNA* library [p.261]

Boldfaced, Page-Referenced Terms

[p.258] PCR, polymerase chain reaction _____

_____

[p.258] primers _____

_____

[p.259] DNA fingerprint _____

_____

[p.259] tandem repeats _____

_____

[p.259] gel electrophoresis _____

_____

[p.260] automated DNA sequencing _____

_____

[p.260] gene library _____

_____

[p.261] probe _____

_____

[p.261] nucleic acid hybridization _____

_____

Complete the Table

1. Complete the following table, which summarizes some of the basic tools and procedures used in recombinant DNA technology.

Tool/Procedure	Definition and Role in Recombinant DNA Technology
a. Automated DNA sequencing [p.260]	
b. Primers [p.258]	
c. cDNA library [p.261]	
d. Gel electrophoresis [p.259]	
e. PCR [p.258]	
f. DNA sequencing [p.260]	
g. Probe [p.261]	
h. DNA fingerprint [p.259]	

Matching

Match the most appropriate letter with each numbered partner.

2. ___ DNA polymerase [p.258]

3. ___ A*, T*, C*, G* [p.260]

4. ___ A, T, C, G [p.260]

5. ___ RFLPs [p.259]

6. ___ tandem repeats [p.259]

7. ___ nucleic acid hybridization [p.261]

A. Identical short lengths of DNA arranged in sequence that differ from one person to the next
B. Recognizes primers as START tags and assembles complementary sequences using the template as a guide
C. DNA fragments of different lengths from a long strand of DNA cleaved by restriction enzymes
D. Nitrogen bases labeled with a molecule that fluoresces a specific color when passing through a laser beam
E. Unlabeled nitrogen bases
F. Pairing of nitrogen bases that occurs between DNA (or RNA) from different sources

16.6. USING THE GENETIC SCRIPTS [p.262]

16.7. DESIGNER PLANTS [pp.262–263]

16.8. GENE TRANSFERS IN ANIMALS [pp.264–265]

16.9. *Focus on Bioethics:* WHO GETS ENHANCED? [p.265]

16.10. SAFETY ISSUES [p.266]

Selected Words: diabetics [p.262], insulin [p.262], human somatotropin [pp.262,264], hemoglobin [p.262], interferon [p.262], *Southern corn blight* [p.262], Ti plasmid [p.263], CFTR protein [p.264], TPA [p.264], human collagen [p.264], *Dolly* [p.264], human serum albumin [p.264], Human Genome Initiative [p.264], EST [p.265], TIGR Assembler [p.265], *eugenic engineering* [p.265], "fail-safe" genes [p.266], *hok* gene [p.266], "ice-minus bacteria" [p.266]

Complete the Table

1. Complete the following table by providing examples of genetically engineered beneficial organisms for uses in:

a. medicine [p.262]	
b. industry [p.262]	
c. agriculture [p.262]	
d. environmental remediation [p.262]	

Short Answer

2. Explain how knowing about the genetic makeup of Earth's organisms can help us reconstruct the evolutionary history of life. [p.262] _____

_____

_____

_____

3. Explain how knowing the composition of genes can help scientists derive counterattacks against rapidly mutating pathogens. [p.262] _____

_____

_____

4. Explain why it is important for farmers to plant many varieties of food and fiber plants instead of just a few varieties. [p.262] _____

_____

_____

5. Describe how genetic engineers transfer specific genes into plant cells. [p.263] _____

_____

_____

_____

6. Researchers are trying to insert the gene for human serum albumin into the chromosomes of dairy cattle. Why might that achievement be useful? [p.264] _____

_____

_____

7. Explain the goal of the Human Genome Initiative. [pp.264–265] _____
_____

In exercises 8–13, summarize the results of the given experimentation dealing with genetic modifications of plants and animals.

8. Cattle may soon be producing human collagen: [p.264] _____
_____

9. The bacterium *Agrobacterium tumefaciens:* [p.263] _____
_____

10. Cotton plants: [p.263] _____
_____

11. Introduction of the rat and human somatotropin gene into fertilized mouse eggs: [p.264] _____
_____

12. "Ice-minus" bacteria and strawberry plants [p.266] _____
_____

13. Dolly, the cloned lamb: [p.264] _____
_____

Fill-in-the-Blanks

The gene harmful to strawberries is called the "ice-forming" gene, and the bacteria are known as

(14) _____ - _____ [p.266] bacteria; genetic engineers were able to remove the harmful gene and

test the modified bacterium on strawberry plants with no adverse effects. Inserting one or more genes into

the (15) _____ _____ [p.254] of an organism for the purpose of correcting genetic defects is known

as (16) _____ _____ [p.254]. Attempting to modify a human trait by inserting genes into sperm or

eggs is called (17) _____ _____ [p.265].

Self-Quiz

____ 1. Small, circular molecules of DNA in bacteria are called _____. [p.256]
   a. plasmids
   b. desmids
   c. pili
   d. F particles
   e. transferins

____ 2. Enzymes used to cut genes in recombinant DNA research are _____. [p.256]
   a. ligases
   b. restriction enzymes
   c. transcriptases
   d. DNA polymerases
   e. replicases

____ 3. The total DNA in a haploid set of chromosomes of a species is its _____. [p.256]
   a. plasmid
   b. enzyme potential
   c. genome
   d. DNA library
   e. none of the above

____ 4. An enzyme that heals random base-pairing of chromosomal fragments and plasmids is _____. [p.256]
   a. reverse transcriptase
   b. DNA polymerase
   c. cDNA
   d. DNA ligase

___ 5. Any DNA molecule that is copied from mRNA is known as _____. [p.257]
   a. cloned DNA
   b. cDNA
   c. DNA ligase
   d. hybrid DNA

___ 6. The most commonly used method of DNA amplification is _____. [p.258]
   a. polymerase chain reaction
   b. gene expression
   c. genome mapping
   d. RFLPs

___ 7. Amplification results in _____. [p.258]
   a. plasmid integration
   b. bacterial conjugation
   c. copies of chromosome DNA or cDNA
   d. production of DNA ligase

___ 8. Tandem repeats are valuable because _____. [p.259]
   a. they reduce the risks of genetic engineering
   b. they provide an easy way to sequence the human genome

   c. they allow fragmenting DNA without enzymes
   d. they provide DNA fragment sizes unique to each person

___ 9. _____ _____ _____ can rapidly reveal the base sequence of cloned DNA or PCR-amplified DNA fragments. [p.260]
   a. Polymerase chain reaction
   b. Nucleic acid hybridization
   c. Recombinant DNA technology
   d. Automated DNA sequencing

___10. A cDNA library is _____. [p.261]
   a. a collection of DNA fragments derived from mRNA and free of introns
   b. cDNA plus the required restriction enzymes
   c. mRNA-cDNA
   d. composed of mature mRNA transcripts

Chapter Objectives/Review Questions

1. List the means by which natural genetic recombination occurs. [p.255]
2. Define *recombinant DNA technology*. [p.255]
3. _____ are small, circular, self-replicating molecules of DNA or RNA within a bacterial cell. [p.256]
4. Some bacteria produce _____ enzymes that cut apart DNA molecules injected into the cell by viruses; such DNA fragments or "_____ ends" often have staggered cuts capable of base-pairing with other DNA molecules cut by the same _____ enzymes. [p.256]
5. Base-pairing between chromosomal fragments and cut plasmids is made permanent by DNA _____. [p.256]
6. Define *cDNA*. [p.257]
7. A special viral enzyme, _____ _____, presides over the process by which mRNA is transcribed into DNA. [p.257]
8. Why do researchers prefer to work with cDNA when working with human genes? [p.257]
9. Multiple, identical copies of DNA fragments produced by restriction enzymes are known as _____ DNA. [pp.257–258]
10. List and define the two major methods of DNA amplification. [p.258]
11. Polymerase chain reaction is the most commonly used method of DNA _____. [p.258]
12. Explain how gel electrophoresis is used to sequence DNA. [pp.259–260]
13. How is a cDNA probe used to identify a desired gene carried by a modified host cell? [p.261]
14. Be able to explain what a DNA library is; review the steps used in creating such a library. [p.261]
15. List some practical genetic uses of RFLPs. [p.262]
16. Tell about the Human Genome Initiative and its implications. [pp.264–265]
17. Define *gene therapy* and *eugenic engineering*. [pp.254,265]

Integrating and Applying Key Concepts

How could scientists guarantee that *Escherichia coli*, the human intestinal bacterium, will not be transformed into a severely pathogenic form and released into the environment if researchers use the bacterium in recombinant DNA experiments?

17

EMERGENCE OF EVOLUTIONARY THOUGHT

Interactive Exercises

Legends of the Flood [pp.270-271]

17.1. EARLY BELIEFS, CONFOUNDING DISCOVERIES [pp.272-273]

Selected Words: "species" [p.272], *sequences* of fossils [p.273]

Boldfaced, Page-Referenced Terms

[p.271] catastrophism _____

_____

[p.272] biogeography _____

_____

[p.272] comparative morphology _____

_____

[p.273] fossils _____

_____

[p.273] evolution _____

_____

Matching

Choose the most appropriate answer for each term.

1. ____comparative anatomy [pp.272–273]
2. ____biogeography [p.272]
3. ____fossils [p.273]
4. ____school of Hippocrates [p.272]
5. ____evolution [p.273]
6. ____great Chain of Being [p.272]
7. ____Buffon [p.273]
8. ____species [p.272]
9. ____Aristotle [p.272]
10. ____sedimentary beds [p.273]

A. Rock layers deposited at different times; may contain fossils
B. Came to view nature as continuum of organization, from lifeless matter through complex forms of plant and animal life
C. Each kind of being that represents a link in the great Chain of Being
D. Modification of species over time
E. Extended from the lowest forms of life to humans and on to spiritual beings
F. Suggested that perhaps species originated in more than one place and perhaps had been modified over time
G. Studies of body structure comparisons and patterning
H. Studies of the world distribution of plants and animals
I. Suggested that the gods were not the cause of the sacred disease
J. One type of direct evidence that organisms lived in the past

17.2. A FLURRY OF NEW THEORIES [pp.274–275]

17.3. DARWIN'S THEORY TAKES FORM [pp.276–277]

17.4. *Focus on Science*: REGARDING THE FIRST OF THE "MISSING LINKS" [p.278]

Selected Words: Cuvier [p.274], inheritance of *acquired* characteristics [p.274], Lamarck [p.274], H.M.S. *Beagle* [p.274], Darwin [p.274], *artificial* selection [p.277], *natural* selection [p.277], Wallace [p.277], *On the Origin of Species* [p.277], "missing link" [p.278], *transitional forms* [p.278], *Archaeopteryx* [p.278]

Boldfaced, Page-Referenced Terms

[p.275] theory of uniformity _____

_____

Choice

For questions 1–12, choose from the following:

a. Georges Cuvier   b. Jean-Baptiste Lamarck

1. ____ Stretching directed "fluida" to the necks of giraffes, which lengthened permanently [p.274]
2. ____ Acknowledged there were abrupt changes in the fossil record that corresponded to discontinuities between certain layers of sedimentary beds; thought this was evidence of change in populations of ancient organisms [p.274]
3. ____ The force for change in organisms is the drive for perfection [p.274]
4. ____ There was only one time of creation that populated the world with all species [p.274]
5. ____ Catastrophism [p.274]
6. ____ Permanently stretched giraffe necks were bestowed on offspring [p.274]

7. ___ When a global catastrophe destroyed many organisms, a few survivors repopulated the world [p.274]

8. ___ Theory of inheritance of acquired characteristics [p.274]

9. ___ The survivors of a global catastrophe were not new species (many differed from related fossils) but naturalists had not discovered them yet [p.274]

10. ___ During a lifetime, environmental pressures and internal "desires" bring about permanent changes [p.274]

11. ___ The force for change is a drive for perfection, up the Chain of Being [p.274]

12. ___ The drive to change is centered in nerves that direct an unknown "fluida" to body parts in need of change [p.274]

Complete the Table

13. Several key players and events in the life of Charles Darwin led him to his conclusions about natural selection and evolution. Summarize these influences by completing the following table.

Event/Person	Importance to Synthesis of Evolutionary Theory
[p.274] a.	Botanist at Cambridge University who perceived Darwin's real interests and arranged for Darwin to become a ship's naturalist
[p.274] b.	Where Darwin earned a degree in theology but also developed his love for natural history
[p.274] c.	British ship that carried Darwin (as a naturalist) on a five-year voyage around the world
[p.274] d.	Wrote *Principles of Geology*; advanced the theory of uniformity; suggested that Earth was much older than 6,000 years
[p.276] e.	Wrote an influential essay (read by Darwin) on human populations asserting that people tend to produce children faster than food supplies, living space, and other resources can be sustained
[p.276] f.	Volcanic islands 900 kilometers from the South American coast where Darwin correlated differences in various species of finches with their environmental challenges
[p.277] g.	The key point in Darwin's theory of evolution; involves reproductive capacity, heritable variations, and adaptive traits
[p.277] h.	English naturalist contemporary with Darwin; independently developed Darwin's theory of evolution before Darwin published
[p.277] i.	Unearthed in 1861; the first transitional fossil (between reptiles and birds); provided evidence for Darwin's theory

Sequence

Read ideas A–G through first, then put them in an order that logically develops the Darwin–Wallace theory of evolution in correct sequence. Number 14 is the first, most fundamental idea. Idea number 15 sets the stage for and underlies the remaining ideas.

14. ___   The first idea in the series [pp.276–277]

15. ___   The second idea in the series [pp.276–277]

16. ___   The third idea in the series [pp.276–277]

17. ___   The fourth idea in the series [pp.276–277]

18. ___   The fifth idea in the series [pp.276–277]

19. ___   The sixth idea in the series [pp.276–277]

20. ___   The concluding idea in the series [pp.276–277]

A. Nature "selects" those individuals with traits that allow them to obtain the resources they need. They live longer and produce more offspring than others in the population who cannot get the resources they need to live and reproduce.

B. As population size increases, available resources dwindle.

C. A population is evolving when the forms of its heritable traits are changing over successive generations.

D. Animal populations tend to reproduce faster than food supplies, living space, and other resources can sustain the populations.

E. The struggle for existence intensifies.

F. There is genetic variation in all sexually reproducing populations; variations in traits might affect individuals' abilities to get resources, and therefore to survive and reproduce in particular environments.

G. Over time the more successful phenotypes will dominate the population that exists in that particular environment.

Self-Quiz

___ 1. An acceptable definition of evolution is _____. [p.273]
   a. changes in organisms that are extinct
   b. changes in organisms since the flood
   c. changes in organisms over time
   d. changes in organisms in only one place

___ 2. The two scientists most closely associated with the concept of evolution are _____. [p.277]
   a. Lyell and Malthus
   b. Henslow and Cuvier
   c. Henslow and Malthus
   d. Darwin and Wallace

___ 3. Studies of the distribution of organisms on the earth is known as _____. [p.272]
   a. a great Chain of Being
   b. stratification
   c. geological evolution
   d. biogeography

___ 4. Buffon suggested that _____. [p.273]
   a. tail bones in a human have no place in a perfectly designed body
   b. perhaps species originated in more than one place and may have been modified over time
   c. the force for change in organisms was a built-in drive for perfection, up the Chain of Being
   d. gradual processes now molding the Earth's surface had also been at work in the past

5. In Lyell's book, *Principles of Geology*, he suggested that _____. [pp.274–275]
   a. tail bones in a human have no place in a perfectly designed body
   b. perhaps species originated in more than one place and may have been modified over time
   c. the force for change in organisms was a built-in drive for perfection, up the Chain of Being
   d. gradual processes now molding the Earth's surface had also been at work in the past

6. One of the central ideas of Lamarck's theory of desired evolution was that _____. [p.274]
   a. tail bones in a human have no place in a perfectly designed body
   b. perhaps species originated in more than one place and may have been modified over time
   c. the force for change in organisms was a built-in drive for perfection, up the Chain of Being
   d. gradual processes now molding the Earth's surface had also been at work in the past

7. The theory of catastrophism is associated with _____. [p.274]
   a. Darwin
   b. Cuvier
   c. Buffon
   d. Lamarck

8. The idea that any population tends to outgrow its resources, and its members must compete for what is available, belonged to _____. [p.276]
   a. Malthus
   b. Darwin
   c. Lyell
   d. Henslow

9. The ideas that all natural populations have the reproductive capacity to exceed the resources required to sustain them and members of a natural population show great variation in their traits are associated with _____. [p.276]
   a. catastrophism
   b. desired evolution
   c. natural selection
   d. Malthus's idea of survival

10. *Archaeopteryx* represents evidence for _____. [p.278]
    a. birds descending from reptiles
    b. evolution
    c. an evolutionary "link" between two major groups of organisms
    d. the existence of fossils
    e. all of the above

Chapter Objectives/Review Questions

1. Relate Aristotle's view of nature and describe what is meant by the great "Chain of Being." [p.272]
2. It was the study of _____ that raised this question: How did so many species get from the center of creation to islands and other isolated places? [p.272]
3. Scientists working in a field known as _____ _____ wondered why animals as different as humans, whales, and bats had so many similarities. [pp.272–273]
4. Give two examples of the type of evidence found by comparative anatomists that suggested living things may have changed with time. [pp.272–273]
5. _____ _____ consist of distinct, multiple layers of sedimentary rock that were originally deposited at different time periods. [p.273]
6. A _____ is defined as the recognizable remains or body impressions of organisms that lived in the past. [p.273]
7. The modification of species over time is called _____. [p.273]
8. Discuss Cuvier's theory of catastrophism. [p.274]
9. What was the main force for change in organisms as described in Lamarck's theory of desired evolution? [p.274]

10. State the significance of the following to the theory of evolution: Henslow, H.M.S. *Beagle*, Lyell, Darwin, Malthus, Wallace, and *Archaeopteryx*. [pp.274–278]
11. To what conclusion was Darwin led when he considered the various species of finches living on the separate islands of the Galapágos? [pp.276–277]
12. Outline the key observations and inferences of Darwin's theory of evolution by natural selection. [pp.276–277]

Integrating and Applying Key Concepts

An Austrian biologist named Paul Kammerer (1880–1926) once reported that he had found solid evidence that the nuptial pads on the forelimbs of the male midwife toad (used to grip the female during mating) were *acquired* by the "energy and diligence" of the parent toad and were then transmitted to offspring for the benefit of future generations. When he presented his evidence, several observing scientists suspected his specimens were forgeries. Scientists attempted to repeat Kammerer's experiments but they were unrepeatable. Kammerer incurred the wrath of the international scientific community, and, in disgrace, he eventually ended his life by his own hand. Considering what you have learned in this chapter, what is your assessment of Kammerer's claim?

18

MICROEVOLUTION

Interactive Exercises

Designer Dogs [pp.280–281]

18.1. INDIVIDUALS DON'T EVOLVE—POPULATIONS DO [pp.282–283]

18.2. *Focus on Science*: WHEN IS A POPULATION *NOT* EVOLVING? [pp.284–285]

Selected Words: "evolution" [p.281], *morphological, physiological,* and *behavioral* traits [p.282], *qualitatively different, quantitatively different* [p.282], morpho- [p.282], continuous variation [p.282], *natural selection, gene flow,* and *genetic drift* [p.283], phenotypic variation [p.283], p, q, p^2, q^2 and $2pq$ [p.284]

Boldfaced, Page-Referenced Terms

[p.281] evolution _____

_____

[p.282] population _____

_____

[p.282] polymorphism _____

_____

[p.282] gene pool _____

_____

[p.282] alleles _____

_____

[p.282] allele frequencies _____

_____

[p.282] genetic equilibrium _____

_____

[p.283] microevolution _____

_____

[p.283] mutation rate _____

_____

[p.283] lethal mutation _____

_____

[p.283] neutral mutation _____

_____

[p.284] Hardy–Weinberg rule _____

_____

Choice

The traits of individuals in a population are often classified as being morphological, physiological, or behavioral traits. In the list of traits below, enter "M" if the trait seems to be morphological, "P" if the trait is physiological, and "B" if the trait is behavioral. [p.282]

1. ___ Frogs have a three-chambered heart.

2. ___ The active transport of Na^+ and K^+ ions is unequal.

3. ___ Humans and orangutans possess an opposable thumb.

4. ___ An organism periodically seeks food.

5. ___ Some animals have a body temperature that fluctuates with the environmental temperature.

6. ___ The platypus is a strange mammal. It has a bill like a duck and lays eggs.

7. ___ Some vertebrates exhibit greater parental protection for offspring than others.

8. ___ During short photoperiods, the pituitary gland releases small quantities of gonadotropins.

9. ___ Red grouse defend large, multipurpose territories where they forage, mate, nest, and rear young.

10. ___ Lampreys have an elongated cylindrical body without scales.

Matching

Select the one most appropriate answer to match the sources of genetic variation. [p.282]

11. ___fertilization

12. ___changes in chromosome structure or number

13. ___crossing over at meiosis

14. ___gene mutation

15. ___independent assortment at meiosis

A. Leads to mixes of paternal and maternal chromosomes in gametes
B. Produces new alleles
C. Leads to the loss, duplication, or alteration of alleles
D. Brings together combinations of alleles from two parents
E. Leads to new combinations of alleles in chromosomes

Short Answer

16. Briefly discuss the role of the environment and genetics in the production of an organism's phenotype. [p.282 and Section 10.8] _____

_____

_____

17. List the conditions (in any order) that must be met before genetic equilibrium (or nonevolution) will occur. [pp.282–284] _____

_____

_____

18. For the following situation, assume that the conditions listed in question 17 do exist; therefore, there should be no change in gene frequency, generation after generation. Consider a population of hamsters in which dominant gene B produces black coat color and recessive gene b produces gray coat color (two alleles are responsible for color). The dominant gene has a frequency of 80 percent (or .80). It would follow that the frequency of the recessive gene is 20 percent (or .20). From this, the assumption is made that 80 percent of all sperm and eggs have gene B. Also, 20 percent of all sperm and eggs carry gene b. [p.284 in the text]

a. Calculate the probabilities of all possible matings in the Punnett square.
b. Summarize the genotype and phenotype frequencies of the F_1 generation.

Genotypes	Phenotypes
____ BB	
____ Bb	____ % black
____ bb	____ % gray

	Sperm 0.80 B	0.20 b
Eggs 0.80 B	BB	Bb
0.20 b	Bb	bb

c. Further assume that the individuals of the F_1 generation produce another generation and the assumptions of the Hardy–Weinberg rule still hold. What are the frequencies of the sperm produced? [p.284]

Parents (F_1)	B sperm	b sperm
___ BB	___	___
___ Bb	___	___
___ bb	___	___
Totals =	___	___

The egg frequencies may be similarly calculated. Note that the gamete frequencies of the F_2 are the same as the gamete frequencies of the last generation. Phenotype percentage also remains the same. Thus, the gene frequencies did not change between the F_1 and the F_2 generation. Again, given the assumptions of the Hardy–Weinberg equilibrium, gene frequencies do not change generation after generation.

19. In a population, 81 percent of the organisms are homozygous dominant, and 1 percent are homozygous recessive. Find the following. [p.284]

   a. the percentage of heterozygotes _____

   b. the frequency of the dominant allele _____

   c. the frequency of the recessive allele _____

20. In a population of 200 individuals, determine the following for a particular locus if $p = 0.80$. [p.284]

   a. the number of homozygous dominant individuals _____

   b. the number of homozygous recessive individuals _____

   c. the number of heterozygous individuals _____

21. If the percentage of gene D is 70 percent in a gene pool, find the percentage of gene d. [p.284]

_____

_____

22. If the frequency of gene R in a population is 0.60, what percentage of the individuals are heterozygous Rr? [p.284]

_____

_____

Matching

Select the most appropriate answer for each.

23. ___gene pool [p.282]

24. ___allele frequencies [p.282]

25. ___neutral mutations [p.283]

26. ___microevolution [p.283]

27. ___lethal mutation [p.283]

28. ___alleles [p.282]

29. ___mutation [p.283]

30. ___polymorphism [p.282]

31. ___genetic equilibrium [p.282]

32. ___natural selection [p.285]

A. A heritable change in DNA
B. Zero evolution
C. Different molecular forms of a gene
D. All of the alleles shared by members of a population and transmissable to the next generation
E. Change or stabilization of allele frequencies due to differences in survival and reproduction among variant members of a population
F. Two or more distinct forms in a population that are qualitatively different
G. Change in which expression of mutated gene always leads to the death of the individual
H. The abundance of each kind of allele in the entire population
I. Neither harmful nor helpful to the individual
J. Changes in allele frequencies brought about by mutation, genetic drift, gene flow, and natural selection

Fill-in-the-Blanks

A(n) (33) _____ [p.282] is a group of individuals of the same species that occupy a given area at a specific time. The (34) _____-_____ [p.284] rule allows researchers to establish a theoretical reference point (baseline) against which changes in allele frequency can be measured. Variation can be expressed in terms of (35) _____ _____ [p.282], which means the relative abundance of different alleles carried by the individuals in that population. The stability of allele ratios that would occur if all individuals had equal probability of surviving and reproducing is called (36) _____ _____ [pp.282–283]. Over time, allele frequencies tend to change through infrequent but inevitable (37) _____ [p.283], which are the original source of genetic variation.

18.3. NATURAL SELECTION REVISITED [p.285]

18.4. DIRECTIONAL CHANGE IN THE RANGE OF VARIATION [pp.286–287]

18.5. SELECTION AGAINST OR IN FAVOR OF EXTREME PHENOTYPES [pp.288–289]

18.6. SPECIAL TYPES OF SELECTION [pp.290–291]

18.7. GENE FLOW [p.291]

18.8. GENETIC DRIFT [pp.292–293]

Selected Words: *Biston betularia* [p.286], *mark-release-recapture* [p.286], *pest resurgence* [p.287], *biological controls* [p.287], antibiotics [p.287], gallmaking fly [p.288], parasitic wasp [p.288], downy woodpecker [p.288], black-bellied seedcracker [p.289], sexual dimorphism [p.290], *sickle-cell anemia* [p.290], *malaria* [p.290], *Plasmodium* [p.290], *emigration* [p.291], *immigration* [p.291], elephant seals [p.293], *Ellis–van Creveld syndrome* [p.293], *endangered species* [p.293], *feline infectious peritonitis* [p.293]

Boldfaced, Page-Referenced Terms

[p.285] fitness _____

_____

[p.285] natural selection _____

_____

[p.286] directional selection _____

_____

[p.288] stabilizing selection _____

_____

[p.289] disruptive selection _____

_____

[p.290] sexual selection _____

_____

[p.290] balanced polymorphism _____

_____

[p.291] gene flow _____

_____

[p.292] genetic drift _____

_____

[p.292] sampling error _____

_____

[p.292] fixation _____

_____

[p.292] bottleneck _____

_____

[p.293] founder effect _____

_____

[p.293] inbreeding _____

_____

Complete the Table

1. Complete the following table, which defines three types of natural selection effects.

Effect	Characteristics	Example
a. Stabilizing selection [p.288]		
b. Directional selection [p.286]		
c. Disruptive selection [p.289]		

Labeling

2. Identify the three curves below as stabilizing selection, directional selection, or disruptive selection.

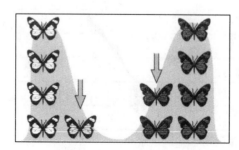

a. _____   b. _____   c. _____

Fill-in-the-Blanks

When individuals of different phenotypes in a population differ in their ability to survive and reproduce, their alleles are subject to (3) _____ [p.285] selection. (4) _____ [p.288] selection favors the intermediate forms of a trait in the population. (5) _____ [p.286] selection occurs when a specific change in the environment causes a heritable trait to occur with increasing frequency and the whole population tends to shift in a parallel direction. (6) _____ [p.289] selection favors the development of two or more distinct polymorphic varieties such that they become increasingly represented in a population and the population splits into different phenotypic variations. (7) _____ _____ [p.288] provide an excellent example of stabilizing selection, because extreme phenotypes of the gallfly are eliminated by birds and wasps. (8) _____-_____ _____ [p.290], a genetic disorder that produces an abnormal form of hemoglobin, is an example of (9) _____ [p.290] selection that maintains a high frequency of the harmful (10) _____ [p.291] along with the normal gene. Such a genetic juggling act is called a (11) _____ _____ [p.290]. Differences in appearance between males and females of a species are known as (12) _____ _____ [p.290]. Among birds and mammals, females act as agents of (13) _____ [p.290] when they choose their mates. (14) _____ [p.290] selection is based on any trait that gives the individual a competitive edge in mating and producing offspring. The more colorful, showier, and larger appearance of male pheasants when compared with females is the result of (15) _____ [p.290] selection. Any alleles in an individual that help it obtain resources, survive and reproduce increase its (16) _____ [p.290] and tend to be in phenotypes that become more numerous in the population as it adapts to its environment.

Choice

For questions 17–30, choose from the following categories of natural selection; in some cases, two letters may be correct.

> a. directional selection   b. stabilizing selection   c. disruptive selection
> d. balanced polymorphism   e. sexual selection

17. ___ May result even when heterozygotes are selected against; bill size in the African finches population provides an example [pp.289–290]

18. ___ Sexual dimorphism [p.290]

19. ___ Phenotypic forms at both ends of the variation range are favored and intermediate forms are selected against [p.289]

20. ___ May account for the persistence of certain phenotypes over time; horsetail plants still bear strong resemblance to their ancient relatives [p.288]

21. ___ Selection maintains two or more alleles for the same trait in steady fashion, generation after generation [p.290]

22. ___ The most common forms of a trait in a population are favored; over time, alleles for uncommon forms are eliminated [p.288]

23. ___ Allele frequencies shift in a steady, consistent direction; this may be due to a change in the environment [p.286]

24. ___ Often results when environmental conditions favor homozygotes over heterozygotes [pp.289–290]

25. ___ Counters the effects of mutation, genetic drift, and gene flow [p.288]

26. ___ Human newborns weighing an average of 7 pounds are favored [p.288]

27. ___ In West Africa, finches known as black-bellied seedcrackers have either large or small bills [p.289]

28. ___ When a trait gives an individual an advantage in reproductive success [p.290]

29. ___ The most frequent wing color of peppered moths shifted from a light form to a dark form as tree trunks became soot-darkened because coal was used for fuel during the industrial revolution [p.286]

30. ___ Females of a species choosing mates to directly affect reproductive success [p.290]

Choice

For questions 31–40, choose from the following microevolutionary forces that can change gene frequency:

a. genetic drift   b. gene flow

31. ___ A random change in allele frequencies over the generations, brought about by chance alone [p.292]

32. ___ Emigration [p.291]

33. ___ Prior to the turn of the century hunters killed all but twenty of a large population of northern elephant seals [p.293]

34. ___ When allele frequencies change due to individuals leaving a population or new individuals enter it [p.291]

35. ___ Long ago, seabirds, winds, or ocean currents carried a few seeds from the Pacific Northwest to the Hawaiian Islands to establish plant populations [p.293]

36. ___ Sometimes result in endangered species [p.293]

37. ___ Founder effects and bottlenecks are two extreme cases [pp.292–293]

38. ___ Immigration [p.291]

39. ___ Only 20,000 cheetahs have survived to the present [p.293]

40. ___ Blue jays make hundreds of round trips carrying acorns from oak trees as much as a mile away to soil in their home territories for winter storage; this introduces new alleles to oak forests [p.291]

Short Answer

41. Distinguish between the two extreme cases of genetic drift: the founder effect and bottlenecks. [pp.292–293]

_____

_____

Fill-in-the-Blanks

Random fluctuations in allele frequencies over time due to chance occurrence alone are called (42) _____

_____; [p.292] it is more pronounced in small populations than in large ones. (43) _____ [p.291]

flow associated with immigration and/or emigration also changes allele frequencies. (44) _____

_____ [p.285] is the differential survival and reproduction of individuals of a population that differ in

one or more traits. (45) _____ _____ [p.285] is the most important microevolutionary process.

Self-Quiz

For questions 1–3, choose from these answers:
- a. gene pool
- b. genetic variation
- c. gene flow
- d. allele frequency

___ 1. Differences in the combinations of alleles carried by the individuals of a population would be its _____. [p.282]

___ 2. For sexually reproducing species, the _____ is a source of potentially enormous variation in traits. [p.282]

___ 3. The relative abundance of each type of allele in a population is the _____. [p.282]

___ 4. p plus q = 1 expresses the _____ of a population; p^2 plus $2pq$ plus q^2 expresses the _____ of a population. [p.284]
- a. genotype frequency; allele frequency
- b. genotype frequency; genotype frequency
- c. allele frequency; genotype frequency
- d. allele frequency; allele frequency

___ 5. The unit used in studying evolution is the _____. [p.282]
- a. population
- b. individual
- c. fossil
- d. missing link

___ 6. Changes in allele frequencies brought about by mutation, genetic drift, gene flow, and natural selection are called _____ processes. [p.283]
- a. genetic equilibrium
- b. microevolution
- c. founder effects
- d. independent assortment

For questions 7–9, choose from these answers:
- a. disruptive selection
- b. genetic drift
- c. directional selection
- d. stabilizing selection

___ 7. Cockroaches becoming increasingly resistant to pesticides is an example of _____. [p.286]

___ 8. The founder effect is a special case of _____. [pp.292–293]

___ 9. Different alleles that code for different forms of hemoglobin have led to _____ in the middle of the malaria belt that extends through West and Central Africa. [p.290]

For questions 10–12, choose from these answers:
- a. balanced polymorphism
- b. sexual dimorphism
- c. genetic equilibrium
- d. sexual selection

___10. Hardy and Weinberg invented an ideal population that was in _____ and used it as a baseline against which to measure evolution in real populations. [p.282]

___11. The term _____ is applied when a population has nonidentical alleles for a trait that are maintained at frequencies greater than 1 percent. [p.290]

___12. Phenotypic differences between the males and females of a species are known as _____. [p.290]

Chapter Objectives/Review Questions

1. A _____ is a group of individuals occupying a given area and belonging to the same species. [p.282]
2. Distinguish among morphological, physiological, and behavioral traits. [p.282]
3. All the genes of an entire population belong to a _____ _____. [p.282]
4. Each kind of gene usually exists in one or more molecular forms, called _____. [p.282]
5. Review the five categories through which genetic variation occurs among individuals. [p.282]
6. The abundance of each kind of allele in the whole population is referred to as *allele* _____. [p.283]
7. A point at which allele frequencies for a trait remain stable through the generations is called genetic _____. [p.283]

8. List the five conditions that must be met for the Hardy–Weinberg rule to apply. [p.283]
9. Changes in allele frequencies brought about by mutation, genetic drift, gene flow, and natural selection are called _____. [p.283]
10. A _____ is a random, heritable change in DNA. [p.283]
11. Distinguish a lethal mutation from a neutral mutation. [p.283]
12. Calculate allele and other genotype frequencies when provided with the homozygous recessive genotype frequency. (For example, assume $q^2 = .36$.) [pp.284–285]
13. _____ _____ is defined as the change or stabilization of allele frequencies due to differences in survival and reproduction among variant members of a population. [p.285]
14. Define and provide an example of directional selection, stabilizing selection, and disruptive selection. [pp.286–289]
15. Define balanced polymorphism. [p.290]
16. The occurrence of phenotypic differences between males and females of a species is called _____ _____. [p.290]
17. _____ selection is based on any trait that gives an individual a competitive edge in mating and producing offspring. [p.290]
18. Allele frequencies change as individuals leave or enter a population; this is gene _____. [p.291]
19. Random fluctuations in allele frequencies over time, due to chance, is called _____ _____. [p.292]
20. Distinguish the founder effect from a bottleneck. [pp.292–293]
21. Relate bottlenecks to endangered species. [p.293]

Integrating and Applying Key Concepts

Can you imagine any way in which directional selection may have occurred or may be occurring in humans? Which factors do you suppose are the driving forces that sustain the trend? Do you think the trend could be reversed? If so, by what factor(s)?

19

SPECIATION

Interactive Exercises

The Case of the Road-Killed Snails [pp.296–297]

19.1. ON THE ROAD TO SPECIATION [pp.298–299]

19.2. REPRODUCTIVE ISOLATING MECHANISMS [pp.300–301]

Selected Words: *Helix aspersa* [p.296], *prezygotic* isolation [p.300], *postzygotic* isolation [p.300], "sibling species" [p.300], ecological, temporal, behavioral, and mechanical isolation [p.300], gametic and zygotic mortality [p.300], hybrid inviability [p.300], hybrid infertility [p.300]

Boldfaced, Page-Referenced Terms

[p.297] speciation _____

_____

[p.298] species _____

[p.298] biological species concept _____

[p.299] gene flow_____

[p.299] genetic divergence _____

_____

[p.299] reproductive isolating mechanisms _____

_____

Choice

For questions 1–10, choose from the following:

a. Mayr's biological species   b. genetic divergence   c. isolating mechanisms

1. ___ Heritable aspect of body form, physiology, or behavior that prevents gene flow between genetically divergent populations [p.299]

2. ___ Groups of interbreeding natural populations [p.298]

3. ___ May occur as a result of stopping gene flow [p.299]

4. ___ Phenotypically very different but can interbreed and produce fertile offspring [p.299]

5. ___ Defined as the buildup of differences in the separated pools of alleles of populations [p.299]

6. ___ A "kind" of living thing [p.298]

7. ___ Factors that prevent horses and donkeys from mating in the wild [p.301]

8. ___ Prezygotic [pp.300–301]

9. ___ Not applicable to organisms that reproduce solely by asexual means or those known only from the fossil record [p.299]

10. ___ Postzygotic [p.301]

Matching

Select the most appropriate answer to match the isolating mechanisms; complete the exercise by writing "pre" in the parentheses if the mechanism is prezygotic and "post" if the mechanism is postzygotic. [all from p.300]

11. ___ecological (   )

12. ___temporal (   )

13. ___hybrid inviability (   )

14. ___mechanical (   )

15. ___zygote mortality (   )

16. ___gametic mortality (   )

17. ___behavioral (   )

18. ___hybrid offspring (   )

A. Potential mates occupy overlapping ranges but reproduce at different times.

B. The first-generation hybrid forms but shows very low fitness.

C. Potential mates occupy different local habitats within the same area.

D. Potential mates meet but cannot figure out what to do about it.

E. Sperm is transferred, but the egg is not fertilized (gametes die or gametes are incompatible).

F. The hybrid is sterile or partially so.

G. Potential mates attempt engagement, but sperm cannot be successfully transferred.

H. The egg is fertilized, but the zygote or embryo dies.

Choice

For questions 19–25, choose from the following isolating mechanisms:

a. temporal   b. behavioral   c. mechanical   d. gametic mortality
e. ecological   f. postzygotic

19. ___ Sterile mules [p.301]

20. ___ Two sage species, each has its flower petals arranged as a "landing platform" for a different pollinator [p.301]

21. ___ Two species of cicada, one matures, emerges, and reproduces every thirteen years, the other every seventeen years [p.300]

22. ___ Populations of the manzanita shrub *Arctostaphylos patula* and *A. viscida* demonstrate differing tolerances to water stress at varying distances from water sources; speciation may be underway [p.300]

23. ___ Pollen grains from one flowering plant species molecularly mismatched with gametes of a different flowering plant species [p.301]

24. ___ Before copulation, male and female birds engage in complex courtship rituals recognized only by birds of their own species [p.301]

Identification

25. The following illustration depicts the divergence of one species as time passes. Each horizontal line represents a population of that species and each vertical line, a point in time. Answer the questions below the illustration. [p.299]

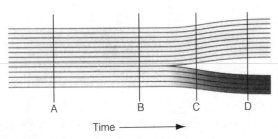

Time ⟶

a. How many species are represented at time A? _____

b. How many species exist at time D? _____

c. Between what times (letters) does divergence begin? _____

d. What letter represents the time that complete divergence is reached? _____

e. Between what two times (letters) is divergence clearly underway but not complete? _____

19.3. SPECIATION IN GEOGRAPHICALLY ISOLATED POPULATIONS [pp.302–303]

19.4. MODELS FOR OTHER SPECIATION ROUTES [pp.304–305]

Selected Words: Panamanian wrasses [p.302], "Pacific" enzymes [p.302], "Atlantic" enzymes [p.302], Hawaiian honeycreepers [p.303], cichlids [p.304], *Triticum aestivum* [p.305], *subspecies* [p.305], "secondary contact" [p.305]

Boldfaced, Page-Referenced Terms

[p.302] allopatric speciation _____

_____

[p.303] archipelago _____

_____

[p.304] sympatric speciation _____

_____

[p.304] polyploidy _____

_____

[p.304] dosage compensation _____

_____

[p.305] parapatric speciation _____

_____

[p.305] hybrid zone _____

_____

Complete the Table

1. Complete the following table, which defines three speciation models.

Speciation Model	Description
a. [p.304]	Suggests that species can form within the range of an existing species, in the absence of physical or ecological barriers; a controversial model; evidence comes from crater lake fishes
b. [p.302]	Emphasizes disruption of gene flow; perhaps the main speciation route; through divergence and evolution of isolating mechanisms, separated populations become reproductively incompatible and cannot interbreed
c. [p.305]	Occurs where populations share a common border; the borders may be permeable to gene flow; in some, gene exchange is confined to a common border, the hybrid zone

Dichotomous Choice

Circle one of two possible answers given between parentheses in each statement.

2. Instant speciation by polyploidy in many flowering plant species represents (parapatric/sympatric) speciation. [p.304]
3. In the past, hybrid orioles existed in a hybrid zone between eastern Baltimore orioles and western Bullock orioles; today, hybrid orioles are becoming less frequent in this example of (allopatric/parapatric) speciation. [p.305]
4. In the absence of gene flow between geographically separate populations, genetic divergence may become sufficient to bring about (allopatric/parapatric) speciation. [p.302]
5. The uplifted ridge of land we now call the Isthmus of Panama divides the Atlantic Ocean from the Pacific Ocean and forms a natural laboratory for biologists to study (allopatric/sympatric) speciation. [p.302]

6. In the 1800s, a major earthquake changed the course of the Mississippi River; this isolated some populations of insects that could not swim or fly; and set the stage for (parapatric/allopatric) speciation. [p.302]

Short Answer

7. Study the following illustration, which shows the possible course of wheat evolution; similar genomes (a genome is a complete set of chromosomes) are designated by the same letter. Answer the questions after the diagram. [All from p.305]

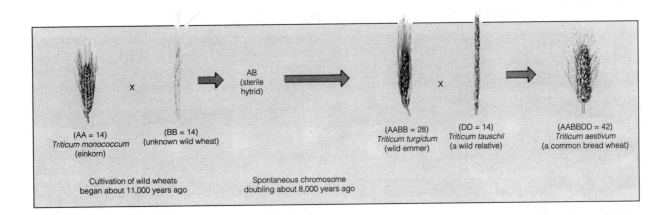

(AA = 14) *Triticum monococcum* (einkorn)	(BB = 14) (unknown wild wheat)	AB (sterile hytrid)	(AABB = 28) *Triticum turgidum* (wild emmer)

(DD = 14) *Triticum tauschii* (a wild relative)

(AABBDD = 42) *Triticum aestivum* (a common bread wheat)

Cultivation of wild wheats began about 11,000 years ago

Spontaneous chromosome doubling about 8,000 years ago

a. How many pairs of chromosomes are found in *Triticum monococcum*? _____

b. How are the *T. monococcum* chromosomes designated? _____

c. How many pairs of chromosomes are found in the unknown species of wheat? _____

d. How are the chromosomes in 7C designated? _____

e. Why is the hybrid designated AB sterile? _____

f. What cellular event must have occurred to create the *T. turgidum* genome? _____

g. Describe the genome of the plant arising from the cross of *T. turgidum* and *T. tauschii*. _____

h. What cellular event must have occurred to create the *T. aestivum* genome? _____

i. What is the source of the A genome in *T. aestivum*? the source of the B genome in *T. aestivum*? the source of the D genome in *T. aestivum*? _____

19.5. PATTERNS OF SPECIATION [pp.306–307]

Selected Words: pattern of evolutionary change [p.306], *branch* [p.306], *branch point* [p.306], *physical access* [p.307], *evolutionary access* [p.307], *ecological access* [p.307]

Boldfaced, Page-Referenced Terms

[p.306] cladogenesis _____

_____

[p.306] anagenesis _____

_____

[p.306] evolutionary tree _____

_____

[p.306] gradual model of speciation _____

_____

[p.306] punctuation model of speciation _____

_____

[p.306] adaptive radiation _____

_____

[p.306] adaptive zone _____

_____

[p.307] key innovation _____

_____

[p.307] extinction _____

_____

[p.307] mass extinction _____

_____

Complete the Table

1. Complete the following table to summarize information to interpret evolutionary tree diagrams. Refer to the text. [p.306]

Branch Form	Interpretation
a.	Traits changed rapidly around the time of speciation
b.	An adaptive radiation occurred
c.	Extinction
d.	Speciation occurred through gradual changes in traits over geologic time
e.	Traits of the new species did not change much thereafter
f.	Evidence of this presumed evolutionary relationship is only sketchy

Matching

Select the most appropriate answer for each.

2. ___adaptive radiation [p.306]

3. ___adaptive zones [pp.306–307]

4. ___anagenesis [p.306]

5. ___cladogenesis [p.306]

6. ___evolutionary tree diagrams [p.306]

7. ___gradual model of speciation [p.306]

8. ___key innovation [p.307]

9. ___mass extinction [p.307]

10. ___physical, evolutionary, and ecological access [p.307]

11. ___punctuation model of speciation [p.306]

A. Modification of some structure or function that permits a lineage to exploit the environment in new or more efficient ways

B. Conditions allowing a lineage to radiate into an adaptive zone

C. Speciation in which most morphological changes are compressed into a brief period of hundreds or thousands of years when populations first start to diverge; evolutionary tree branches make abrupt 90-degree turns

D. Ways of life such as "burrowing in the seafloor"

E. Speciation with slight angles on the evolutionary tree, which conveys many small changes in form over long time spans

F. An abrupt rise in the rates of species disappearance above background level

G. Speciation occurring within a single, unbranched line of descent

H. A method of summarizing information about the continuities of relationships among species

I. A branching speciation pattern; isolated populations diverge

J. A burst of microevolutionary activity within a lineage; results in the formation of new species in a wide range of habitats

Fill-in-the-Blanks

(12) _____ [p.297] is the process whereby species are formed. A (13) _____ [p.298] is composed of one or more populations of individuals who can interbreed under natural conditions and produce fertile, reproductively isolated offspring. (14) _____ [p.299] can be described as a process in which differences in alleles accumulate between populations. Any aspect of structure, function, or behavior that prevents interbreeding is a (15) _____ _____ [p.299] mechanism. Physical barriers that prevent gene flow, as in the case of large rivers changing course or forests giving way to grasslands, are (16) _____ [p.302] barriers. Wheat is an example of a species formed by (17) _____ [p.305] and (18) _____ [p.305].

Self-Quiz

___ 1. Of the following, _____ is (are) not a major consideration of the biological species concept. [p.299]
a. groups of populations
b. reproductive isolation
c. asexual reproduction
d. interbreeding natural populations
e. production of fertile offspring

___ 2. When something prevents gene flow between two populations or subpopulations, _____ may occur. [p.299]
a. genetic drift, natural selection, and mutation
b. genetic divergence
c. a buildup of differences in the separated gene pools of alleles
d. all of the above

____ 3. When potential mates occupy different local habitats within the same area, it is termed _____ isolation. [p.300]
   a. temporal
   b. behavioral
   c. mechanical
   d. ecological

____ 4. "Potential mates occupy overlapping ranges but reproduce at different times" is a description of _____ isolation. [p.300]
   a. temporal
   b. behavioral
   c. mechanical
   d. ecological

____ 5. Of the following, _____ is a postzygotic isolating mechanism. [p.300]
   a. behavioral isolation
   b. gametic mortality
   c. hybrid inviability
   d. mechanical isolation

For questions 6–8, choose from the following answers:
   a. parapatric speciation
   b. sympatric speciation
   c. allopatric speciation

____ 6. When the Isthmus of Panama was formed, it provided contemporary scientists with an ideal natural laboratory to study _____. [p.302]

____ 7. Polyploidy and cichlid fishes of crater lakes provide evidence for _____. [p.304]

____ 8. The identification of a hybrid zone between the ranges of eastern Baltimore orioles and western Bullock orioles is an example of _____. [p.305]

____ 9. The branching speciation pattern revealed by the fossil record is called _____. [p.306]
   a. anagenesis
   b. the gradual model
   c. cladogenesis
   d. the punctuation model

____ 10. Species formation by many small changes over long time spans fits the _____ model of speciation; rapid speciation with morphological changes compressed into a brief period of population divergence describes the _____ model of speciation. [p.306]
   a. anagenesis; punctuation
   b. punctuation; gradual
   c. gradual; cladogenesis
   d. gradual; punctuation

____ 11. Adaptive radiation is _____. [pp.306–307]
   a. a burst of microevolutionary activity within a lineage
   b. the formation of new species in a wide range of habitats
   c. the spreading of lineages into unfilled adaptive zones
   d. a lineage radiating into an adaptive zone when it has physical, ecological, or evolutionary access to it
   e. all of the above

____ 12. A catastrophe during which major groups disappear is called a _____. [p.307]
   a. mass extinction
   b. reverse background radiation
   c. background extinction
   d. speciation extinction

For questions 13–14, choose from the following answers:
   a. stabilizing selection
   b. divergence
   c. a reproductive isolating mechanism
   d. polyploidy

____ 13. Two species of sage plants with differently shaped floral parts that prevent pollination by the same pollinator serve as an example of _____. [p.301]

____ 14. _____ occurs when isolated populations accumulate allele frequency differences between them over time. [p.299]

Chapter Objectives/Review Questions

1. The process by which species are formed is known as _____. [p.297]
2. State the biological species concept as phrased by Ernst Mayr. [pp.298–299]
3. _____ _____ is a buildup of differences in separated pools of alleles. [p.299]
4. The evolution of reproductive _____ _____ paves the way for genetic divergence and speciation. [pp.299–301]
5. Briefly define the major categories of prezygotic and postzygotic isolating mechanisms (see text, Table 19.1). [p.300]
6. _____ speciation occurs when daughter species form gradually by divergence in the absence of gene flow between geographically separate populations. [p.302]
7. In _____ speciation, daughter species arise, sometimes rapidly, from a small proportion of individuals within an existing population. [p.304]
8. Explain why sympatric speciation by polyploidy is a rapid method of speciation. [pp.304–305]
9. When daughter species form from a small proportion of individuals along a common border between two populations, it is called _____ speciation. [p.305]
10. In some cases of parapatric speciation, gene exchange between two species is confined to a _____ zone. [p.305]
11. Distinguish cladogenesis from anagenesis. [p.306]
12. Explain the use of evolutionary tree diagrams and the symbolism used. [p.306]
13. Explain why branches on an evolutionary tree that have slight angles indicate the _____ model of speciation. [p.306]
14. Branches on an evolutionary tree that turn abruptly with 90-degree turns are consistent with the _____ model of speciation. [p.306]
15. An adaptive radiation is a burst of _____ activity within a lineage that results in the formation of new species in a variety of habitats. [p.307]
16. Cite examples of adaptive zones. [p.307]
17. How does a key innovation differ from mass extinction? [p.307]

Integrating and Applying Key Concepts

Systematics is defined as the practice of describing, naming, and classifying living things; this includes the comparative study of organisms and all relationships among them. Plant systematists who work with flowering plants readily accept Ernst Mayr's definition of a "biological species" as described in this chapter. However, in real practice systematists often must also work with a concept known as the "morphological species." For example, the statement is sometimes made that "two plant specimens belong to the same morphological species but not to the same biological species." Explain the meaning of that statement. What type of experimental evidence would be necessary as a basis for this statement? From your study of this chapter, can you suggest a reason that application of the term "morphological species" is sometimes necessary? What difficulties and inaccuracies, in terms of identifying a species, might the use of both species concepts present?

20

THE MACROEVOLUTIONARY PUZZLE

He Sees Seashells . . . So Where's the Sea?

FOSSILS—EVIDENCE OF ANCIENT LIFE
    Fossilization
    Interpreting the Geologic Tombs
    Interpreting the Fossil Record

EVIDENCE FROM COMPARATIVE MORPHOLOGY
    Morphological Divergence and Homologous
        Structures
    Potential Confusion From Analogous Structures

EVIDENCE FROM PATTERNS OF DEVELOPMENT
    Developmental Program of Larkspurs
    Developmental Program of Vertebrates

EVIDENCE FROM COMPARATIVE BIOCHEMISTRY
    Protein Comparisons
    Nucleic Acid Comparisons
    Molecular Clocks

IDENTIFYING SPECIES, PAST AND PRESENT
    Assigning Names to Species
    Groupings of Species—The Higher Taxa
    Putting Classification Schemes to Work

FINDING EVOLUTIONARY RELATIONSHIPS
    AMONG SPECIES
    The Systematics Approach
    Actually, There Is No Single Systematics
    Approach

Focus on Science: CONSTRUCTING A CLADOGRAM

HOW MANY KINGDOMS?
    A Five-Kingdom Scheme
    Along Came the Archaebacteria

Focus on Science: ALONG CAME DEEP GREEN

EVIDENCE OF A CHANGING EARTH
    An Outrageous Hypothesis
    Drifting Continents, Changing Seas

Focus on Science: DATING PIECES OF THE
MACROEVOLUTIONARY PUZZLE

Interactive Exercises

He Sees Seashells . . . So Where's the Sea? [pp.310–311]

20.1. FOSSILS—EVIDENCE OF ANCIENT LIFE [pp.312–313]

20.2. EVIDENCE FROM COMPARATIVE MORPHOLOGY [pp.314–315]

20.3. EVIDENCE FROM PATTERNS OF DEVELOPMENT [pp.316–317]

20.4. EVIDENCE FROM COMPARATIVE BIOCHEMISTRY [pp.318–319]

Selected Words: species [p.311], *trace* fossils [p.312], *Cooksonia* [p.312], Cenozoic, Mesozoic, Paleozoic, Proterozoic, Archean [all p.313], sedimentary layers [p.313], Carboniferous [p.313], Permian [p.313], Jurassic [p.313], Cretaceous [p.313], "stem" reptile [p.314], key innovation [p.314], *Delphinium* (larkspur) [p.316], vertebrate [p.316], cytochrome *c* [p.318], "hybrid" molecule [p.318], giant panda [p.319]

Boldfaced, Page-Referenced Terms

[p.311] lineages _it connects all species, past and present_

[p.311] macroevolution _of large scale patterns, trends, and rates of change among families and other more inclusive groups of species._

[p.312] fossils _stone hard evidence of ancient life._

[p.312] fossilization _a slow process that starts when an organism becomes burried and it's remains become infused with dissolved metal ions and other organic componds._

[p.313] stratification _the layering of sedimentary deposits._

[p.313] geologic time scale _the chronological chart of Earth's history that dates fossils_

[p.314] comparative morphology _comparing body forms and structures of major lineages._

[p.314] morphological divergence _the change from the body form of a common ancestor_

[p.314] homology _a similarity in one or more body parts in different organisms that can be attributed to decent from a common ancestor._

[p.315] morphological convergence _The evolution of similar patterns, even though there is no common past lineages._

[p.315] analogy _body parts of different animals that seem similar but evolved in different ways (fins, flippers)_

[p.318] neutral mutations _slight structural differences in highly conserved genes of different lineages_

[p.318] DNA–DNA hybridization

[p.319] molecular clock _the accumulation of neutral mutations in highly conserved genes accumulated at a regular rate._

Matching

Select the most appropriate answer.

1. ___fossilization [p.312]
2. ___stratification [p.313]
3. ___trace fossils [p.312]
4. ___continuity of relationship [p.311]
5. ___macroevolution [p.311]
6. ___fossils [p.312]
7. ___biased fossil record [p.313]
8. ___formation of sedimentary deposits [p.313]
9. ___oldest fossils [p.313]
10. ___rupture or tilt of sedimentary layers [p.313]

A. Evolutionary connectedness between organisms, past to present
B. Found in the deepest rock layers
C. Affected by movements of the Earth's crust, hardness of organisms, population sizes, and certain environments
D. Produced by gradual deposition of silt, volcanic ash, and other materials
E. Refers to large-scale patterns, trends, and rates of change among groups of species
F. A process favored by rapid burial in the absence of oxygen
G. Evidence of later geologic disturbance
H. Recognizable, physical evidence of ancient life; the most common are bones, teeth, shells, spore capsules, seeds, and other hard parts
I. Includes the indirect evidence of coprolites and imprints of leaves, stems, tracks, trails, and burrows
J. A layering of sedimentary deposits

Sequence

Earth history has been divided into four great eras that are based on four abrupt transitions in the fossil record. The oldest era has been subdivided. Arrange the eras in correct chronological sequence from the oldest to the youngest. [p.313]

11. _4_   A. Mesozoic
12. _5_   B. Cenozoic
13. _2_   C. Proterozoic
14. _1_   D. Archean
15. _3_   E. Paleozoic

Matching

Choose the one most appropriate answer for each.

16. ___Archean [p.313]
17. ___Mesozoic [p.313]
18. ___Proterozoic [p.313]
19. ___Paleozoic [p.313]
20. ___Cenozoic [p.313]
21. ___molecular clock [p.318]
22. ___geologic time scale [p.313]

A. The "modern era" of geologic time
B. Source of the most ancient fossils prior to subdivision of this era
C. The boundaries mark the times of mass extinctions
D. The first era of the geologic time scale; a recent subdivision of the Proterozoic
E. An era that follows the Paleozoic
F. An era whose fossils followed those of the Proterozoic
G. A method used to estimate the timing of divergences

Complete the Table

In the following geologic time scale table, fill in the missing information. [p.313]

	Period	Epoch	Millions of Years Ago (mya)
(23) ERA	QUATERNARY	Recent	0.01–
		Pleistocene	1.55
	TERTIARY	Pliocene	5
		Miocene	25
		Oligocene	38
		Eocene	54
		Paleocene	**(32)**
(24) ERA	**(28)**	Late	100
		Early	138
	(29)		205
	TRIASSIC		**(33)**
(25) ERA	**(30)**		290
	CARBONIFEROUS		360
	DEVONIAN		410
	SILURIAN		435
	ORDOVICIAN		505
	(31)		**(34)**
(26) EON			**(35)**
(27) EON			**(36)**

True–False

If the statement is true, write a T in the blank. If the statement is false, correct it by changing the underlined word(s) and writing the correct word(s) in the answer blank.

_____ 37. The extent to which a strand of DNA or RNA isolated from individuals of one species will base-pair with a comparable strand from individuals of a different species is a rough measure of the evolutionary <u>distance</u> between them. [p.318]

_____ 38. The amino acid sequence of cytochrome *c* is very similar in such distantly related organisms as (1) yeast, (2) wheat, and (3) humans; the probability that this striking resemblance resulted from chance alone is very <u>high</u>. [p.318]

_____ 39. Analogous structures in two different lineages indicate strong evidence of morphological <u>divergence</u>. [p.315]

_____ 40. The same body parts that became modified in different ways in different lineages yet also develop from similar tissues in their embryos are homologous structures that provide very strong evidence of morphological <u>divergence</u>. [pp.314–315]

_____ 41. Biochemical <u>similarities</u> are fewest among the most closely related species and most numerous among the most distantly related. [p.319]

_____ 42. All species that ever evolved are related to one another, by way of <u>descent</u>. [p.311]

Choice

For questions 43–46, choose from the following:

                a. analogy   b. homology   c. an example of a lineage
d. a specific example of morphological convergence   e. a specific example of morphological divergence

43. ___ Ancestral ape-human, *Homo erectus*, and modern humans (*Homo sapiens*). [p.311]

44. ___ Body parts that were once quite different among distantly related lineages but are now quite similar in function and in apparent structure. [p.315]

45. ___ The human arm and hand, the porpoise's front flipper, and the bird's wing. [p.314]

46. ___ Body parts of different organisms that develop from similar embryonic parts yet might result in quite different-appearing adult structures that would probably not have similar functions. [p.314]

Labeling

Evidence for macroevolution comes from comparative morphology, patterns of development and from comparative biochemistry. For each of the following listed items, place an "M" in the blank if the evidence comes from comparative morphology, a "B" if the evidence comes from comparative biochemistry, and a "D" if the evidence is from patterns of development.

47. ___ Early in the vertebrate developmental program, embryos of different lineages proceed through strikingly similar stages. [p.316]

48. ___ Comparison of the developmental rates of changes in a chimpanzee skull and a human skull. [p.317]

49. ___ The evolution of different rates of flower development in two different larkspur species also led to the coevolution of two different pollinators, honeybees and hummingbirds. [p.316]

50. ___ Using accumulated neutral mutations to date the divergence of two species from a common ancestor. [pp.318–319]

51. ___ The amino acid sequence in the cytochrome *c* of humans precisely matches the sequence in chimpanzees. [p.318]

52. ___ Establishing a rough measure of evolutionary distance between two organisms by use of DNA hybridization studies that demonstrate the extent to which the DNA from one species base-pairs with another. [p.318]

53. ___ Regulatory genes control the rate of growth of different body parts and can produce large differences in the development of two very similar embryos. [p.317]

54. ___ Results from nucleic acid hybridization studies supported an earlier view that giant pandas, bears, and red pandas are related by common descent. [p.319]

55. ___ Distantly related vertebrates, such as sharks, penguins, and porpoises, show a similarity to one another in their proportion, position, and function of body parts. [p.314]

56. ___ Finding similarities in vertebrate forelimbs when comparing the wings of pterosaurs, birds, bats, and porpoise flippers. [p.314]

20.5. IDENTIFYING SPECIES, PAST AND PRESENT [pp.320–321]

20.6. FINDING EVOLUTIONARY RELATIONSHIPS AMONG SPECIES [pp.322–323]

20.7. *Focus on Science:* CONSTRUCTING A CLADOGRAM [pp.324–325]

20.8. HOW MANY KINGDOMS? [pp.326–327]

20.9. *Focus on Science:* ALONG CAME DEEP GREEN [p.327]

Selected Words: Linnaeus [p.320], *classical taxonomy* [p.322], evolutionary tree diagram [p.322], *cladistic taxonomy* [p.322], ingroup [p.324], outgroup [p.324], monophyletic group [p.324], ancestor [p.325], Whittaker's five-kingdom classification scheme [p.326], Woese [p.326], Bult [p.326], *Methanococcus jannaschii* [p.326], *three-domain scheme* [p.326], *species* [p.327], Deep Green [p.327]

Boldfaced, Page-Referenced Terms

[p.320] taxonomy _identifying, naming, and classifying species_

[p.320] binomial system _the way species are assigned a two part Latin name_

[p.320] genus, genera (pl.) _the first part of a species name. more generic_

[p.320] specific name _the second part of a species name. In combination with the genus identifies the organism_

[p.321] classification schemes _organized ways of retrieving information about particular species._

[p.321] higher taxa, taxon (sing.) _more inclusive groupings that are meant to reflect relationships among species_

[p.321] phylogeny _the evolutionary history of organisms, extinct or living_

[p.322] evolutionary systematics _the branch of biology that applies evolutionary theory to the task of identifying patterns of diversity over time and in the environment._

[p.322] derived trait _a novel feature that evolved only once and is shared only among descendants of the ancestral species in which it evolved._

[p.325] cladograms _evolutionary tree diagrams based on a cladistic approach_

[p.326] Monera _____

[p.326] Protista _____

[p.326] Fungi _____

[p.326] Plantae _____

[p.326] Animalia _____

[p.326] Archaebacteria _____

_____

[p.326] Eubacteria _____

_____

[p.327] six-kingdom classification scheme _____

_____

Matching

Choose the single most appropriate letter.

1. ___black bear [p.321]

2. ___phylogeny [p.321]

3. ___binomial system [p.320]

4. ___*Pinus strobus* and *Pinus banksiana* [p.320]

5. ___species [p.320]

6. ___genus [p.320]

7. ___family [p.321]

8. ___taxa [p.321]

9. ___*Ursus americanus* (black bear) and *Homarus americanus* (Atlantic lobster) [p.321]

10. ___Linnaeus [p.320]

A. A group of similar species
B. The taxon most often studied
C. Evolutionary relationships among species, reflected in classification schemes
D. The organizing units of classification schemes
E. Classification level that includes similar genera
F. Originated the modern practice of providing two scientific names for organisms
G. An example of a common name
H. Two species belonging to one genus
I. An example of two very different organisms having the same specific epithet (species name)
J. System of assigning a two-part Latin name to a species

Short Answer

11. Arrange the following jumbled taxa in proper order, with the most inclusive first: family, class, species, kingdom, genus, phylum (or division), and order. [p.320] _____

_____

_____

_____

12. Arrange the following jumbled taxa and their categories in proper order, the most inclusive first: genus: *Archibaccharis*; kingdom: Plantae; order: Asterales; species: *lineariloba*; class: Dicotyledonae; division: Anthophyta; family: Asteraceae. [p.320]_____

_____

_____

_____

13. List three methods employed by systematics to elucidate the patterns of diversity in an evolutionary context. [p.322]_____

_____

_____

Complete the Table

14. Complete the following table to define two different approaches to the problem of reconstructing the history of life.

School of Systematics	Definition
a. Evolutionary systematics [p.322]	
b. Cladistics [p.322]	

Short Answer

15. What are monophyletic lineages? [p.324] _____

_____

16. Describe the construction of a cladogram. [p.324] _____

_____

17. What is an outgroup? [p.324] _____

_____

18. What is meant by a derived trait? [p.322] _____

_____

19. What is the essential information portrayed by a cladogram? [p.323] _____

_____

Cladogram Interpretation

20. Study the following cladogram of seven taxa. The vertical bars on the stem of the cladogram represent shared derived traits. Various taxa are indicated by letters. Answer the questions following the cladogram.

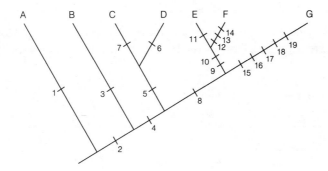

a. On the cladogram, what types of information could be represented by the numbered traits? [p.324]

_____

_____

b. Which derived trait is shared by taxa EFG? [p.325] _____

c. Which derived trait is shared by taxa CDEFG? [p.325] _____

d. Which is the unique derived trait shared by taxa C and D? [p.324] _____

_____

e. What does it mean if some taxa are closer together on the cladogram than others? [p.325]

_____

f. Which taxon on the cladogram represents the outgroup condition? [p.324] _____

_____

g. What is a monophyletic taxon? [p.324]_____

_____

h. For this cladogram, list the monophyletic taxa shown as well as shared derived traits. [p.324]

_____

Matching

Choose the one best answer for each. [All from p.326]

21. ___Monera   A. Multicelled heterotrophs that feed by extracellular digestion and absorption
22. ___Protista   B. Single-celled prokaryotes, some autotrophs, others heterotrophs
                   C. Diverse multicelled heterotrophs, including predators and parasites
23. ___Fungi   D. Multicelled photosynthetic autotrophs
24. ___Plantae   E. Diverse single-celled eukaryotes, some photosynthetic autotrophs, many
                      heterotrophs
25. ___Animalia

20.10. EVIDENCE OF A CHANGING EARTH [pp.328–329]

20.11. *Focus on Science:* DATING PIECES OF THE MACROEVOLUTIONARY PUZZLE
   [p.330]

Selected Words: sedimentary layers [p.328], continental drift [p.328], mantle [p.328], seafloor spreading [p.328], crust [p.328], *Glossopteris* [p.328], Gondwana [p.328], upwelling plumes [p.329], *superplumes* [p.329], "hot spots" [329]

Boldfaced, Page-Referenced Terms

[p.328] theory of uniformity _the changing of the Earth through time._

_____

[p.328] Pangea _____

_____

[p.328] plate tectonics _____

_____

[p.330] radiometric dating _____

_____

[p.330] half-life _____

_____

Fill-in-the-Blanks

The theory of (1) _____ [p.328] describes the mountain building and (2) _____ [p.328] that had repeatedly changed Earth's surface gradually over immense periods of time; (3) _____ [p.328] layers were evidence of these processes. By using the (4) _____ [p.328] record of North American and European rocks formed 200 million years ago, geologists concluded that these two continents (and others) had once been joined together as a supercontinent: (5) _____ [p.328], which included all of Earth's land masses. In addition, deep-sea probes and submersibles had revealed evidence of (6) _____ [p.328] spreading that contributed to movements of Earth's crust. All this evidence was combined by geologists and paleontologists to describe a new theory of (7) _____ _____ [p.328], which helps scientists to understand how new land masses are built by (8) _____ [p.329] action; when and where earthquakes are likely to occur; and how, in the past, the movements of land masses profoundly influenced the (9) _____ [p.329] of life.

Although specific types of fossils are generally associated with particular rock layers (strata), no one was able to assign firm dates along the continuum of the geologic time scale until the discovery of (10) _____ _____ [p.330], which made it possible to convert (11) _____ [p.330] ages to absolute ones. By a procedure called (12) _____ _____ [p.330], researchers measure proportions of (1) an unstable radioisotope in a mineral that became trapped long ago in a new rock and (2) a more stable daughter isotope that formed from it in the same rock. (12) works most reliably for (13) _____ [p.330] rocks or ashes. Carbon 14 (^{14}C) has a (14) _____ - _____ [p.330] of 5,730 years. Suppose you had three pieces of the same kind of wood, identical in weight, but differing in age. Piece A was cut yesterday and releases 80 emissions per minute; Piece B was cut from a 5,730-year-old wooded enclosure of a mummy and registers (15) _____ [p.330] emissions per minute. Piece C was found in a sealed-off cave, registered half of (15)'s emission count, and was therefore judged to be (16) _____ [p.330] years old.

Self-Quiz

_d_ 1. The patterns, trends, and rates of change among lineages over geologic time are known as _____. [p.311]
    a. taxonomy
    b. classification
    c. a binomial system
    d. macroevolution

For questions 2–3, choose from the following answers. [pp.314–315]
    a. convergence
    b. divergence

_b_ 2. The wings of pterosaurs, birds, and bats serve as examples of _____.

_a_ 3. Penguins and porpoises serve as examples of _____.

For questions 4–8, choose from the following answers.
    a. nucleic acid hybridization
    b. homologous
    c. macroevolution
    d. analogous
    e. regulatory genes

_c_ 4. _____ refers to changes in groups of species. [p.311] b 5. _____ structures resemble one another due to common descent. [p.314]

d 6. Examples of _____ structures are the shark fin and the penguin wing. [p.315]

e 7. _____ control the rate of growth in different body parts. [p.317]

a 8. A rough measure of the evolutionary distance between two species is provided by _____. [p.318]

a 9. Early embryos of vertebrates strongly resemble one another *because* _____. [p.316]
   a. the genes that guide early embryonic development are the same (or similar) in all vertebrates
   b. the analogous structures they each contain are evidence of morphological convergence
   c. the homologous structures they each contain provide very strong evidence of morphological divergence
   d. nucleic acid hybridization techniques reveal many structural alterations that have resulted from gene mutations in early embryos

b 10. The significant contribution by Linnaeus was to develop _____. [p.320]
   a. the idea that organisms should be placed in two kingdoms
   b. the binomial system
   c. the theory of evolution
   d. the idea that a species could have several common names

c 11. Phylogeny is _____. [p.321]
   a. the identification of organisms and assigning names to them
   b. producing a "retrieval system" of several levels
   c. an evolutionary history of organism(s), both living and extinct
   d a study of the adaptive responses of various organisms

c 12. Lineages sharing a common evolutionary heritage are termed _____. [p.324]
   a. phenetic
   b. polyphyletic
   c. monophyletic
   d. homologous

d 13. All living organisms have eukaryotic cell structure except for members of the _____. [p.326]
   a. Animalia
   b. Plantae
   c. Fungi
   d. Archae- and Eubacteria
   e. Protista

Chapter Objectives/Review Questions

1. _____ refers to large-scale patterns, trends, and rates of change among groups of species. [p.311]
2. Give a generalized definition of a "fossil"; cite examples. [p.312]
3. _____ begins with rapid burial in sediments or volcanic ash. [p.312]
4. What is the name given to the layering of sedimentary deposits? [p.313]
5. Are the oldest fossils found in the deepest layers of sedimentary rocks or nearer the surface? [p.313]
6. The macroevolutionary pattern of change from a common ancestor is known as morphological _____. [p.314]
7. Distinguish homologies from analogies as applied to organisms. [pp.314–315]
8. A pattern of long-term change in similar directions among remotely related lineages is called morphological _____. [p.315]
9. Explain why two species of larkspurs have flowers with different shapes that attract different pollinators. [p.316]
10. Early in the vertebrate developmental program, _____ of different lineages proceeds through strikingly similar stages. [p.316]

11. Mutation of _____ genes explains the differences in the timing of growth sequences of chimps and humans (even though they have nearly identical genes). [p.317]
12. Some nucleic acid comparisons are based on _____ _____ hybridization. [p.318]
13. Define *neutral mutations* and discuss their use as molecular clocks. [pp.318–319]
14. Describe the binomial system as developed by Linnaeus; why is Latin used as the language of binomials? [p.320]
15. Arrange the classification groupings correctly, from most to least inclusive. [p.320]
16. Modern classification schemes reflect _____, the evolutionary relationships among species, starting with the most ancestral and including all the branches leading to their descendants. [p.321]
17. _____ _____ is a branch of biology that deals with patterns of diversity in an evolutionary context. [p.322]
18. Explain the difficulties taxonomists may encounter identifying species. [pp.320–323]
19. Describe the construction, appearance, and use of cladograms. [pp.322–325]
20. List the five-kingdoms of life in the Whittaker system, and cite examples of the organisms in each. [p.326]
21. Present the evidence that has caused geologists to amend the theory of uniformity and accept the ideas of plate tectonics as they attempt to decipher Earth's evolutionary history. [pp.313,328–329]

Integrating and Applying Key Concepts

Water crowfoot plants (family Ranunculaceae) often have two or more distinct leaf types within the same species and on the same plant. One type is capillary (hairlike), found submerged in the water or growing well above the water. The other type is laminate (flat and expanded), and is found floating on the water or submerged. In some *Ranunculus* species, three different leaf shapes form on a plant in a sequence. Suppose two taxonomists describe the same crowfoot plant as being two different and distinct species due to these two different leaf shapes. Can you think of difficulties that might arise when other researchers are required to refer to these "two species" in the course of their work?

21

THE ORIGIN AND EVOLUTION OF LIFE

Interactive Exercises

In the Beginning . . . [pp.334–335]

21.1. CONDITIONS ON THE EARLY EARTH [pp.336–337]

21.2. EMERGENCE OF THE FIRST LIVING CELLS [pp.338–339]

Selected Words: galaxy [p.334], nebula [p.334], core [p.336], template [p.337], "protenoids" [p.337], *metabolism* [p.338], photosynthesis [p.338], chlorophylls [p.338], porphyrins [p.338], cytochrome [p.338], coenzymes [p.338], selectively permeable [p.339]

Boldfaced, Page-Referenced Terms

[p.334] big bang _____

_____

[p.336] crust _____

_____

[p.336] mantle _____

_____

[p.339] RNA world _____

_____

[p.339] proto-cells _____

_____

Choice

Answer questions 1–20 by choosing from the following probable stages of the physical and chemical evolution of life.

a. formation of Earth   b. early Earth and the first atmosphere   c. the synthesis of organic compounds
d. origin of agents of metabolism   e. origin of self-replicating systems
f. origin of the first plasma membranes

1. ___ Most likely, the proto-cells were little more than membrane sacs protecting information-storing templates and various metabolic agents from the environment. [p.339]

2. ___ Simple systems (of this type) of RNA, enzymes, and coenzymes have been created in the laboratory. [p.338]

3. ___ Four billion years ago, Earth was a thin-crusted inferno. [p.336]

4. ___ Sunlight, lightning, or heat escaping from Earth's crust could have supplied the energy to drive their condensation into complex organic molecules. [pp.336–337]

5. ___ During the first 600 million years of Earth history, enzymes, ATP, and other molecules could have assembled spontaneously at the same locations. [p.338]

6. ___ Was gaseous oxygen (O_2) present? Probably not. [p.336]

7. ___ Sidney Fox heated amino acids under dry conditions to form protein chains, which he placed in hot water. The cooled chains self-assembled into small, stable spheres. The spheres were selectively permeable. [p.339]

8. ___ Stanley Miller mixed hydrogen, methane, ammonia, and water in a reaction chamber that recirculated the mixture and bombarded it with a spark discharge. Within a week, amino acids and other small organic compounds had been formed. [p.337]

9. ___ Without an oxygen-free atmosphere, the organic compounds that started the story of life never would have formed on their own. Free oxygen would have attacked them. [p.336]

10. ___ Much of Earth's inner rocky material melted. [p.336]

11. ___ Imagine an ancient sunlit estuary, rich in clay deposits. Countless aggregations of organic molecules stick to the clay. Some of these may have been porphyrin molecules that are part of the chlorophylls and cytochromes. [p.338]

12. ___ In other experiments, fatty acids and glycerol combined to form long-tail lipids under conditions that simulated evaporating tidepools. The lipids self-assembled into small, water-filled sacs, which in many were like cell membranes. [p.339]

13. ___ We still don't know how DNA entered the picture. [p.339]

14. ___ Rocks collected from Mars, meteorites, and the moon contain chemical precursors that must have been present on the early Earth. [p.336]

15. ___ This differentiation resulted in the formation of a crust of basalt, granite, and other types of low-density rock, a rocky region of intermediate density (the mantle), and a high-density, partially molten core of nickel and iron. [p.336]

16. ___ When the crust finally cooled and solidified, water condensed into clouds and the rains began. For millions of years, runoff from rains stripped mineral salts and other compounds from Earth's parched rocks. [p.336]

17. ___ The close association of enzymes, ATP, and other molecules would have promoted chemical interactions. Sunlight energy alone could have driven the spontaneous formation of RNA molecules. [p.338]

18. ___ Even if amino acids did form in the early seas, they wouldn't have lasted long. In water, the favored direction of most spontaneous reactions is toward hydrolysis, not condensation. [p.337]

19. ___ Between 4.6 and 4.5 billion years ago, the outer regions of the cloud cooled. [p.336]

20. ___ At first, it probably consisted of gaseous hydrogen (H_2), nitrogen (N_2), carbon monoxide (CO), and carbon dioxide (CO_2). [p.336]

21.3. ORIGIN OF PROKARYOTIC AND EUKARYOTIC CELLS [pp.340–341]

21.4. *Focus on Science:* WHERE DID ORGANELLES COME FROM? [pp.342–343]

Selected Words: Laurentia [p.341], "Precambrian" times [p.341], *Euglena* [p.342], Lynn Margulis [p.343], "mitochondrial code" [p.343], *Cyanophora paradoxa* [p.343]

Boldfaced, Page-Referenced Terms

[p.340] Archean _____

_____

[p.340] prokaryotic cells _____

_____

[p.340] eubacteria _____

_____

[p.340] archaebacteria _____

_____

[p.340] eukaryotic cells _____

_____

[p.340] stromatolites _____

_____

[p.341] Proterozoic _____

_____

[p.343] endosymbiosis _____

_____

[p.343] protistans _____

_____

Choice

For questions 1–15, choose from the following.

a. Archean   b. Proterozoic

1. ___ By 2.5 billion years ago, the noncyclic pathway of photosynthesis had evolved in some eubacterial species. [p.341]

2. ___ By 1.2 billion years ago, eukaryotes had originated. [p.341]

3. ___ The first prokaryotic cells emerged. [p.340]

4. ___ Oxygen, one of the by-products of photosynthesis, began to accumulate. [p.341]

5. ___ There was an absence of free oxygen. [p.340]

6. ___ There was a divergence of the original prokaryotic lineage into three evolutionary directions. [p.340]

7. ___ About 800 million years ago, stromatolites began a dramatic decline. [p.341]

8. ___ Between 3.5 and 3.2 billion years ago, the cyclic pathway of photosynthesis evolved in some species of eubacteria. [p.340]

9. ___ Fermentation pathways were the most likely sources of energy. [p.340]

10. ___ An oxygen-rich atmosphere stopped the further chemical origin of living cells. [p.341]

11. ___ Food was available, predators were absent, and biological molecules were free from oxygen attacks. [p.340]

12. ___ Aerobic respiration became the dominant energy-releasing pathway. [p.341]

13. ___ Archaebacteria and eubacteria arose. [p.340]

14. ___ The first cells may have originated in tidal flats or on the seafloor, in muddy sediments warmed by heat from volcanic vents. [p.340]

15. ___ Near the shores of the supercontinent Laurentia, small, soft-bodied animals were leaving tracks and burrows on the seafloor. [p.341]

Fill-in-the-Blanks

(See *Focus on Science:* Where Did Organelles Come From? [pp.342–343])

The most important feature of eukaryotic cells is the abundance of membrane-bound (16) _____ [p.342] in the cytoplasm. The origin of these structures remains a mystery. Some probably evolved gradually, through (17) _____ [p.342] mutations and natural selection. Some extant prokaryotic cells do possess infoldings of the (18) _____ [p.342] membrane, a site where enzymes and other metabolic agents are embedded. In prokaryotic cells that were ancestral to eukaryotic cells, similar membranous infoldings may have served as a (19) _____ [p.342] from the surface to deep within the cell. These membranes may have evolved into (20) _____ [p.342] channels and into an envelope around the DNA. The advantage of such membranous enclosures may have been to protect (21) _____ [p.342] and their products from foreign invader cells. Evidence for this is that yeasts, simple nucleated eukaryotic cells, have an abundance of (22) _____ [p.342] which they transfer from cell to cell much as prokaryotic cells do.

As the evolution of eukaryotes proceeded, accidental partnerships between different (23) _____ [p.342] species must have formed countless times. Some partnerships resulted in the origin of mitochondria, chloroplasts, and other organelles. This is the theory of (24) _____ [p.343], developed in greatest detail by Lynn Margulis. According to this theory, (25) _____ [p.343] arose after the noncyclic pathway of photosynthesis emerged and oxygen had accumulated to significant levels. (26) _____ [p.343] transport systems had been expanded in some bacteria to include extra cytochromes to donate electrons to (27) _____ [p.343] in a system of aerobic respiration. By 1.2 billion years ago or earlier, amoebalike ancestors of eukaryotes were engulfing aerobic (28) _____ [p.343] and perhaps forming endocytic vesicles around the food for delivery to the cytoplasm for digestion. Some aerobic bacteria resisted digestion and thrived in a new, protected, nutrient-rich environment. In time they released extra (29) _____ [p.343]—which the hosts came to depend on for growth, greater activity, and assembly of other structures. The guest aerobic bacteria came to depend on the metabolic functions of their (30) _____ [p.343] cell. The aerobic and anaerobic cells were now (31) _____ [p.343] of independent existence. The guests had become (32) _____ [p.343], supreme suppliers of (33) _____ [p.343].

Mitochondria are bacteria-size and replicate their own (34) _____ [p.343] dividing independently of the host cell's division process. In addition, (35) _____ [p.343] may be descended from aerobic eubacteria that engaged in oxygen-producing photosynthesis. Such cells may have been engulfed, resisted digestion, and provided their respiring (36) _____ [p.343] cells with needed oxygen. Chloroplasts resemble some (37) _____ [p.343] in metabolism and overall nucleic acid sequence. Chloroplasts have self-replicating DNA, and they divide independently of the host cell's division.

New cells appeared on the evolutionary stage that were equipped with a nucleus, ER membranes, and mitochondria or chloroplasts (or both); they were the first eukaryotic cells, the first (38) _____ [p.343].

Identification

For questions 39–53, fill in the blanks. Choose from the following: Animals, Archaebacterial, Thermophiles, Chloroplasts, Eubacterial, Eukaryote, Eukaryotes, Fungi, Mesozoic, Mitochondria, Paleozoic, Plants

39. _____ [p.340]

40. _____ [p.340]

41. _____ [p.340]

42. _____ [p.341]

43. _____ [p.341]

44. _____ [p.341]

45. _____ [p.341]

46. _____ [p.341]

47. _____ [p.341]

48. _____ [p.341]

49. _____ [p.341]

50. _____ [p.341]

51. _____ [p.341]

52. _____ [p.341]

53. _____ [p.341]

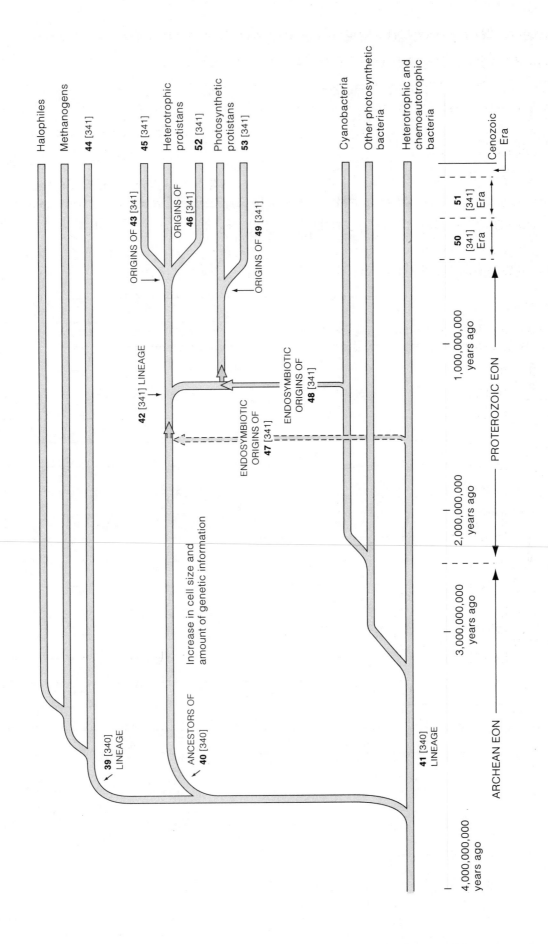

21.5. LIFE IN THE PALEOZOIC ERA [pp.344–345]

21.6. LIFE IN THE MESOZOIC ERA [pp.346–347]

21.7. *Focus on Science:* HORRENDOUS END TO DOMINANCE [p.348]

21.8. LIFE IN THE CENOZOIC ERA [p.349]

21.9. SUMMARY OF EARTH AND LIFE HISTORY [pp.350–351]

Selected Words: Cambrian [p.344], trilobites [p.344], Ordovician [p.344], nautiloids [p.344], Silurian [p.344], armor-plated fishes [p.344], Devonian [p.345], *Psilophyton* [p.345], Carboniferous [p.345], Permian [p.345], *Dimetrodon* [p.345], Pangea [p.345], Tethys Sea [p.345], *Lystrosaurus* [p.346], *Triceratops* [p.346], *Velociraptor* [p.346], Alvarez and Alvarez [p.348], Melosh [p.348], *Smilodon* [p.349]

Boldfaced, Page-Referenced Terms

[p.344] Paleozoic _____

_____

[p.344] Ediacarans _____

_____

[p.346] Mesozoic _____

_____

[p.344] gymnosperms _____

_____

[p.344] angiosperms _____

_____

[p.348] dinosaurs _____

_____

[p.348] asteroids _____

_____

[p.348] K-T asteroid impact theory _____

_____

[p.348] global broiling hypothesis _____

_____

[p.349] Cenozoic _____

_____

Sequence

Refer to Figure 21.20 in the text. Study of the geologic record reveals that as the major events in the evolution of Earth and its organisms occurred, there were periodic major *extinctions* of organisms followed by major *radiations* of organisms. Using the following list, arrange the letters of the extinctions and radiations in the approximate order in which they occurred, from most recent to most ancient. [All from p.350]

1. ___   A. Pangea, worldwide ocean forms; shallow seas squeezed out. Major radiations of reptiles, gymnosperms.

2. ___   B. Land masses dispersed near equator. Simple marine communities only. Origin of animals with hard parts.

3. ___   C. Mass extinction of many marine invertebrates, most fishes.

4. ___   D. Pangea breakup begins. Rich marine communities. Major radiations of dinosaurs.

5. ___   E. Asteroid impact? Mass extinction of all dinosaurs and many marine organisms.

6. ___   F. Recovery, radiations of marine invertebrates, fishes, dinosaurs. Gymnosperms the dominant land plants. Origin of mammals.

7. ___   G. Major glaciations. Modern humans emerge and begin what may be the greatest mass extinction of all time on land, starting with ice age hunters.

8. ___   H. Unprecedented mountain building as continents rupture, drift, collide. Major climatic shifts; vast grasslands emerge. Major radiations of flowering plants, insects, birds, mammals. Origin of earliest human forms.

9. ___   I. Oxygen accumulating in atmosphere. Origin of aerobic metabolism.

10. ___   J. Laurasia forms. Gondwana crosses South Pole. Mass extinction of many marine species. Vast swamplands, early vascular plants. Radiation of fishes continues. Origin of amphibians.

11. ___   K. Pangea breakup continues, broad inland seas form. Major radiations of marine invertebrates, fishes, insects, dinosaurs, origin of angiosperms (flowering plants).

12. ___   L. Asteroid impact? Mass extinction of many organisms in seas, some on land; dinosaurs, mammals survive.

13. ___   M. Mass extinction. Nearly all species in seas and on land perish.

14. ___   N. Tethys Sea forms. Recurring glaciations. Major radiations of insects, amphibians. Spore-bearing plants dominant; gymnosperms present; origin of reptiles.

Chronology of Events—Geologic Time

Refer to Figure 21.20 in the text. From the list of evolutionary events above (A–N), select the letters occurring in a particular era of time by circling the appropriate letters. [p.351]

15. ___ Cenozoic: A – B – C – D – E – F – G – H – I – J – K – L – M – N

16. ___ Border of Cenozoic–Mesozoic: A – B – C – D – E – F – G – H – I – J – K – L – M – N

17. ___ Mesozoic: A – B – C – D – E – F – G – H – I – J – K – L – M – N

18. ___ Border of Mesozoic–Paleozoic: A – B – C – D – E – F – G – H – I – J – K – L – M – N

19. ___ Paleozoic: A – B – C – D – E – F – G – H – I – J – K – L – M – N

Complete the Table

20. Refer to Figure 21.20 in the text. To review some of the events that occurred in the geologic past, complete the following table by entering the geologic era (Archean, Proterozoic, Paleozoic, Mesozoic, and Cenozoic) and the *approximate* time in millions of years since the time of the events. [p.351]

Era/Eon	Time	Events
a.		A few reptile lineages give rise to mammals and the dinosaurs.
b.		Formation of Earth's crust, early atmosphere, oceans; chemical evolution leading to the origin of life.
c.		Origin of amphibians.
d.		Origin of animals with hard parts.
e.		Rocks 3.5 billion years old contain fossils of well-developed prokaryotic cells that probably lived in tidal mud flats.
f.		Flowering plants emerge, gymnosperms begin their decline.
g.		In the Carboniferous, there were major radiations of insects and amphibians; gymnosperms present, origin of reptiles.
h.		Oxygen accumulated in the atmosphere.
i.		Before the close of this era, the first photosynthetic bacteria had evolved.
j.		Insects, amphibians, and early reptiles flourished in the swamp forests of the Permian.
k.		Most of the major animal phyla evolved in rather short order.
l.		Dinosaurs ruled.
m.		Origin of aerobic metabolism; origin of protistans, algae, fungi, animals.
n.		Grasslands emerge and serve as new adaptive zones for plant-eating mammals and their predators.
o.		Humans destroy habitats and many species.
p.		The first ice age initiated the first global mass extinction; reef life everywhere collapsed.
q.		The invasion of land begins; small stalked plants establish themselves along muddy margins, and the lobe-finned fishes ancestral to amphibians move onto land.

Crossword

ACROSS

2. A conifer, cycad, or *Ginkgo*, for example [p.346]
5. The most recent period of the Paleozoic, at the end of which nearly all species on land and in the sea perish. [p.345]
7. Concerning all of Earth [p.348]
10. Mesozoic reptile with a "three-toed" foot print [p.346]
11. Ancestor of a cell [p.339]
12. Mat of calcium-hardened photosynthetic eubacteria [p.340]
13. A gymnosperm that resembles a squat palm tree [p.345]
15. A layer lying between the crust and core of Earth [p.336]
16. The most recent period of the Mesozoic era [p.346]
18. A massive continent that included the "entire Earth" [p.345]
20. An era during which fish, amphibians, and reptiles first appeared [pp.344–345]
25. Single-celled eukaryote [p.343]
26. A fossil of an organism that lived 610 to 550 million years ago [p.344]
27. The most recent era [p.349]
29. A rocky, metallic body hurtling through space [p.348]
30. A statement about how events occur in the natural world [p.342]
31. A halophile, methanogen, or thermophile [p.341]
32. to hit; affect [p.348]

DOWN

1. Theory of _____, best developed by Lynn Margulis [pp.342–343]
2. Southern supercontinent of the Paleozoic era [p.344]
3. Small, leafy, nonflowering plant [p.345]
4. A long period of time [pp.344,346,349]
6. An epoch of the Tertiary Period of the Cenozoic Era [p.349]
8. Cooking on a very hot surface [p.348]
9. A supercontinent that became part of Pangea and later formed the continents of Earth's northern hemisphere [p.350]
11. O_2 accumulated in Earth's atmosphere during the eon. [p.341]
13. Hardened, crunchy surface [p.336]
14. _____ plants radiated dramatically during the Cenozoic [p.351]
17. Photosynthetic bacteria originated during this eon [p.340]
19. Flowering plant [p.346]
21. Supercontinent of Proterozoic [p.350]
22. Dinosaurs and gymnosperms dominated this era [p.346]
23. A longer period of time than 4 down [pp.344–341]
24. The big _____ [p.334]
28. The _____ world was based on ribonucleic acid template use. [p.339]

Self-Quiz

___ 1. Between 4.6 and 4.5 billion years ago, _____. [p.336]
   a. Earth was a thin-crusted inferno
   b. the outer regions of the cloud from which our solar system formed was cooling
   c. life originated
   d. the atmosphere was laden with oxygen

___ 2. More than 3.8 billion years ago, _____. [p.336]
   a. Earth was a thin-crusted inferno
   b. the outer regions of the cloud from which our solar system formed were cooling
   c. life originated
   d. the atmosphere was laden with oxygen

For questions 3–10, choose from these answers:
   a. Archean
   b. Cenozoic
   c. Mesozoic
   d. Paleozoic
   e. Proterozoic

___ 3. Dinosaurs and gymnosperms were the dominant forms of life during the _____ era. [p.346]

___ 4. The Alps, Andes, Himalayas, and Cascade Range were born during major reorganization of land masses early in the _____ era. [p.349]

___ 5. The composition of Earth's atmosphere changed during the _____ eon from one that was anaerobic to one that was aerobic. [p.341]

___ 6. Invertebrates, primitive plants, and primitive vertebrates were the principal groups of organisms on Earth during the _____ era. [pp.344–345]

___ 7. Before the close of the _____ eon, the first photosynthetic bacteria had evolved. [p.340]

___ 8. The _____ era ended with the greatest of all extinctions, the Permian extinction. [p.345]

___ 9. Late in the _____ era, flowering plants arose and underwent a major radiation. [p.346]

___ 10. The _____ era included adaptive zones into which plant-eating mammals and their predators radiated. [p.349]

Chapter Objectives/Review Questions

1. Cloudlike remnants of stars are mostly _____, but they also contain water, iron, silicates, hydrogen cyanide, methane, ammonia, formaldehyde, and many other simple inorganic and organic substances. [p.336]
2. Describe the formation of early Earth before the formation of the first atmosphere. [p.336]
3. List the probable chemical constituents of Earth's first atmosphere. [p.336]
4. _____ oxygen was probably not present in early Earth's atmosphere. [p.336]
5. Describe experimental evidence provided by Stanley Miller (and others) that the formation of biological molecules from simple precursor molecules might have occurred on early Earth. [p.337]
6. _____ might have been the medium on which condensation reactions yielding complex organic compounds occurred. [p.337]
7. During the first _____ years of Earth's history, enzymes, ATP, and other molecules could have assembled at the same location; their close association would have promoted chemical interactions—and the beginning of _____ pathways. [p.338]
8. Although simple, self-replicating systems of RNA, enzymes, and coenzymes have been created in the laboratory, the chemical ancestors of _____ and DNA remain unknown. [p.339]

9. Describe the experiments performed by Sidney Fox that aided understanding of the origin of plasma membranes. [pp.337,339]
10. The first eon of the geologic time scale is now called the _____, "the beginning." [p.340]
11. List the three evolutionary directions of the original prokaryotic lineage during the early Archean era. [p.340]
12. Populations of some early anaerobic eubacteria used the cyclic photosynthetic pathway; their populations formed huge mats called _____. [p.340]
13. Describe how the endosymbiosis theory may help explain the origin of eukaryotic cells. [pp.342–343]
14. Generally discuss the important geological and biological events occurring throughout the Archean and Proterozoic eons, and the Paleozoic, Mesozoic, and Cenozoic eras. [pp.344–349]
15. List the three geologic eras and two geologic eons in proper order, from oldest to youngest. [pp.350–351]

Integrating and Applying Key Concepts

As Earth becomes increasingly loaded with carbon dioxide and various industrial waste products, how do you think living forms on Earth will evolve to cope with these changes?

22

BACTERIA AND VIRUSES

Interactive Exercises

The Unseen Multitudes [pp.354–355]

22.1. CHARACTERISTICS OF BACTERIA [pp.356–357]

22.2. BACTERIAL GROWTH AND REPRODUCTION [pp.358–359]

Selected Words: nanometer [p.354], photoautotrophic [p.356], chemoautotrophic [p.356], photoheterotrophic [p.356], chemoheterotrophic [p.356], parasite [p.356], saprobic [p.356], peptidoglycan [p.357], Gram-positive [p.357], Gram-negative [p.357], *Streptococcus aureus* [p.357], *Escherichia coli* [p.357], *Salmonella* [p.359], *Streptococcus* [p.359]

Boldfaced, Page-Referenced Terms

[p.354] microorganisms _____

_____

[p.354] pathogens _____

_____

[p.356] coccus, cocci _____

_____

[p.356] bacillus, bacilli _____

_____

[p.356] spirillum, spirilla _____

[p.356] prokaryotic cells _____

[p.357] cell wall _____

[p.357] Gram stain _____

[p.357] glycocalyx _____

[p.357] bacterial flagella (sing., flagellum) _____

[p.357] pilus, pili _____

[p.358] bacterial chromosome _____

[p.358] prokaryotic fission _____

[p.359] plasmid _____

[p.359] bacterial conjugation _____

Choice

Choose from the lettered terms; write the letter in the blank next to the best description.

a. archaebacteria   b. chemoautotrophic eubacteria   c. chemoheterotrophic eubacteria
d. photoautotrophic eubacteria   e photoheterotrophic eubacteria

1. ___ Use CO_2 from the environment as the usual source of carbon atoms and use electrons, hydrogen, and energy released from chemical reactions to assemble chains of carbon (food storage) [p.356]

2. ___ Use CO_2 and H_2O from the environment as sources of carbon, hydrogen, and oxygen atoms and use sunlight to power the assembly of food storage molecules [p.356]

3. ___ Cannot use CO_2 from the environment to construct own carbon chains; instead obtain nutrients from the products, wastes, or remains of other organisms; can break down glucose to pyruvate and follow it with fermentation of some sort [p.356]

4. ___ Cannot use CO_2 from the environment to construct own cellular molecules but can absorb sunlight and transfer some of that energy to the bonds of ATP; must obtain food molecules, (carbon chains) produced by other organisms to construct their own molecules [p.356]

5. ___ Are either parasites or saprobes, but are not self-feeders [p.356]

Fill-in-the-Blanks

Bacteria are microscopic, (6) _____ [p.356] cells having one bacterial chromosome and, often, a number of smaller (7) _____ [p.359]. The cells of nearly all bacterial species have a(n) (8) _____ [p.357] around the plasma membrane, and a(n) (9) _____ [p.357] or slime layer surrounding the cell wall. Typically, the width or length of these cells falls between 1 and 10 (10) (choose one) ❏ millimeters ❏ nanometers ❏ centimeters ❏ micrometers [p.356]. Most bacteria reproduce by (11) _____ _____ [p.358]. Spherical bacteria are (12) _____ [p.356], rod-shaped bacteria are (13) _____ [p.356], and helical bacteria are (14) _____ [p.356]. (15) Gram-_____ [p.356] bacteria retain the purple stain when washed with alcohol.

22.3. BACTERIAL CLASSIFICATION [p.359]
22.4. MAJOR BACTERIAL GROUPS [p.360]
22.5. ARCHAEBACTERIA [pp.360–361]
22.4. EUBACTERIA [pp.362–363]

Selected Words: *Methanobacterium* [p.360], *Halobacterium* [p.360], *Thermoplasma* [p.360], anaerobic electron transport [p.360], *Nostoc* [p.362], *Nitrobacter* [p.360], bacteriorhodopsin [p.361], *Sulfolobus* [p.361], *Thermus aquaticus* [p.361], peptidoglycan [p.362], cyanobacteria [p.362], *Anabaena* [p.362], *Lactobacillus* [p.362], *E. coli* [p.362], *Azospirillum* [p.362], *Rhizobium* [p.362], *Clostridium botulinum* [p.362], botulism [p.362], *Clostridium tetani* [p.362], tetanus [p.362], *Borrelia burgdorferi* [p.363], Lyme disease [p. 363], *Rickettsia rickettsii* [p.363], Rocky Mountain spotted fever [p.363]

Boldfaced, Page-Referenced Terms

[p.359] numerical taxonomy _____

_____

[p.359] eubacteria _____

_____

[p.359] archaebacteria _____

_____

[p.360] strain _____

_____

[p.360] methanogens _____

_____

[p.361] halophiles _____

_____

[p.361] extreme thermophiles _____

_____

[p.362] heterocysts _____

_____

[p.362] endospores _____

_____

[p.363] fruiting bodies _____

_____

Fill-in-the-Blanks

In many respects, their cell structure, metabolism, and nucleic acid sequences are unique to the (1) _____ [p.360]; for example, none has a cell wall that contains (2) _____ [p.362]. On the other hand, (3) _____ [p.362] are far more common than the three rather unusual types of (1). The most common photoautotrophic bacteria are the (4) _____ [p.362] (also called the blue-green algae). *Anabaena* and others of the (4) group produce oxygen during photosynthesis. Heterocysts are cells in *Anabaena* that carry out (5) _____ - _____ [p.362]. Many species of chemoautotrophic eubacteria affect the global cycling of nitrogen, (6) _____ [p.362], and other nutrients. Sugarcane and corn plants benefit from a (7) _____ [p.362]-fixing spirochete, *Azospirillum*. Beans and other legumes benefit from the nitrogen-fixing activities of (8) _____ [p.362], which dwells in their roots.

When environmental conditions become adverse, many bacteria form (9) _____ [pp.362–363], which resist moisture loss, irradiation, disinfectants, and even acids. Endospore formation by bacteria can kill humans if they enter the food supply and are not killed by high (10) _____ [p.363] and high pressure. Anaerobic bacteria can live in canned food, reproduce, and produce deadly toxins. Two examples of pathogenic (disease-causing) bacteria that form endospores harmful to humans are (11) _____ _____ [p.362] and (12) _____ _____ [p.362]. (13) _____ _____ [p.363] may be the most common tick-borne disease in the United States by now; tick bites deliver (14) _____ [p.363] from one host to another. Bacterial behavior depends on (15) _____ _____ [p.363], which change shape when they absorb or connect with chemical compounds. Cyanobacteria require (16) _____ [recall p.356] as their source of energy that drives their metabolic activities. Most of the world's bacteria are (17) (choose one) ❏ producers ❏ consumers ❏ decomposers [p.362], so we think of them as "good" heterotrophs. (18) _____ [p.362] include two of the major producers of antibiotics: *Actinomyces* and (19) _____ [p.360]. *Escherichia coli*, which dwells in our gut, synthesizes vitamin (20) _____ [p.362] and substances useful in digesting fats.

Matching

Match each of the items below with a lowercase letter designating its principal bacterial group and an upper-case letter denoting its best descriptor from the right-hand column.

a. archaebacteria   b. chemoautotrophic eubacteria   c. chemoheterotrophic eubacteria
d. photoautotrophic eubacteria   e. photoheterotrophic eubacteria

21. ___, ___*Anabaena* [pp.360,362]

22. ___, ___*Bacillus, Clostridium* [pp.362–363]

23. ___, ___*Escherichia coli* [pp.357,360,362]

24. ___, ___*Halobacterium* [pp.360–361]

25. ___, ___*Lactobacillus* [pp.360,362]

26. ___, ___*Methanobacterium* [pp.360–361]

27. ___, ___*Nitrobacter, Nitrosomonas* [p.360]

28. ___, ___*Rhizobium, Agrobacterium* [pp.360,362]

29. ___, ___*Rhodospirillum* [p.360]

30. ___, ___*Salmonella* [pp.359,360]

31. ___, ___*Spirochaeta, Treponema* [p.360]

32. ___, ___*Staphylococcus, Streptococcus* [pp.357,360]

33. ___, ___*Streptomyces, Actinomyces* [p.360]

34. ___, ___*Thermoplasma, Sulfolobus* [pp.360–361]

A. Lives in anaerobic sediments of swamps and in animal gut; chemosynthetic; used in sewage treatment facilities

B. Purple; generally in anaerobic sediments of lakes or ponds; do not produce oxygen, do not use water as a source of electrons

C. Endospore-forming rods and cocci that live in the soil and in the animal gut; some major pathogens

D. Gram-positive cocci that live in the soil and in the skin and mucous membranes of animals; some major pathogens

E. Gram-positive nonsporulating rods that ferment plant and animal material; some are important in dairy industry; others contaminate milk, cheese

F. In acidic soil, hot springs, hydrothermal vents on seafloor; may use sulfur as a source of electrons for ATP formation

G. Live in extremely salty water; have a unique form of photosynthesis

H. Gram-negative aerobic rods and cocci that live in soil or aquatic habitats or are parasites of animals and/or plants; some fix nitrogen

I. Nitrifying bacteria that live in the soil, fresh water, and marine habitats; play a major role in the nitrogen cycle

J. Gram-negative anaerobic rod that inhabits the human colon where it produces vitamin K

K. Major Gram-negative pathogens of the human gut that cause specific types of food poisoning

L. Mostly in lakes and ponds; cyanobacteria; produce O_2 from water as an electron donor

M. Major producer of antibiotics; an actinomycete that lives in soil and some aquatic habitats

N. Helically coiled, motile parasites of animals; some are major pathogens

22.7. THE VIRUSES [pp.364–365]

22.8. VIRAL MULTIPLICATION CYCLES [pp.366–367]

22.9. *Focus on Health:* **INFECTIOUS PARTICLES TINIER THAN VIRUSES** [p.367]

22.10. *Focus on Health:* **THE NATURE OF INFECTIOUS DISEASES** [pp.368–369]

Selected Words: *helical, polyhedral, enveloped,* and *complex* viruses [p.364], *AIDS* [p.364], *Herpes simplex, type I* [p.366], *cold sores* [p.366], kuru [p.367], Creutzfeldt-Jakob disease [p.367], scrapie [p.367], mad cow disease (BSE) [p.367], *contagious, sporadic,* and *endemic* diseases [p.368], *Ebola* virus [p.368], monkeypox [p.368], *Streptococcus pneumoniae* [p.369], *Salmonella enteridis* [p.369]

Boldfaced, Page-Referenced Terms

[p.364] virus _____

_____

[p.364] bacteriophages _____

_____

[p.364] lytic pathway _____

_____

[p.366] lysis _____

_____

[p.364] lysogenic pathway _____

_____

[p.367] viroids _____

_____

[p.367] prions _____

_____

[p.368] infection _____

_____

[p.368] disease _____

_____

[p.368] epidemic _____

_____

[p.368] pandemic _____

_____

[p.368] emerging pathogens _____

_____

Short Answer

1. a. State the principal characteristics of viruses. [p.364] _____

_____

   b. Describe the structure of viruses. [pp.364–367]_____

_____

   c. Distinguish among the ways viruses replicate themselves. [pp.366–367]_____

_____

2. a. List five specific viruses that cause human illness. [pp.364–365] _____

_____

   b. Describe how each virus in (2a) does its dirty work. [pp.364–368] _____

_____

_____

_____

_____

_____

_____

Fill-in-the-Blanks

A(n) (3) _____ [p.364] is a noncellular, nonliving infectious agent, each of which consists of a central
(4) _____ _____ [p.364] core surrounded by a protective (5) _____ _____ [p.364].
(6) _____ [p.364] contain the blueprints for making more of themselves but cannot carry on metabolic
activities. Chickenpox and shingles are two infections caused by DNA viruses from the (7) _____
[p.365] category. Naked strands or circles of RNA that lack a protein coat are called (8) _____ [p.367].
(9) _____ [p.365] are RNA viruses that infect animal cells, cause AIDS, and follow (10) _____
[pp.366–367] pathways of replication. (11) _____ [p.364] are the usual units of measurement with
which to measure viruses, whereas microbiologists measure bacteria and protistans in terms of
(12) _____ [pp.362–363]. A bacterium 86 micrometers in length is (13) _____ [p.354] nanometers
long. During a period of (14) _____ [p.366], viral genes remain inactive inside the host cell and any of
its descendants. Pathogenic protein particles are called (15) _____ [p.367].

Identification

Identify the virus that causes the following illnesses by writing the name of the virus in the blank preceding
the disease. Then tell whether it is a DNA virus or an RNA virus. [p.365]

16. _____ , _____ Common colds

17. _____ , _____ AIDS, leukemia

18. _____ , _____ Cold sores, chickenpox

Matching

Match each item below with the correct lettered description.

19. ___antibiotic [p.369,glossary]
20. ___bacteriophage [p.364]
21. ___endemic [p.368]
22. ___epidemic [p.368]
23. ___lysogenic pathway [p.366]
24. ___lytic pathway [p.366]
25. ___microorganism [p.354]
26. ___pathogen [p.354]
27. ___prion [p.367]
28. ___sporadic [p.368]
29. ___viroid [p.367]
30. ___virus [p.364]

A. "Naked" RNA bits that resemble introns
B. Disease that breaks out irregularly, affects few organisms
C. Chemical substance that interferes with gene expression or other normal functions of bacteria
D. Disease abruptly spreads through large portions of a population
E. Disease that occurs continuously, but is localized to a relatively small portion of the population
F. Small proteins that are altered products of a gene; linked to 8 degenerative diseases of the nervous system
G. Any organism too small to be seen without a microscope
H. A virus that infects a bacterium
I. Damage and destruction to host cells occurs quickly
J. Noncellular infectious agent that must take over a living cell in order to reproduce itself
K. Viral nucleic acid is integrated into the nucleic acid system of the host cell and replicated during this time
L. Any disease-causing organism or agent

Self-Quiz

Multiple Choice

___ 1. Which of the following diseases is *not* caused by a virus? [p.365]
   a. hepatitis B
   b. polio
   c. influenza
   d. syphilis

___ 2. Bacteriophages are _____. [p.364]
   a. viruses that parasitize bacteria
   b. bacteria that parasitize viruses
   c. bacteria that phagocytize viruses
   d. composed of a protein core surrounded by a nucleic acid coat

Matching

Match all applicable letters with the appropriate terms. A letter may be used more than once, and a blank may contain more than one letter.

3. ___*Anabaena, Nostoc* [pp.360,362]
4. ___*Clostridium botulinum* [pp.360,362]
5. ___*Escherichia coli* [pp.357,360–361]
6. ___*Herpes simplex* [p.366]
7. ___*HIV* [pp.364–365]
8. ___*Lactobacillus* [pp.360,362]
9. ___*Staphylococcus* [p.357]

A. Bacteria
B. Virus
C. Cyanobacteria
D. Gram-positive eubacteria
E. Cause cold sores and a type of venereal disease
F. Associated with AIDS

Matching

Match the following pictures with their names.

10. ___[p.365]   A. *Bacillus*
11. ___[p.364]   B. bacteriophage
                    C. *Clostridium tetani*
12. ___[p.362]   D. Cyanobacterium
                    E. *Influenza virus*
13. ___[p.364]   F. HIV
14. ___[p.363]
15. ___[p.354]

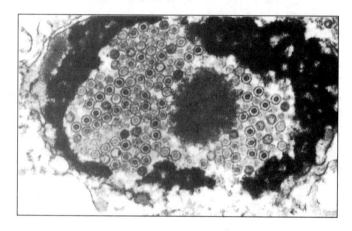

10.

11.

12.

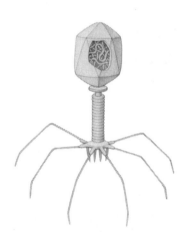

13.

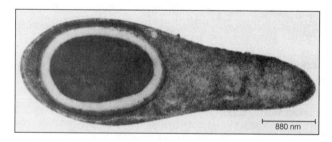

14.

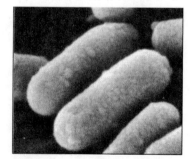

15.

ACROSS

2. _____ can assemble their own food molecules by using light energy to join together carbon, hydrogen and oxygen atoms. [p.362]
3. Prokaryotic _____ is the process most bacteria use to reproduce. [p.358]
6. Disease causers [p.354]
7. Bacterial _____ transfers a plasmid from a donor to a recipient cell. [p.359]
8. A chemical substance produced by *actinomycetes* and other microorganisms that kills or inhibits the growth of other microorganisms [p.362]
10. A whiplike organelle of locomotion [p.356]
12. Boy, young man
14. Group that includes blue-green algae [p.362]
16. Sticky mesh of polysaccharides, polypeptides, or both; alternate name of either a capsule or slime layer [p.357]
17. _____ cells existed before the origin of the nucleus and of eukaryotic cells. [p.356]
19. Cannot synthesize their own food molecules; *can* use sunlight as an energy source to make ATP, but cannot fix CO_2 from their surroundings; must obtain carbon compounds from other organisms [p.356]

20. Viruses that infect bacterial host cells [p.364]
23. Any organism that can be seen most clearly by using a microscope [p.354]
24. A short filamentous protein that projects above a cell wall; helps tether a bacterium to another bacterium or to a surface. [p.357]
25. A(n) _____ occurs when, over the entire world, a disease spreads through large portions of many populations for a limited period. [p.368]
26. _____ are structures that resist heat, drying, boiling, and radiation; can give rise to new bacterial cells. [p.362]
27. A(n) _____ occurs when a disease abruptly spreads through large portions of a single population for a limited period. [p.368]
28. Salt-loving archaebacterium [p.361]
29. Methane-producing archaebacterium [p.360]
30. Spherical bacterium [p.356]

DOWN

1. Rod-shaped bacterium [p.356]
2. Substance not in archaebacterial cell walls [p.362]
4. Helical [p.356]
5. _____-positive bacterial cell walls have an affinity for crystal violet stain. [p.357]
9. _____ include Earth's major decomposers, nitrogen-fixers and fermenters of milk. [p.362]
11. The _____ pathway of bacteriophage multiplication assembles new viral particles very soon after infecting the host cell; no integration into host cell's DNA. [p.366]
13. This group includes the most ancient of Earth's bacteria. [p.360]
15. Bacteria that can synthesize their own food molecules by using energy released from specific chemical reactions [p.356]
18. Heat-tolerant bacterial type [p.361]
21. The _____ pathway of bacteriophage multiplication includes integrating viral DNA into its bacterial host's chromosome. [p.366]
22. Small loops of DNA in addition to the main bacterial "chromosome" [p.359]

Chapter Objectives/Review Questions

1. Distinguish chemoautotrophs from photoautotrophs. [p.356]
2. Describe the principal body forms of monerans (inside and outside). [pp.356–357]
3. Explain how, with no nucleus, or few if any, membrane-bound organelles, bacteria reproduce themselves and obtain energy to carry on metabolism. [pp.358–359]
4. State the ways in which archaebacteria differ from eubacteria. [pp.360–362]
5. Describe the shapes of various viral types and explain the ways in which viruses infect their hosts. [pp.364–365]
6. Distinguish the lytic and lysogenic patterns of viral replication. [pp.366–367]

Integrating and Applying Key Concepts

The textbook (Figure 51.16) identifies natural gas as a nonrenewable fuel resource, yet there is a group of archaebacteria that produce methane, the burning of which can serve as a fuel for heating and/or cooking. Recall or imagine how these bacteria could be incorporated into a system that could serve human societies by generating methane in a cycle that is renewable. Why did your text categorize natural gas as a nonrenewable resource? Is methane a constituent of natural gas? Why or why not?

23

PROTISTANS

Interactive Exercises

Kingdom at the Crossroads [pp.372–373]

23.1. PARASITIC AND PREDATORY MOLDS [pp.374–375]

Selected Words: antheridium [p.374], oogonium [p.374], downy mildew [p.374], *Plasmopara viticola* [p.374], *Phytophthora infestans* [p.374], late blight [p.374], cellular slime molds [p.375], *Dictyostelium discoideum* [p.375], plasmodial slime molds [p.375]

Boldfaced, Page-Referenced Terms

[p.372] protistans _____

_____

[p.374] chytrids _____

[p.374] water molds _____

[p.374] slime molds _____

[p.374] saprobes _____

[p.374] mycelium (pl., mycelia) _____

_____

[p.374] hyphae (sing., hypha) _____

_____

Dichotomous Choice

1. For each structure, indicate with a P if it is a prokaryotic (bacterial) characteristic, and indicate with an E if it is a eukaryotic characteristic.

a. double-membraned nucleus [p.372]	
b. mitochondria present [p.372]	
c. lack membrane-bound organelles [recall ch.22]	
d. engage in mitosis [p.372]	
e. circular chromosome present [recall ch.22]	
f. endoplasmic reticulum present [p.372]	
g. cilia or flagella with 9+2 core [p.372]	

Complete the Table

2. Fill in each box with letters selected from the following grouped items. Choose from the choices in group I on p.241 for the boxes in column I, from group II for the boxes in column II and from group III for the boxes in column III.

Protistan Group of Funguslike Organisms	I. Structural Features	II. Behaviors and Typical Habitats	III. Representatives
chytrids (Chytridiomycota) [p.374]	a.	b.	c.
cellular slime molds (Acrasiomycota) [p.375]	d.	e.	f.
plasmodial slime molds (Myxomycota) [p.375]	g.	h.	i.
water molds (Oomycota) [p.374]	j.	k.	l.

I.	II.	III.
A. Mycelium present in most multicelled species B. Free-living amoebalike cells present C. Rhizoids present in many species as adults D. Single-celled types are globular/spherical E. Produce flagellated asexual spores	F. Most are saprobic decomposers G. Some are parasites on or in living organisms H. Phagocytic predators I. Crawl on rotting plant parts during part of their life cycle J. Some dwell in marine habitats K. Some dwell in freshwater habitats L. Most are terrestrial—living in the soil, on rotten logs or other vegetation	M. *Phytophthora infestans* (causes late blight in potatoes and tomatoes) N. *Plasmopara viticola* (causes downy mildew in grapes) O. *Dictyostelium discoideum* (commonly used for laboratory studies of development) P. *Physarum* Q. *Saprolegnia* (causes "ick" on aquarium fishes)

Fill-in-the-Blanks

(3) _____ [p.374] and (4) _____ _____ [p.374] are two groups of fungi that produce motile

spores; this is a primitive trait that may resemble ancestral fungi that lived several hundred million years

ago in watery habitats. Water molds are only distantly related to other fungi and are seen to have evolved

from yellow-green or (5) _____ [p.374] algae. The cells of some (6) _____ _____ [p.375]

differentiate and form (7) _____ _____ [p.375]: stalked structures bearing spores at their tips.

Unlike fungi, the slime molds are (8) _____ _____ [p.374]. Some slime-mold spores resemble the

spores of many (9) _____ [p.375]. Slime molds also spend part of their life creeping about like

(10) _____ [p.375] and engulfing food.

　　Many fungi have cells merged lengthwise, forming tubes that have thin transparent walls reinforced

with (11) _____ [p.374]; so do some (12) _____ [p.374]. Some fungal species, such as late blight,

are (13) [choose one] ❑ parasitic, ❑ saprophytic [p.374]. The vegetative body of most true fungi is a

(14) _____ [p.374], which is a mesh of branched, tubular filaments. Their metabolic activities enable

them to act as (15) _____ [p.374] in ecosystems. "Saprobes" secrete (16) _____ [p.374] into their

surroundings, where large organic molecules are broken down into smaller components that the fungal cells

then absorb. Although most protistans that resemble fungi are (17) _____ [p.374] (obtaining their

nutrients from nonliving organic matter), some are (18) _____ [p.374] and get their nutrients directly

from their living host's tissues. An example of a parasitic water mold is (19) _____ [p.374], which

causes cottony growths on some aquarium fish. (20) _____ _____ [p.374] is a major plant

pathogen on grapes.

23.2. THE ANIMAL-LIKE PROTISTANS [p.376]

23.3. AMOEBOID PROTOZOANS [pp.376–377]

23.4. CILIATED PROTOZOANS [pp.378–379]

23.5. ANIMAL-LIKE FLAGELLATES [p.380]

23.6. SPOROZOANS [p.381]

23.7. *Focus on Health:* MALARIA AND THE NIGHT-FEEDING MOSQUITOES [p.382]

Selected Words: *Entamoeba histolytica* [p.376], amoebic dysentery [p.376], multiple fission [p.376], *Amoeba proteus* [p.376], foraminiferans [p.377], radiolarians [p.377], heliozoans [p.377], *Paramecium* [p.378], hypotrichs [p.378], *Balantidium coli* [p.379], *micro*nucleus [p.379], *macro*nucleus [p.379], *Giardia lamblia* [p.380], giardiasis [p.380], *Trypanosoma brucei* [p.380], African sleeping sickness [p.380], *Trypanosoma cruzi* [p.380], *Chagas disease* [p.380], *Trichomonas vaginalis* [p.380], *Cryptosporidium* [p.381], cryptosporidiosis [p.381], *Pneumocystis carinii* [p.381], *Toxoplasma* [p.381], *toxoplasmosis* [p.381], malaria [p.382], *Plasmodium* [p.382], vaccine [p.382]

Boldfaced, Page-Referenced Terms

[p.376] protozoans _____

_____

[p.376] multiple fission _____

_____

[p.376] cyst _____

_____

[p.376] amoeboid protozoans _____

_____

[p.376] pseudopods _____

_____

[p.376] rhizopods _____

_____

[p.377] actinopods _____

_____

[p.377] plankton _____

_____

[p.378] ciliated protozoans _____

_____

[p.378] pellicle _____

_____

[p.379] contractile vacuoles _____

_____

[p.379] conjugation _____

_____

[p.380] animal-like flagellates _____

_____

[p.381] sporozoan _____

_____

Fill-in-the-Blanks

Amoebas move by sending out (1) _____ [p.376], which surround food and engulf it. (2) _____
[p.377] secrete a hard exterior covering of calcareous material that is peppered with tiny holes through
which sticky, food-trapping pseudopods extend. Needle-like (3) _____ [p.377] often support the
pseudopods in heliozoans. Accumulated shells of (4) _____ [p.377], which generally have a skeleton of
silica (glass), and foraminiferans are key components of many oceanic sediments.

 Paramecium is a ciliate that lives in (5) _____ [p.378] environments and depends on (6) _____
_____ [p.379] for eliminating the excess water constantly flowing into the cell. *Paramecium* has a
(7) _____ [p.378], a cavity that opens to the external watery world. Once inside the cavity, food
particles become enclosed in (8) _____-_____ _____ [p.378], where digestion takes place.

 Examples of flagellated protozoans that are parasitic include the (9) _____ [p.380], two species of
which cause African sleeping sickness and Chagas disease.

 (10) _____ [p.382] is a famous sporozoan that causes malaria. When a particular (11) _____
[p.382] draws blood from an infected individual, (12) _____ [p.382] of the parasite fuse to form zygotes,
which eventually develop within the mosquito.

Matching

Put as many letters in each blank as are applicable.

13. ___*Amoeba proteus* [p.376]

14. ___*Entamoeba histolytica* [p.376]

15. ___foraminiferans [p.377]

16. ___*Paramecium* [p.378]

17. ___*Plasmodium* [p.381]

18. ___*Trichomonas vaginalis* [p.380]

19. ___*Trypanosoma brucei* [p.380]

A. Ciliophora
B. Mastigophora
C. Sarcodina
D. Apicomplexa
E. Amoeboid protozoans
F. Animal-like flagellates
G. African sleeping sickness
H. Malaria
I. Amoebic dysentery
J. Sporozoans
K. Primary component of many ocean sediments
L. Ciliated protozoans

Matching

Match the following pictures with the names below.

20. ___[p.376]
21. ___[p.377]
22. ___[p.377]
23. ___[p.380]
24. ___[p.378]

A. *Amoeba proteus*
B. Flagellated protozoans
C. Foraminiferans
D. Heliozoans
E. *Paramecium*

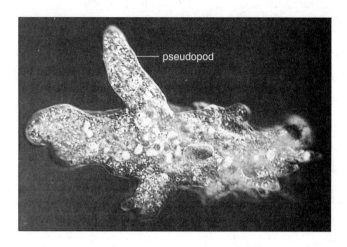

pseudopod

20.

21.

22.

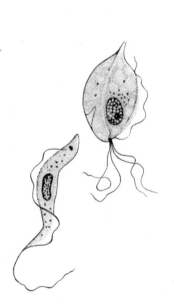

23.

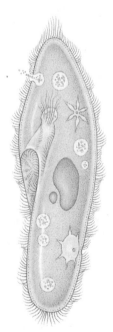

24.

23.8. REGARDING "THE ALGAE" [p.383]

23.9. EUGLENOIDS [p.383]

23.10. CHRYSOPHYTES AND DINOFLAGELLATES [pp.384–385]

23.11. RED ALGAE [p.386]

23.12. BROWN ALGAE [p.387]

23.13. GREEN ALGAE [pp.388–389]

Selected Words: *"the algae"* [p.383], Euglenophyta [p.383], *heterotrophs* [p.383], photosynthetic [p.384], *Euglena* [p.383], *"eyespot"* [p.383], Chrysophyta [p.384], diatoms [p.384], *Pfiesteria piscicida* [p.384], Rhodophyta [p.386], *nori* cultivation [p.386], phycobilins [p.386], agar [p.386], carrageenan [p.386], Phaeophyta [p.387], *Sargassum* [p.387], fucoxanthin [pp.384,387], algins [p.387], Chlorophyta [p.388], *Volvox* [p.388], *Spirogyra* [p.389], *Chlamydomonas* [p.389]

Boldfaced, Page-Referenced Terms

[p.383] phytoplankton _____

_____

[p.383] euglenoids _____

_____

[p.384] chrysophytes _____

_____

[p.384] algal bloom _____

_____

[p.384] golden algae _____

_____

[p.384] yellow-green algae _____

_____

[p.384] diatoms _____

_____

[p.384] coccolithophores _____

_____

[p.384] dinoflagellates _____

_____

[p.384] red tides _____

_____

[p.386] red algae _____

_____

[p.387] brown algae _____

_____

[p.388] green algae _____

_____

Fill-in-the-Blanks

Most euglenoids contain (1) _____ [p.383], which enable them to carry out photosynthesis. A(An) (2) _____ [p.383] of carotenoid pigment granules partly shields a light-sensitive receptor and enables *Euglena* to remain where light is optimal for its activities. Some strains of *Euglena* can be converted from photosynthetic, chloroplast-containing forms to strains that are (3) _____ [p.383].

The term (4) "_____" [p.383] no longer has formal classification significance, because organisms once lumped under that term are now assigned to a variety of different groups. (5) _____ [p.384] include 600 species of "yellow-green algae," about 500 species of "golden algae," and more than 5,600 existing species of golden-brown (6) _____ [p.384]. Except in yellow-green algae, photosynthetic chrysophytes contain xanthophylls and (7) _____ [p.384]; those pigments mask the green color of chlorophyll in golden algae and diatoms. Diatom cells have external thin, overlapping "shells" of (8) _____ [p.384] that fit together like a pillbox. 270,000 metric tons of (9) _____ _____ [p.384] are extracted annually from a quarry near Lompoc, California, and are used to make abrasives, (10) _____ [p.384] materials, and insulating materials. Dinoflagellates are mostly photosynthetic members of marine (11) _____ [p.384] and freshwater ecosystems; some forms are also heterotrophic. (12) _____ [p.384] undergo explosive population growth and color the seas red or brown, causing a red tide that may kill hundreds or thousands of fish and, occasionally, people.

Several species of red algae secrete (13) _____ [p.386] (used in culture media) as part of their cell walls. Most red algae live in (14) _____ [p.386] habitats. Some red algae have cell (15) _____ [p.386] hardened with calcium carbonate, participate in coral reef building, and are major producers. The (16) _____ [p.387] algae live offshore or in intertidal zones and have many representatives with large sporophytes known as kelps; some species produce (17) _____ [p.387], a valuable thickening agent. Green algae are thought to be ancestral to more complex plants, because they have the same types and proportions of (18) _____ [p.388] pigments, have (19) _____ [p.388] in their cell walls, and store their carbohydrates as (20) _____ [p.388].

Complete the Table

21. Complete the following table.

Type of Alga	Typical Pigments	May Originate From or Be Related to	Uses by Humans	Representatives
Red algae (Rhodophyta) [p.386]	a.	b.	c.	d.
Brown algae (Phaeophyta) [p.387]	e.	f.	g.	h.
Green algae (Chlorophyta) [pp.388–389]	i.	j.	k.	l.

Labeling and Matching

Identify each indicated part of the following illustration by entering its name in the appropriate numbered blank. Choose from the following terms: *cytoplasmic fusion, asexual reproduction, resistant zygote, fertilization, zygote, meiosis and germination, spore mitosis, gametes meet*. Complete the exercise by matching from the list below, entering the correct letter in the parentheses following each label. [All from p.389] Consult the illustration on p.248.

22. _____ ( )

23. _____ _____ ( )

24. _____ and _____ ( )

25. _____ _____ ( )

26. _____ _____ ( )

27. _____ _____ ( )

28. _____ _____ ( )

29. _____ ( )

A. Fusion of two gametes of different mating types
B. A device to survive unfavorable environmental conditions
C. More spore copies are produced
D. Fusion of two haploid nuclei
E. Haploid cells form smaller haploid gametes when nitrogen levels are low
F. Formed after fertilization
G. Two haploid gametes coming together
H. Reduction of the chromosome number

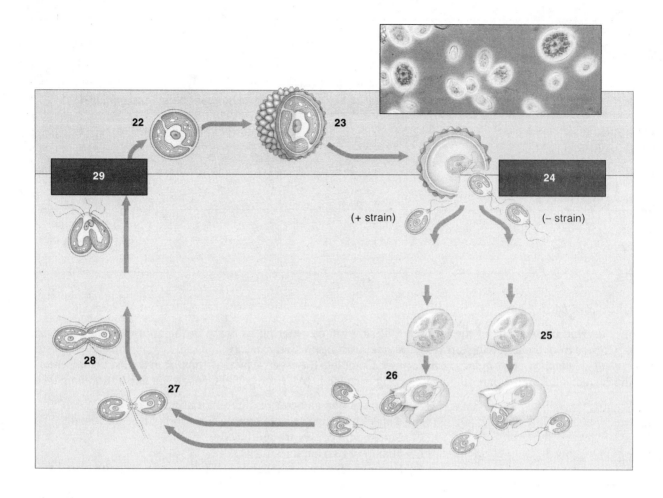

More Choice

30. For each of the following groups, indicate with a "+" if it has chloroplasts, and a "−" if members of that group lack chloroplasts and the ability to do photosynthesis.

a. brown algae [p.387]	f. protozoans [p.376]
b. chytrids [p.374]	g. red algae [p.386]
c. chrysophytes [p.384]	h. slime molds [p.375]
d. dinoflagellates [p.384]	i. sporozoans [p.381]
e. green algae [pp.388–389]	j. water molds [p.374]

Self-Quiz

___ 1. Which is *not* regarded as one of the three major protistan lineages? [p.373]
   a. red, brown and green algae
   b. blue-green algae (cyanobacteria)
   c. euglenoids, chrysophytes, and dinoflagellates
   d. chytrids, water molds, protozoans, sporozoans, and slime molds

___ 2. Which of the following specialized structures is not correctly paired with a function? [p.379]
   a. gullet—ingestion
   b. cilia—food gathering
   c. contractile vacuole—digestion
   d. chloroplast—food production

___ 3. Chytrids, water molds, and slime molds resemble the fungi in all the following characteristics *except:* [p.374]
   a. They are heterotrophs.
   b. They produce motile spores during their life cycle; fungi don't.
   c. Many are saprobic decomposers or parasites.
   d. They produce spore-bearing structures.

___ 4. Population "blooms" of _____ cause "red tides" and extensive fish kills. [p.384]
   a. *Euglena*
   b. specific dinoflagellates
   c. diatoms
   d. *Plasmodium*

___ 5. Which of the following is generally *not* a protistan characteristic? [p.391]
   a. cells usually 1–10 micrometers
   b. eukaryotic with lots of membrane-bound organelles
   c. usually undergo mitosis for cell division to occur
   d. complex chromosomes within a nucleus

___ 6. Which of the following protistans does *not* cause great misery to humans? [p.375]
   a. *Dictyostelium discoideum*
   b. *Entamoeba histolytica*
   c. *Plasmodium*
   d. *Trypanosoma brucei*

___ 7. Which of the following is not a sporozoan (parasitic protistans that must complete part of their life cycles inside specific cells of host organisms)? [p.378]
   a. *Cryptosporidium*
   b. *Pneumocystis carinii*
   c. *Plasmodium*
   d. *Paramecium*

___ 8. Red, brown, and green "algae" are found in the kingdom _____. [p.383]
   a. Plantae
   b. Monera
   c. Protista
   d. all of the above

___ 9. Red algae _____. [p.386]
   a. are primarily marine organisms
   b. are thought to have developed from green algae
   c. contain xanthophyll as their main accessory pigments
   d. all of the above

___10. Stemlike structure, leaflike blades, and gas-filled floats are found in the species of _____. [p.387]
   a. red algae
   b. brown algae
   c. bryophytes
   d. green algae

___11. Because of pigmentation, cellulose walls, and starch storage similarities, the _____ algae are thought to be ancestral to more complex plants. [p.388]
   a. red
   b. brown
   c. blue-green
   d. green

Matching

Match all applicable letters with the appropriate terms. A letter may be used more than once, and a blank may contain more than one letter.

12. _____ *Amoeba proteus* [p.376]

13. _____ diatoms [p.384]

14. _____ *Dictyostelium* [p.375]

15. _____ foraminifera [p.377]

16. _____ *Pfiesteria piscicida* [pp.384–385]

17. _____ *Paramecium* [pp.378–379]

18. _____ *Plasmodium* [p.382]

19. _____ *Phytophthora infestans* [p.374]

A. Protista
B. Slime mold
C. Water mold
D. Dinoflagellates
E. Obtain food by using pseudopodia
F. Causes malaria
G. A sporozoan
H. A ciliate
I. Live in "glass" houses
J. Live in hardened shells that have thousands of tiny holes, through which pseudopodia protrude

Matching

20. ___ [p.384]

21. ___ [p.378]

22. ___ [p.377]

23. ___ [p.376]

24. ___ [p.383]

A. *Amoeba proteus*
B. diatoms
C. *Euglena*
D. foraminiferans
E. *Paramecium*

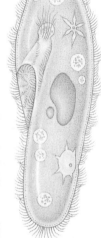

21.

20.

22.

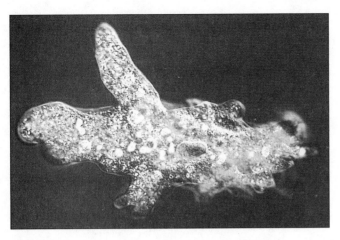

23. _____

24. _____

Chapter Objectives/Review Questions

1. Explain how heterotrophic protistans could have acquired the capacity for photosynthesis, and state the evidence to support your explanation. [p.372]
2. Discuss the contributions that protistans make to Earth's ecosystems and the ways that humans use protistans to make specific products. [pp.374–389]
3. State the principal characteristics of the amoebas, radiolarians, and foraminiferans. Indicate how they generally move from one place to another and how they obtain food. [pp.376–377]
4. Two flagellated protozoans that cause human misery are _____ and _____. [p.380]
5. Characterize the sporozoan group, identify the group's most prominent representative, and describe the life cycle of that organism. [pp.381–382]
6. List the features common to most ciliated protozoans. [pp.378–379]
7. How do golden algae resemble diatoms? [p.384]
8. Explain what causes red tides. [p.384]
9. State the outstanding characteristics of organisms of the red, brown, and green algae divisions. [pp.386–389]

Integrating and Applying Key Concepts

Explain why totally submerged aquatic plants that live in deep water never developed heterosporous life cycles.

24

FUNGI

Interactive Exercises

Ode to the Fungus Among Us [pp.392–393]

24.1. CHARACTERISTICS OF FUNGI [p.394]

Selected Words: *Lobaria* [p.392], *Polyporus* [p.392], "fungus-root" [p.392], "imperfect fungi" [p.394]

Boldfaced, Page-Referenced Terms

[p.392] symbiosis _____

_____

[p.392] mutualism _____

_____

[p.392] lichen _____

_____

[p.392] mycorrhiza (plural, mycorrhizae) _____

_____

[p.392] decomposers _____

_____

[p.392] extracellular digestion and absorption _____

_____

[p.394] fungi _____

_____

[p.394] saprobes _____

_____

[p.394] parasites _____

_____

[p.394] zygomycetes _____

_____

[p.394] sac fungi _____

_____

[p.394] club fungi _____

_____

[p.394] spores _____

_____

[p.394] mycelium (plural, mycelia) _____

_____

[p.394] hypha (plural, hyphae) _____

_____

Fill-in-the-Blanks

(1) _____ [p.392] refers to species that live closely together. In cases of (1) called (2) _____ [p.392], their interaction benefits both partners or does one of them no harm. A(n) (3) _____ [p.392] is a vegetative body in which a fungus has become intertwined with one or more photosynthetic organisms. Fungi that enter into mutualistic interactions with young tree roots form a(n) (4) _____ [p.392], which means "fungus-root." Plants benefit from fungi because they are premier (5) _____ [p.392] that break down organic compounds in their surroundings.

Matching

Choose the most appropriate answer for each term.

6. ___saprobes [p.394]

7. ___parasites [p.394]

8. ___mycelium [p.394]

9. ___fungi [p.394]

10. ___hypha [p.394]

11. ___zygomycetes, sac fungi, club fungi [p.394]

12. ___spores [p.394]

13. ___extracellular digestion and absorption [p.392]

A. Nonmotile reproductive cells or multicelled structures; often walled and germinate following dispersal from the parent body

B. Represent major lineages of fungal evolution

C. A mesh of branching fungal filaments that grows over and into organic matter, secretes digestive enzymes, and functions in food absorption

D. Fungi that obtain nutrients from nonliving organic matter and so cause its decay

E. Fungi that extract nutrients from tissues of a living host

F. External secretion of enzymes prior to nutrient intake

G. Each filament in a mycelium; consists of cells of interconnecting cytoplasm and chitin-reinforced walls

H. A richly diverse group of heterotrophs that are premier decomposers

24.2. CONSIDER THE CLUB FUNGI [pp.394–395]

Selected Words: *Agaricus brunnescens* [p.394], *Armillaria bulbosa* [p.394], *Amanita muscaria* [p.395], *dikaryotic* mycelium [p.395]

Boldfaced, Page-Referenced Terms

[p.395] mushrooms _____

_____

[p.395] basidiospores _____

_____

Matching

Choose the most appropriate answer for each term.

1. ___fungal rusts and smuts [p.394]

2. ___*Agaricus brunnescens* [p.394]

3. ___*Armillaria bulbosa* [p.394]

4. ___*Amanita muscaria* [p.395]

5. ___*Amanita phalloides* [p.395]

A. Fly agaric mushroom, causes hallucinations when eaten; ritualistic use

B. Among the oldest and largest of the club fungi

C. Destroys entire fields of wheat, corn, and other major crops

D. Common cultivated mushroom; multimillion-dollar business

E. Death cap mushroom; kills humans by liver and kidney degeneration

Fill-in-the-Blanks

The numbered items in the following illustration of a club fungus life cycle represent missing information; complete the blanks in the following narrative to supply that missing information.

The mature mushroom is actually a short-lived (6) _____ [p.395] body; its living (7) _____ [p.395] is buried in soil or decaying wood. Each mushroom consists of a cap and a stalk. Spore-producing (8) _____ [p.395] -shaped structures develop on the gills. Each bears two haploid ($n + n$) nuclei. (9) _____ [p.395] fusion occurs within the club-shaped structures, which yields a(n) (10) _____ [p.395] stage. (11) _____ [p.395] occurs within the club-shaped structures, and four haploid (12) _____ [p.395] emerge at the tip of each. The spores are released, and each may germinate into a haploid (n) mycelium. When hyphae of two compatible mating strains meet, (13) _____ [p.395] fusion occurs. Following this, a(n) (14) "_____" [p.395] ($n + n$) mycelium forms and gives rise to the spore-bearing mushrooms.

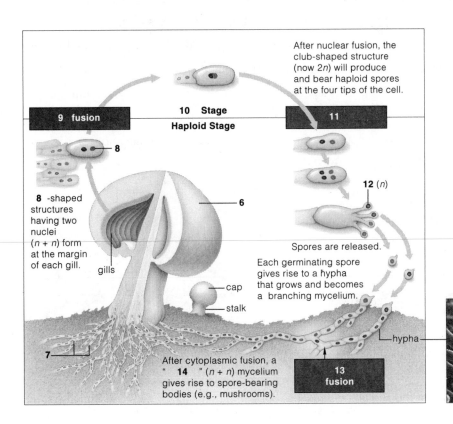

After nuclear fusion, the club-shaped structure (now $2n$) will produce and bear haploid spores at the four tips of the cell.

9 fusion

10 Stage
Haploid Stage

11

8

8 -shaped structures having two nuclei ($n + n$) form at the margin of each gill.

gills

cap

stalk

6

12 (n)

Spores are released.

Each germinating spore gives rise to a hypha that grows and becomes a branching mycelium.

hypha

7

After cytoplasmic fusion, a " 14 " ($n + n$) mycelium gives rise to spore-bearing bodies (e.g., mushrooms).

13 fusion

hypha in mycelium

24.3. SPORES AND MORE SPORES [pp.396–397]

Selected Words: *Rhizopus stolonifer* [p.396], *asci* [p.396], *Candida albicans* [p.397], *Penicillium* [p.397], *Aspergillus* [p.397], *Neurospora crassa* [p.397]

Boldfaced, Page-Referenced Terms

[p.396] zygosporangium _____

_____

[p.396] ascospores _____

_____

Fill-in-the-Blanks

The numbered items in the following illustration (*Rhizopus* life cycle) represent missing information; complete the blanks in the following narrative to supply missing information about zygomycetes.

The sexual phase begins when haploid (1) _____ [p.396] of two different mating strains (+ and –) grow into each other and fuse due to a chemical attraction. Two (2) _____ [p.396] form between two hyphae, and several haploid nuclei are produced inside each. Later, their nuclei fuse, forming a zygote with a thick protective wall, called a(n) (3) _____ [p.396]. This structure may remain dormant for several months. Meiosis proceeds, and haploid sexual (4) _____ [p.396] are produced when this structure germinates. Each gives rise to stalked structures, each with a spore sac on its tip, that can produce many spores, each of which can be the start of an extensive (5) _____ [p.396].

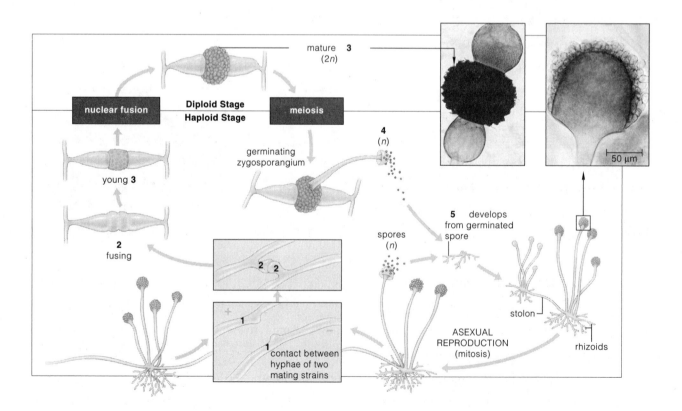

Matching

Choose the most appropriate answer for each term.

6. ___truffles and morels [p.397]
7. ___yeasts [p.397]
8. ___*Candida albicans* [p.397]
9. ___*Penicillium* spp. [p.397]
10. ___*Aspergillis* spp. [p.397]
11. ___*Neurospora crassa* [p.397]

A. "Flavor" Camembert and Roquefort cheeses; produce antibiotics
B. Has uses in genetic research
C. Fermenting by-products put to use by bakers and vintners
D. Causes vexing infections in humans
E. Highly prized edibles
F. Make citric acid for candies and soft drinks; ferment soybeans for soy sauce

24.4. THE SYMBIONTS REVISITED [pp.398–399]

24.5. *Focus on Science:* A LOOK AT THE UNLOVED FEW [p.400]

Selected Words: lichen [p.398], *myco*biont [p.398], *photo*biont [p.398], mycorrhizae [p.398], *ecto*mycorrhiza [p.398], *endo*mycorrhizae [p.398], *Cryphonectria parasitica* [p.400], *Ajellomyces capsulatus* [p.400], histoplasmosis [p.400], *Claviceps purpurea* [p.400], ergotism [p.400], *Epidermophyton floccosum* [p.400], *Venturia inaequalis* [p.400]

Fill-in-the-Blanks

(1) _____ [p.398] refers to species that live in close ecological association. In cases of symbiosis called

(2) _____ [p.398], interaction benefits both partners or does one of them no harm. A (3) _____

[p.398] is a mutualistic interaction between a fungus and one or more photosynthetic species. The fungal

part of a lichen is known as the (4) _____ [p.398], and the photosynthetic component is the

(5) _____ [p.398]. Of about 13,500 known types of lichens, nearly half incorporate (6) _____ [p.398]

fungi. A lichen forms after the tip of a fungal (7) _____ [p.398] binds with a suitable host cell. Both lose

their wall, and their (8) _____ [p.398] fuses or the hypha induces the host cell to cup around it. The

(9) _____ [p.398] and the (10) _____ [p.398] grow and multiply together. The lichen commonly has

distinct (11) _____ [p.398]. The overall pattern of growth may be leaflike, flattened, pendulous, or

erect. Lichens typically colonize places that are too (12) _____ [p.398] for most organisms. Almost

always, the (13) _____ [p.399] is the largest component of the lichen. The fungus benefits by having a

long-term source of nutrients that it absorbs from cells of the (14) _____ [p.399]. Nutrient withdrawals

affect the (15) _____ [p.399] growth a bit, but the lichen may help (16) _____ [p.399] it. If more

than one fungus is present in the lichen, it might be a mycobiont, a (17) _____ [p.399], or even an

opportunist that is using the lichen as a substrate.

Fungi also are mutualists with young tree roots, as (18) _____ [p.399]. Without (18), plants cannot

grow as (19) _____. [p.399]. In (20) _____ [p.399], hyphae form a dense net around living cells in

the roots but do not penetrate them. Ectomycorrhizae are common in (21) _____ [p.399] forests and

help trees withstand seasonal changes in temperature and rainfall. About 5,000 fungal species, mostly

(22) _____ [p.399] fungi, enter into such associations. The more common (23) _____ [p.399] form

in about 80 percent of all vascular plants. These fungal hyphae (24) _____ [p.399] plant cells, as they do in lichens. Fewer than 200 species of (25) _____ [p.399] serve as the fungal partner. Their hyphae branch extensively, forming tree-shaped (26) _____ [p.399] structures in plant cells. Hyphae also extend several centimeters into the (27) _____ [p.399].

Complete the Table

28. After reading *Focus on Science*, "A Look at the Unloved Few" [text p.400], complete the following table, which deals with five pathogenic and toxic fungi.

Fungus Latin Name	Description
[p.400] a.	Turned most of eastern North America's chestnut trees into stubby versions of their former selves
[p.400] b.	Histoplasmosis
[p.400] c.	Athlete's foot
[p.400] d.	Apple scab
[p.400] e.	Ergotism

Self-Quiz

Labeling and Matching

In the blank beneath each of the following illustrations (1–8), identify the organism by common name (or scientific name if a common name is unavailable). Then match each organism with the appropriate item (may be used more than once) from the list after the illustrations by entering the letter in the parentheses.

1. _____ ( )
    [p.394]

2. _____ ( )
    [p.401]

    A. Sac fungi
    B. Zygomycetes
    C. Club fungi
    D. Imperfect fungi
    E. Algae and fungi

3. _____ ( )
   [p.395]

4. _____ ( )
   [p.397]

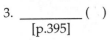

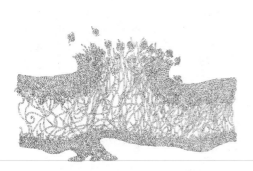

5. _____ ( )
   [p.398]

6. _____ ( )
   [p.398]

7. _____ ( )
   [p.393]

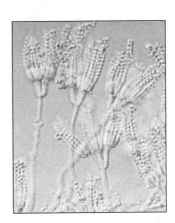

8. _____ ( )
   [p.397]

Multiple Choice

___ 9. Most true fungi send out cellular filaments called _____. [p.394]
   a. mycelia
   b. hyphae
   c. mycorrhizae
   d. asci

___ 10. Heterotrophic species of fungi can be _____. [pp.394, 398]
   a. saprobic
   b. parasitic
   c. mutualistic
   d. all of the above

Choice

For questions 11–20, choose from the following:
   a. club fungi
   b. imperfect fungi
   c. sac fungi
   d. zygomycetes

___11. The group that includes *Rhizopus stolonifer*, the notorious black bread mold, is the _____. [p.396]

___12. The group that includes delectable morels and truffles but also includes bakers' and brewers' yeasts is the _____ _____. [p.397]

___13. The group that includes shelf fungi, which decompose dead and dying trees, and mycorrhizal symbionts that help trees extract mineral ions from the soil, is the _____ _____. [pp.394,399]

___14. The group that includes the commercial mushroom, *Agaricus brunnescens*, as well as the death cap mushroom, *Amanita phalloides*, is the _____ _____. [pp.394–395]

___15. The group that includes *Penicillium*, which has a variety of species that produce penicillin and substances that flavor Camembert and Roquefort cheeses, is the _____ _____. [p.397]

___16. The group whose spore-producing structures (asci) are shaped like flasks, globes, and shallow cups is the _____ _____. [p.396]

___17. The group that forms a thin, clear covering around the zygote, the zygosporangium, is the _____. [p.396]

___18. The group whose spore-producing structures are club-shaped is the _____ _____. [p.395]

___19. The groups that are symbiotic with young roots of shrubs and trees in mycorrhizal associations are _____ and _____. [p.399]

___20. A group of puzzling kinds of fungi whose members are lumped together but do not form a formal taxonomic group. [p.397]

Chapter Objectives/Review Questions

1. Fungi are heterotrophs; most are _____ and obtain nutrients from nonliving organic matter and so cause its decay. [p.394]
2. Other fungi are _____; they extract nutrients from tissues of a living host. [p.394]
3. The common names of the three major lineages of fungi are the _____, the _____ _____ fungi, and the _____ _____ fungi. [p.394]
4. Distinguish among the meanings of the following terms: *hypha, hyphae; mycelium,* and *mycelia.* [p.394]
5. Describe the diverse appearances of the fungi classified as club fungi. [p.394]
6. List the reproductive modes of fungi. [p.394]
7. The short-lived reproductive bodies of most club fungi are known as _____. [p.395]
8. The spores produced by members of the club fungi are known as _____. [p.395]
9. A _____ mycelium is one in which the hyphae have undergone cytoplasmic fusion but not nuclear fusion. [p.395]
10. Review the generalized life cycle of a club fungus. [p.395]
11. What is the importance of the club fungi in the genus *Amanita*? [p.395]
12. Zygomycetes form sexual spores by way of _____. [p.396]
13. Review the life cycle of *Rhizopus*. [p.396]
14. Most sac fungi produce sexual spores called _____. [p.396]
15. What are some typical shapes of the reproductive structures enclosing the asci? [p.396]
16. Flavoring cheeses, producing citric acid for candies and soft drinks, food spoilage, genetic research, uses by bakers and vintners, are all activities associated with fungi known as the _____ fungi. [p.397]
17. Define *mutualism* and explain why a lichen fits that definition. [p.398]
18. Distinguish the mycobiont from the photobiont. [p.398]
19. Describe the fungus–plant root association known as mycorrhizae. [p.399]
20. Distinguish ectomycorrhizae from endomycorrhizae. [p.399]
21. What is the effect of pollution on mycorrhizae? [p.399]
22. Give the name of the fungus that causes the disease known as ergotism; list the symptoms of ergotism. [p.400]
23. Name the cause and describe the symptoms of histoplasmosis. [p.400]

Integrating and Applying Key Concepts

Suppose humans acquired a few well-placed fungal genes that caused them to reproduce in the manner of a "typical" fungus [text Figure 24.4, p.395]. Try to imagine the behavioral changes that humans would likely undergo. Would their food supplies necessarily be different? table manners? stages of their life cycle? courtship patterns? habitat? Would the natural limits to population increase be the same? Would their body structure change? Would there necessarily have to be separate sexes? Compose a descriptive science-fiction tale about two mutants that find each other and set up "housekeeping" together.

25

PLANTS

Interactive Exercises

Pioneers in a New World [pp.402–403]

25.1. EVOLUTIONARY TRENDS AMONG PLANTS [pp.404–405]

*Selected Words: non*vascular plants [p.404], *seedless* vascular plants [p.404], *seed-bearing* vascular plants [p.404], *haploid (n)* phase [p.404], diploid (2*n*) phase [p.404], *hetero*spory [p.405], *homo*spory [p.405]

Boldfaced, Page-Referenced Terms

[p.404] vascular plants _____

_____

[p.404] bryophytes _____

_____

[p.404] gymnosperms _____

_____

[p.404] angiosperms _____

_____

[p.404] root systems _____

_____

[p.404] shoot systems _____

_____

[p.404] lignin _____

_____

[p.404] xylem _____

_____

[p.404] phloem _____

_____

[p.404] cuticle _____

_____

[p.404] stomata _____

_____

[p.404] gametophytes _____

_____

[p.404] sporophyte _____

_____

[p.404] spores _____

_____

[p.405] pollen grains _____

_____

[p.405] seed _____

_____

Matching

Choose the most appropriate answer for each term.

1. ___ seedless vascular plants [p.404]
2. ___ angiosperms [p.404]
3. ___ vascular plants [p.404]
4. ___ bryophytes [p.404]
5. ___ gymnosperms [p.404]

A. Liverworts, hornworts, and mosses
B Seed-bearing plants that include cycads, ginkgo, gnetophytes, and conifers
C. Whisk ferns, lycophytes, horsetails, and ferns
D. A group of plants that bear flowers and seeds
E. In general, a large number of plants having internal tissues that conduct water and solutes through the plant body

Complete the Table

6. As plants evolved, several key evolutionary developments occurred that solved the problems of living in new land environments. Complete the following table to summarize these events. Choose from heterospory, xylem and phloem, seed, shoot systems, sporophytes, spores, lignin, gametophytes, root systems, stomata, pollen grains, and cuticle.

Evolutionary Trends	Survival Problem Solved
[p.404] a.	Provides a large surface area for rapidly taking up soil water and dissolved mineral ions; often anchors the plant
[p.404] b.	Consist of stems and leaves that function efficiently in the absorption of sunlight energy and CO_2 from the air
[p.404] c.	Allows extensive growth of stems and branches; a very hard organic substance that strengthens cell walls, thus allowing erect plant parts to display leaves to sunlight
[p.404] d.	Provides cellular pipelines to distribute water, dissolved ions, and solutes such as dissolved sugars to all living plant cells
[p.404] e.	A waxy coat on many plant organs that helps conserve water on hot, dry days
[p.404] f.	Tiny openings across the surfaces of stems and some leaves that help control the absorption of CO_2 and restrict evaporative water loss
[p.404] g.	The haploid (n) phase that dominates the life cycles of algae
[p.404] h.	The diploid ($2n$) phase that dominates the life cycles of most plants
[p.404] i.	Haploid cells produced by meiosis in sporophyte plants; later divide by mitosis to give rise to the gametophytes
[p.405] j.	Condition in some seedless plant species and seed-bearing plants where two kinds of spores are produced
[p.405] k.	Developed from one type of spore in gymnosperms and angiosperms; in turn develops into mature, sperm-bearing male gametophytes
[p.405] l.	Consists of an embryo sporophyte, nutritive tissues, and a protective coat

25.2. BRYOPHYTES [pp.406–407]

Selected Words: Polytrichum [p.406], *Sphagnum* [p.407]

Boldfaced, Page-Referenced Terms

[p.406] mosses _____

_____

[p.406] liverworts _____

_____

[p.406] hornworts _____

_____

[p.407] peat bogs _____

_____

True–False

If the statement is true, write a T in the blank. If the statement is false, make it correct by changing the underlined word(s) and writing the correct word(s) in the answer blank.

_____ 1. Mosses are especially sensitive to <u>water</u> pollution. [p.406]

_____ 2. Bryophytes <u>have</u> leaflike, stemlike, and rootlike parts although they do not contain xylem or phloem. [p.406]

_____ 3. Most bryophytes have <u>rhizomes</u>, elongated cells or threads that attach gametophytes to soil and serve as absorptive structures. [p.406]

_____ 4. Bryophytes are the simplest plants to exhibit a cuticle, cellular jackets around parts that produce sperms and eggs, and large gametophytes that do not depend on <u>sporophytes</u> for nutrition. [p.406]

_____ 5. The true <u>liverworts</u> are the most common bryophytes. [p.406]

_____ 6. Following fertilization, zygotes give rise to <u>gametophytes</u>. [p.407]

_____ 7. Each <u>sporophyte</u> consists of a stalk and a jacketed structure in which spores will develop. [p.407]

_____ 8. <u>Club</u> moss is a bog moss whose large, dead cells in their leaflike parts soak up five times as much water as cotton does. [p.407]

_____ 9. Bryophyte sperm reach eggs by movement through <u>air</u>. [p.407]

_____10. Bryophytes are <u>vascular</u> plants with flagellated sperm. [p.407]

Fill-in-the-Blanks

The numbered items in the following illustration represent missing information about a typical moss life cycle; complete the blanks in the following narrative to supply that missing information.

The gametophytes are the green leafy "moss plants." Sperm develop in jacketed structures at the shoot tip of the male (11) _____ [pp.406–407], and eggs develop in jacketed structures at the shoot tip of the female (12) _____ [pp.406–407]. Raindrops or a film of water on plant surfaces transport (13) _____ [pp.406–407] to the egg-producing structure. (14) _____ [pp.406–407] occurs within the egg-producing structure. The (15) _____ [pp.406–407] grows and develops into a mature (16) _____ [pp.406–407] (with sporangium and stalk) while attached to the gametophyte. (17) _____ [pp.406–407] occurs within the sporangium of the sporophyte where haploid (18) _____ [pp.406–407] form, develop, and are released. The released spores germinate with some growing and developing into male or female (19) _____ [pp.406–407].

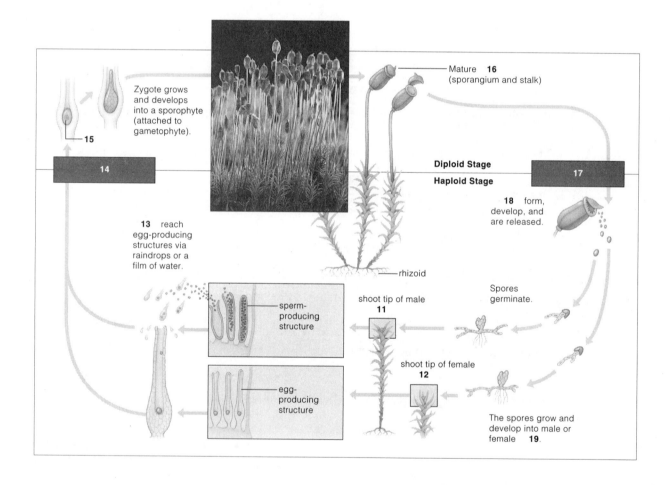

25.3. EXISTING SEEDLESS VASCULAR PLANTS [pp.408–409]

25.4. *Focus on the Environment:* ANCIENT CARBON TREASURES [p.410]

Selected Words: *Cooksonia* [p.408], *Psilotum* [p.408], *Lycopodium* [p.408], *Selaginella* [p.408], *Equisetum* [p.408], *epiphyte* [p.409], lycophyte trees [p.410], *Calamites* [p.410], "fossil fuels" [p.410]

Boldfaced, Page-Referenced Terms

[p.408] whisk ferns _____

_____

[p.408] lycophytes _____

_____

[p.408] horsetails _____

_____

[p.408] ferns _____

_____

[p.408] rhizomes _____

_____

[p.409] strobilus (plural, strobili) _____

_____

[p.410] coal _____

_____

Choice

For questions 1–20, choose from the following.

       a. whisk ferns   b. lycophytes   c. horsetails   d. ferns   e. applies to a, b, c, and d

1. ___ A group in which only the genus *Equisetum* survives [p.408]
2. ___ Familiar club mosses growing as mats on forest floors [p.408]
3. ___ Seedless vascular plants [p.408]
4. ___ *Psilotum* [p.408]
5. ___ Rust-colored patches, the sori, are on the lower surface of their fronds. [p.409]
6. ___ Some tropical species are the size of trees. [p.409]
7. ___ Ancestral plants living in the Carboniferous; through time, heat, and pressure, these plants became peat and coal, the "ancient carbon treasures." [p.410]
8. ___ Stems were used by pioneers of the American West to scrub cooking pots. [p.409]
9. ___ The sporophytes have no roots or leaves. [p.408]
10. ___ Mature leaves are usually divided into leaflets. [p.409]
11. ___ The sporophyte has vascular tissues. [pp.408–409]
12. ___ When the spore chamber snaps open, spores catapult through the air. [p.409]

13. ___ Grow in mud soil of streambanks and in disturbed habitats, such as roadsides and vacant lots [p.408]

14. ___ Sporophytes have rhizomes and hollow, photosynthetic, aboveground stems with scale-shaped leaves. [p.409]

15. ___ The sporophyte is the larger, longer-lived phase of the life cycle. [p.408]

16. ___ *Lycopodium* [p.408]

17. ___ The young leaves are coiled into the shape of a fiddlehead. [p.409]

18. ___ A germinating spore develops into a small, green, heart-shaped gametophyte. [p.409]

19. ___ *Selaginella* is heterosporous and produces two spores. [p.408]

20. ___ "Amphibians" of the plant kingdom; life cycles require water [p.408]

Fill-in-the-Blanks

The numbered items on the following illustration of a generalized fern life cycle represent missing information; complete the blanks in the following narrative to supply that missing information.

Fern leaves (fronds) of the sporophyte are usually divided into leaflets. The underground stem of the sporophyte is termed a (21) _____ [p.409]. On the undersides of many fern fronds, rust-colored patches, each of which is called a (22) _____ [p.409], is composed of spore chambers. (23) _____ [p.409] of diploid cells within each sporangium produces haploid (24) _____ [p.409]. The (25) _____ [p.409] are catapulted into the air when each spore chamber snaps open. A spore may germinate and grow into a (26) _____ [p.409] that is small, green, and heart-shaped. Jacketed structures develop on the underside of the mature (27) _____ [p.409]. Each male jacketed structure produces many (28) _____ [p.409], whereas each female jacketed structure produces a single (29) _____ [p.409]. These gametes meet in (30) _____ [p.409]. The diploid (31) _____ [p.409] is first formed inside the female jacketed structure; it divides to form the developing (32) _____ [p.409], still attached to the gametophyte.

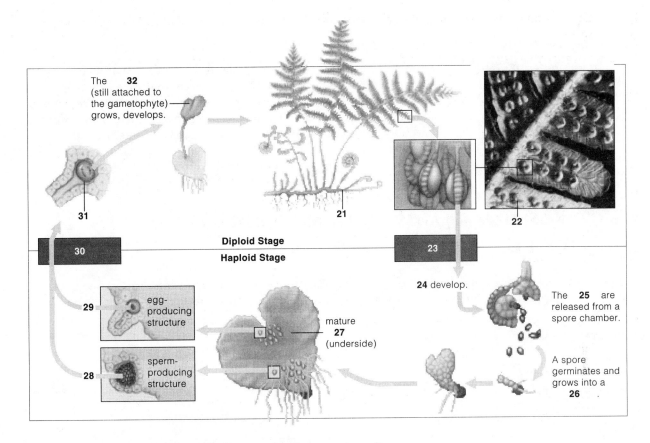

The **32** (still attached to the gametophyte) grows, develops.

31

30 **Diploid Stage**

Haploid Stage

21

22

23

24 develop.

29 egg-producing structure

mature **27** (underside)

28 sperm-producing structure

The **25** are released from a spore chamber.

A spore germinates and grows into a **26**.

22.5. THE RISE OF THE SEED-BEARING PLANTS [p.411]

Selected Words: *Pinus* [p.411], *Prunus* [p.411]

Boldfaced, Page-Referenced Terms

[p.411] microspores _____

_____

[p.411] pollination _____

_____

[p.411] megaspores _____

_____

[p.411] ovules _____

_____

[p.411] seed ferns _____

_____

[p.411] progymnosperms _____

_____

Matching

Choose the most appropriate answer for each term.

1. ___microspores [p.411]
2. ___pollination [p.411]
3. ___megaspore [p.411]
4. ___ovule [p.411]
5. ___seed ferns [p.411]

A. Spore type found in seed-bearing plants that develop within ovules
B. Female reproductive parts, which are seeds at maturity
C. Spore type found in seed-bearing plants that gives rise to pollen grains
D. Formerly dominant seed-bearing plant group along with gymnosperms and the later angiosperms; replaced by cycads, conifers, and other gymnosperms
E. The name for the arrival of pollen grains on the female reproductive structures

25.6. GYMNOSPERMS—PLANTS WITH "NAKED" SEEDS [pp.412–413]

25.7. A CLOSER LOOK AT THE CONIFERS [pp.414–415]

Selected Words: gymnos [p.412], sperma [p.412], evergreen [p.412], deciduous [p.412], Zamia [p.412], Ginkgo biloba [p.413], Gnetum [p.413], Ephedra [p.413], Welwitschia mirabilis [p.413]

Boldfaced, Page-Referenced Terms

[p.412] conifers _____

_____

[p.412] cones _____

_____

[p.412] cycads _____

_____

[p.413] ginkgos _____

_____

[p.413] gnetophytes _____

_____

[p.415] deforestation _____

_____

Choice

For questions 1–12, choose from the following:

    a. cycads   b. ginkgos   c. gnetophytes   d. conifers   e. gymnosperms (includes a, b, c, d)

1. ___ Fleshy-coated seeds of female trees produce an awful stench. [p.413]

2. ___ Includes pines and redwoods [p.412]

3. ___ Seeds and a flour made from the trunk are edible after removal of poisonous alkaloids. [p.412]

4. ___ Only a single species survives, the maidenhair tree. [p.413]

5. ___ Includes *Welwitschia* of hot deserts of south and west Africa, *Gnetum* of humid tropical regions, and *Ephedra* of California deserts and other arid regions [p.413]

6. ___ Form pollen-bearing cones and seed-bearing cones on separate plants; leaves superficially resemble those of palm trees [p.412]

7. ___ Seeds are mature ovules. [pp.412–413]

8. ___ The favored male trees are now widely planted; they have attractive, fan-shaped leaves and are resistant to insects, disease, and air pollutants. [p.413]

9. ___ Their ovules and seeds are not covered; they are borne on surfaces of spore-producing reproductive structures. [pp.412–413]

10. ___ Some sporophyte plants in this group mainly have a deep-reaching taproot; the exposed part is a woody, disk-shaped stem bearing cones and one or two strap-shaped leaves that split lengthwise repeatedly as the plant ages. [p.413]

11. ___ Most species are "evergreen" trees and shrubs with needlelike or scalelike leaves. [p.412]

12. ___ Includes conifers, cycads, ginkgos, and gnetophytes [pp.412–413]

Fill-in-the-Blanks

The numbered items in the following illustration represent missing information; complete the blanks in the following narrative to supply that missing information.

The familiar pine tree, a conifer, represents the mature (13) _____ [p.414]. Pine trees produce two kinds of spores in two kinds of cones. Pollen grains are produced in male (14) _____ [p.414]. Ovules are produced in young female (15) _____ [p.414]. Inside each (16) _____ [p.414], (17) _____ [p.414] occurs to produce haploid megaspores; one develops into a many-celled female (18) _____ [p.414] that contains haploid (19) _____ [p.414]. Diploid cells within pollen sacs of male cones undergo (20) _____ [p.414] to produce haploid microspores. Microspores develop into (21) _____ [p.414] grains. (22) _____ [p.414] occurs when spring air currents deposit pollen grains near ovules of female cones. A pollen (23) _____ [p.414] representing a male gametophyte grows toward the female gametophyte. (24) _____ [p.414] nuclei form within the pollen tube as it grows toward the egg. (25) _____ [p.414] follows, and the ovule becomes a seed that is composed of an outer seed coat, the (26) _____ [p.414] diploid sporophyte plant, and nutritive tissue.

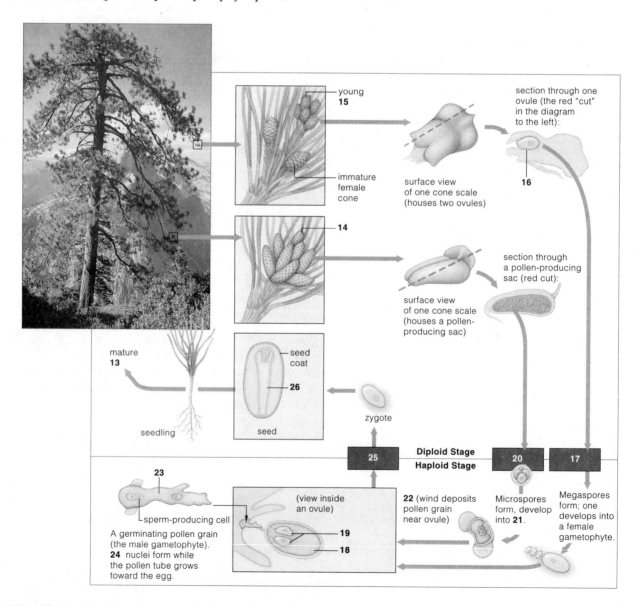

25.8. ANGIOSPERMS—THE FLOWERING, SEED-BEARING PLANTS [p.416]
25.9. VISUAL OVERVIEW OF FLOWERING PLANT LIFE CYCLES [p.417]
25.10. *Commentary:* SEED PLANTS AND PEOPLE [pp.418–419]

Selected Words: angeion [p.416], *Nymphaea* [p.416], *Arceuthobium* [p.416], *Monotropa uniflora* [p.416], *Eucalyptus* trees [p.416], large sporophyte [p.416], endosperm [p.416], seeds [p.416], fruits [p.416], *Lilium* [p.417], "double" fertilization [p.417], embryo [p.417], *Homo erectus* [p.418], *Agave* [p.418], Mexican cockroach plants [p.418], neem tree leaves [p.418], *Digitalis* [p.418], *Triticum* [p.418], *Secale* [p.418], *Saccharum* [p.418], *Theobroma* [p.418], *Aloe vera* [p.419], *Nicotiana* [p.419], *Cannabis* [p.419], *Hyoscyamus* [p.419]

Boldfaced, Page-Referenced Terms

[p.416] flowers _____

_____

[p.416] pollinators _____

_____

[p.416] dicots _____

_____

[p.416] monocots _____

_____

Matching

Choose the most appropriate answer for each term.

1. ___examples of monocot plants [p.416]
2. ___flowers [p.416]
3. ___pollinators [p.416]
4. ___double fertilization [p.417]
5. ___fruits [p.416]
6. ___endosperm [p.416]
7. ___examples of dicot plants [p.416]
8. ___seed [p.416]

A. Nutritive tissue found within a seed
B. Orchids, palms, lilies, and grasses, including rye, sugarcane, corn, rice, and wheat
C. Unique angiosperm reproductive structures
D. Mature ovaries; protect and help disperse plant embryos
E. Insects, bats, birds, and other animals that withdraw nectar or pollen from a flower and, in so doing, transfer pollen to its female reproductive parts
F. Among all plants, unique to flowering plant life cycles; one sperm fertilizes the egg, the other sperm fertilizes a cell that gives rise to endosperm
G. Packaged in fruits; each covered by a protective tissue and containing an embryo and nutritive tissue
H. Most herbaceous plants, such as cabbages and daisies; most flowering shrubs and trees, such as oaks and apple trees; water lilies and cacti

Matching

Choose the most appropriate answer for each term.

9. ___Agave [p.418]
10. ___Digitalis [p.418]
11. ___Secale [p.418]
12. ___Triticum [p.418]
13. ___Theobroma [p.418]
14. ___Saccharum officinarum [p.418]
15. ___Aloe vera [p.419]
16. ___Nicotiana [p.419]
17. ___Cannabis sativa [p.419]
18. ___Hyoscyamus niger [p.419]
19. ___neem tree [p.419]

A. tobacco
B. common bread wheat
C. leaf extracts kill nematodes, insects, and mites but not the natural predators of these common pests
D. source of marijuana and other mind-altering substances
E. sugarcane
F. extracts stabilize the heartbeat and blood circulation
G. used to make twine and ropes from century plants
H. henbane, source of belladonna and other alkaloids
I. wheat and rye
J. cocoa butter or chocolate
K. juices soothe sun-damaged skin

Self-Quiz

Complete the Table

1. Complete the following table to compare the plant groups studied in this chapter.

Plant Group	Dominant Generation	Vascular Tissue Present?	Seeds Present?
[pp.406–407] a. Bryophytes			
[p.408] b. Lycophytes			
[pp.408–409] c. Horsetails			
[p.409] d. Ferns			
[pp.412–414] e. Gymnosperms			
[pp.416–417] f. Angiosperms			

Multiple Choice

___ 2. The plants around you today are descendants of ancient species of _____ [p.402]
a. brown algae
b. green algae
c. bryophytes
d. red algae

___ 3. The _____ is *not* a trend in the evolution of plants. [pp.404–405]
a. evolution of roots, stems, and leaves
b. shift from homospory to heterospory
c. shift from diploid to haploid dominance
d. development of xylem and phloem
e. development of cuticles and stomata

___ 4. Existing nonvascular plants do *not* include _____. [pp.406–407]
   a. horsetails
   b. mosses
   c. liverworts
   d. hornworts

___ 5. Plants possessing xylem and phloem are called _____ plants. [p.404]
   a. gametophyte
   b. nonvascular
   c. vascular
   d. seedless

___ 6. Bryophytes _____. [p.406]
   a. have vascular systems that enable them to live on land
   b. include lycopods, horsetails, and ferns
   c. have true roots but not stems
   d. include mosses, liverworts, and hornworts

___ 7. _____ are *not* seedless vascular plants. [p.408]
   a. Lycophytes
   b. Gymnosperms
   c. Horsetails
   d. Whisk ferns
   e. Ferns

___ 8. In horsetails, lycopods, and ferns, _____. [pp.408–409]
   a. spores give rise to gametophytes
   b. the dominant plant body is a gametophyte
   c. the sporophyte bears sperm- and egg-producing structures
   d. all of the above

___ 9. _____ are seed plants. [pp.411–413]
   a. Cycads and ginkgos
   b. Conifers
   c. Angiosperms
   d. all of the above

___10. In complex land plants, the diploid stage is resistant to adverse environmental conditions such as dwindling water supplies and cold weather. The diploid stage progresses through this sequence: _____. [p.411]
   a. gametophyte → male and female gametes
   b. spores → sporophyte
   c. zygote → sporophyte
   d. zygote → gametophyte

___11. Monocots and dicots are groups of _____. [p.416]
   a. gymnosperms
   b. club mosses
   c. angiosperms
   d. horsetails

Chapter Objectives/Review Questions

1. Every plant you see today is a descendant of ancient species of _____ algae that lived near the water's edge or made it onto land. [p.402]
2. Most of the members of the plant kingdom are _____ plants, with internal tissues that conduct water and solutes through roots, stems, and leaves. [p.404]
3. Distinguish between the seed-bearing vascular plants known as gymnosperms and those known as angiosperms. [p.404]
4. State the general functions of the root systems and shoot systems of vascular plants. [p.404]
5. Explain the significance of "cells with lignified walls." [p.404]
6. _____ tissue distributes water and dissolved ions to all of the plant's living cells; _____ tissue distributes dissolved sugars and other photosynthetic products. [p.404]
7. Name the plant structures that protect leaves and young stems from water loss, and also name the plant structures serving as routes for absorbing carbon dioxide and controlling evaporative water loss. [p.404]
8. Give the reasons that diploid dominance allowed plants to successfully exploit the land environment. [pp.404–405]
9. As plants evolved two spore types (heterospory), one spore type develops into pollen grains that become mature sperm-bearing male _____; the other spore type develops into female _____, where eggs form and later become fertilized. [p.405]
10. The combination of a plant embryo, nutritive tissues, and protective tissues constitutes a _____. [p.405]
11. Mosses, liverworts, and hornworts belong to a plant group called the _____. [p.406]

12. Describe and state the functions of rhizoids. [p.406]
13. What group of plants first displayed cuticles, cellular jackets around the parts that produce sperms and eggs, and large gametophytes that retain sporophytes? [p.406]
14. The remains of peat mosses accumulate into compressed, excessively moist mats called _____ _____. [p.407]
15. List the four groups of seedless vascular plants. [p.408]
16. The _____ is the larger, longer-lived phase of the seedless vascular plants. [p.408]
17. Describe structural characteristics of *Psilotum*, *Lycopodium*, *Equisetum*, and a fern; and generally describe their life cycles. [pp.408–409]
18. _____ are underground, branching, short, mostly horizontal absorptive stems. [p.408]
19. Some sporophytes of club mosses have nonphotosynthetic, cone-shaped clusters of leaves known as a _____ that bears spore-producing structures. [p.408]
20. Explain the meaning of the general term *epiphyte*. [p.409]
21. Describe the sources of and the formation of coal. [p.410]
22. Why is coal said to be a nonrenewable source of energy? [p.410]
23. Microspores give rise to _____ _____. [p.411]
24. Define the term *pollination*. [p.411]
25. An _____ contains the female gametophyte, surrounded by nutritive tissue, and a jacket of cell layers. [p.411]
26. _____ develop inside ovules. [p.411]
27. _____ may have been the first seed-bearing plants, although their seeds may have been seedlike structures. [p.411]
28. Conifers, cycads, ginkgos, and gnetophytes are all members of the _____ lineage. [p.412]
29. Describe the structure of a typical conifer cone. [p.412]
30. Briefly characterize plants known as cycads, ginkgos, and gnetophytes. [pp.412–413]
31. List reasons why *Ginkgo biloba* is a unique plant. [p.413]
32. *Gnetum*, *Ephedra*, and *Welwitschia* represent genera of _____. [p.413]
33. Describe the life cycle of *Pinus*, a somewhat typical gymnosperm. [p.414]
34. _____ is the removal of all trees from large tracts, as by clear-cutting. [p.415]
35. Only angiosperms produce reproductive structures known as _____. [p.416]
36. Most flowering plants coevolved with _____ such as insects, bats, birds, and other animals that withdraw nectar or pollen from a flower. [p.416]
37. Name and cite examples of the two classes of flowering plants. [p.416]
38. _____, a nutritive tissue, surrounds embryo sporophytes inside the seeds of flowering plants. [p.416]
39. As seeds develop, tissues of most ovaries and other structures mature into _____. [p.416]
40. List human uses for the following plants: *Agave*, neem tree, *Digitalis*, *Secale*, *Triticum*, *Theobroma cacao*, *Saccharum officinarum*, *Aloe vera*, *Nicotiana*, *Cannabis sativa*, and *Hyoscyamus*.

Integrating and Applying Key Concepts

Considering the pine life cycle, how many haploid (*n*) and diploid (2*n*) phases are represented within a pine seed?

26

ANIMALS: THE INVERTEBRATES

Interactive Exercises

Madeleine's Limbs [pp.422–423]

26.1. OVERVIEW OF THE ANIMAL KINGDOM [pp.424–425]
26.2. PUZZLES ABOUT ORIGINS [p.426]
26.3. SPONGES—SUCCESS IN SIMPLICITY [pp.426–427]

Selected Words: anterior end [p.425], *posterior* end [p.425], *dorsal* surface [p.425], *ventral* surface [p.425], pseudocoel [p.425], *Paramecium* [p.426], *Volvox* [p.426], *Trichoplax adhaerens* [p.426], *Euplectella* [p.427]

Boldfaced, Page-Referenced Terms

[p.424] animals _____

[p.424] ectoderm _____

_____

[p.424] endoderm _____

_____

[p.424] mesoderm _____

_____

[p.424] vertebrates _____

_____

[p.424] invertebrates _____

_____

[p.424] bilateral symmetry _____

_____

[p.425] cephalization _____

_____

[p.425] gut _____

_____

[p.425] coelom _____

_____

[p.426] placozoan _____

_____

[p.426] sponges _____

_____

[p.426] collar cells _____

_____

[p.427] larva (plural, larvae) _____

_____

Matching

Choose the most appropriate answer for each term.

1. ___animals [p.424]
2. ___ventral surface [p.425]
3. ___ectoderm, endoderm, mesoderm [p.424]
4. ___anterior end [p.425]
5. ___vertebrates [p.424]
6. ___invertebrates [p.424]
7. ___radial symmetry [p.424]
8. ___bilateral symmetry [p.424]
9. ___dorsal surface [p.425]
10. ___gut [p.425]
11. ___coelom [p.425]
12. ___thoracic cavity [p.425]
13. ___abdominal cavity [p.425]
14. ___posterior end [p.425]
15. ___segmentation [p.425]
16. ___cephalization [p.425]

A. All animals whose ancestors evolved before backbones did
B. An evolutionary process whereby sensory structures and nerve cells became concentrated in a head
C. The back surface
D. Animal body cavity lined with a peritoneum—found in most bilateral animals; some worms lack this cavity, other worms have a false cavity
E. Head end
F. Animals having body parts arranged regularly around a central axis, like spokes of a bike wheel
G. Upper coelom cavity holding a heart and lungs
H. Primary tissue layers that give rise to all adult animal tissues and organs
I. Surface opposite the dorsal surface
J. Region inside animal body in which food is digested
K. Animals having right and left halves that are mirror images of each other
L. Series of animal body units that may or may not be similar to one another
M. End opposite the anterior end
N. Lower coelom cavity holding a stomach, intestines, and other organs
O. Multicellular organisms with tissues forming organs and organ systems, diploid body cells, heterotrophic, aerobic respiration, sexual reproduction, sometimes asexual, most are motile in some part of the life cycle, and the life cycle shows embryonic development
P. Animals with a "backbone"

Complete the Table

17. Complete the following table by filling in the appropriate phylum or representative group name.

Phylum	Some Representative	Number of Known Species
a. [p.424]	*Trichoplax*; simplest animal	1
b. Porifera [p.424]		8,000
c. [p.424]	Hydrozoans, jellyfishes, corals, sea anemones	11,000
d. Platyhelminthes [p.424]		15,000
e. [p.424]	Pinworms, hookworms	20,000
f. [p.424]	Tiny body with crown of cilia, great internal complexity; "wheel animals"	1,800
g. Mollusca [p.424]		110,000
h. [p.424]	Leeches, earthworms, polychaetes	15,000
i. Arthropoda [p.424]		1,000,000
j. [p.424]	Sea stars, sea urchins	6,000

Choice

For questions 18–27, answer questions about animal origins by choosing from the following:

a. *Paramecium*   b. *Volvox*   c. *Trichoplax adhaerens*
d. different animal lineages arose from more than one group of protistanlike ancestors

18. ___ The only known placozoan [p.426]

19. ___ Similar ciliate forerunners may have had multiple nuclei within a single cell [p.426]

20. ___ Similar to colonies that became flattened and crept on the seafloor [p.426]

21. ___ The answer to animal origins might require more than one answer [p.426]

22. ___ As simple as an animal can get [p.426]

23. ___ By another hypothesis, multicelled animals arose from flagellated cells that live in similar hollow, spherical colonies [p.426]

24. ___ A soft-bodied marine animal, shaped a bit like a tiny pita bread [p.426]

25. ___ By one hypothesis, the animal forerunners were ciliates, much like this organism [p.426]

26. ___ Has no symmetry and no mouth [p.426]

27. ___ In a similar organism, the division of labor characterizing multicellularity might have begun [p.426]

Matching

Choose the most appropriate answer for each term.

28. ___fragmentation [p.427]

29. ___sponge phylum [p.426]

30. ___adult [p.427]

31. ___larva [p.427]

32. ___amoeboid cells [pp.426–427]

33. ___sponge skeletal elements [p.426]

34. ___Trichoplax [p.426]

35. ___microvilli [pp.426–427]

36. ___collar cells [pp.426–427]

37. ___water entering a sponge body [pp.426–427]

38. ___gemmules [p.427]

A. Flagellated cells that absorb and move water through a sponge as well as engulf food
B. Reside in a gelatinous substance between inner and outer cell linings
C. An organism whose two cell layers resemble those of a sponge
D. Sexually mature form of a species
E. Form the "collars" of collar cells
F. Clusters of sponge cells capable of germinating and establishing new colonies
G. Microscopic pores and chambers
H. Random chunks of sponge tissue break off and grow into more sponges
I. Spongin fibers, glasslike spicules of silica or calcium carbonate or both
J. Sexually immature form preceding the adult
K. Porifera

26.4. CNIDARIANS—TISSUES EMERGE [pp.428–429]

26.5. VARIATIONS ON THE CNIDARIAN BODY PLAN [pp.430–431]

26.6. COMB JELLIES [p.431]

Selected Words: *Hydra* [p.428], *Chrysaora* [p.428], gastrodermis [p.428], *Obelia* [p.430], *Physalia* [p.431], *Cestum veneris* [p.431], *Mnemiopsis* [p.431]

Boldfaced, Page-Referenced Terms

[p.428] Cnidaria _____

_____

[p.428] nematocysts _____

_____

[p.428] medusa _____

_____

[p.428] polyp _____

_____

[pp.428–429] epithelium (plural, epithelia) _____

_____

[p.429] nerve cells _____

_____

[p.429] contractile cells _____

_____

[p.429] hydrostatic skeleton _____

_____

[p.430] gonads _____

_____

[p.430] planulas _____

_____

[p.431] comb jellies _____

_____

Fill-in-the-Blanks

All members of the phylum (1) _____ [p.428] are radial animals; they include jellyfishes, tentacled sea anemones, corals, and animals such as *Hydra*. Most of these animals live in the sea and they alone produce (2) _____ [p.428], capsules capable of discharging threads that entangle or pierce prey. Cnidarians have two common body plans, the (3) _____ [p.428] that look like bells or upside down saucers, and the (4) _____ [p.428] that has a tubelike body with a tentacle-fringed mouth at one end. The saclike cnidarian gut processes food with its (5) _____ [p.428], a sheetlike lining with glandular cells that secrete digestive enzymes. An (6) _____ [p.428] lines the rest of the body's surfaces. Each of these linings is referred to as a(n) (7) _____ [p.428], a tissue with a free surface that faces the environment or some type of fluid inside the body. Cnidaria epithelia house (8) _____ [p.429] cells, which receive signals from receptors that sense changes in the surroundings and send signals to (9) _____ [p.429] cells that can carry out suitable responses. The nerve cells interact as a (10) "_____ net" [p.429], a simple nervous tissue, to control movement and changes in shape.

The (11) _____ [p.429] is a layer of gelatinous secreted material that lies between the epidermis and gastrodermis. (12) _____ [p.429] contain enough mesoglea to impart buoyancy and serve as a firm yet deformable skeleton against which contractile cells act.

Any fluid-filled cavity or cell mass against which contractile cells can act is a (13) _____ [p.429] skeleton. The contractile cells of most polyps, which have little (14) _____ [p.429], act against water in their gut.

Many cnidarian species have only a polyp or a medusa stage in the life cycle with the medusa being the sexual form. However, some cnidarians such as *Obelia* and *Physalia* exhibit (15) _____ [p.430] body forms in their life cycles. They have simple (16) _____ [p.430] that rupture and release gametes. Zygotes formed at fertilization develop into (17) _____ [p.430], a kind of swimming or creeping larva that usually possesses ciliated epidermal cells. In time, a mouth opens at one end transforming the larva into a polyp or a medusa, and the cycle begins anew.

The (18) _____ [p.430]-forming corals and other colonial anthozoans are a fine example of variation on the basic cnidarian body plan. The (18)-formers thrive in clear, warm water that is at least (19) _____ (temperature) [p.430]. The colonies consist of polyps that have secreted (20) _____ [p.430]-reinforced skeletons, which interconnect with one another. Over time, the (21) _____ [p.431] accumulate and so become the main building materials for reefs such as the Great Barrier Reef. Reef-building corals receive

(22) _____ [p.431] inputs from the changing tides and from dinoflagellate mutualists living in their tissues.

As a final example of cnidarian diversity, consider *Physalia*, informally called the (23) _____ [p.431] man-of-war. The (24) _____ in the nematocysts of this infamous hydrozoan poses a danger to bathers and fishermen as well as to prey organisms (fish). Although *Physalia* lives mainly in warm waters, currents sometimes move it up to the (25) _____ [p.431] coasts of North America and Europe. A blue, gas-filled float that develops from the (26) _____ [p.431] keeps the colony near the water's surface, where winds move it about. Under the float, groups of polyps and medusae interact as (27) "_____" [p.431] in feeding, reproduction, defense, and other specialized tasks.

Labeling

Identify each indicated part of the following illustration:

28. _____ _____ [p.430]

29. _____ _____ [p.430]

30. _____ _____ [p.430]

31. _____ [p.430]

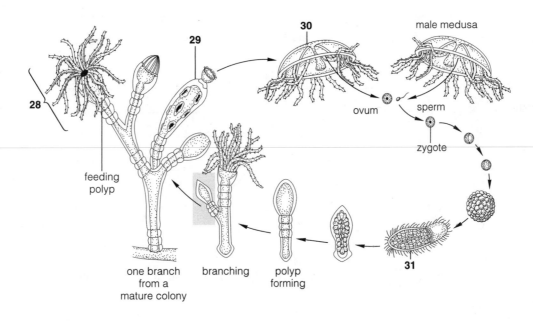

Dichotomous Choice

Circle one of two possible answers given between parentheses in each statement.

32. Evolutionarily, comb jellies are of interest because they possess cells with (single/multiple) cilia as in many complex animals. [p.431]
33. Comb jellies are classified in the phylum (Ctenophora/Cnidaria). [p.431]
34. The comb jellies are the simplest animals with embryonic tissues that are like (mesoderm/endoderm). [p.431]
35. If comb jellies are sliced into two equal halves, then the halves are sliced into quarters, two of the quarters will be (mirror/identical) images of the other two. [p.431]

36. All comb jellies are weak-swimming predators in planktonic communities, and all show modified (bilateral/radial) symmetry. [p.431]
37. A comb jelly has (eight/ten) rows of comblike structures made of thick, fused, cilia. [p.431]
38. In a comb jelly, all the combs in a row beat in waves and propel the animal (backward/forward), usually mouth first. [p.431]
39. Comb jellies (do/do not) produce nematocysts. [p.431]
40. Some comb jellies use their sticky (lips/combs) to capture prey. [p.431]
41. Some species of comb jellies have (four/two) muscular tentacles with branches that are equipped with sticky cells. [p.431]

26.7. ACOELOMATE ANIMALS—AND THE SIMPLEST ORGAN SYSTEMS [pp.432–433]

Selected Words: *organ-system* level of construction [p.432], *transverse* fission [p.432], *definitive* host [p.432], *intermediate* host [p.432], scolex [p.433]

Boldfaced, Page-Referenced Terms

[p.432] organ _____

_____

[p.432] organ system _____

_____

[p.432] flatworms _____

_____

[p.432] hermaphrodites _____

_____

[p.433] proglottids _____

_____

[p.433] ribbon worms _____

_____

Choice

For questions 1–25, choose from the following:

           a. turbellarians   b. flukes   c. tapeworms   d. ribbon worms

1. ___ Parasitic worms [pp.432–433]

2. ___ Flame cells [p.432]

3. ___ Parasitize the intestines of vertebrates [p.433]

4. ___ Planarians that reproduce asexually by transverse fission [p.432]

5. ___ Possess a scolex [p.433]

6. ___ They swallow or suck tissue fluids from worms, mollusks, and crustaceans. [p.433]

7. ___ Ancestral forms probably had a gut but later lost it during their evolution in animal intestines. [p.433]

8. ___ Phylum Nemertea [p.433]

9. ___ Water-regulating systems have one or more tiny, branched tubes called protonephridia. [p.433]

10. ___ Only planarians and a few others live in freshwater habitats. [p.432]

11. ___ Their life cycles have sexual and asexual phases and at least two kinds of hosts. [p.432]

12. ___ May be related to flatworms but have a complete gut, circulatory system, and other traits that are notable departures from them [p.433]

13. ___ Flame cells, each with a tuft of cilia, that drives out excess water to the surroundings [p.432]

14. ___ Thrive in predigested food in vertebrate intestines [p.433]

15. ___ Resemble flatworms in their tissue organization [p.433]

16. ___ After division, each half regenerates the missing parts. [p.432]

17. ___ Proglottids are new units of the body that bud just behind the head. [p.433]

18. ___ These parasites attach to the intestinal wall by a scolex. [p.433]

19. ___ They have a proboscis, which is a tubular, prey-piercing, venom-delivering device tucked inside the head end. [p.433]

20. ___ Possess a structure equipped with suckers, hooks, or both [p.433]

21. ___ They reach sexual maturity in an animal that serves as their primary host; larval stages use an intermediate host, in which they either develop or become encysted. [pp.432–433]

22. ___ Older proglottids store fertilized eggs; they break off and leave the body in feces. [p.433]

23. ___ They differ from flatworms in having a circulatory system, a complete gut, and separate sexes. [p.433]

24. ___ Similar to flatworms in having ciliated surfaces, secreting mucus, and moving by beating their cilia through it [p.433]

25. ___ They grow and reach sexual maturity in an animal that serves as their definitive host. [pp.432–433]

Labeling

Identify the parts of the animal shown dissected in the accompanying drawings.

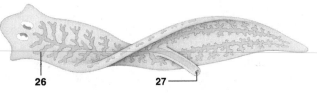

26. _____ _____ [p.432]

27. _____ [p.432]

28. _____ [p.432]

29. _____ _____ [p.432]

30. _____ [p.432]

31. _____ [p.432]

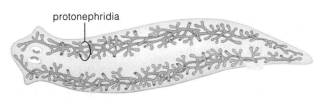

Short Answer

Answer exercises 32–35 for the drawing of the accompanying dissected animal.

32. What is the common name (or genus) of the animal dissected? _____ [p.432]

33. Is the animal parasitic? _____ [p.432]

34. Is the animal hermaphroditic? _____ [p.432]

35. Name the coelom type exhibited by this animal. _____ [p.432]

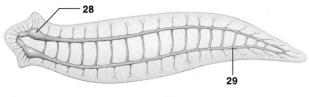

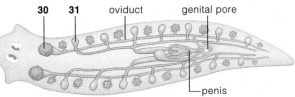

26.8. ROUNDWORMS [p.434]

26.9. *Focus on Health:* A ROGUE'S GALLERY OF WORMS [pp.434–435]

26.10. ROTIFERS [p.436]

Selected Words: *Paratylenchus* [p.434], *Caenorhabditis elegans* [p.434], *schistosomiasis* [p.434], *Schistosoma japonicum* [p.434], *Enterobius vermicularis* [p.434], *Taenia saginata* [p.435], *Trichinella spiralis* [p.435], *Wuchereria bancrofti* [p.435], *elephantiasis* [p.435], *Philodina roseola* [p.436]

Boldfaced, Page-Referenced Terms

[p.434] roundworms _____

_____

[p.436] rotifers _____

_____

Short Answer

Answer exercises 1–3 for the following drawing of a dissected animal.

1. What is the common name of the animal dissected? _____ [p.434]

2. Is the animal hermaphroditic? _____ [p.434]

3. Name the coelom type exhibited by this animal. _____ [p.434]

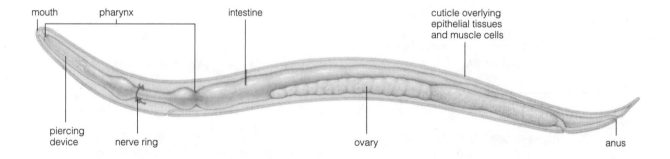

Answer exercises 4–8 for the following drawing of a dissected animal.

4. What is the common name of this animal? _____ [p.436]

5. What type of symmetry does this animal possess? _____ [p.436]

6. Is this animal cephalized? _____ [p.436]

7. Name the type of coelom this animal possesses. _____ [p.436]

8. Why is this animal said to be a "wheel animal"? _____ [p.436]

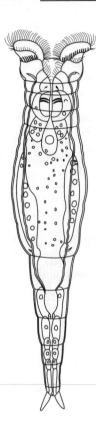

Choice

For questions 9–26, choose from the following.

a. roundworms   b. rotifers

9. ___ All but about 5 percent live in freshwater, such as lakes, ponds, and even within films of water on mosses and other plants. [p.436]

10. ___ At night, the centimeter-long females migrate to the host's anal region to lay eggs; itching and scratching occurs that transfers eggs. [pp.434–435]

11. ___ Hookworms, pinworms, and other types parasitize plants and animals. [pp.434–435]

12. ___ Probably the most abundant of all multicelled animals alive today [p.434]

13. ___ All have a crown of cilia used in swimming and wafting food to the mouth. [p.436]

14. ___ All have a bilateral, cylindrical body, usually tapered at both ends, and protected by a tough cuticle. [p.434]

15. ___ Parasitic forms can do extensive damage to their hosts, which include humans, cats, dogs, cows, and sheep as well as valued crop plants. [p.434]

16. ___ Most species are less than a millimeter long, yet they have a pharynx, an esophagus, digestive glands, a stomach, usually an intestine and anus, and protonephridia. [p.436]

17. ___ Adult forms become lodged in the body's lymph nodes; elephantiasis occurs, an enlargement of legs and other body regions due to blockage of lymph flow. [p.435]

18. ___ Become infected when a juvenile may penetrate the bare skin; the parasite then travels the bloodstream to the lungs [p.435]

19. ___ Some have "eyes." [p.436]

20. ___ Two "toes" exude substances that attach free-living species to substrates feeding time. [p.436]

21. ___ A mosquito is *Wuchereria*'s intermediate host. [p.435]

22. ___ Its rhythmic motions reminded early microscopists of a turning wheel. [p.436]

23. ___ Humans become infected by *Trichinella spiralis* mostly by eating insufficiently cooked meat from pigs or certain game animals. [p.435]

24. ___ They eat bacteria and microscopic algae. [p.436]

25. ___ *Enterobius vermicularis* lives in the large intestine of humans. [pp.434–435]

26. ___ Thousands of scavenging types may occupy a handful of rich topsoil. [p.434]

Fill-in-the-Blanks

The numbered items on the following illustration (blood fluke life cycle) represent missing information; complete the corresponding numbered blanks in the narrative to supply missing information about flukes.

The life cycle of the Southeast Asian blood fluke (*Schistosoma japonicum*) requires a human primary host standing in water in which the fluke larvae can swim. This life cycle also requires an aquatic snail as an intermediate host. Flukes reproduce (27) _____ [p.434], and eggs mature in a human body. Eggs leave the body in feces, then hatch into ciliated, swimming (28) _____ [p.434] that burrow into a (29) _____ [p.434] and multiply asexually. In time, many fork-tailed (30) _____ [p.434] develop. These leave the snail and swim until they contact (31) _____ [p.434] skin. They bore inward and migrate to thin-walled intestinal veins, and the cycle begins anew. In infected humans, white blood cells that defend the body attack the masses of fluke eggs, and grainy masses form in tissues. In time, the liver, spleen, bladder, and kidneys deteriorate.

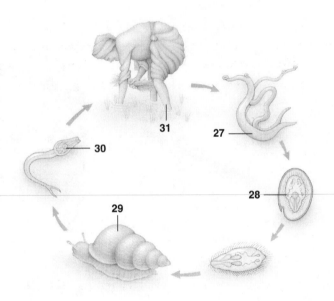

Fill-in-the-Blanks

The numbered items on the following illustration (beef tapeworm life cycle) represent missing information; complete the corresponding numbered blanks in the narrative to supply missing information about the beef tapeworm.

(32) _____ [p.435], each with an inverted scolex of a future tapeworm, become encysted in (33) _____ [p.435] host tissues (such as skeletal muscle). A (34) _____ [p.435], a definitive host, eats infected and undercooked beef containing tapeworm cysts. The scolex of a larva turns inside out, attaches to the wall of the host's (35) _____ _____ [p.435] and begins to absorb host nutrients. Many (36) _____ [p.435] form, by budding; each of these segments becomes sexually mature and has both male and female reproductive (37) _____ [p.435]. Ripe (38) _____ [p.435] containing fertilized eggs leave the host in (39) _____ [p.435], which may contaminate water and vegetation. Inside each fertilized egg, an embryonic (40) _____ [p.435] form develops. Cattle may ingest embryonated eggs or ripe proglottids and so become (41) _____ [p.435] hosts.

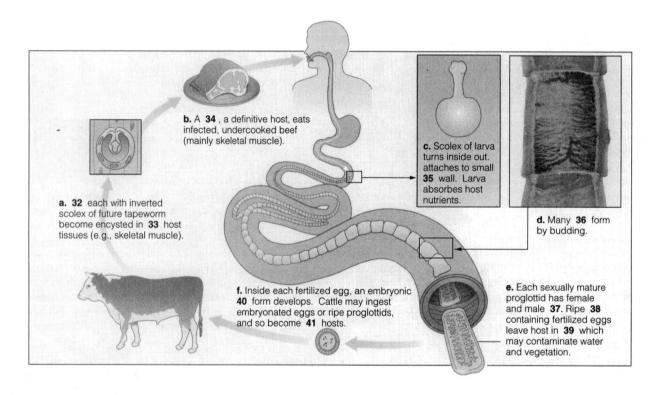

b. A **34**, a definitive host, eats infected, undercooked beef (mainly skeletal muscle).

c. Scolex of larva turns inside out. attaches to small **35** wall. Larva absorbes host nutrients.

a. **32** each with inverted scolex of future tapeworm become encysted in **33** host tissues (e.g., skeletal muscle).

d. Many **36** form by budding.

f. Inside each fertilized egg, an embryonic **40** form develops. Cattle may ingest embryonated eggs or ripe proglottids, and so become **41** hosts.

e. Each sexually mature proglottid has female and male **37**. Ripe **38** containing fertilized eggs leave host in **39** which may contaminate water and vegetation.

26.11. TWO MAJOR DIVERGENCES [p.436]

26.12. A SAMPLING OF MOLLUSCAN DIVERSITY [p.437]

26.13. EVOLUTIONARY EXPERIMENTS WITH MOLLUSCAN BODY PLANS [pp.438–439]

Selected Words: *spiral* cleavage [p.436], *radial* cleavage [p.436], *molluscus* [p.437], *Aplysia* [p.438], *Dosidiscus* [p.439], *jet propulsion* [p.439]

Boldfaced, Page-Referenced Terms

[p.436] protostomes _____

_____

[p.436] deuterostomes _____

_____

[p.437] mollusks _____

_____

[p.437] mantle _____

_____

[p.438] torsion _____

_____

Choice

For questions 1–10, choose from the following:

a. protostomes   b. deuterostomes

1. ___ The first external opening in these embryos become the anus; the second becomes the mouth [p.436]
2. ___ Animals having a developmental pattern in which the early cell divisions are parallel and perpendicular to the axis [p.436]
3. ___ A coelom arises from spaces in the mesoderm [p.436]
4. ___ Radial cleavage [p.436]
5. ___ The first external opening in these embryos becomes the mouth [p.436]
6. ___ A coelom forms from outpouchings of the gut wall [p.436]
7. ___ Spiral cleavage [p.436]
8. ___ Animal having a developmental pattern in which early cell divisions are at oblique angles relative to the genetically prescribed body axis [p.436]
9. ___ Echinoderms and chordates [p.436]
10. ___ Mollusks, annelids, and arthropods [p.436]

Matching

Identify the animals pictured below by matching each with the appropriate description.

11. _____ [p.438] Animal A   I. Bivalve

12. _____ [p.439] Animal B   II. Cephalopod

13. _____ [p.439] Animal C   III. Gastropod

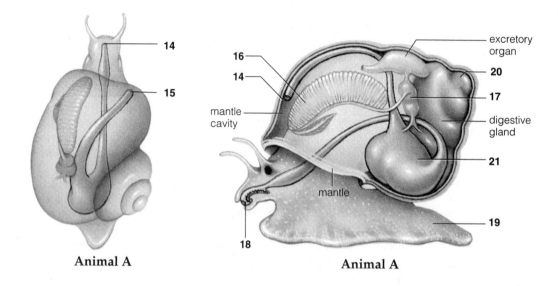

Animal A   **Animal A**

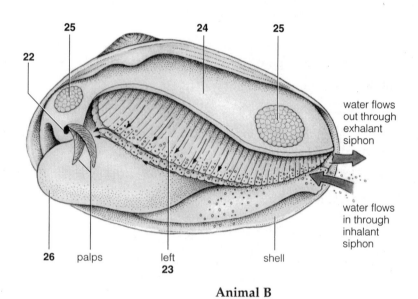

Animal B

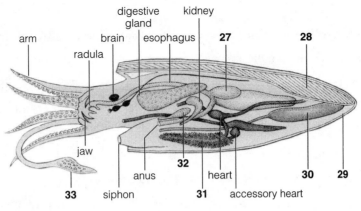

arm
radula
brain
digestive gland
esophagus
kidney
27
28
jaw
33
siphon
anus
32
31
heart
accessory heart
30
29

Animal C

Labeling

Identify each numbered part in the preceding drawings by writing its name in the appropriate blank.

14. _____ [p.438]

15. _____ [p.438]

16. _____ [p.438]

17. _____ [p.438]

18. _____ [p.438]

19. _____ [p.438]

20. _____ [p.438]

21. _____ [p.438]

22. _____ [p.439]

23. _____ [p.439]

24. _____ [p.439]

25. _____ [p.439]

26. _____ [p.439]

27. _____ [p.439]

28. _____ _____ [p.439]

29. _____ [p.439]

30. _____ _____ [p.439]

31. _____ [p.439]

32. _____ _____ [p.439]

33. _____ [p.439]

Fill-in-the-Blanks

A (34) _____ [p.437] is a bilateral animal having a small coelom and a fleshy soft body. Most have a (35) _____ [p.437] of calcium carbonate and protein, which were secreted from cells of a tissue that drapes like a skirt over the body mass. This tissue, the (36) _____ [p.437], is unique to mollusks. Special respiratory organs, the (37) _____ [p.437] contain thin-walled leaflets for gas exchange. Most mollusks have a fleshy (38) _____ [p.437]. Many have a (39) _____ [p.437], a tonguelike, toothed organ that by rhythmic protractions and retractions, rasps small algae and other food from substrates and draws it into the mouth. Mollusks with a well-developed head have (40) _____ [p.437] and (41) _____ [p.437], but not all have a head. There is great diversity in the phylum, and here we review four classes: chitons, gastropods, bivalves, and cephalopods.

Choice

For questions 42–65, choose from the following classes of the phylum Mollusca:

a. chitons   b. gastropods   c. bivalves   d. cephalopods

42. ___ Class with the swiftest invertebrates, the squids [p.437]
43. ___ The "belly foots" [p.437]
44. ___ The largest class, with 90,000 species [p.437]
45. ___ Possess a dorsal shell divided into eight plates [p.437]
46. ___ Class with the smartest invertebrates, the octopuses [p.437]
47. ___ So named because their soft foot spreads out as they crawl [p.437]
48. ___ Animals having a "two-valved shell" [p.437]
49. ___ Class with the largest invertebrates, the giant squids [p.437]
50. ___ Coiling compacts the organs into a mass that can be balanced above the body, rather like a backpack [p.437]
51. ___ Class with the most complex invertebrates in the world, the octopuses and squids, with a memory and a capacity for learning [p.437]
52. ___ Highly active predators of the seas [p.437]
53. ___ Many have spirally coiled or conical shells. [p.437]
54. ___ The only mollusks with a closed circulatory system [p.439]
55. ___ Includes clams, scallops, oysters, and mussels [p.437]
56. ___ Includes aquatic snails, land snails, and sea slugs [p.437]
57. ___ Move rapidly with a system of jet propulsion [pp.437,439]
58. ___ Water is drawn into the mantle cavity through one siphon and leaves through the other, carrying wastes. [p.439]
59. ___ Most can discharge dark fluid from an ink sac, perhaps to confuse predators. [p.439]
60. ___ The anus dumps wastes near the mouth. [p.438]
61. ___ Being highly active, they have great demands for oxygen. [p.439]

62. ___ The only class of mollusks where torsion operates [p.438]

63. ___ Class containing the chambered nautilus [p.439]

64. ___ The shells of many are lined with iridescent mother-of-pearl. [p.437]

65. ___ As the embryo develops, a cavity between the mantle and shell twists 180 degrees counterclockwise, as does nearly all of the visceral mass. [p.438]

26.14. ANNELIDS—SEGMENTS GALORE [pp.440–441]

Selected Words: setae or chaetae [p.440], *Hirudo medicinalis* [p.440]

Boldfaced, Page-Referenced Terms

[p.440] annelids _____

_____

[p.441] nephridia (singular, nephridium) _____

_____

[p.441] brain _____

_____

[p.441] nerve cords _____

_____

Matching

Choose the appropriate answer for each term.

1. ___cuticle [p.441]
2. ___annelids [p.440]
3. ___earthworms [pp.440–441]
4. ___nephridia with cells similar to flame cells [p.441]
5. ___marine polychaetes [p.440]
6. ___brain [p.441]
7. ___nephridia [p.441]
8. ___nerve cords [p.441]
9. ___setae or chaetae [p.440]
10. ___advantage of segmentation [p.440]
11. ___hydrostatic skeleton [p.441]
12. ___leeches [p.440]
13. ___earthworm locomotion [p.441]
14. ___ganglion [p.441]

A. Fluid-cushioned coelomic chambers
B. Oligochaete scavengers with a closed circulatory system and few bristles per segment
C. Paired, each a bundle of extensions of nerve cell bodies leading away from the brain
D. Possess many bristles per segment
E. Muscle contraction with protraction and retraction of segment bristles
F. Secreted wrapping around the body surface of most annelids that permits respiratory exchange
G. A rudimentary aggregation of nerve cell bodies that integrate sensory input and muscle responses for the whole body
H. Lack bristles
I. Implies an evolutionary link between flatworms and annelids
J. Different body parts can evolve separately and specialize in different tasks
K. Except for leeches, chitin-reinforced bristles on each side of the body on nearly all segments
L. The term means "ringed forms"
M. An enlargement of the nerve cord in each segment of an earthworm
N. Regulates volume and composition of body fluids; often begins with a funnel-shaped structure in each segment

Labeling

Identify each indicated part of the following illustrations.

15. _____ [p.441]

16. _____ [p.441]

17. _____ _____ [p.441]

18. _____ [p.441]

19. _____ [p.441]

20. _____ _____ [p.441]

21. _____ [p.441]

22. _____ [p.441]

23. _____ [p.441]

24. _____ [p.441]

25. _____ _____ [p.441]

26. Name this animal. _____ [p.441]

27. Name this animal's phylum. _____ [p.440]

28. Name two distinguishing characteristics of this group _____

_____ [pp.440–441]

29. Is this animal segmented? (   ) yes (   ) no [pp.440–441]

30. Symmetry of adult: (   ) radial (   ) bilateral [p.440]

31. Does this animal have a true coelom? (   ) yes (   ) no [p.440]

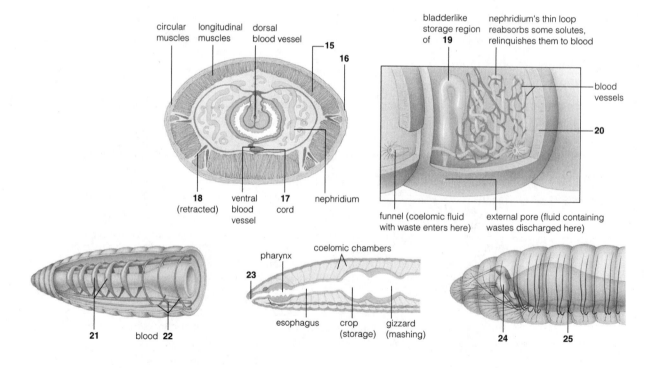

26.15. ARTHROPODS—THE MOST SUCCESSFUL ORGANISMS ON EARTH [p.442]

Selected Words: *juvenile* [p.442], *division of labor* [p.442]

Boldfaced, Page-Referenced Terms

[p.442] arthropods _____

_____

[p.442] exoskeleton _____

_____

[p.442] molting _____

_____

[p.442] metamorphosis _____

_____

Choice

For questions 1–13, choose from the following six adaptations that contributed to the success of arthropods:

    a. hardened exoskeletons   b. fused and modified segments   c. jointed appendages
    d. respiratory structures   e. specialized sensory structures   f. division of labor

1. ___ Intricate eyes and other sensory organs that contributed to arthropod success [p.442]

2. ___ The new individual is a *juvenile*, a miniaturized form of the adult that simply changes in size and proportion until reaching sexual maturity. [p.442]

3. ___ A cuticle of chitin, proteins, and surface waxes that may be impregnated with calcium carbonate [p.442]

4. ___ In most of their existing descendants, however, the serial repeats of the body wall and organs are masked, for many fused-together, modified segments perform more specialized functions. [p.442]

5. ___ Metamorphosis from embryo to adult forms [p.442]

6. ___ Might have evolved as defenses against predation [p.442]

7. ___ Immature stages, such as caterpillars, specialize in feeding and growing in size. [p.442]

8. ___ Gills of aquatic arthropods [p.442]

9. ___ In the ancestors of insects, different segments became combined into a head, a thorax, and an abdomen. [p.442]

10. ___ A jointed exoskeleton was a key innovation that led to appendages as diverse as wings, antennae, and legs. [p.442]

11. ___ Numerous species have a wide angle of vision and can process visual information from many directions. [p.442]

12. ___ Air-conducting tubes evolved among insects and other land-dwellers. [p.442]

13. ___ Their waxy surfaces restrict evaporative water loss and can support a body deprived of water's buoyancy. [p.442]

14. Why are the arthropods said to be the most biologically successful organisms on Earth? [p.442]

_____

_____

_____

_____

26.16. A LOOK AT SPIDERS AND THEIR KIN [p.443]

26.17. A LOOK AT THE CRUSTACEANS [pp.444–445]

26.18. HOW MANY LEGS? [p.445]

Selected Words: *open* circulatory system [p.443], *Scutigera* [p.445]

Boldfaced, Page-Referenced Terms

[p.445] millipedes _____

_____

[p.445] centipedes_____

_____

Matching

Select the most appropriate answer for each term.

1. ____ticks [p.443]
2. ____arachnid forebody appendages [p.443]
3. ____arachnids [p.443]
4. ____arachnid hindbody appendages [p.443]
5. ____chelicerates [p.443]
6. ____spider, internal organs [p.443]
7. ____efficient predatory arachnids [p.443]

A. Scorpions and spiders that sting, bite, and may subdue prey with venom
B. An open circulatory system and book lungs
C. A group including mites, horseshoe crabs, sea spiders, spiders, ticks, and chigger mites
D. Spin out silk thread for webs and egg cases
E. Some transmit bacterial agents of Rocky Mountain spotted fever or Lyme disease to humans
F. A familiar group including scorpions, spiders, ticks, and chigger mites
G. Four pairs of legs, a pair of pedipalps that have mainly sensory functions, and a pair of chelicerae that can inflict wounds and discharge venom

Dichotomous Choice

Circle one of two possible answers given between parentheses in each statement.

8. Nearly all arthropods possess (strong claws/an exoskeleton). [p.444]
9. The giant crustaceans are (lobsters and crabs/barnacles and pillbugs). [p.444]
10. The simplest crustaceans have many pairs of (different/similar) appendages along their length. [p.444]
11. (Barnacles/Lobsters and crabs) have strong claws that collect food, intimidate other animals, and sometimes dig burrows. [p.444]
12. (Barnacles/Lobsters and crabs) have feathery appendages that comb microscopic bits of food from the water. [p.444]

13. (Barnacles/Copepods) are the most numerous animals in aquatic habitats, maybe even in the world. [p.445]
14. Of all arthropods, only (lobsters and crabs/barnacles) have a calcified "shell." [p.445]
15. Adult (barnacles/copepods) cement themselves to wharf pilings, rocks, and similar surfaces. [p.445]
16. As is true of other arthropods, crustaceans undergo a series of (rapid feedings/molts) and so shed the exoskeleton during their life cycle. [p.445]
17. As (millipedes/centipedes) develop, pairs of segments fuse, so each segment in the cylindrical body ends up with two pairs of legs. [p.445]
18. Adult (millipedes/centipedes) have a flattened body, are fast-moving, and have a pair of walking legs on every segment except two. [p.445]
19. (Millipedes/Centipedes) mainly scavenge for decaying vegetation in soil and forest litter. [p.445]
20. (Millipedes/Centipedes) are fast-moving, aggressive predators, outfitted with fangs and venom glands. [p.445]

Labeling

Identify each numbered part of the animal pictured at right.

21. _____ [p.444]

22. _____ [p.444]

23. _____ [p.444]

24. _____ [p.444]

25. _____ [p.444]

26. _____ [p.444]

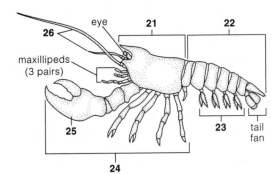

Answer questions 27–32 for the animal pictured above and to the right.

27. Name the animal pictured. _____ [p.443]

28. Name the subgroup of arthropods to which this animal belongs. _____ [p.443]

29. Name two distinguishing characteristics of this group. _____ [p.443]

30. Is this animal segmented? ( ) yes ( ) no [p.443]

31. Symmetry of adult: ( ) radial ( ) bilateral [p.443]

32. Does this animal have a true coelom? ( ) yes ( ) no [p.443]

Identify each numbered body part in this illustration, then answer question 38.

33. _____ _____ [p.443]

34. _____ [p.443]

35. _____ [p.443]

36. _____ [p.443]

37. _____ _____ [p.443]

38. Name the subgroup of arthropods to which the animal belongs. _____ [p.443]

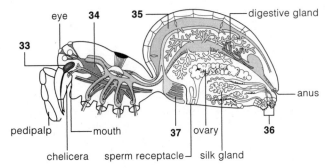

26.19. A LOOK AT INSECT DIVERSITY [pp.446–447]

26.20. *Focus on Health:* UNWELCOME ARTHROPODS [pp.448–449]

Selected Words: incomplete metamorphosis [p.446], *complete* metamorphosis [p.446], *Loxosceles* [p.448], *Ixodes* [p.448], Lyme disease [p.448], *Ixodes dammini* [p.448], *Borrelia burgdorferi* [p.448], *Centruroides sculpuratus* [p.449], *Diabrotica virgifera* [p.441]

Boldfaced, Page-Referenced Terms

[p.446] Malpighian tubules _____

_____

[p.446] nymphs _____

_____

[p.446] pupae _____

_____

Matching

Choose the most appropriate answer for each term.

1. ___complete metamorphosis [p.446]

2. ___the most successful species of insects [p.446]

3. ___incomplete metamorphosis [p.446]

4. ___insect success based on aggressive competition with humans [pp.446–447]

5. ___Malpighian tubules [p.446]

6. ___metamorphosis [p.446]

7. ___a variation of insect headparts [p.446]

8. ___insect activities considered beneficial to humans [p.447]

9. ___insect life-cycle stages [p.446]

10. ___shared insect adaptations [p.446]

A. Post-embryonic resumption of growth and transformation into an adult form

B. Destruction of crops, stored food, wool, paper, and timber, drawing blood from humans and their pets, transmitting pathogenic microorganisms

C. Head, thorax, and abdomen, paired sensory antennae and mouthparts, three pairs of legs, and two pairs of wings

D. Chewing, sponging up, siphoning, piercing, and sucking

E. Involves gradual, partial change from the first immature form until the last molt

F. Winged insects, also the only winged invertebrates

G. Pollination of flowering plants and crop plants; parasitizing or attacking plants humans would rather do without

H. Larva–nymph–pupa–adult

I. Structures that collect nitrogen-containing wastes from blood and convert them to harmless crystals of uric acid that are eliminated with feces

J. Tissues of immature forms are destroyed and replaced before emergence of the adult

Choice

For questions 11–20, choose from the following:

> a. spiders   b. ticks   c. scorpions   d. beetles

11. ___ Corn rootworm [p.449]

12. ___ Lyme disease [pp.448–449]

13. ___ The poisonous brown recluse [p.448]

14. ___ *Centruroides sculpturatus* from Arizona, the most dangerous species in the United States [p.449]

15. ___ About five people die each year in the United States as a result of black widow bites [p.448]

16. ___ *Ixodes dammini* [p.448]

17. ___ Possess large, prey-seizing pincers and a venom-dispensing stinger at the tip of a narrowed, jointed abdomen [p.449]

18. ___ Rocky Mountain spotted fever, scrub typhus, tularemia, babesiasis, and encephalitis [p.449]

19. ___ Cucurbitacin is being used effectively against corn rootworms

20. ___ Do not jump or fly; they crawl onto grasses and shrubs, then onto animals that brush past

26.21. THE PUZZLING ECHINODERMS [pp.450–451]

Selected Terms: *Echinodermata* [p.450]

Boldfaced, Page-Referenced Terms

[p.450] echinoderms _____

_____

[p.451] water-vascular system _____

_____

Fill-in-the-Blanks

The second lineage of coelomate animals is referred to as the (1) _____ [p.450]. The major invertebrate members of this lineage include (2) _____ [p.450]. The body wall of all echinoderms has protective spines, spicules, or plates made rigid with (3) _____ _____ [p.450]. Most echinoderms also have a well-developed internal (4) _____ [p.450], which is composed of calcium carbonate and other substances secreted from specialized cells. Oddly, adult echinoderms have (5) _____ [p.450] symmetry with some bilateral features, but some produce larvae with (6) _____ symmetry [p.450]. Adult echinoderms have no (7) _____ [p.450] but a decentralized (8) _____ [p.450] system allows them to respond to information about food, predators, and so forth that is coming from different directions. For example, any (9) _____ [pp.450–451] of a sea star that senses the shell of a tasty scallop can become the leader, directing the remainder of the body to move in a direction suitable for prey capture.

The (10) _____ [p.451] feet of sea stars are used for walking, burrowing, clinging to a rock, or gripping a meal of clam or snail. These "feet" are parts of a (11) _____ [p.451] vascular system unique to echinoderms. Each foot contains a(n) (12) _____ [p.451] that acts like a rubber bulb on a medicine dropper as it contracts and forces fluid into a foot that then lengthens. Tube feet change shape constantly as (13) _____ [p.451] action redistributes fluid through the water vascular system. Some sea stars are simply able to swallow their prey (14) _____ [p.451]; others can push part of their stomach outside the mouth and around their prey, then start (15) _____ [p.451] their meal even before swallowing it. Coarse, indigestible remnants are regurgitated back through the mouth. Their small (16) _____ [p.451] is of no help in getting rid of empty clam or snail shells.

Labeling

Identify each indicated part of the two following illustrations.

17. _____ _____ [p.451]   21. _____ [p.451]

18. _____ _____ [p.451]   22. _____ _____ [p.451]

19. _____ [p.451]   23. _____ [p.451]

20. _____ [p.451]   24. _____ _____ [p.451]

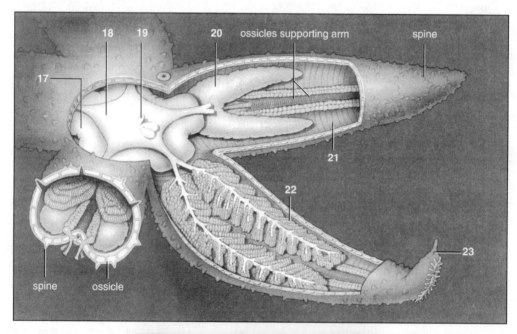

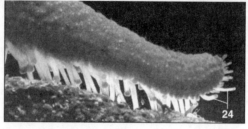

Short Answer

Answer questions 25–26 for the animal just pictured.

25. Name the animal shown. _____ [p.451]

26. Protostome ( ) or deuterostome ( ) [p.450]

Identifying

27. Name the system shown at the right. _____ _____ system [p.451]

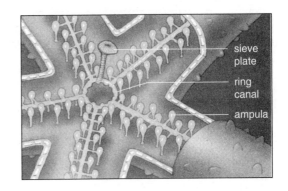

28. Identify each creature below by its common name. Write the name in the blank below the picture, a–d.

29. Name the phylum of the animals pictured below. _____ [p.450]

30. Name two distinguishing characteristics of this group. _____ _____ [pp.450–451]

31. Symmetry of adults represented below is: ( ) radial ( ) bilateral [p.450]

a. _____ [p.450]

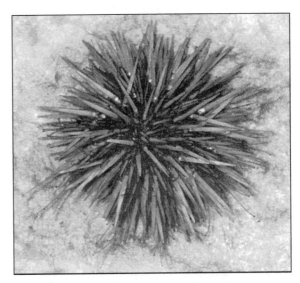

b. _____ [p.450]

c. _____ [p.450]

d. _____ [p.450]

Self-Quiz

___ 1. Which of the following is *not* true of sponges? They have no _____.
[pp.426–427]
a. distinct cell types
b. nerve cells
c. muscles
d. gut

___ 2. Which of the following is *not* a protostome? [p.450]
a. earthworm
b. crayfish or lobster
c. sea star
d. squid

___ 3. Deuterostomes undergo _____ cleavage; protostomes undergo _____ cleavage. [p.436]
a. radial; spiral
b. radial; radial
c. spiral; radial
d. spiral; spiral

___ 4. Bilateral symmetry is characteristic of _____. [p.432]
a. cnidarians
b. sponges
c. jellyfish
d. flatworms

___ 5. Flukes and tapeworms are parasitic _____. [pp.432–433]
a. leeches
b. flatworms
c. jellyfish
d. roundworms

___ 6. Insects include _____. [pp.446–447]
a. spiders, mites, ticks
b. centipedes and millipedes
c. termites, aphids, and beetles
d. all of the above

___ 7. The simplest animals to exhibit the organ-system level of construction are the _____. [p.432]
a. sponges
b. arthropods
c. cnidarians
d. flatworms

___ 8. Torsion is a process characteristic of _____. [p.438]
a. chitons
b. bivalves
c. gastropods
d. cephalopods
e. echinoderms

___ 9. The _____ body plan is characterized by bilateral symmetry, a flattened body, cephalization, and a digestive system with a pharynx for feeding, and hermaphroditism. [p.432]
a. annelid
b. roundworm
c. echinoderm
d. flatworm

___10. The _____ have bilateral symmetry, cylindrical bodies tapered on both ends, a tough protective cuticle, a false coelom, and present the simplest example of a complete digestive system. [p.434]
a. roundworms
b. cnidarians
c. flatworms
d. echinoderms

___11. A _____ with a peritoneum separates the gut and body wall of most bilateral animals. [p.436]
a. coelom
b. mesoderm
c. mantle
d. water-vascular system

___12. In some annelid groups, a _____ is composed of cells similar to flatworm flame cells. [p.441]
a. trachea
b. nephridium
c. mantle
d. parapodium

Matching

Match each of the following phyla with the corresponding characteristics (letters a–k) and representatives (A–O). A phylum may match with more than one letter from the group of representatives.

___, ___ 13. Annelida [pp.440–441]

___, ___ 14. Arthropoda [p.442]

___, ___ 15. Cnidaria [pp.428–429]

___, ___ 16. Echinodermata [pp.450–451]

___, ___ 17. Mollusca [pp.437–439]

___, ___ 18. Nematoda [p.434]

___, ___ 19. Ctenophora [p.431]

___, ___ 20. Platyhelminthes [pp.432–433]

___, ___ 21. Porifera [pp.426–427]

___, ___ 22. Rotifera [p.436]

___, ___ 23. Nemertea [p.433]

a. choanocytes (= collar cells) + spicules
b. jointed legs + an exoskeleton
c. pseudocoelomate + wheel organ + soft body
d. soft body + mantle; may or may not have radula or shell
e. bilateral symmetry + blind-sac gut
f. radial symmetry + blind-sac gut; stinging cells
g. body compartmentalized into repetitive segments; coelom containing nephridia (= primitive kidneys)
h. bilateral, soft-bodied, ciliated, elongated predators; circulatory system, a complete gut, separate sexes, and a tubular prey-piercing proboscis
i. comb-bearing, weak-swimming predators, cells with multiple cilia
j. tube feet + calcium carbonate structures in skin
k. complete gut + bilateral symmetry + cuticle; includes many parasitic species, some of which are harmful to humans

A. Comb jellies
B. Corals, sea anemones, and *Hydra*
C. Tapeworms and planaria
D. Insects
E. Jellyfish and the Portuguese man-of-war
F. Sand dollars and starfishes
G. Earthworms and leeches
H. Lobsters, shrimp, and crayfish
I. Organisms with spicules and collar cells
J. Scorpions and millipedes
K Octopuses and oysters
L. Ribbon worms
M. Flukes
N. Hookworm, trichina worm
O. Small animals with a crown of cilia and two "exuding toes"

Chapter Objectives/Review Questions

1. List the six general characteristics that define an "animal." [p.424]
2. _____ cells are the forerunners of the primary tissue layers, the ectoderm, endoderm, and, in most species, mesoderm. [p.424]
3. List the primary characteristic that separates vertebrates from invertebrates. [p.424]
4. Distinguish radial symmetry from bilateral symmetry, and generally describe various animal gut types. [pp.424–425]
5. Describe meanings of the following terms relating to aspects of an animal body: *anterior* and *posterior* ends, *dorsal* and *ventral* surfaces, and *cephalization*. [p.425]
6. The _____ is a tubular or saclike region in the body in which food is digested, then absorbed into the internal environment. [p.425]
7. List two benefits that the development of a coelom brings to an animal. [p.425]
8. Define *pseudocoel*, and describe what types of animals this term is applied to. [p.425]
9. What is meant by a "segmented animal?" [p.425]

10. *Trichoplax* is as simple as an animal can get, having only _____ distinct layers of cells. [p.426]
11. List the characteristics that distinguish sponges from other animal groups. [pp.426–427]
12. Describe the processes involved in sponge reproduction, both asexual and sexual. [p.427]
13. State what nematocysts are used for and explain how they function. [p.428]
14. Two cnidarian body types are the _____ and the _____. [p.428]
15. Describe the structure typical of a cnidarian, using terms such as *epithelium, nerve cells, contractile cells, nerve net, mesoglea,* and *hydrostatic skeleton.* [pp.428–429]
16. Describe the life cycle of *Obelia.* [p.430]
17. Describe a "comb jelly" in terms of its characteristics. Explain why the symmetry of a comb jelly is not truly bilateral. [p.431]
18. Comb jellies are the simplest animals with embryonic tissues that are like _____. [p.431]
19. Define *organ-system* level of construction and relate this to flatworms. [p.432]
20. List the three main types of flatworms and briefly describe each; name the groups that are parasitic. [pp.432–433]
21. Members of the phylum Nemertea are known as _____ _____; describe them and the habitats they live in. [p.433]
22. Describe the body plan of roundworms, comparing its various systems with those of the flatworm body plan. [p.434]
23. Southeast Asian blood flukes, tapeworms, and pinworms are examples of _____ parasites. [pp.434–435]
24. Describe the size, structure, and the environment of the rotifers. [p.436]
25. Define, by their characteristics, *protostome* and *deuterostome* lineages, and cite examples of animal groups belonging to each lineage. [p.436]
26. List and generally describe the major groups of mollusks and their members. [p.437]
27. Define *mantle,* and tell what role it plays in the molluscan body. [p.437]
28. Describe the process of torsion that occurs only in gastropods. [p.438]
29. Explain why cephalopods came to have such well-developed sensory and motor systems and are able to learn. [p.439]
30. Describe the advantages of segmentation, and tell how this relates to the development of specialized internal organs. [p.440]
31. Generally describe the external and internal structure of a typical annelid, the earthworm. [pp.440–441]
32. List four different lineages of arthropods; briefly describe each. [p.442]
33. List the six arthropod adaptations that led to their success. [p.442]
34. List the groups of organisms known as the familiar chelicerates. [p.443]
35. Describe the characteristics of and the significance of the arachnid lifestyle. [p.443]
36. Name the most obvious characteristic shared by most crustaceans. [p.444]
37. Name some common types of crustaceans. [pp.444–445]
38. Compare and contrast the structural characteristics of millipedes and centipedes. [p.445]
39. Name the characteristics that all insects share, even though they may appear very dissimilar. [p.446]
40. The developmental process of many insects proceeds through very different postembryonic stages, and then to an adult form by a process known as _____. [p.446]
41. List members of the following groups that are unwelcome to humans: spiders, mites, scorpions, and beetles; tell why each is unwelcome. [pp. 448–449]
42. The major invertebrate members of the _____ lineage are the echinoderms; list their major characteristics. [pp.450–451]
43. List five examples of animals known as echinoderms. [pp.450–451]
44. Describe how locomotion and eating occur in sea stars. [p.451]

Integrating and Applying Key Concepts

Scan text Figure 26.45, on p.452, to verify that most highly evolved invertebrate animals have bilateral symmetry, a complete gut, a true coelom, and segmented bodies. Why do you suppose that having a true coelom and a segmented body is considered more highly evolved than the condition of lacking a coelom or possessing a false coelom, and having an unsegmented body? Cite evidence in the chapter that would support or not support the information found in Figure 26.45.

27

ANIMALS: THE VERTEBRATES

Interactive Exercises

Making Do (Rather Well) With What You've Got [pp.454–455]

27.1. THE CHORDATE HERITAGE [p.456]

27.2. INVERTEBRATE CHORDATES [pp.456–457]

27.3. EVOLUTIONARY TRENDS AMONG THE VERTEBRATES [pp.458–459]

27.4. EXISTING JAWLESS FISHES [p.459]

27.5. EXISTING JAWED FISHES [pp.460–461]

Selected Words: duck-billed platypus [p.454], "invertebrate chordates" [p.456], "sea squirts" [p.456], larva [p.456]

Boldfaced, Page-Referenced Terms

[p.456] chordates _____

_____

[p.456] vertebrates _____

_____

[p.456] notochord _____

_____

[p.456] nerve cord _____

_____

[p.456] pharynx _____

_____

[p.456] tunicates _____

_____

[p.456] filter feeders _____

_____

[p.456] gill slits _____

_____

[p.456] lancelets _____

_____

[p.458] vertebrae _____

_____

[p.458] jaws _____

_____

[p.458] fins _____

_____

[p.458] gills _____

_____

[p.459] lungs _____

_____

[p.459] ostracoderms _____

_____

[p.459] placoderms _____

_____

[p.459] hagfishes _____

_____

[p.459] lampreys _____

_____

[p.460] swim bladder _____

_____

[p.460] cartilaginous fishes _____

_____

[p.460] scales _____

_____

[p.460] bony fishes _____

_____

[p.461] lobe-finned fishes _____

_____

Fill-in-the-Blanks

Each animal is an assortment of (1) _____ [p.455]; many have been conserved from remote ancestors and others are unique to its branch on the animal family tree. Four major features distinguish the embryos of chordates from those of all other animals: a hollow dorsal (2) _____ _____ [p.455], a (3) _____ [p.455] with slits in its wall, a (4) _____ [p.455], and a tail that extends past the anus at least during part of its life. In some chordates, the (5) _____ [p.456] chordates, the notochord is *not* divided into a skeletal column of separate, hard segments; in others, the (6) _____ [p.456], it is. Invertebrate chordates living today are represented by tunicates and (7) _____ [p.456], which obtain their food by (8) _____ - _____ [p.456]; they draw in plankton-laden water through the mouth and pass it over sheets of mucus, which trap the particulate food before the water exits through the (9) _____ _____ [p.456] in the pharynx. (10) _____ [p.456] are among the most primitive of all living chordates; when they are tiny, they look and swim like (11) _____ [p.456]. A rod of stiffened tissue, the (12) _____ [p.456], cooperates with muscles to act like a torsion bar propelling the larva forward. Most (10) remain attached to rocks or hard substrates in marine habitats after their larvae undergo (13) _____ [p.456].

Even though the adult forms of hemichordates, echinoderms, and chordates look very different, they have similar embryonic developmental patterns. Chordates may have developed from a mutated ancestral deuterostome (14) _____ [p.458] of an attached, filter-feeding adult. A larva is an immature, motile form of an organism, but if a(n) (15) _____ [p.458] occurred that caused (16) _____ _____ [p.458] to become functional in the larval body, then a motile larva that could reproduce would have been more successful in finding food than a sessile adult; in time, the attached adult stage in the species would be eliminated.

The ancestors of the vertebrate line may have been mutated forms of their closest relatives, the (17) _____ [p.455], in which the notochord became segmented and the segments became hardened (18) _____ [p.458]. The vertebral column was the foundation for fast-moving (19) _____ [p.458], some of which were ancestral to all other vertebrates. The evolution of (20) _____ [p.458] intensified the competition for prey and the competition to avoid being preyed on; animals in which mutations expanded the nerve cord into a (21) _____ [p.458] that enabled the animal to compete effectively survived more frequently than their duller-witted fellows and passed along their genes into the next generations. Fins became (22) _____ [p.458]; in some fishes, those became (23) _____ [p.458] and equipped with skeletal supports; these forms set the stage for the development of legs, arms, and wings in later groups.

As the ancestors of land vertebrates began spending less time immersed and more of their lives exposed to air, use of gills declined and (24) _____ [p.459] evolved; more elaborate and efficient (25) _____ [p.459] systems evolved along with more complex and efficient lungs.

Exercises 26–27 refer to the illustrations below. [Both on p.457]

26. Name the creature in illustration B. _____

27. Name the creature in illustration C. _____

Labeling

Name the structures numbered in illustrations A–C.
Also, name the groups numbered in Figure X.

28. _____ [p.457]

29. _____ with _____ _____ [p.457]

30. _____ [p.455]

31. _____ [p.455]

32. _____ _____ [p.455]

33. _____ [p.457] with _____ _____

34. _____ [p.457]

35. _____ _____ [p.457]

36. _____ _____ [p.457]

37. _____ _____ _____ _____ [p.457]

38. _____ [p.457]

39. _____ [p.457]

40. _____ _____ _____ _____ [p.457]

28 — nerve cord
"heart" — atrium
29 — attachments points

A

CARTILAGINOUS FISHES
32 — JAWLESS FISHES
31 — LANCELETS
30 — TUNICATES

X

36
35
mantle — genital duct
34
33 — atrial cavity
tunic
gonad
circulatory system (with heart)

26. [p.457] **B**

40   39
tentacles   37   atrial cavity   pore of atrial cavity   38   segmental muscles

27. [p.457] **C**

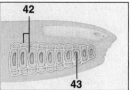

41. [p.458]   D

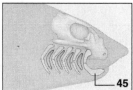

44. [p.458]   E

Short Answer

Answer the following questions in the blanks.

41. Name the creature whose head is shown in illustration D. [p.458] _____

_____

42. Name the structures. [p.458] _____ _____

43. Name the structures. [p.458] _____ _____

44. Name the creature shown in illustration E. [p.458] _____

45. What structures occupy the front of its head space? [p.458] _____

...mes wherever an asterisk (*) appears, and provide the class name wherever a box ❐

...lata	Phylum: Chordata		
...s (or you)	Subphylum: Urochordata	Subphylum: Cephalochordata	Subphylum: Vertebrata
	47.* [p.456]	48.* [p.456]	Class: *Agnatha* 49.* [p.456]
			Class: *Placodermi* 50.* [p.456]
			Class: *Chondrichthyes* 51.* [p.456]
			Class: *Osteichthyes* 52.* [p.456]
			Class: 53. ❐ amphibians [p.456]
			Class: 54. ❐ reptiles [p.456]
			Class: 55. ❐ birds [p.456]
			Class: 56. ❐ mammals [p.456]

Fill-in-the-Blanks

Cartilaginous fishes include about 850 species of rays, skates, (57) _____ [p.460], and chimaeras. They have conspicuous fins and five to seven (58) _____ _____ [p.460] on both sides of the pharynx. Ninety-six percent of the existing species of fishes are (59) _____ [p.460]. Their ancestors arose during the Silurian period perhaps as early as 450 million years ago and soon gave rise to three lineages: the (60) _____-_____ [p.460] fishes, the lobe-finned fishes, and the lungfishes.

Matching

Match the numbered item with its letter. One letter is used twice. [All from p

61. ___
62. ___
63. ___
64. ___
65. ___
66. ___
67. ___
68. ___
69. ___
70. ___
71. ___
72. ___
73. ___
74. ___
75. ___
76. ___

A. Anal fin
B. Anus
C. Brain
D. Caudal fin
E. Dorsal fin
F. Gallbladder
G. Heart
H. Intestine
I. Kidney
J. Liver
K. Pectoral fin (paired)
L. Pelvic fin (paired)
M. Stomach
N. Swim bladder
O. Urinary bladder

Analysis and Short Answer

Figure 27.11 mentions features that distinguish the amphibian-lungfish lineage from the lineage that leads to sturgeons and other bony fishes. At least one of those features may have developed near ⑤ in the following evolutionary tree (see p.314).

Exercises 77–87 refer to this evolutionary diagram.

77. What single feature do the lampreys, hagfishes, and extinct ostracoderms have in common that is different from the placoderms? [p.459] _____

78. How did ostracoderms feed? [p.459] _____

79. A mutation in ostracoderm stock led to the development of what feature in all organisms that descended from ①? [p.459] _____

80. A genetic event at ② led to the development of an endoskeleton made of what? [p.460]_____

81. Mutations at ③ led to an endoskeleton of what? [p.460]_____

82. Mutations at ④ led to which spectacularly diverse fishes that have delicate fins originating from the dermis? [pp.460–461] _____

ns incorporate fleshy extensions from the body? [pp.461–462]

_____ibians? [p.462] _____

_____ fishes? [p.460] _____

____ nt lineages of *bony* fishes first appear in the fossil record?

_____

_____

____ d the fork in the evolutionary path that led to the amphib-
_____ary diagram] _____

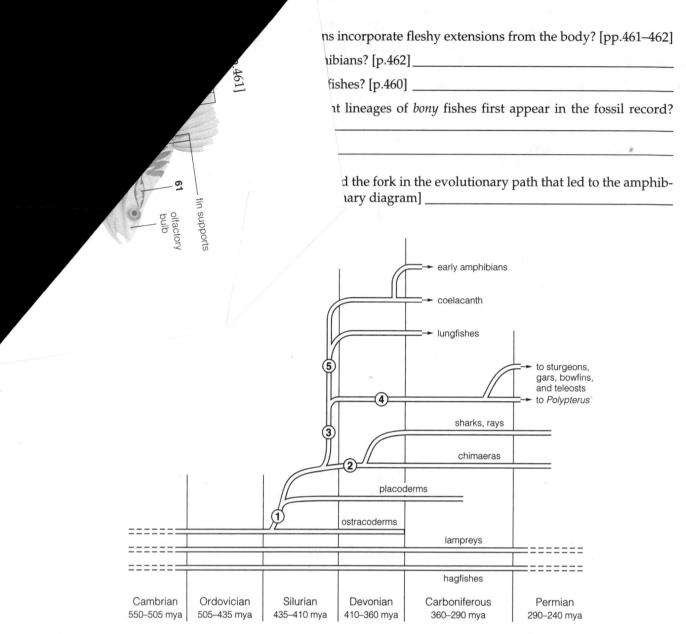

mya = million years ago

27.6. AMPHIBIANS [pp.462–463]

27.7. THE RISE OF REPTILES [pp.464–465]

27.8. MAJOR GROUPS OF EXISTING REPTILES [pp.466–467]

Selected Words: Amphibia [p.462], *Ichthyostega* [p.462], *salamander* [p.462], newt [p.462], frog [p.462], toad [p.462], caecilian [p.462], Carboniferous [p.464], Reptilia [p.464], crocodile [p.464], *Maiasaura* [p.465], Cretaceous [p.465], dinosaur [p.465], *Velociraptor* [p.466], turtle [p.466], lizard [p.466], snake [p.466], tuatara [p.467]

Boldfaced, Page-Referenced Terms

[p.462] amphibian _____

_____

[p.464] reptiles _____

_____

[p.464] amniote egg _____

_____

Fill-in-the-Blanks

Natural selection acting on lobe-finned fishes during the Devonian period favored the evolution of ever more efficient (1) _____ [p.462] used in gas exchange and stronger (2) _____ [p.462] used in locomotion. Without the buoyancy of water, an animal traveling over land must support its own weight against the pull of gravity. The (3) _____ [p.462] of early amphibians underwent dramatic modifications that involved evaluating incoming signals related to vision, hearing, and (4) _____ [p.462]. Although fish have (5) _____ [p.669]-chambered hearts, amphibians have (6) _____ [p.669]-chambered hearts (see Fig. 39.4). Early in amphibian evolution, mutations may have created the third chamber, which added a second (7) _____ [p.669] in addition to the already existing atrium and ventricle. The Devonian period brought humid, forested swamps with an abundance of aquatic invertebrates and (8) _____ [p.462]—ideal prey for amphibians.

There are three groups of existing amphibians: (9) _____ [p.463], frogs and toads, and caecilians. Amphibians require free-standing (10) _____ [p.462] or at least a moist habitat to (11) _____ [p.462]. Amphibian skin generally lacks scales but contains many glands, some of which produce (12) _____ [p.463].

Several features helped: modification of (13) _____ [p.464] bones that supported the trunk favored swiftness, modification of teeth and jaws enabled them to feed efficiently on a variety of prey items, and the development of a (14) _____ [p.464] egg protected the embryo inside from drying out, even in dry habitats.

(For questions 15–26, consult Figure 27.14 of the main text, the following time line, and the figure for the Analysis and Short Answer exercise on page 314 of this Study Guide.)

Today's reptiles include (15) _____ [p.466], crocodilians, snakes, and (16) _____ [p.466]. All rely on (17) _____ [p.464] fertilization, and most lay leathery-shelled eggs. Although amphibians originated during Devonian times, ancestral "stem" reptiles appeared during the (18) _____ [p.465] period, about 340 million years ago. Reptilian groups living today that have existed on Earth longest are the (19) _____ [p.465]; their ancestral path diverged from that of the "stem" reptiles during the (20) _____ [p.465] period. Crocodilian ancestors appeared in the early (21) _____ [Study Guide, p.317] period, about 220 million years ago. Snake ancestry diverged from lizard stocks during the early (22) _____ [Study Guide, p.317] period, about 140 million years ago. (23) _____ [p.466] are more closely related to extinct dinosaurs and crocodiles than to any other existing vertebrates; they, too, have a (24) _____ [p.466] - chambered heart. Mammals have descended from therapsids, which in turn are descended from the (25) _____ [Fig. 27.14, p.465] group of reptiles, which diverged earlier from the stem reptile group during the (26) _____ [p.465] period, approximately 320 million years ago.

Sequence

In blanks 27–33, arrange the following groups in sequence, from earliest to latest, according to their appearance in the fossil record: [All from p.465]

A. Birds   B. Crocodilians   C. Dinosaurs   D. Early ancestors of mammals (= therapsid reptiles)
E. Early ancestors of turtles (= anapsid reptiles)   F. Snakes   G. "Stem" reptiles

27. ____ 28. ____ 29. ____ 30. ____ 31. ____ 32. ____ 33. ____

Choice

In blanks 34–40, select, from the following choices, the geologic period in which each group (from items 27–33) first appeared and write it in its corresponding space. For example, blank 27 contains the letter of the earliest group to appear on Earth; blank 34 should contain the letter of its geologic period, chosen from the following. [All from p.465]

Paleozoic Era		Mesozoic Era		
A. Carboniferous	B. Permian	C. Triassic	D. Jurassic	E. Cretaceous
360–290 mya	290–240 mya	240–205 mya	205–138 mya	138–65 mya

34. ____ 35. ____ 36. ____ 37. ____ 38. ____ 39. ____ 40. ____

Analysis and Short Answer

41. This chapter mentions the features that distinguish, say, the mammals from all the remaining groups on the right-hand side of the diagram; therefore, at some time during the evolution of mammals, mutations that produced those features appeared. Consult the following evolutionary diagram and imagine what sort of mutation(s) occurred at the numbered places. [Diagram adapted from Figure 27.14.]

   a. What may have occurred at ①? _____

   b. What may have occurred at ②? _____

   c. What may have occurred at ③? _____

   d. What may have occurred at ④? _____

   e. What may have occurred at ⑤? _____

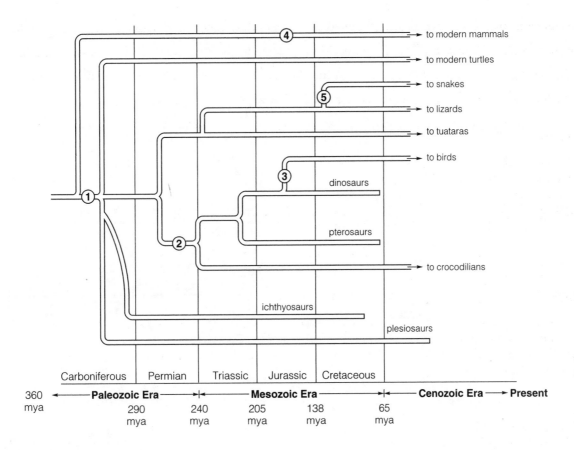

Carboniferous | Permian | Triassic | Jurassic | Cretaceous

360 mya ← **Paleozoic Era** → | ← **Mesozoic Era** → | ← **Cenozoic Era** → **Present**

290 mya   240 mya   205 mya   138 mya   65 mya

27.9. BIRDS [pp.468–469]

27.10. THE RISE OF MAMMALS [pp.470–471]

27.11. A PORTFOLIO OF EXISTING MAMMALS [pp.472–473]

Selected Words: Aves [p.468], *Archaeopteryx* [p.468], animal migration [p.468], keeled sternum [p.469], Mammalia [p.470], incisors, canines, premolars, and molars [p.470], egg-laying mammals [p.471], pouched mammals [p.471], placental mammals [p.471], mammary glands [p.471], spiny anteater [p.472], koala [p.472]

Boldfaced, Page-Referenced Terms

[p.468] birds _____

_____

[p.468] feathers _____

_____

[p.470] mammals _____

_____

[p.470] behavioral flexibility _____

_____

[p.470] dentition _____

_____

[p.471] therapsids _____

_____

[p.471] therians _____

_____

[p.471] monotremes _____

_____

[p.471] marsupials _____

_____

[p.471] eutherians _____

_____

[p.472] convergent evolution _____

_____

[p.472] placenta _____

_____

Labeling

Label the structures pictured at the right.

1. _____ [p.468]
2. _____ [p.468]
3. _____ _____ [p.468]

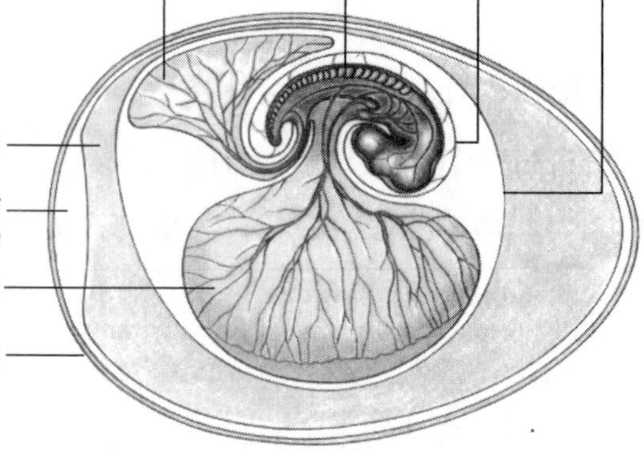

Fill-in-the-Blanks

Birds descended from (4) _____ [p.468] that ran around on two legs some 160 million years ago. All birds have (5) _____ [p.468] that insulate and help get the bird aloft. Generally, birds have a greatly enlarged (6) _____ [p.469] to which flight muscles are attached. Bird bones contain (7) _____ _____ [p.469], and air flows through sacs, not into and out of them, for gas exchange in the lungs. Birds also lay (8) _____ [recall p.464] eggs, have complex courtship behaviors, and generally nurture their offspring. Birds are warm blooded: able to regulate their body (9) _____ [use logic], which is generally higher than that of mammals. Flight muscles attach to a large (10) _____ sternum (breastbone) [p.469].

Flight demands (11) ❏ high ❏ low [p.469] metabolic rates, which require an abundant supply of (12) _____ [p.469] pumped to all body parts by means of a large, durable (13) ❏ 2 ❏ 3 ❏ 4 ❏ 6 – [p.469] chambered heart. Almost 9,000 species of birds show amazing variation in body structure. The smallest adult bird is a(n) (14) _____ [p.469] that weighs 2.25 grams. The largest existing bird is the (15) _____ [p.469] which weighs about 150 kilograms, can sprint fast but cannot fly.

During the Cenozoic Era (65 million years ago to the present), birds, (16) _____ [p.471], and flowering plants evolved to dominate Earth's assemblage of organisms. (16) are warm-blooded vertebrates with (17) _____ [p.471] that began their evolution more than 200 million years ago.

There are three groups of existing mammals: those that lay eggs (examples are the (18) _____ [p.472] and the spiny anteater), those that are (19) _____ [p.471] (examples are the opossum and the kangaroo), and those that are (20) _____ [p.471] mammals. Mammals regulate their body temperature; like birds, they have a (21) _____ [see Chapter 39] -chambered heart, and show a high degree of parental nurture. Most mammals have (22) _____ [p.470] as a means of insulation, and mammalian mothers generally suckle their young with milk. Mammals differ from reptiles and birds in (23) _____ [p.470] (the type, number, and size of teeth). Most mammalian females produce one (24) _____ [p.472] per developing offspring inside the uterus.

The ancestors of today's mammals diverged from small, hairless reptiles called (25) _____ [p.471] more than 200 million years ago during the Triassic. Through mutation and natural selection during Jurassic times they had developed hair and major changes in the body form, jaws and teeth; now they were called (26) _____ [p.471], which coexisted with the diverse groups of dominant (27) _____ [p.471] through the Cretaceous. When the (27)s became extinct, diverse adaptive zones awaited exploitation by the three principal lineages, which, during the subsequent 65 million years, have blossomed into 4,500 known mammalian species. (28) _____ _____ [p.472] has occurred as evolutionarily distinct, geographically isolated lineages evolved in similar ways in similar habitats.

27.12. EVOLUTIONARY TRENDS AMONG THE PRIMATES [pp.474–475]
27.13. FROM PRIMATES TO HOMINIDS [pp.476–477]
27.14. EMERGENCE OF EARLY HUMANS [pp.478–479]
27.15. *Focus on Science:* OUT OF AFRICA—ONCE, TWICE, OR . . . [p.480]

Selected Words: prosimians [p.474], tarsioids [p.474], anthropoids [p.474], arboreal [p.474], monkey [p.474], ape [p.474], lemur [p.474], hominoids [p.474], savannas [p.474], prehensile [p.475], opposable [p.475], Paleocene [p.476], *Plesiadapis* [p.476], *Aegyptopithecus* [p.476], dryopiths [p.476], Miocene [p.476], *Ardipithecus ramidus* [p.477], "Lucy" [p.477], *Australopithecus* [p.477], gracile [p.477], robust [p.477], *Homo habilis* [p.478], *Homo erectus* [p.479], *Homo sapiens* [p.479], cultural [p.479], "race" [p.480], multiregional model [p.480], African emergence model [p.480]

Boldfaced, Page-Referenced Terms

[p.474] Primates _____

_____

[p.474] hominids _____

_____

[p.475] bipedalism _____

_____

[p.475] culture _____

_____

[p.477] australopiths _____

_____

[p.478] humans _____

_____

Fill-in-the-Blanks

During the Cenozoic Era (65 million years ago to the present), birds, (1) _____ [recall p.471], and flowering plants evolved to dominate Earth's assemblage of organisms. (1) are warm-blooded vertebrates with (2) _____ [p.470] that began their evolution more than 200 million years ago. There are many groups of (3) _____ [p.474] within the class Mammalia; each order has its distinctive array of characteristics. Humans, apes, monkeys, and prosimians are all (4) _____ [p.474]; members of this order have excellent (5) _____ [p.474] perception as a result of their forward-directed eyes, hands that are (6) _____ [p.475] and opposable; they are adapted for (7) _____ [p.475] instead of running, and primates rely less on their sense of (8) _____ [p.474] and more on daytime vision. Their (9) _____ [p.475] became larger and more complex; this trend was accompanied by refined technologies and the development of (10) _____ [p.475]: the collection of behavior patterns of a social group, passed from generation to generation by learning and by symbolic behavior (language). Modification in the (11) _____ [p.475] led to increased dexterity and manipulative skills. Changes in primate (12) _____ [p.475] indicate that there was a shift from eating insects to fruit and leaves and on to a mixed diet. Primates began to evolve from ancestral mammals more than (13) _____ [p.476] million years ago, during the Paleocene. The first primates resembled small (14) _____ [p.476] or tree shrews; they foraged at night for (15) _____ [p.476], seeds, buds, and eggs on the forest floor, and they could clumsily climb trees searching for safety and sleep.

During the (16) _____ [p.476] (25–5 million years ago), the continents began to assume their current positions, and climates became cooler and (17) _____ [p.476]. Under these conditions, forests began to give way to mixed woodlands and (18) _____ [p.476]; an adaptive radiation of apelike forms—the first hominoids—took place as subpopulations of apes became reproductively isolated within the shrinking stands of trees. The chimpanzee-sized "tree apes," or (19) _____ [p.476], originated during this time; eventually, they ranged throughout Africa, Europe, and Southern (20) _____ [p.476]. Between (21) _____ [p.477] million and (22) _____ [p.477] million years ago, two lineages branched from a common ancestor; one lineage gave rise to the great apes, and the other gave rise to the first hominids.

Matching

Fill in the blanks of the following evolutionary diagram with the appropriate letter from the choice list. Consult Table 27.2. Choose from the lettered groups below to match boldfaced numbers. [All from p.474]

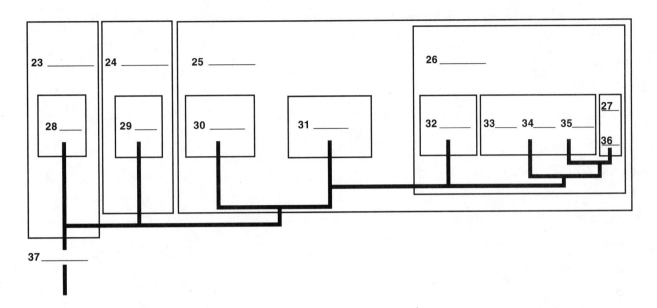

A. Anthropoids
B. Chimpanzee
C. Gibbon, siamang
D. Gorilla
E. Hominids
F. Hominoids
G. Humans and their most recent ancestors
H. Lemurs, lorises
I. New World monkeys (spider monkeys, etc.)
J. Old World monkeys (baboons, etc.)
K. Orangutan
L. Prosimians
M. Tarsiers
N. Tarsioids
O. Rodentlike primate of the Paleocene

Choice

Choose from the following choices for 38–47. Choose the single *smallest* group to which each belongs. [All from p.474]

a. anthropoids   b. hominids   c. hominoids   d prosimians   e tarsioids

38. __ Australopiths
39. __ Chimpanzees
40. __ Dryopiths
41. __ Gibbons
42. __ Gorillas
43. __ Humans
44. __ Lemurs
45. __ New World monkeys
46. __ Old World monkeys
47. __ Orangutans

Fill-in-the-Blanks

(48) _____ [p.477] include all lineages that descended from the ancestral groups that had diverged from the ancestors of chimpanzees approximately (49) _____ [p.477] million years ago during the Miocene epoch. (50) _____ [p.477] such as Lucy were apelike in many skeletal details, but they walked upright like humans. Other human features include: a large (51) _____ [p.478] that provides for distinctive analytical and verbal skills, complex social behavior, and technological innovation. Changes in the bones of the hand led to greater (52) _____ _____ [p.478] to manipulate the simple tools that the earliest humans, (53) *Homo* _____ [p.478] of Olduvai Gorge, created as they foraged for fruits, leaves, insects, roots, and the remains of animal carcasses. The oldest fossils of the genus *Homo* date from approximately (54) _____ [p.478] million years ago. Between 2 million and until at least 53,000 years ago, during the Pleistocene periods of glaciation, there were also intermittent periods of warming; during these interglacial times, a larger-brained human species, (55) _____ _____ [p.479], migrated out of Africa and into China, Southeast Asia, and Europe. Over time, (55) _____ _____ became better at toolmaking and learned how to control (56) _____ [p.479] as their people adapted to a wide range of habitats and were associated with abundant (57) _____ [p.479] artifacts. The (58) _____ [p.479] were a distinct hominid population that appeared 200,000 years ago in southern France, central Europe, and the Near East; their cranial capacity was indistinguishable from our own, they were massively built and they had a complex culture; their disappearance coincided with the appearance of anatomically modern humans.

Anatomically modern humans evolved from (59) _____ _____ [p.480]. Analyses of (60) _____ [p.480] and (61) _____ [p.480] studies support a model of human origins hypothesizing that *Homo erectus* migrated out of Africa approximately two million years ago, then formed distinctive subpopulations of modern humans as an outcome of genetic divergence in different geographic regions.

Another model states that *H. sapiens* originated in Sub-Saharan Africa between 200,000 and 100,000 years ago and later emigrated from Africa. The oldest known fossils of early modern humans (*H. sapiens*) are from (62) _____ [p.480]; they are (63) _____ [p.479] years old. From (64) _____ [p.479] years ago to the present, human evolution has been almost entirely cultural rather than biological. By 30,000 years ago, there was only one remaining hominid species: (65) _____ _____ [p.479].

Labeling

Identify each indicated part of the accompanying illustrations.

66. _____ [p.476,p.482]

67. _____ [p.476,p.482]

68. _____ [p.477,p.482]

69. _____ [p.474,p.482]

70. _____ [p.474,p.482]

71. _____ _____ [p.477,p.482]

72. _____ _____ [p.479,p.482]

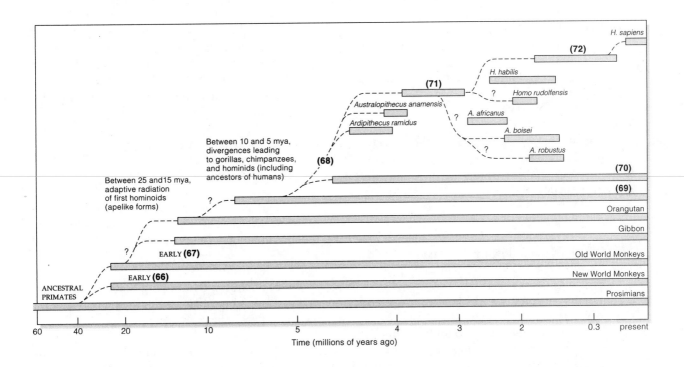

Self-Quiz

___ 1. Filter-feeding chordates rely on _____, which have cilia that create water currents and mucous sheets that capture nutrients suspended in the water. [p.456]
   a. notochords
   b. differentially permeable membranes
   c. filiform tongues
   d. gill slits

___ 2. In true fishes, the gills serve primarily _____ function. [p.458]
   a. a gas-exchange
   b. a feeding
   c. a water-elimination
   d. both a feeding and a gas-exchange

___ 3. The heart in amphibians _____. [p.669, referred from p.465]
   a. pumps blood more rapidly than the heart of fish
   b. is efficient enough for amphibians but would not be efficient for birds and mammals
   c. has three chambers (ventricle and two atria)
   d. all of the above

___ 4. The feeding behavior of true fishes selected for highly developed _____. [p.458]
   a. parapodia
   b. notochords
   c. sense organs
   d. gill slits

___ 5. Which of the following is *not* considered to have been a key character in early primate evolution? [pp.474–475]
   a. eyes adapted for discerning color and shape in a three-dimensional field
   b. body and limbs adapted for tree climbing
   c. bipedalism and increased cranial capacity
   d. eyes adapted for discerning movement in a three-dimensional field

___ 6. Primitive primates generally live _____. [p.474]
   a. in tropical and subtropical forest canopies
   b. in temperate savanna and grassland habitats
   c. near rivers, lakes, and streams in the East African Rift Valley
   d. in caves where there are abundant supplies of insects

___ 7. The earliest fossils of *Homo habilis*, *Homo erectus*, and *Homo sapiens* date from approximately _____, _____, and _____ years ago, respectively. [pp.478–479]
   a. 5, 4, and 2 million
   b. 2.5, 2, and 0.1 million
   c. 25,000, 2,000, and 100 thousand
   d. 5,000, 4,000, and 20 thousand

___ 8. The hominid evolutionary line stems from a divergence (fork in a phylogenetic tree) from the ape line that apparently occurred _____. [p.477]
   a. somewhere between 6 million and 4 million years ago
   b. about 3 million years ago
   c. during the Pliocene epoch
   d. less than 2 million years ago

___ 9. _____ was an Oligocene anthropoid that probably predated the divergence leading to Old World monkeys and the apes, with dentition more like that of dryopiths and less like that of the Paleocene primates with rodentlike teeth. [p.476]
   a. *Aegyptopithecus*
   b. *Australopithecus*
   c. *Homo erectus*
   d. *Plesiadapis*

___10. Donald Johanson, from the University of California at Berkeley, discovered Lucy (named for the Beatles tune), who was a(n) _____. [p.477]
   a. dryopith
   b. australopith
   c. member of *Homo*
   d. prosimian

___11. A hominid of Europe and Asia that became extinct nearly 30,000 years ago was _____. [p.479]
   a. a dryopith
   b. *Australopithecus*
   c. *Homo erectus*
   d. Neandertal

Matching

Choose the one most appropriate answer for each.

12. ___anthropoids [p.474]

13. ___australopiths [p.477]

14. ___Cenozoic [p.476; recall geologic time scale]

15. ___hominids [p.474]

16. ___hominoids [p.474]

17. ___Miocene [p.476]

18. ___primates [p.474]

19. ___prosimians [p.474]

A. A group that includes apes and humans
B. Organisms in a suborder that includes New World and Old World monkeys, apes, and humans
C. An era that began 65 to 63 million years ago; characterized by the evolution of birds, mammals, and flowering plants
D. A group that includes humans and their most recent ancestors
E. An epoch of the Cenozoic era lasting from 25 million to 5 million years ago; characterized by the appearance of primitive apes, whales, and grazing animals of the grasslands
F. Organisms in a suborder that includes tree shrews, lemurs, and others
G. A group that includes prosimians, tarsioids, and anthropoids
H. Bipedal organisms living from about 4 million to 1 million years ago, with essentially human bodies and ape-shaped heads; brains no larger than those of chimpanzees

Matching

Match the following groups and classes with the corresponding characteristics (a–i) and representatives (A–I).

___ , ___20. Amphibians [p.462]

___ , ___21. Birds [p.468]

___ , ___22. Bony fishes [pp.460–461]

___ , ___23. Cartilaginous fishes [p.460]

___ , ___24. Cephalochordates [pp.456–457]

___ , ___25. Jawless fishes [p.459]

___ , ___26. Mammals [p.470]

___ , ___27. Reptiles [p.464]

___ , ___28. Urochordates [p.456]

a. hair + vertebrae
b. feathers + hollow bones
c. jawless + cartilaginous skeleton (in existing species)
d. two pairs of limbs (usually) + glandular skin + "jelly"-covered eggs
e. amniote eggs + scaly skin + bony skeleton
f. invertebrate + sessile adult that cannot swim
g. jaws + cartilaginous skeleton + vertebrae
h. in adult, notochord stretches from head to tail; mostly burrowed-in, adult can swim
i. bony skeleton + skin covered with scales and mucus

A. lancelet
B. loons, penguins, and eagles
C. tunicates, sea squirts
D. sharks and manta rays
E. lampreys and hagfishes (and ostracoderms)
F. true eels and sea horses
G. lizards and turtles
H. caecilians and salamanders
I. platypuses and opossums

Identification

Provide the common name and the major chordate group for each creature pictured as follows and on pages 457–458, 460–462, 466, 469, and 473.

29. [p.462] _____, _____

30. [p.473] _____, _____

31. [p.461] _____, _____

32. [p.458] _____, _____

33. [p.469] _____, _____

34. [p.466] _____, _____

35. [p.460] _____, _____

36. [p.461, Fig. 27.9e] _____, _____

37. [p.460] _____, _____

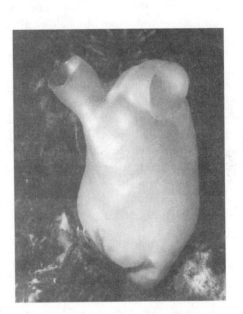

38. [p.457] _____, _____

39. [p.457] _____, _____

Chapter Objectives/Review Questions

1. Name each class of vertebrates and state the distinguishing features of each. [pp.456,459–460,462,465, 468,470]
2. List three characteristics found only in chordates. [p.456]
3. Describe the adaptations that sustain the sessile or sedentary life-style seen in primitive chordates such as tunicates and lancelets. [pp.456-457]
4. State what sort of changes occurred in the primitive chordate body plan that could have promoted the emergence of vertebrates. [pp.458–459]
5. Describe the differences between primitive and advanced fishes in terms of skeleton, jaws, special senses, and brain. [pp.458–461]
6. Describe the changes that enabled aquatic fishes to give rise to land dwellers. [pp.461–462]
7. Discuss the effects that increased parental nurture of offspring in birds and mammals has had on courtship behavior and reproductive physiology. [pp.468–472]
8. Where do tarsier survivors dwell today on Earth? [p.474]
9. Beginning with the primates most closely related to humans, list the main groups of primates in order by decreasing closeness of relationship to humans. [p.474]
10. Five key characters of primate evolution are _____, _____, _____, _____ and _____. [pp.474–475]
11. Describe the general physical features and behavioral patterns attributed to early primates. [p.476]
12. Trace primate evolutionary development through the Cenozoic era. Describe how Earth's climates were changing as primates changed and adapted. Be specific about times of major divergence. [pp.476–479]
13. State which anatomical features underwent the greatest changes along the evolutionary line from early anthropoids to humans. [pp.474–475]
14. Explain how you think *Homo sapiens* arose. Make sure your theory incorporates existing paleontological (fossil), biochemical, and morphological data. [pp.479–480]

Integrating and Applying Key Concepts

Birds and mammals both have four-chambered hearts, high metabolic rates, and regulate their body temperatures efficiently. Both groups evolved from reptiles, so one would think that those same traits would have developed in ancestral reptiles. Data suggest that most reptiles have a heart intermediate between three and four chambers, lower metabolic rates, and body temperatures that are not well regulated and tend to rise and fall in accord with the environmental temperature. If the three traits mentioned in the first sentence had developed in reptilian groups, how might their lives have been different?

Suppose someone told you that some time between 12 million and 6 million years ago dryopiths were forced by larger predatory members of the cat family to flee the forests and take up residence in estuarine, riverine, and sea coastal habitats where they could take refuge in the nearby water to evade the tigers. Those that, through mutations, became naked, developed an upright stance, developed subcutaneous fat deposits as insulation, and developed a bridged nose that had advantages in watery habitats (features that other dryopiths that remained inland never developed) survived and expanded their populations. As time went on, predation by the big cats and competition with other animals for available food caused most of the terrestrial dryopiths to become extinct, but the water-habitat varieties survived as scattered remnant populations, adapting to easily available shellfish and fish, wild rice and oats, and various tubers, nuts, and fruits. It was in these aquatic habitats that the first food-getting tools (baskets, nets, and pebble tools) were developed, as well as the first words that signified different kinds of food. How does such a story fit with current speculations about and evidence of human origins? How could such a story be shown to be true or false?

Crossword Puzzle: Primate Evolution

Across

1. A group that includes prosimians, tarsioids, and anthropoids [p.474]
3. A group that includes apes and humans [p.474]
7. Small, daisylike flowers _____
10. A kind of tall, showy flower; usually lavender or purple
11. A cheek tooth that crushes and grinds
12. All the behavior patterns of a social group passed by learning and language from generation to generation [p.475]
13. Pay great homage to; worship
14. Forest apes that lived 13 million years ago in Africa, Europe, and southern Asia [p.476]
17. Pointed teeth that enable the tearing of flesh [p.470]
18. *Homo* _____ was the first out-of-Africa hominid to control and use fire for heating and cooking [p.479]
19. A group of primates intermediate between lemurs and monkeys [p.474]

Down

2. Southern ape-humans, the fossils of which have dates from 3.7 to 1.25 million years ago [p.477]
4. A vertebrate with hair [p.470]
5. A flat chisel or conelike tooth that nips or cuts food [p.470]
6. A group that includes modern humans and their direct-line ancestors since divergence from the ape line [p.470]
8. The specific (species) name of modern humans [p.479]
9. Habitual two-legged method of locomotion [p.475]
15. _____ *Homo* used simple stone tools. [p.478]
16. Hard structures that can provide clues about what an animal typically eats [p.470]

28

BIODIVERSITY IN PERSPECTIVE

Interactive Exercises

The Human Touch [pp.484–485]

28.1. ON MASS EXTINCTIONS AND SLOW RECOVERIES [pp.486–487]

28.2. THE NEWLY ENDANGERED SPECIES [pp.488–489]

28.3. *Focus on the Environment:* CASE STUDY: THE ONCE AND FUTURE REEFS [pp.490–491]

28.4. *Focus on Science:* RACHEL'S WARNING [p.492]

Selected Words: Easter Island [p.484], biodiversity [p.485], taxa [p.486], mosasaur [p.487], dodo [p.487], *endemic* [p.488], E. O. Wilson [p.488], *exotic* species [p.489], *Corallina* [p.490], Rachel Carson [p.492], *Silent Spring* [p.492]

Boldfaced, Page-Referenced Terms

[p.488] endangered species _____

_____

[p.488] habitat loss _____

_____

[p.488] habitat fragmentation _____

_____

[p.489] habitat islands _____

_____

[p.489] indicator species _____

[p.490] coral reefs _____

Fill-in-the-Blanks

The full range of (1) _____ [p.486] is greater now than it has ever been in the past, even though an estimated 99 percent of all species that have ever lived are now extinct. Each global episode that brought about a (2) _____ _____ [p.486] that reduced the total number of different species also acted to encourage evolutionary change and (3) _____ _____ [p.486] into newly vacated adaptive zones. Recovery was exceedingly slow, requiring 20 to 100 (4) (choose one) ❏ hundred ❏ million ❏ billion [p.486] years to reach the same level again. Each lineage had become either a winner or loser when environmental conditions changed drastically.

Approximately 65 million years ago, an event caused the (5) _____ [p.487] and (6) _____ [p.486] to become extinct; the Mesozoic Era ended with the (7) _____ [p.487] boundary, and the (8) _____ [p.486] Era began after the event ended. More than 300 years ago, Dutch sailors clubbed to death the last (9) _____ [p.487], a flightless bird that lived only on Mauritius Island.

No biodiversity-shattering asteroids have hit Earth for 65 million years, yet the (10) _____ [number] [p.488] major extinction event is underway, and it is being driven largely by (11) _____ [p.488] activities. Currently, (12) _____ [p.488] species of mammals are considered to be endangered, which means they are endemic and extremely vulnerable to extinction.

We are just one species among millions, but because of our rapid population growth, we threaten the existence of others by way of habitat losses, species introductions, (13) _____ [p.488], and illegal wildlife trading.

The physical reduction in suitable places to live, as well as the poisoning of places to live as a result of chemical or radioactive pollution are what E. O. Wilson defines as (14) _____ _____ [p.488], which is one of the major threats to more than 90 percent of endangered species. (1) is greatest in the tropics, where habitats on land are being lost daily on a grand scale due to (15) _____ [p.488]. In the United States alone (choose from the following: ❏ 50% ❏ 85–95% ❏ 98%) (16) _____ [p.488] percent of the wetlands, (17) _____ [p.488] percent of the old growth forests, and (18) _____ [p.488] percent of the tall-grass prairie are gone because of human destruction of these ecosystems. When large exposures of ecosystems become fragmented into isolated patches, there are more edges to the habitat so that populations are more vulnerable to (19) _____ [p.488], winds, fires, temperature changes, and (20) _____ [p.488]. Also, the new areas may not be large enough to contain enough resources for successful (21) _____ [p.488] and survival.

About half of the plant and animal populations that have become extinct since 1600 were native to (22) _____ [p.489] because they could not escape to anywhere else easily. (22) need not be surrounded by water to be considered (22); national parks, tropical forests, (23) _____ [p.489] and reserves are examples. In general according to MacArthur and Wilson, who developed a model of island (24) _____ [p.489], a 90 percent loss of habitat will drive about (25) _____ [p.489] percent of endemic populations to extinction. About (26) _____ [p.489] percent of the 9,600 known species of birds are now facing habitat loss. (27) _____ [p.489] species are those introduced into new habitats; they adapt and displace endemic species. (27) are a major factor in almost 70 percent of the situations that are driving endemic species to extinction.

Predation by humans has driven most of the great (28) _____ [p.489] to the brink of extinction; instead of being regarded as exploitable commodities to be made into corset stays and fertilizer, (28) should be conserved for their intelligence, their roles in healthy ecosystems, and their complex communication systems.

Of all the marine ecosystems, (29) _____ _____ [p.490] show the most spectacular biodiversity. Abnormal, widespread (30) _____ [p.490] of (29) began in the 1980s when sea surface temperatures increased and the colorful symbiotic dinoflagellates were expelled by the polyps. Commercial fishermen from Japan, Indonesia, and Kenya routinely dynamite (29) and squirt sodium cyanide into hiding places to stun fish, which float to the surface.

In 1962, (31) _____ _____ [p.492] published her book, *Silent Spring*, which documented the harmful effects of pesticides on wildlife. She suggested that indiscriminate use of pesticides could endanger human populations as well as kill robins, catbirds, doves, jays, and wrens. She died from cancer without knowing that she had been instrumental in starting the (32) _____ [p.492] movement.

28.5. CONSERVATION BIOLOGY [pp.492–493]
28.6. RECONCILING BIODIVERSITY WITH HUMAN DEMANDS [pp.494–495]

Selected Words: Zea diploperennis [p.492], systematics [p.492], bioeconomic analysis [p.493], sustainable development [p.493], Gary Hartshorn [p.494], Jack Turnell [p.495]

Boldfaced, Page-Referenced Terms

[p.492] conservation biology _____

_____

[p.492] hot spots _____

_____

[p.493] ecoregion _____

_____

[p.494] strip logging _____

_____

[p.495] riparian zone _____

_____

Fill-in-the-Blanks

"An awful symmetry . . . binds the rise of (1) _____ [p.492] to the fall of biodiversity." In the next 50 years, Earth's human population size may reach (2) _____ [p.492] billion, with most growth occurring in the developing countries, where scientific knowledge of their biological treasure trove is little and biodiversity is greatest. Awareness of the impending extinction crisis gave rise to the field of pure and applied research known as (3) _____ _____ [p.492]. Scientists in this field perform a systematic (4) _____ [p.492] and description of the full range of biological diversity, try to understand the evolutionary and ecological origins of diversity, and try to identify (5) _____ [p.492] that might maintain populations at levels high enough to keep Earth's ecosystems functioning well with enough abundance that could also sustain the burgeoning human populations. (6) _____ _____ [p.492] are habitats that have the greatest number of endemic species found nowhere else that are in the greatest danger of extinction due to human activities.

Many plants are the sources of medicines and other chemical products. Quite possibly, large (7) _____ [p.493] companies will pay royalties to countries that have large numbers of different plant species that might serve as sources of medicines or herbal tonics.

The richest sources of biodiversity on land are the (8) _____ _____ _____ [p.494]. Most environmentalists would say that this (9) _____ [p.493] should not be logged at all, because when the logs are taken out of this kind of habitat most of the productivity is lost and is difficult to regain if traditional clear-cutting methods are used. However, Gary Hartshorn has proposed the method of (10) _____ _____ [p.494] as a way to harvest sustainably a limited supply of wood for local economies and limited amounts of diverse, exotic woods prized by developed countries. Where terrain is sloped and features drainage streams, a narrow corridor that follows the land's (11) _____ [p.494] is cleared; the upper part of it is used to build a (12) _____ [p.494] to haul away the logs taken from the lower part of the corridor. After a few years, (13) _____ [p.494] seeded from the native trees start to grow in the original corridor and another corridor is cleared above the road. Precious (14) _____ [p.494] leached from the exposed soil trickle down into the first corridor and are taken up by (13), which benefit by more rapid (15) _____ [p.494] and begin a profitable logging cycle that the region can sustain over time.

Plants associated with (16) _____ [p.495] zones along streams or rivers afford a line of defense against damage by (17) _____ [p.495] because they "drink up" water from spring (18) _____ [p.495] and summer storms. In the western part of the United States, (16) zones shelter 67 to 75 percent of the (19) _____ [p.495] species that must spend at least part of their life cycles there. Unfortunately, compared with wild horned (20) _____ [p.495], cattle drink a lot more water, so they tend to congregate at (16) zones, trampling and feeding on the grasses and tender shrubs until they are gone, leaving bare banks that erode easily. Developing feeding and (21) _____ [p.495] sites away from (16) zones and providing supplemental feed at different grazing areas can restore (16) zones and sustain endemic wildlife diversity by providing natural sources of food, shelter, or shade. If people want to eat beef, perhaps part of the cost of buying beef should be reserved for ranchers like Jack Turnell who are willing to build the fences necessary to restrict cattle access to (16) zones and develop (21) and feeding sites elsewhere.

· Throughout much of the world, the explosive rate of human (22) _____ _____ [p.495] is the elephant in the room that few people want to acknowledge and do something about.

Self-Quiz

___ 1. All the following statements are apparently true of Easter Island *except:* [pp.484–485]
   a. It was settled by voyagers from Hawaii around A.D. 350, possibly after being blown off course by a series of storms.
   b. Dense palm forests, hau hau trees, toromino shrubs, and grasses were abundant on the island.
   c. The settlers planted crops of taro, bananas, sugarcane, and sweet potatoes.
   d. Edible seafood species vanished from the protected waters around the island so fishermen had to build larger boats to sail far out on the open sea.
   e. The islanders ate rats and each other before the population vanished sometime after 1724, when Captain James Cook visited Easter Island.

___ 2. The most important lesson for humans worldwide that can be understood from the story of Easter Island is _____. [p.485]
   a. that you shouldn't cut down the last tree on the island, because then you won't be able to make a boat with which to escape to another place
   b. that preserving biodiversity generally sustains societies and allows them to enjoy abundant resources as long as they keep their human population in check
   c. that arable land grows richer with nutrients as time passes if intelligent planting methods are used
   d. that by 1400, at least 10,000 humans were living on an island that consisted of approximately 165 square kilometers (64 square miles) of surface land
   e. that it resulted in humans cannibalizing other humans

___ 3. Human worldwide population is expected to reach _____ billion by 2050, and human demands for food, materials, and living space are threatening biodiversity around the world. [p.492]
   a. 3
   b. 6
   c. 9
   d. 12
   e. 15

___ 4. Conservation biology includes all the following *except* _____. [p.492]
   a. identifying methods to maintain and use biodiversity for the benefit of humans
   b. surveying the full range of biodiversity in the world ecoregions
   c. encouraging the use of genetically altered crops planted as monocultures to reduce competition with native varieties of plants
   d. analyzing the evolutionary origins of ecoregions
   e. analyzing the ecological origins of ecoregions

For questions 5–8, choose from these possible answers:
   a. Strip logging
   b. Habitat fragmentation
   c. Overharvesting
   d. Extinction
   e. Relocating watering and feeding stations

___ 5. _____ can play a major role in the restoration of riparian zones. [p.495]

___ 6. _____ can play a major role in the sustainable use of tropical rain forests. [p.494]

_____ 7. In North America, _____ made it easier for skunks, opossums, raccoons, and blue jays to feast on eggs and juveniles of migratory songbirds. [pp.488–489]

_____ 8. In marine environments, _____ has led to edible species vanishing from safe, inshore waters and moving to other deeper, offshore regions; on land, it has led to the extinction of dodos and bison. [pp.484,488]

_____ 9. Biodiversity is a measure of _____. [p.485]
   a. the total number of organisms associated with a defined amount of living room
   b. evolutionary success
   c. indicator species that are thriving in a defined area
   d. the number of different species associated with a defined amount of living room
   e. comparing exotic species with endemic species

_____10. All of the following are true of coral reefs *except* _____. [p.490]
   a. Coral reefs generally develop between latitudes 25° north and south.
   b. Increases in the surface temperatures of the tropical and subtropical oceans have increased since the 1980s.
   c. Some unscrupulous reef fishermen use dynamite and/or sodium cyanide to make reef animals come out of their hiding places.
   d. Since the 1980s, Indonesian fisherman have been causing widespread bleaching of living corals by squirting laundry bleach on them to prepare them for sale to pet stores and gift shops.
   e. Reef formations off Florida's key largo have declined by one-third, mostly since 1970.

Chapter Objectives/Review Questions

1. Describe the history of human occupation of Easter Island. Note the human choices and behaviors that inexorably led to a dramatic crash in biodiversity on the island. [pp.484–485]
2. Explain how it can be that 99 percent of all species that have ever lived are extinct, but biodiversity is greater now than it has ever been in the past. [pp.486–487]
3. Describe the supposed relationship between the dodo or land tortoise and a tree (*Calvaria major*) on the island of Mauritius. [p.487]
4. Earth's sixth major extinction event is underway. State its principal cause and then list four secondary causes brought about by the principal cause. [pp.488–492]
5. Describe (a) the coral reef and (b) tropical rain forest, which are the marine and land-based ecoregions of greatest biodiversity. List the principal threats to each, and suggest what needs to be done to keep these ecosystems functioning on a sustainable, healthy basis. [pp.490,494]
6. Identify the contributions that Rachel Carson, E. O. Wilson, Gary Hartshorn, the World Wildlife Fund, Jack Turnell, Lucy Bunkley-Williams, and Ernest Williams have made to our understanding of conservation biology. [pp.488,490,492,493,495]

Integrating and Applying Key Concepts

Think about your life and identify five specific things that you could do to encourage biodiversity in ecosystems, some of which are far away from your home and some of which are very close to you. Are there any streams, rivers or forests undergoing changes for the worse near where you live? How could you find information about how these places were before your region was settled by the human ancestors of its current inhabitants? What can you do to help these places maintain high biodiversity? Why should you bother to put your personal energy into such activities?

29

PLANT TISSUES

Interactive Exercises

Plants Versus the Volcano [pp.498–499]

29.1. OVERVIEW OF THE PLANT BODY [pp.500–501]

Selected Words: "annuals" [p.500], *nonwoody* [p.500], "biennials" [p.500], "perennials" [p.500], *Lycopersicon* [p.500], *primary* meristems [p.501], *lengthening* of stems [p.501], periderm [p.501], *thickening* of stems [p.501]

Boldfaced, Page-Referenced Terms

[p.500] shoots _____

_____

[p.500] roots _____

_____

[p.500] ground tissue system _____

_____

[p.500] vascular tissue system _____

_____

[p.500] dermal tissue system _____

_____

[p.501] meristems _____

_____

[p.501] apical meristem _____

_____

[p.501] vascular cambium _____

_____

[p.501] cork cambium _____

_____

Labeling and Matching

Identify each part of the accompanying illustration. Choose from dermal tissues (epidermis), root system, ground tissues, shoot system, and vascular tissues. Complete the exercise by matching and entering the letter of the proper description in the parentheses following each label.

1. _____ _____ ( ) [p.500]
2. _____ _____ ( ) [p.500]
3. _____ _____ ( ) [p.500]
4. _____ _____ ( ) [p.500]
5. _____ _____ ( ) [p.500]

A. Typically consists of stems, leaves, and flowers (reproductive shoots)
B. Typically grows below ground, anchors aboveground parts, absorbs soil water and minerals, stores and releases food, and anchors the aboveground parts
C. Tissues that make up the bulk of the plant body
D. Covers and protects the plant's surfaces
E. Two conducting tissues that distribute water and solutes through the plant body

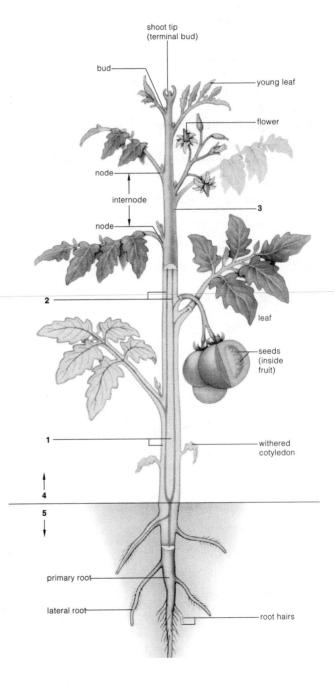

Matching

Choose the most appropriate answer for each term.

6. ___secondary growth [p.501]

7. ___vascular cambium and cork cambium [p.501]

8. ___meristems [p.501]

9. ___apical meristems [p.501]

10. ___primary growth [p.501]

11. ___transitional meristems [p.501]

A. Represented by a lengthening of stems and roots originating from cell divisions at apical meristems
B. Localized regions of dividing cells
C. Cell populations forming from apical meristems; include protoderm, ground meristem, and procambium
D. Lateral meristems giving rise to, respectively, secondary vascular tissues, and a sturdier plant covering that replaces epidermis
E. Represented by a thickening of stems and roots caused by activity of the lateral meristems
F. Located in the dome-shaped tips of all shoots and roots; responsible for lengthening those organs

Labeling

It is important to understand the terms that identify the thin sections (slices) of plant organs and tissues that are prepared for study. Label each of the following three diagrams. Choose from radial section (cut along the radius of the organ), transverse or cross section (cuts perpendicular to the long axis of the organ), or tangential section (cuts made at right angles to the radius of the organ). *Note:* The dark area indicates the slice.

12. _____ section [p.501]

13. _____ section [p.501]

14. _____ section [p.501]

Labeling and Matching

Identify each part of the accompanying illustration. Choose from shoot apical meristem, root apical meristem, root transitional meristems, shoot transitional meristems, and lateral meristems. Complete the exercise by matching and entering the letter of the proper description in the parentheses after each label.

15. _____ _____ _____ (   ) [p.501]

16. _____ _____ _____ (   ) [p.501]

17. _____ _____ _____ (   ) [p.501]

18. _____ _____ _____ (   ) [p.501]

19. _____ _____ (   ) [p.501]

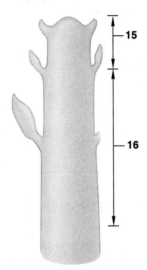

A. Near all root tips, also gives rise to three transitional meristems from which the root's primary tissue systems develop (lengthening)

B. These embryonic descendants of apical meristem (protoderm, ground meristem, and procambium) divide, grow, and differentiate into a shoot's primary tissue systems

C. Found only inside the older stems and roots of woody plants; sources of secondary growth (increases in diameter)

D. A region of embryonic cells near the dome-shaped tip of all shoots; the source of primary growth (lengthening)

E. Embryonic descendants of the root apical meristem (the protoderm, ground meristem, and procambium) that divide, grow, and differentiate into the root's primary tissue systems

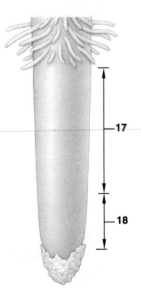

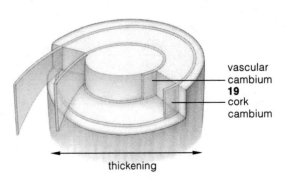

Complete the Table

20. The cells of the transitional meristems and the lateral meristems give rise to tissues, either primary or secondary. Complete the following table, which summarizes the activity of these meristems.

Meristem	Tissue(s)	Primary or Secondary
[p.501] a. Protoderm		
[p.501] b. Ground meristem		
[p.501] c. Procambium		
[p.501] d. Vascular cambium		
[p.501] e. Cork cambium		

29.2. TYPES OF PLANT TISSUES [pp.502–503]

Selected Words: mesophyll [p.502], *fibers* [p.502], *sclereids* [p.502], *vessel members* [p.503], *tracheids* [p.503], *sieve-tube members* [p.503], *companion cells* [p.503]

Boldfaced, Page-Referenced Terms

[p.502] parenchyma _____

_____

[p.502] collenchyma _____

_____

[p.502] sclerenchyma _____

_____

[p.503] xylem _____

_____

[p.503] phloem _____

_____

[p.503] epidermis _____

_____

[p.503] cuticle _____

_____

[p.503] stoma (plural, stomata) _____

_____

[p.503] periderm _____

_____

[p.503] dicots _____

_____

[p.503] monocots _____

_____

Labeling

Label the following illustrations as either collenchyma (uneven wall thickenings, often appearing like thickened corners; parenchyma (thin-walled with intercellular spaces); or sclerenchyma (very thick walls in cross-section, includes fibers and sclereids such as stone cells).

1. _____
   [p.502]

2. _____
   [p.502]

3. _____
   [p.502]

thick secondary wall

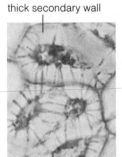

4. _____
   [p.502]

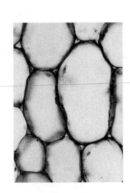

5. _____
   [p.502]

6. _____
   [p.502]

Choice

For questions 7–17, choose from the following:

a. parenchyma   b. collenchyma   c. sclerenchyma

7. ___ Patches or cylinders of this tissue are found near the surface of lengthening stems. [p.502]

8. ___ The cells have thick, lignin-impregnated walls. [p.502]

9. ___ Cells are alive at maturity and retain the capacity to divide. [p.502]

10. ___ Some types specialize in storage, secretion, and other tasks. [p.502]

11. ___ Cells are mostly elongated, with unevenly thickened walls. [p.502]

12. ___ Provides flexible support for primary tissues [p.502]

13. ___ Abundant air spaces are found around the cells. [p.502]

14. ___ Supports mature plant parts and often protects seeds [p.502]

15. ___ Cells are thin-walled, pliable, and many-sided. [p.502]

16. ___ Mesophyll is specialized for photosynthesis. [p.502]

17. ___ Fibers and sclereids [p.502]

Labeling

Identify the cell types and cellular structures in the accompanying illustrations by entering the correct name in the blanks. Complete the exercise by entering the name of the complex tissue (xylem or phloem) in which that cell or structure is found.

18. _____ (_____) [p.502]

19. _____ (_____) [p.502]

20. _____ (_____) [p.502]

21. _____ (_____) [p.502]

22. _____ (_____) [p.502]

23. _____ (_____) [p.502]

24. _____ (_____) [p.502]

25. _____ (_____) [p.502]

26. _____ (_____) [p.502]

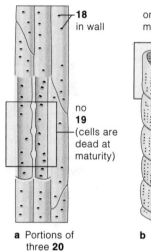

18 in wall

no **19** (cells are dead at maturity)

a Portions of three **20**

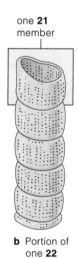

one **21** member

b Portion of one **22**

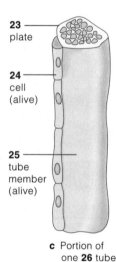

23 plate

24 cell (alive)

25 tube member (alive)

c Portion of one **26** tube

Complete the Table

27. The two types of vascular tissues, xylem and phloem, occur in strands called *vascular bundles* within the plant body (each of these complex tissues also has fibers and parenchyma). Complete the following table, which summarizes important information about these vascular tissues.

Vascular Tissue	Major Conducting Cells	Cells Alive?	Function(s)
[p.503] a. Xylem			
[p.503] b. Phloem			

28. Two types of dermal tissues cover the plant body. Complete the following table, which summarizes information about the dermal tissues.

Dermal Tissue	Primary or Secondary Plant Body	Function(s)
[p.503] a. Epidermis		
[p.503] b. Periderm		

29. There are two classes of flowering plants, monocots and dicots. Complete the following table, which summarizes information about the two groups of flowering plants.

Class	Number of Cotyledons	Number of Floral Parts	Leaf Venation	Pollen Grains	Vascular Bundles
[p.503] a. Monocots					
[p.503] b. Dicots					

29.3. PRIMARY STRUCTURE OF SHOOTS [pp.504–505]

Selected Words: Coleus [p.412], *Medicago* [p.413], *Zea mays* [p.413]

Boldfaced, Page-Referenced Terms

[p.504] bud _____

_____

[p.504] vascular bundles _____

_____

[p.504] cortex _____

_____

[p.504] pith _____

_____

Labeling

Name the structures numbered in the following illustrations of shoot primary development.

1. _____ [p.504]
2. _____ _____ [p.504]
3. _____ [p.504]
4. _____ [p.504]
5. _____ [p.504]
6. _____ [p.504]
7. _____ [p.504]
8. _____ [p.504]
9. _____ _____ [p.504]
10. _____ [p.504]
11. _____ [p.504]
12. _____ [p.504]
13. _____ [p.504]
14. _____ _____ [p.504]
15. _____ _____ [p.504]

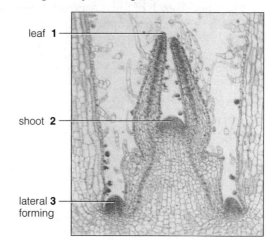

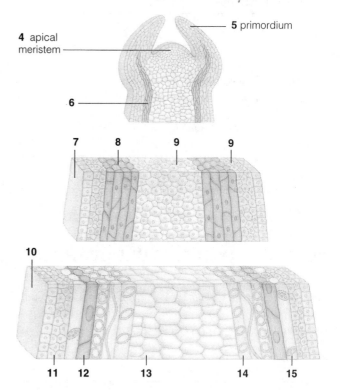

Labeling

Name the structures numbered in the following illustrations of the monocot stem. Choose from air space, epidermis, sclerenchyma cells, vessel, vascular bundle, sieve tube, ground tissue, and companion cell.

16. _____ [p.505]

17. _____ _____ [p.505]

18. _____ _____ [p.505]

19. _____ _____ [p.505]

20. _____ _____ [p.505]

21. _____ _____ [p.505]

22. _____ _____ [p.505]

23. _____ _____ [p.505]

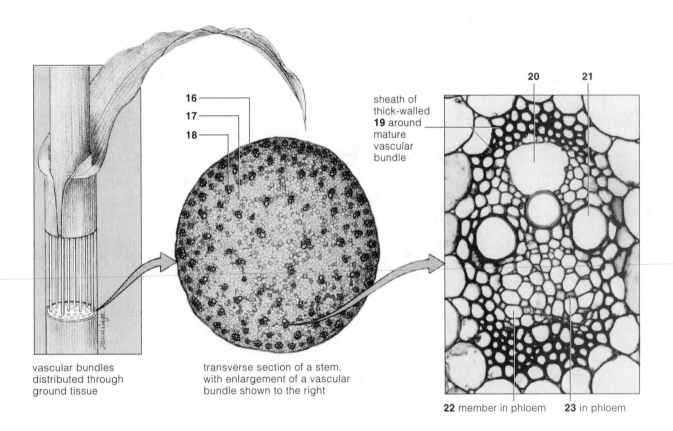

16

17

18

vascular bundles distributed through ground tissue

transverse section of a stem, with enlargement of a vascular bundle shown to the right

sheath of thick-walled **19** around mature vascular bundle

20   **21**

22 member in phloem   **23** in phloem

Labeling

Name the structures numbered in the following illustrations of the herbaceous dicot stem. Choose from sieve tube and companion cells in phloem, vascular bundle, xylem vessels, cortex, vascular cambium, pith, phloem fibers, and epidermis.

24. _____ [p.505]

25. _____ [p.505]

26. _____ _____ [p.505]

27. _____ [p.505]

28. _____ [p.505]

29. _____ _____ [p.505]

30. _____ _____ [p.505]

31. _____ [p.505]

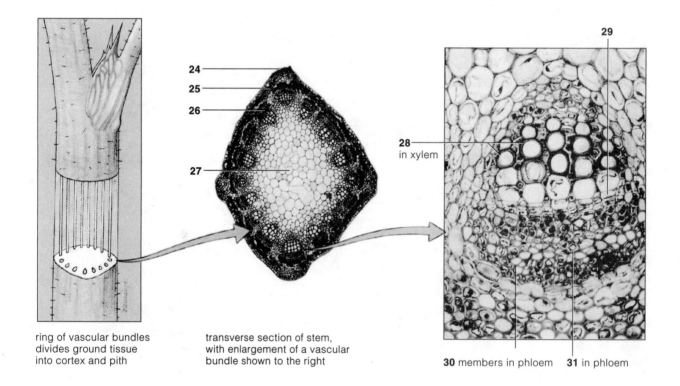

ring of vascular bundles divides ground tissue into cortex and pith

transverse section of stem, with enlargement of a vascular bundle shown to the right

30 members in phloem   **31** in phloem

29.4. A CLOSER LOOK AT LEAVES [pp.506–507]

Selected Words: deciduous [p.506], evergreen [p.506], *simple* leaves [p.506], *compound* leaves [p.506], *Agapanthus* [p.506], *Phaseolus* [p.507], *palisade* mesophyll [p.507], *spongy* mesophyll [p.507]

Boldfaced, Page-Referenced Terms

[p.506] leaf _____

_____

[p.507] mesophyll _____

_____

[p.507] veins _____

_____

Labeling

Name the structures numbered in the illustrations of leaf development and leaf forms.

1. _____ [p.506]
2. _____ [p.506]
3. _____ _____ [p.506]
4. _____ [p.506]
5. _____ [p.506]
6. _____ [p.506]

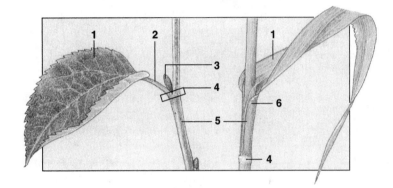

Dichotomous Choice

Circle one of two possible answers given between parentheses in each statement.

7. The leaf illustrated on the left above is a (monocot/dicot). [p.506]

8. The leaf illustrated on the right above is a (monocot/dicot). [p.506]

Labeling and Matching

Identify each numbered part of the accompanying illustration [see text Figure 29.16 on p.507]. Complete the exercise by matching and entering the letter of the proper function description in the parentheses following each label.

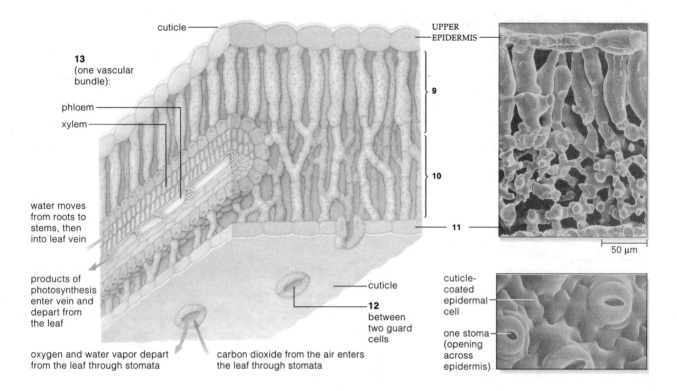

9. _____ _____ ( ) [p.507]

10. _____ _____ ( ) [p.507]

11. _____ _____ ( ) [p.507]

12. _____ ( ) [p.507]

13. _____ _____ ( ) [p.507]

A. Lowermost cuticle-covered cell layer

B. Loosely packed photosynthetic parenchyma cells just above the lower epidermal layer

C. Allows movement of oxygen and water vapor out of leaves and allows carbon dioxide to enter

D. Photosynthetic parenchyma cells just beneath the upper epidermis

E. Move water and solutes to photosynthetic cells and carry products away from them

29.5. PRIMARY STRUCTURE OF ROOTS [pp.508–509]

Selected Words: *adventitious* structures [p.508], *Eschscholzia* [p.508], *Zea mays* [p.508], mucigel [p.508], *Ranunculus* [p.509], *Salix* [p.509], endodermis [p.509], pericycle [p.509]

Boldfaced, Page-Referenced Terms

[p.508] primary root _____

_____

[p.508] lateral roots _____

_____

[p.508] taproot system _____

_____

[p.508] fibrous root system _____

_____

[p.508] root hairs _____

_____

[p.509] vascular cylinder _____

_____

Labeling and Matching

Identify each numbered part of the accompanying illustration. Complete the exercise by matching the letter of the proper description in the parentheses following each label. Some choices are used more than once.

1. _____ _____ ( ) [p.508]
2. _____ ( ) [p.508]
3. _____ ( ) [p.508]
4. _____ ( ) [p.508]
5. _____ ( ) [p.508]
6. _____ _____ _____ ( ) [p.508]
7. _____ _____ ( ) [p.508]
8. _____ ( ) [p.508]
9. _____ ( ) [p.508]
10. _____ _____ ( ) [p.508]

A. Dome-shaped cell mass produced by the apical meristem
B. Part of the vascular cylinder; gives rise to lateral roots
C. Part of the vascular cylinder; transports photosynthetic products
D. Ground tissue region surrounding the vascular cylinder
E. The absorptive interface with the root's environment
F. The region of apical and primary meristems
G. Innermost part of the root cortex; helps control water and mineral movement into the vascular column
H. Epidermal cell extensions; greatly increases the surface available for taking up water and solutes

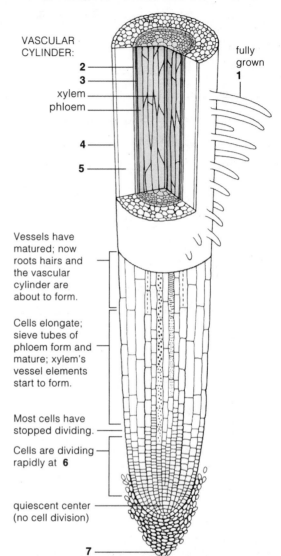

VASCULAR CYLINDER:
2
3
xylem
phloem
4
5

fully grown
1

Vessels have matured; now roots hairs and the vascular cylinder are about to form.

Cells elongate; sieve tubes of phloem form and mature; xylem's vessel elements start to form.

Most cells have stopped dividing.

Cells are dividing rapidly at **6**

quiescent center (no cell division)

7

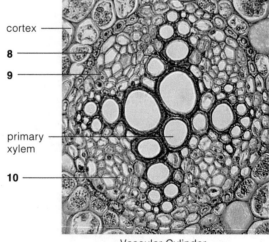

cortex
8
9
primary xylem
10

Vascular Cylinder

Fill-in-the-Blanks

The first part of a seedling to emerge from the seed coat is the (11) _____ [p.508] root. In most dicot seedlings, the primary root increases in diameter while it grows downward. Later, (12) _____ [p.508] roots begin forming in internal tissues at an angle perpendicular to the primary root's axis, then they erupt through epidermis. Oak trees, carrots, poppies, and dandelions are examples of plants whose primary root and lateral branchings represent a(n) (13) _____ [p.508] system. In monocots such as grasses, the primary root is short-lived; in its place, numerous (14) _____ [p.508] roots arise from the stem of the young plant. Such roots and their branches are somewhat alike in length and diameter and form a(n) (15) _____ [p.508] root system.

Some root epidermal cells send out absorptive extensions called root (16) _____ [p.508]. Vascular tissues form a (17) _____ [p.509] cylinder, a central column inside the root consisting of primary xylem and phloem and one or more layers of parenchyma called the (18) _____ [p.509]. Ground tissues surrounding the cylinder are called the root (19) _____ [p.509]. A monocot's vascular cylinder divides the ground tissue system into cortex and (20) _____ [p.509] regions. There are many (21) _____ [p.509] spaces in between cells of the ground tissue system, and (22) _____ [p.509] can easily diffuse through them. Water entering the root moves from cell to cell until it reaches the (23) _____ [p.509], the innermost cell layer of the root cortex. Abutting walls of its cells are waterproof, so they force incoming water to pass through the cytoplasm. This arrangement helps (24) _____ [p.509] the movement of water and dissolved substances into the vascular cylinder. Just inside the endodermis is the (25) _____ [p.509]. This part of the vascular cylinder gives rise to (26) _____ [p.509] roots, which then erupt through the (27) _____ [p.509] and epidermis.

29.6. ACCUMULATED SECONDARY GROWTH—THE WOODY PLANTS [pp.510–511]

29.7. A LOOK AT WOOD AND BARK [pp.512–513]

Selected Words: *nonwoody* plants [p.510], *woody* plants [p.510], *Castanea* [p.510], *fusiform initials* [p.510], *ray initials* [p.511], *inner* face [p.511], *outer* face [p.511], secondary xylem [p.512], *girdling* [p.512], *early* wood [p.513], *late* wood [p.513], "tree rings" [p.513]

Boldfaced, Page-Referenced Terms

[p.510] annuals _____

_____

[p.510] biennials _____

_____

[p.510] perennials _____

_____

[p.512] bark _____

_____

[p.512] cork _____

_____

[p.512] heartwood _____

_____

[p.512] sapwood _____

_____

[p.513] growth rings _____

_____

[p.513] hardwood _____

_____

[p.513] softwood _____

_____

[p.513] compartmentalization _____

_____

Matching

Choose the most appropriate answer for each term.

1. ___tree rings [p.513]
2. ___heartwood [p.512]
3. ___sapwood [p.512]
4. ___early wood [p.513]
5. ___nonwoody plants [p.510]
6. ___woody plants [p.510]
7. ___softwood [p.513]
8. ___girdling [p.512]
9. ___bark [p.512]
10. ___cork [p.512]
11. ___late wood [p.513]
12. ___biennial [p.510]
13. ___annuals [p.510]
14. ___hardwood [p.513]
15. ___perennials [p.510]

A. Produced by cells of the cork cambium
B. Includes all tissues external to the vascular cambium
C. Alternating bands of early and late wood, which reflect light differently; show secondary growth during two or more growing seasons
D. Complete the life cycle in a single growing season and are generally herbaceous
E. Another term for herbaceous plants
F. The first xylem cells produced at the start of the growing season; tend to have large diameters and thin walls
G. Produced by conifers that lack fibers and vessels in their xylem
H. Term applied to the wood of dicot trees that evolved in temperate and tropical regions; possess vessels, tracheids, and fibers in their xylem
I. Produced by some monocots and many dicots; they put on extensive secondary growth over two or more growing seasons
J. Formed in dry summer, has xylem cells with smaller diameters and thicker walls
K. A dumping ground at the center of older stems and roots for metabolic wastes such as resins, oils, gums, and tannins
L. Removal of a band of secondary phloem around a trunk's circumference; the tree will eventually die
M. Plants such as carrots that live for two growing seasons
N. Secondary growth located between heartwood and the vascular cambium; wet, usually pale
O. Plants that add secondary growth during two or more growing seasons

Labeling and Matching

Identify each numbered part of the accompanying illustration. Complete the exercise by matching and entering the letter of the proper description in the parentheses following each label.

16. _____ ( ) [p.513]

17. _____ _____ ( ) [p.513]

18. _____ _____ ( ) [p.513]

19. _____ ( ) [p.513]

20. year _____ ( ) [p.513]

21. and 22. years _____ and _____ ( ) [p.513]

23. _____ _____ ( ) [p.513]

A. All secondary growth
B. Produced later in the growing season; vessels have smaller diameters with thick walls
C. Includes the primary growth and some secondary growth
D. Includes all tissues external to the vascular cambium
E. Large-diameter conducting cell in the xylem
F. Produced early in the growing season; vessels have large diameters and thin walls
G. Produces secondary vascular tissues, xylem, and phloem

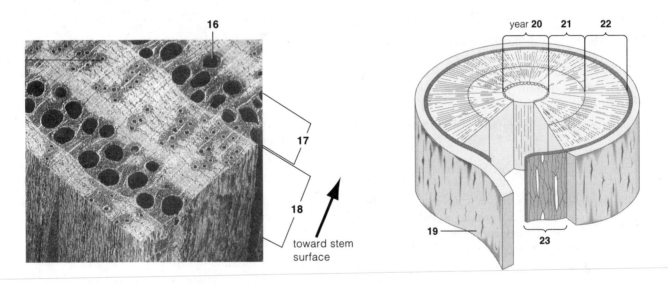

16

17

18

toward stem surface

year 20 21 22

19

23

Self-Quiz

___ 1. _____ covers and protects the plant's surfaces. [p.500]
   a. Ground tissue
   b. Dermal tissue
   c. Vascular tissue
   d. Pericycle

___ 2. Which of the following is *not* considered a type of simple tissue? [p.502]
   a. Epidermis
   b. Parenchyma
   c. Collenchyma
   d. Sclerenchyma

___ 3. Of the following cell types, which one does *not* appear in vascular tissues? [p.503]
   a. vessel members
   b. cork cells
   c. tracheids
   d. sieve-tube members
   e. companion cells

___ 4. The _____ is a leaflike structure that is part of the embryo; monocot embryos have one, dicot embryos have two. [p.503]
   a. shoot tip
   b. root tip
   c. cotyledon
   d. apical meristem

___ 5. Each part of the stem where one or more leaves are attached is a(n) _____. [p.504]
   a. node
   b. internode
   c. vascular bundle
   d. cotyledon

___ 6. Which of the following structures is *not* considered meristematic? [p.501]
   a. vascular cambium
   b. lateral meristem
   c. cork cambium
   d. endodermis

___ 7. Which of the following statements about monocots is *false*? [p.503]
   a. They are usually herbaceous.
   b. They develop one cotyledon in their seeds.
   c. Their vascular bundles are scattered throughout the ground tissue of their stems.
   d. They have a single central vascular cylinder in their stems.

___ 8. New plants grow and older plant parts lengthen through cell divisions at _____ meristems present at root and shoot tips; older roots and stems of woody plants increase in diameter through cell divisions at _____ meristems. [p.501]
   a. lateral; lateral
   b. lateral; apical
   c. apical; apical
   d. apical; lateral

___ 9. Vascular bundles called _____ form a network through a leaf blade. [p.507]
   a. xylem
   b. phloem
   c. veins
   d. cuticles
   e. vessels

___10. A primary root and its lateral branchings represent a(n) _____ system. [p.508]
   a. lateral root
   b. adventitious root
   c. taproot
   d. branch root

___11. Plants whose vegetative growth and seed formation continue year after year are _____ plants. [p.510]
   a. annual
   b. perennial
   c. nonwoody
   d. herbaceous

___12. The _____ layer of a root divides to produce lateral roots. [p.509]
   a. endodermis
   b. pericycle
   c. xylem
   d. cortex
   e. phloem

Chapter Objectives/Review Questions

1. The above ground parts of flowering plants are called _____; the plants' descending parts are called _____. [p.500]
2. Distinguish among the ground tissue system, the vascular tissue system, and the dermal tissue system. [p.500]
3. Plants grow at localized regions of self-perpetuating embryonic cells called _____. [p.501]
4. Lengthening of stems and roots originates at _____ meristems and all dividing tissues derived from them; this is called _____ growth. [p.501]
5. Cell populations of protoderm, ground meristem, and procambium are derived from the apical meristem and are known as _____ meristems. [p.501]
6. Increases in the thickness of a plant originate at _____ meristems. [p.501]
7. Describe the role of vascular cambium and cork cambium in producing secondary tissues of the plant body. [p.501]
8. Visually identify and generally describe the simple tissues called parenchyma, collenchyma, and sclerenchyma. [p.502]

9. Fibers and sclereids are both types of _____ cells. [p.502]
10. Name the cell wall compound that was necessary for the evolution of rigid and erect land plants. [p.503]
11. _____ tissue conducts soil water and dissolved minerals, and it mechanically supports the plant. [p.503]
12. _____ tissue transports sugars and other solutes. [p.503]
13. Name and describe the functions of the conducting cells in xylem and phloem. [p.503]
14. All surfaces of primary plant parts are covered and protected by a dermal tissue system called _____ and a surface coating called a _____. [p.503]
15. What is the general function of guard cells and stomata found within the epidermis of young stems and leaves? [p.503]
16. The cork cells of _____ replace the epidermis of stems and roots showing secondary growth. [p.503]
17. Distinguish between monocots and dicots by listing their characteristics and citing examples of each group. [p.503]
18. A _____ is an undeveloped shoot of mostly meristematic tissue, often protected by scales; it gives rise to new stems, leaves, and flowers. [p.504]
19. The primary xylem and phloem develop as vascular _____. [p.504]
20. Distinguish between the stem's cortex and its pith. [p.504]
21. Visually distinguish between monocot stems and dicot stems, as seen in cross-section. [p.505]
22. Describe the principal difference between deciduous and evergreen plants. [p.506]
23. How does the "simple" leaf type differ from "compound" leaves? [p.506]
24. Describe the structure (cells and layers) and major functions of leaf epidermis, mesophyll, and vein tissue. [pp.506–507]
25. The _____ root is the first to poke through the coat of a germinating seed; later, _____ roots erupt through the epidermis. [p.508]
26. How does a taproot system differ from a fibrous root system? [p.508]
27. Define the term *adventitious*. [p.508]
28. Describe the origin and function of root hairs. [pp.508–509]
29. A _____ _____ consists of primary xylem and primary phloem and one or more layers of parenchyma cells called the pericycle. [p.509]
30. Describe the passage of soil water through root epidermis to the xylem of the vascular cylinder; include the role of the endodermis. [p.509]
31. Define these categories of flowering plants: annuals, biennials, and perennials. [p.510]
32. Distinguish a woody plant from a nonwoody plant in terms of secondary growth. [p.510]
33. Each growing season, new tissues that increase the girth of woody plants originate at their _____ meristems. [p.510]
34. Describe the formation of cork and bark. [p.512]
35. Distinguish heartwood from sapwood, and early wood from late wood. [pp.512–513]
36. Explain the origin of the annual growth layers (tree rings) seen in a cross-section of a tree trunk. [p.513]
37. Hardwood trees possess _____ and _____, but softwood trees lack these cells. [p.513]

Integrating and Applying Key Concepts

Imagine (and then describe) the specific behavioral restrictions that might be imposed if the human body resembled the plant body in having (1) open growth with apical meristematic regions, (2) stomata in the epidermis, (3) cells with chloroplasts, (4) excess carbohydrates stored primarily as starch rather than as fat, and (5) dependence on the soil as a source of water and inorganic compounds.

30

PLANT NUTRITION AND TRANSPORT

Interactive Exercises

Flies for Dinner [pp.516–517]

30.1. SOIL AND ITS NUTRIENTS [pp.518–519]

Selected Words: *Dionaea muscipula* [p.516], *Darlingtonia californica* [p.517], *profile* properties [p.518], *macro*nutrients [p.519], *micro*nutrients [p.519]

Boldfaced, Page-Referenced Terms

[p.516] carnivorous plants _____

_____

[p.516] plant physiology _____

_____

[p.518] soil _____

_____

[p.518] humus _____

_____

[p.518] loams _____

_____

[p.518] topsoil _____

_____

[p.518] nutrients _____

_____

[p.519] leaching _____

_____

[p.519] erosion _____

_____

Matching

Choose the most appropriate answer for each term.

1. ___soil [p.518]
2. ___humus [p.518]
3. ___loams [p.518]
4. ___profile properties [p.518]
5. ___topsoil [p.518]
6. ___nutrients [p.518]
7. ___macronutrients [p.519]
8. ___micronutrients [p.519]
9. ___leaching [p.519]
10. ___erosion [p.519]

A. Elements essential for a given organism because, directly or indirectly, they have roles in metabolism that cannot be fulfilled by any other element
B. The layered characteristics of soils, which are in different stages of development in different places
C. The decomposing organic material in soil
D. The movement of land under the force of wind, running waters, and ice
E. Elements other than the macronutrients that are essential for plant growth
F. Consists of particles of minerals mixed with variable amounts of decomposing organic material
G. Refers to the removal of some of the nutrients in soil as water percolates through it
H. Uppermost part of the soil that is highly variable in depth (A horizon); the most essential layer for plant growth
I. Soils having more or less equal proportions of sand, silt, and clay; best for plant growth
J. Nine of the essential elements required for plant growth

Fill-in-the-Blanks

The three essential elements that plants use as their main metabolic building blocks are oxygen, carbon, and (11) _____ [p.518]. Besides these elements, plants depend on the uptake of at least (number) (12) _____ [p.519] others. These are typically available to plants in (13) _____ [p.519] forms. Plants give up (14) _____ [p.519] ions to the clay in exchange for weakly bound elements such as Ca^{++} and K^+. Nine essential elements are (15) _____ [p.519] and are required in amounts above 0.5 percent of the plant's dry weight. The other elements are (16) _____ [p.519]; they make up traces of the plant's dry weight.

Complete the Table

17. Thirteen essential elements are available to plants as mineral ions. Complete the following table, which summarizes information about these important plant nutrients.

Mineral Element	Macronutrient or Micronutrient	Known Functions
[p.519] a.		Roles in chlorophyll synthesis, electron transport
[p.519] b.		Activation of enzymes, role in maintaining water–solute balance
[p.519] c.		Role in chlorophyll synthesis; coenzyme activity
[p.519] d.		Role in root, shoot growth; role in photolysis
[p.519] e.		Role in chlorophyll synthesis; coenzyme activity
[p.519] f.		Component of enzyme used in nitrogen metabolism
[p.519] g.		Component of proteins, nucleic acids, coenzymes, chlorophyll
[p.519] h.		Component of most proteins, two vitamins
[p.519] i.		Roles in flowering, germination, fruiting, cell division, nitrogen metabolism
[p.519] j.		Role in formation of auxin, chloroplasts, and starch; enzyme component
[p.519] k.		Roles in cementing cell walls, regulation of many cell functions
[p.519] l.		Component of several enzymes
[p.519] m.		Component of nucleic acids, phospholipids, ATP

30.2. HOW DO ROOTS ABSORB WATER AND MINERAL IONS? [pp.520–521]

Selected Words: *Rhizobium* [p.521], *Bradyrhizobium* [p.521]

Boldfaced, Page-Referenced Terms

[p.520] vascular cylinder _____

_____

[p.520] endodermis _____

_____

[p.520] Casparian strip _____

_____

[p.520] exodermis _____

_____

[p.520] root hairs _____

_____

[p.520] mutualism _____

_____

[p.521] nitrogen fixation _____

_____

[p.521] root nodules _____

_____

[p.521] mycorrhizae (singular, mycorrhiza) _____

_____

Labeling and Matching

Identify each numbered part of the following illustration that deals with the control of nutrient uptake by plant roots. Choose from the following: epidermis, water movement, cytoplasm, root hair, vascular cylinder, endodermal cell wall, exodermis, endodermis, cortex, and Casparian strip. Complete the exercise by matching from the following list and entering the correct letter in the parentheses after each label.

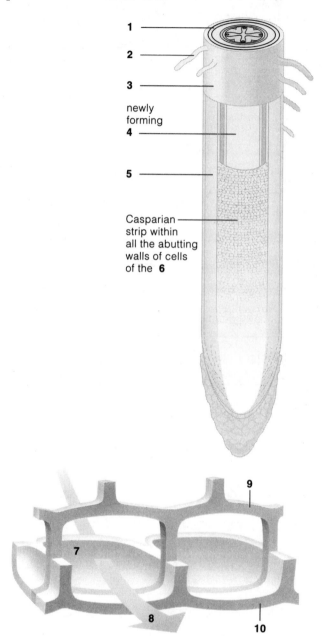

newly
forming

Casparian strip within all the abutting walls of cells of the **6**

1. _____ (   ) [p.520]

2. _____ _____ (   ) [p.520]

3. _____ (   ) [p.520]

4. _____ _____ (   ) [p.520]

5. _____ (   ) [p.520]

6. _____ (   ) [p.520]

7. _____ (   ) [p.520]

8. _____ _____ (   ) [p.520]

9. _____ _____ (   ) [p.520]

10. _____ _____ _____ (   ) [p.520]

A. Cellular area through which water and dissolved nutrients must move, because of Casparian strips

B. A layer of cortex cells just inside the epidermis; also equipped with Casparian strips

C. Cellular area of a root between the exodermis (if present) and the endodermis

D. Waxy band acting as an impermeable barrier between the walls of abutting endodermal cells; forces water and dissolved nutrients through the cytoplasm of endodermal cells

E. Tiny extensions of the root epidermal cells that greatly increase absorptive capacity of a root

F. Specific location of the waxy bands known as Casparian strips

G. Substance whose diffusion occurs through the cytoplasm of endodermal cells because of Casparian strips

H. Outermost layer of root cells

I. A cylindrical layer of cells that wraps around the vascular column

J. Tissues include the xylem, phloem, and pericycle

Sequence

Arrange in correct chronological sequence the path that water and nutrients take from the soil to cells in living plant tissues. Write the letter of the first step next to 11, the letter of the second step next to 12, and so on.

11. ___[p.520]   A. Cortex cells lacking Casparian strips

12. ___[p.520]   B. Endodermis cells with Casparian strips

13. ___[p.520]   C. Root epidermis and root hairs

14. ___[p.520]   D. Exodermis cells with Casparian strips (in some plants)

15. ___[p.520]   E. Vascular cylinder

Dichotomous Choice

Circle one of two possible answers given between parentheses in each statement.

16. A two-way flow of benefits between species is a symbiotic interaction known as (mutualism/parasitism). [p.520]

17. (Gaseous nitrogen/Nitrogen "fixed" by bacteria) represents the chemical form of nitrogen plants can use in their metabolism. [p.521]

18. Nitrogen-fixing bacteria reside in localized swellings on legume plants known as (root hairs/root nodules). [p.521]

19. Mycorrhizae represent symbiotic relationships in which fungi and the roots they cover both benefit; the roots receive (sugars and nitrogen-containing compounds/scarce minerals that the fungus is better able to absorb). [p.521]

20. In a mycorrhizal interaction between a young root and a fungus, the fungus benefits by obtaining (sugars and nitrogen-containing compounds/scarce minerals that the fungus is better able to absorb). [p.521]

30.3. HOW IS WATER TRANSPORTED THROUGH PLANTS? [pp.522–523]

Selected Words: cohesion [p.522], tension [p.522]

Boldfaced, Page-Referenced Terms

[p.522] transpiration _____

_____

[p.522] xylem _____

_____

[p.522] tracheids _____

_____

[p.522] vessel members _____

_____

[p.522] cohesion–tension theory _____

_____

Fill-in-the-Blanks

The cohesion–tension theory explains (1) _____ [p.522] transport to the tops of plants, which are often very high. It travels pipelines in the (2) _____ [p.522], which are formed by water-conducting cells called (3) _____ [p.522] and (4) _____ [p.522]. The process begins with the drying power of air, which causes (5) _____ [p.522], the evaporation of water from plant parts exposed to air. The collective strength of (6) _____ [p.522] bonds between water molecules in the narrow, tubular xylem cells imparts (7) _____ [p.522]. This provides unbroken, fluid columns of water. The xylem water columns, as they are pulled upward, are under (8) _____ [p.522]. This force extends from the veins inside leaves, down through the stems, and on into the young (9) _____ [p.522] where water is being absorbed. As long as water molecules continue to escape from the plant, the continuous tension inside the (10) _____ [p.522] permits more molecules to be pulled upward from the roots to replace them.

30.4. HOW DO STEMS AND LEAVES CONSERVE WATER? [pp.524–525]

Selected Words: inward diffusion [p.524], *outward* diffusion [p.524], *Opuntia* [p.524], *Commelina communis* [p.525]

Boldfaced, Page-Referenced Terms

[p.524] cuticle _____

_____

[p.524] cutin _____

_____

[p.524] stomata (singular, stoma) _____

_____

[p.524] guard cells _____

_____

[p.525] turgor pressure _____

_____

[p.525] CAM plants _____

_____

True–False

If the statement is true, write a T in the blank. If the statement is false, make it correct by changing the underlined word(s) and writing the correct word(s) in the answer blank.

_____ 1. Of the water moving into a leaf, 2 percent or more is lost by evaporation into the surrounding air. [p.524]

_____ 2. When evaporation exceeds water uptake by roots, plant tissues wilt and water-dependent activities are seriously disrupted. [p.524]

_____ 3. The surfaces of all plant epidermal cell walls have an outer layer of waxes embedded in cutin, the cuticle, which restricts water loss, restricts inward diffusion of carbon dioxide, and limits outward diffusion of oxygen by-products. [p.524]

_____ 4. Plant epidermal layers are peppered with tiny openings called <u>guard cells</u> through which water leaves the plant and carbon dioxide enters. [p.524]

_____ 5. When a pair of guard cells swells with turgor pressure, the opening between them <u>closes</u>. [p.525]

_____ 6. In most plants, stomata remain <u>open</u> during the daylight photosynthetic period; water is lost from plants, but they accumulate carbon dioxide. [p.525]

_____ 7. Stomata stay <u>closed</u> at night in most plants. [p.525]

_____ 8. Photosynthesis starts when the sun comes up; as the morning progresses, carbon dioxide levels <u>increase</u> in cells, including guard cells. [p.525]

_____ 9. As the morning progresses, a drop in carbon dioxide level within guard cells triggers an inward active transport of potassium ions; water follows by osmosis and the fluid pressure <u>closes</u> the stoma. [p.525]

_____ 10. When the sun goes down and photosynthesis stops, carbon dioxide levels rise; stomata <u>close</u> when potassium, then water, moves out of the guard cells. [p.525]

_____ 11. CAM plants such as cacti and other succulents open stomata during the <u>day</u> when they fix carbon dioxide by way of a special C4 metabolic pathway. [p.525]

_____ 12. CAM plants use carbon dioxide in photosynthesis the following <u>night</u> when stomata are closed. [p.525]

30.5. HOW ARE ORGANIC COMPOUNDS DISTRIBUTED THROUGH PLANTS?
[pp.526–527]

Selected Words: source [p.526], sink [p.526], Sonchus [p.527]

Boldfaced, Page-Referenced Terms

[p.526] phloem _____

_____

[p.526] sieve tubes _____

_____

[p.526] companion cells _____

_____

[p.526] translocation _____

_____

[p.526] pressure flow theory _____

_____

Fill-in-the-Blanks

Sucrose and other organic products resulting from (1) _____ [p.526] are used throughout the plant organs. Most plant cells store their carbohydrates as (2) _____ [p.526] in plastids. The (3) _____ [p.526] and fats, which cells synthesize from carbohydrates and the amino acids distributed to them, are stored in many (4) _____ [p.526]. Avocados and some other fruits also accumulate (5) _____ [p.526]. (6) _____ [p.526] molecules are too large to cross cell membranes and too insoluble to be transported to other regions of the plant body. Overall, (7) _____ [p.526] are too large and fats are too insoluble to be transported from the storage sites. Plant cells convert storage forms of organic compounds to (8) _____ [p.526] of smaller size that are more easily transported through the phloem. For example, the cells degrade starch to glucose monomers. When one of these monomers combines with fructose, the result is (9) _____ [p.526], an easily transportable sugar. Experiments with phloem-embedded aphid mouthparts revealed that (10) _____ [p.526], an easily transportable sugar, was the most abundant carbohydrate being forced out of those tubes.

Matching

Choose the most appropriate answer for each term.

11. ___translocation [p.526]

12. ___sieve-tube members [p.526]

13. ___companion cells [p.526]

14. ___aphids [p.526]

15. ___source [p.526]

16. ___sink [p.526]

17. ___pressure flow theory [pp.526–527]

A. Any region where organic compounds are being loaded into the sieve tubes

B. Nonconducting cells adjacent to sieve-tube members that supply energy to load sucrose at the source

C. Any region of the plant where organic compounds are being unloaded from the sieve-tube system and used or stored

D. Process occurring in phloem that distributes sucrose and other organic compounds through the plant (apparently under pressure)

E. Internal pressure builds up at the source end of a sieve-tube system and pushes the solute-rich solution toward a sink, where they are removed

F. Passive conduits for translocation within vascular bundles; water and organic compounds flow rapidly through large pores on their end walls

G. Insects used to verify that in most plant species sucrose is the main carbohydrate translocated under pressure

Self-Quiz

___ 1. For plants, the essential elements include _____. [p.518]
   a. oxygen, carbon, and nitrogen
   b. oxygen, hydrogen, and nitrogen
   c. oxygen, carbon, and hydrogen
   d. carbon, nitrogen, and hydrogen

___ 2. Macronutrients are the nine dissolved mineral ions that _____. [p.519]
   a. furnish the basic ingredients for photosynthesis
   b. occur in only small traces in plant tissues
   c. are required in amounts above 0.5 percent of the plant's dry weight
   d. can function only without the presence of micronutrients
   e. both a and c

___ 3. Gaseous nitrogen is converted to a plant-usable form by _____. [p.521]
   a. root nodules
   b. mycorrhizae
   c. nitrogen-fixing bacteria
   d. Venus flytraps

___ 4. _____ prevent(s) inward-moving water from moving past the abutting walls of the root endodermal cells. [p.520]
   a. Cytoplasm
   b. Plasma membranes
   c. Osmosis
   d. Casparian strips

___ 5. Most of the water moving into a leaf is lost through _____. [p.522]
   a. osmotic gradients being established
   b. evaporation of water from plant parts exposed to air
   c. pressure flow forces
   d. translocation

___ 6. Stomata remain _____ during daylight, when photosynthesis occurs, but remain _____ during the night when carbon dioxide accumulates through aerobic respiration. [pp.524–525]
   a. open; open
   b. closed; open
   c. closed; closed
   d. open; closed

___ 7. By control of _____ levels inside the guard cells of stomata, the activity of stomata is controlled when leaves are losing more water than roots can absorb. [p.525]
   a. oxygen
   b. potassium
   c carbon dioxide
   d. ATP

___ 8. Without _____, plants would rapidly wilt and die during hot, dry spells. [p.524]
   a. a cuticle
   b mycorrhizae
   c. phloem
   d. cotyledons

___ 9. The _____ theory of water transport states that hydrogen bonding allows water molecules to maintain a continuous fluid column as water is pulled from roots to leaves. [p.522]
   a. pressure flow
   b. cohesion–tension
   c. evaporation
   d. abscission

___10. Leaves represent _____ regions; growing leaves, stems, fruits, seeds, and roots represent _____ regions. [p.526]
   a. source; source
   b. sink; source
   c. source; sink
   d. sink; sink

Chapter Objectives/Review Questions

1. Explain how carnivorous plants such as Venus flytraps and bladderworts accomplish their nutritional needs. [pp.516–517]
2. _____ _____ is the study of adaptations by which plants function in their environment. [p.517]
3. Define the term *soil*. [p.518]
4. Plants do best in _____, which are the soils having more or less equal proportions of sand, silt, and clay. [p.518]
5. Name the soil layer that is the most essential for plant growth. [p.518]
6. Define the term *nutrients* in terms of plant nutrition. [p.518]
7. Name the three elements considered essential for plant nutrition. [p.518]
8. Distinguish between macronutrients and micronutrients in relation to their role in plant nutrition. [p.519]
9. _____ refers to the removal of some of the nutrients in soil as water percolates through it. [p.519]
10. The movement of land under the force of wind, running water, and ice is called _____. [p.519]
11. Trace the path of water and mineral ions into roots; name the structure and function of plant structures involved. [p.520]
12. Differentiate between the endodermis and the exodermis of the root cortex. [p.520]
13. Because of the presence of the _____ strip in the walls of endodermal cells, water can move into the vascular cylinder only by crossing the plasma membrane and diffusing through the cytoplasm. [p.520]

14. Explain why root hairs are so valuable in root absorption. [p.520]
15. Define the term *mutualism*. [p.520]
16. Describe the roles of root nodules and mycorrhizae in plant nutrition. [p.521]
17. The evaporation of water from leaves as well as from stems and other plant parts is known as _____. [p.522]
18. Henry Dixon's _____–_____ theory explains how water moves upward in an unbroken column through xylem to the tops of tall trees. [p.522]
19. In a plant's vascular tissues, water moves through pipelines called _____. [p.522]
20. Give the key points in Dixon's explanation of upward water transport in plants. [pp.522–523]
21. Even mildly stressed plants would rapidly wilt and die were it not for the _____ covering their parts; describe additional functions of this structure. [p.524]
22. Describe the chemical compounds found in cutin. [p.524]
23. Evaporation from plant parts occurs mostly at _____, tiny epidermal passageways of leaves and stems. [p.524]
24. Explain the mechanism by which stomata open during daylight and close during the night. [p.525]
25. Define *turgor pressure*, and explain its role in controlling water loss at stomata. [p.525]
26. Describe the mechanisms by which CAM plants conserve water. [p.525]
27. Describe the role of phloem sieve-tube members and companion cells in translocation. [p.526]
28. _____ is the main form in which sugars are transported through most plants. [p.526]
29. Define *translocation*. [p.526]
30. According to the _____ _____ theory, pressure builds up at the source end of a sieve-tube system and pushes solutes toward a sink, where they are removed. [pp.526–527]
31. Companion cells supply the _____ that loads sucrose at the source. [p.527]
32. Name the gradients responsible for continuous flow of organic compounds through phloem. [pp.526–527]

Integrating and Applying Key Concepts

How do you think maple syrup is made from maple trees? Which specific systems of the plant are involved, and why are maple trees tapped only at certain times of the year?

31

PLANT REPRODUCTION

Interactive Exercises

A Coevolutionary Tale [pp.530–531]

31.1. REPRODUCTIVE STRUCTURES OF FLOWERING PLANTS [pp.532–533]

31.2. *Focus on the Environment:* POLLEN SETS ME SNEEZING [p.533]

Selected Words: *Angraecum sesquipedale* [p.531], *Prunus* [p.532], *Rosa* [p.533], *perfect* flowers [p.533], *imperfect* flowers [p.533], *allergic rhinitis* [p.533]

Boldfaced, Page-Referenced Terms

[p.530] flower _____

_____

[p.530] coevolution _____

_____

[p.530] pollinator _____

_____

[p.532] sporophyte _____

_____

[p.532] gametophytes _____

_____

[p.533] stamens _____

_____

[p.533] pollen grains _____

_____

[p.533] carpels _____

_____

[p.533] ovary _____

_____

Labeling and Matching

Identify each numbered part of the accompanying illustration. Complete the exercise by matching and entering the letter of the proper description in the parentheses following each label.

1. _____ (   ) [p.532]

2. _____ (   ) [p.532]

3. _____ (   ) [p.532]

4. _____ (   ) [p.532]

5. _____ (   ) [p.532]

6. _____ (   ) [p.532]

A. An event that produces a young sporophyte
B. A reproductive shoot produced by the sporophyte
C. The "plant"; a vegetative body that develops from a zygote
D. Cellular division event occurring within flowers to produce spores
E. Produces haploid eggs by mitosis
F. Produces haploid sperm by mitosis

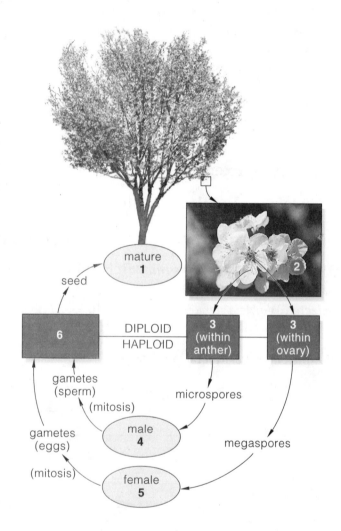

Labeling

Identify each numbered part of the accompanying illustration.

7. _____ [p.532]

8. _____ [p.532]

9. _____ [p.532]

10. _____ [p.532]

11. _____ [p.532]

12. _____ [p.532]

13. _____ [p.532]

14. _____ [p.532]

15. _____ [p.532]

16. _____ [p.532]

17. _____ [p.532]

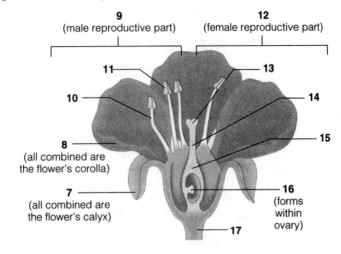

9
(male reproductive part)

12
(female reproductive part)

11

13

10

14

15

8
(all combined are
the flower's corolla)

7
(all combined are
the flower's calyx)

16
(forms
within
ovary)

17

Matching

Choose the most appropriate answer for each term.

18. ___sepals [p.532]

19. ___petals [p.532]

20. ___stamens [p.533]

21. ___ovule [p.532]

22. ___pollen grain [p.533]

23. ___carpel [p.533]

24. ___ovary [p.533]

25. ___perfect flowers [p.533]

26. ___imperfect flowers [p.533]

A. Have both male and female parts
B. Mature haploid spore whose contents develop into a male gametophyte
C. Collectively, the flower's "corolla"
D. Have male or female parts, but not both
E. Female reproductive part; includes stigma, style, and ovary
F. Structure inside the ovary; matures to become a seed
G. Found just inside the flower's corolla, the male reproductive parts
H. Lower portion of the carpel where egg formation, fertilization, and seed development occur
I. Outermost leaflike whorl of floral organs; collectively, the calyx

31.3. A NEW GENERATION BEGINS [pp.534–535]

Boldfaced, Page-Referenced Terms

[p.534] microspores _____

_____

[p.534] ovule _____

_____

[p.534] megaspores _____

_____

[p.534] endosperm _____

_____

[p.534] pollination _____

_____

[p.534] double fertilization _____

_____

Fill-in-the-Blanks

The numbered items in the following illustration represent missing information; complete the blanks in the following narrative to supply that information.

Within each (1) _____ [p.534], mitotic divisions produce four masses of spore-forming cells, each mass forming within a (2) _____ _____ [p.534]. Each one of these diploid cells is known as a (3) _____ _____ [p.534] cell and undergoes (4) _____ [p.534] to produce four haploid (5) _____ [p.534]. Mitosis within each haploid microspore results in a two-celled haploid body, the immature male gametophyte. One of these cells will give rise to a (6) _____ _____ [p.534]; the other cell will develop into a (7) _____-_____ [p.534] cell. Mature microspores are eventually released from the pollen sacs of the anther as (8) _____ [p.534]. Pollination occurs and after the pollen lands on a (9) _____ [p.534] of a carpel, the pollen tube develops from one of the cells in the pollen grain; the other cell within the pollen grain divides to form two sperm cells. As the pollen tube grows through the carpel tissues, it contains the two sperm cells and a tube nucleus. The pollen tube with its two sperm cells and the tube nucleus is known as the mature (10) _____ _____ [p.534].

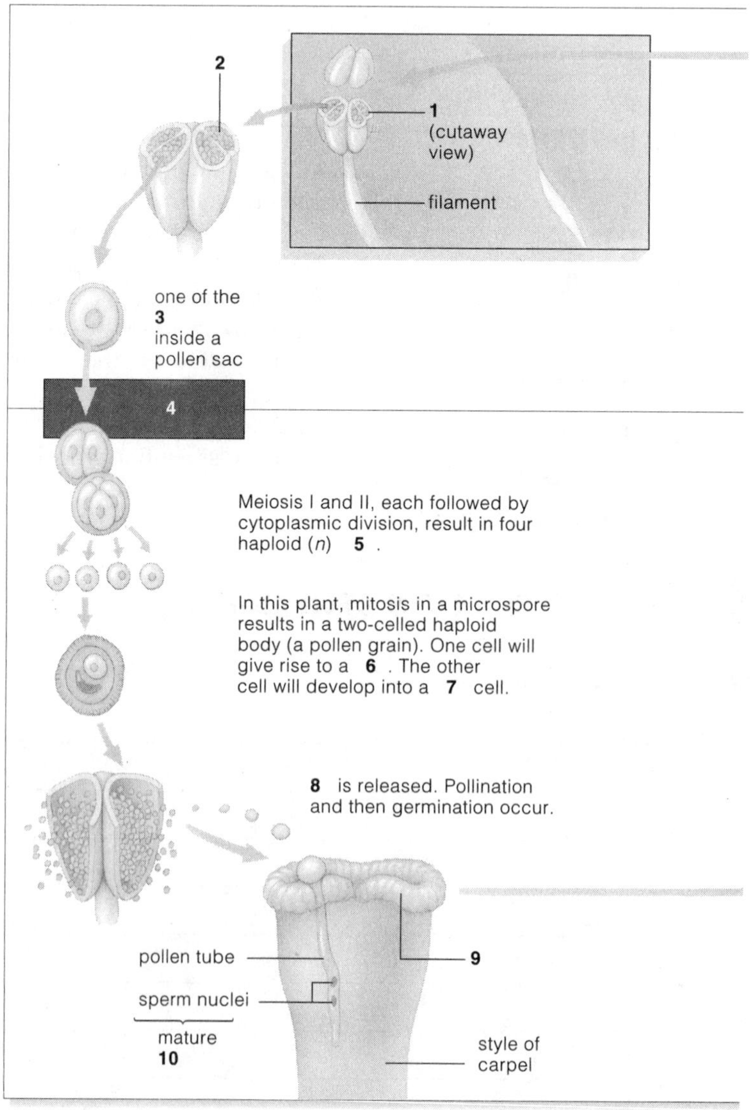

2

1
(cutaway view)

— filament

one of the **3** inside a pollen sac

4

Meiosis I and II, each followed by cytoplasmic division, result in four haploid (*n*) **5** .

In this plant, mitosis in a microspore results in a two-celled haploid body (a pollen grain). One cell will give rise to a **6** . The other cell will develop into a **7** cell.

8 is released. Pollination and then germination occur.

pollen tube —
sperm nuclei —

mature
10

9

style of carpel

Fill-in-the-Blanks

The numbered items on the following illustration represent missing information; complete the numbered blanks in the following narrative to supply that information.

In the carpel of the flower, one or more dome-shaped, diploid tissue masses develop on the inner wall of the ovary. Each mass is the beginning of a(n) (11) _____ [pp.534–535]. A tissue forms inside a domed mass as it grows, and one or two protective layers called (12) _____ [pp.534–535] form around it. Inside each mass, a diploid cell (the endosperm mother cell) divides by (13) _____ [pp.534–535] to form four haploid spores known as (14) _____ [pp.534–535]. Commonly, all but one (15) _____ [pp.534–535] disintegrates. The remaining (16) _____ [pp.534–535] undergoes (17) _____ [pp.534–535] three times without cytoplasmic division. At first, this structure is a cell with (18) _____ [pp.534–535] haploid nuclei. Cytoplasmic division results in a seven-cell (19) _____ _____ [pp.534–535], which

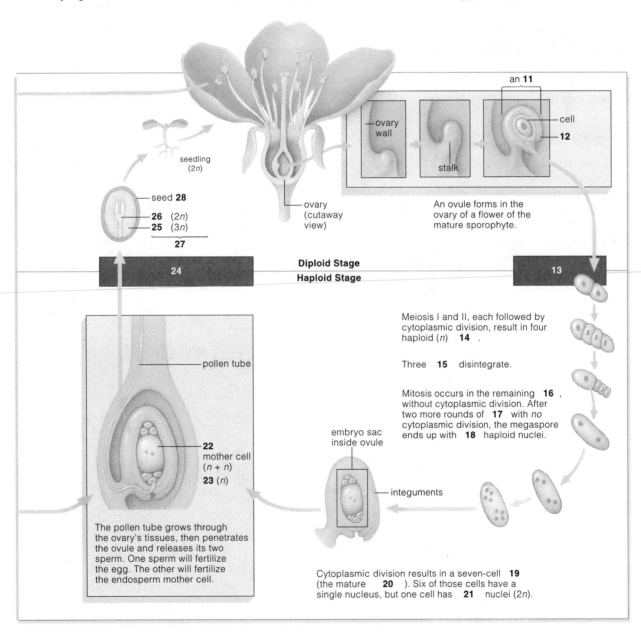

an **11**

ovary wall

stalk

cell

12

An ovule forms in the ovary of a flower of the mature sporophyte.

seedling (2n)

seed **28**

26 (2n)

25 (3n)

27

ovary (cutaway view)

Diploid Stage

24

Haploid Stage

13

Meiosis I and II, each followed by cytoplasmic division, result in four haploid (n) **14**.

Three **15** disintegrate.

Mitosis occurs in the remaining **16**, without cytoplasmic division. After two more rounds of **17** with *no* cytoplasmic division, the megaspore ends up with **18** haploid nuclei.

pollen tube

22 mother cell (n + n)

23 (n)

embryo sac inside ovule

integuments

The pollen tube grows through the ovary's tissues, then penetrates the ovule and releases its two sperm. One sperm will fertilize the egg. The other will fertilize the endosperm mother cell.

Cytoplasmic division results in a seven-cell **19** (the mature **20**). Six of those cells have a single nucleus, but one cell has **21** nuclei (2n).

represents the mature (20) _____ _____ [pp.534–535]. Six of those cells have a single nucleus, but one cell has (21) _____ [pp.534–535] (number) nuclei ($2n$) and represents the (22) _____ [pp.534–535] mother cell ($n + n$). Another haploid cell within the embryo sac is the (23) _____ [pp.534–535]. Following (24) _____ _____ [p.535] with one sperm, the $n + n$ cell will help form the $3n$ (25) _____ [p.535], a nutritive tissue for the forthcoming embryo. The other sperm involved in this unique fertilization process fertilizes the haploid egg; this combination forms the diploid (26) _____ [p.535]. Thus the ovule is transformed to a (27) _____ [p.535] that is composed of three parts, a seed (28) _____ [p.535], an embryo, and nourishment for the embryo, the endosperm.

Short Answer

29. What guides the growth of the pollen tube down through the female floral tissues toward the chamber holding the egg? [p.535] _____

_____

30. Describe the site of double fertilization, an event known only in flowering plants. [p.535] _____

_____

_____

Complete the Table

31. Complete the following table to summarize the unique double fertilization occurring only in flowering plant life cycles.

Double Fertilization Products	Origin	Produces?	Function
[p.535] a. Zygote ($2n$) nucleus			
[p.535] b. Endosperm ($3n$) nucleus			

31.4. FROM ZYGOTE TO SEEDS AND FRUITS [pp.536–537]

Selected Words: *Capsella* [p.536], *simple* fruits [pp.536–537], *aggregate* fruits [pp.536–537], *multiple* fruits [pp.536–537], *Ananas* [p.537], *Fragaria* [p.537], *endo*carp [p.537], *meso*carp [p.537], *exo*carp [p.537]

Boldfaced, Page-Referenced Terms

[p.536] fruit _____

_____

[p.536] cotyledons _____

_____

[p.536] seed _____

_____

[p.537] pericarp _____

_____

Complete the Table

1. Complete the following table, which summarizes concepts associated with seeds and fruits.

Structure	Origin
[p.536] a. Cotyledons	
[p.536] b. Seeds	
[p.536] c. Seed coat	
[pp.536–537] d. Fruit	

Labeling

Identify each numbered part of the accompanying illustration.

2. _____ [p.536]

3. _____ [p.536]

4. _____ [p.536]

5. _____ [p.536]

6. _____ [p.536]

7. _____ _____ [p.536]

8. _____ _____ [p.536]

9. _____ [p.536]

10. _____ [p.536]

11. _____ [p.536]

12. _____ _____ [p.536]

13. _____ [p.536]

14. _____ [p.536]

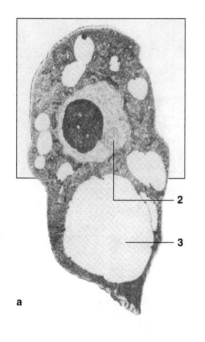

a

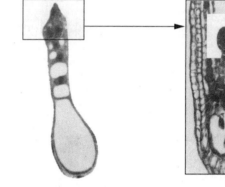

b Upper part of **4**
  gives rise to embryo

c Globular
  5 stage

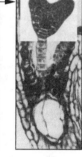

d Heart-shaped
  stage of **6**

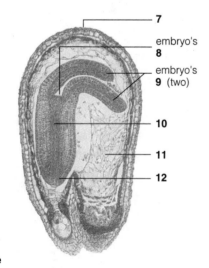

e

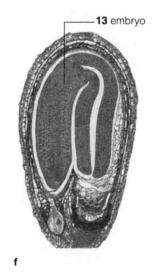

f

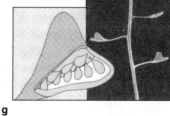

A **14** (a mature ovary)
cut open to show
seeds (the mature
ovules). The embryo
in each seed is at
stage (f).

g

Fill-in-the-Blanks

Following fertilization, the newly formed zygote initiates a course of (15) _____ [p.536] cell divisions that lead to a mature embryo (16) _____ [p.536]. The embryo develops as part of a(n) (17) _____ [p.536] and is accompanied by the formation of a (18) _____ [p.536], a mature ovary.

By the time a *Capsella* embryo is near maturity, two (19) _____ [p.536], or seed leaves, have begun to develop from two lobes of meristematic tissue. Dicot embryos, such as *Capsella*, have (number) (20) _____ [p.536] cotyledon(s) and monocot embryos have (21) (number) _____ [p.536] cotyledon(s). Like most dicots, the *Capsella* embryo has rather thick cotyledons that absorb nutrients from the (22) _____ [p.536] of the seed and stores them in its cotyledons. In corn, wheat, and most other monocots, endosperm is not tapped until the seed (23) _____ [p.536]. Digestive enzymes become stockpiled inside the (24) _____ [p.536] cotyledons of monocot embryos. When the (25) _____ [p.536] do become active, nutrients stored in the endosperm will be released and transferred to the growing (26) _____ [p.536].

From the time a zygote forms until an embryo matures, a parent sporophyte plant transfers nutrients to tissues of the (27) _____ [p.536]. Food reserves accumulate in (28) _____ [p.536] or cotyledons. Eventually, the ovule separates from the (29) _____ [p.536] wall, and its integuments thicken and harden into a seed (30) _____ [p.536]. The embryo, food reserves, and seed coat are a self-contained package—a (31) _____ [p.536], which is defined as a mature (32) _____ [p.536]. While seeds are forming, changes occur in other floral parts and begin to form (33) _____ [p.536]. There are three categories of these. (34) _____ [p.536] (either dry or fleshy, and derived from a single ovary), (35) _____ [p.536] (from many separate ovaries of a single flower), and (36) _____ [p.536] (from many separate ovaries of a single flower, all attached to the same receptacle). An apple is an (37) _____ [p.537] fruit, composed of mainly an enlarged receptacle and calyx.

Botanists often refer to three divisions of fleshy fruits. The (38) _____ [p.537] is the innermost portion around a seed or seeds. (39) _____ [p.537] is the fleshy portion. (40) _____ [p.537] is the skin. Together, the three fruit regions are called a (41) _____ [p.537].

Choice

For questions 42–48, choose from the following fruit types.

a. simple dry dehiscent   b. simple dry indehiscent   c. simple fleshy fruits
d. aggregate fruits   e. multiple fruits   f. accessory fruits

42. ___ pineapple [p.537]

43. ___ nuts, grains [p.537]

44. ___ raspberries [p.537]

45. ___ grapes, tomatoes [p.537]

46. ___ apples, pears, strawberries [p.537]

47. ___ sunflowers and carrots [p.537]

48. ___ oranges [p.537]

49. ___ peaches, cherries, and other drupes [p.537]

31.5. DISPERSAL OF FRUITS AND SEEDS [p.538]

31.6. *Focus on Science:* WHY SO MANY FLOWERS AND SO FEW FRUITS [p.539]

Selected Words: seed dispersal [p.538], *giant saguaro* [p.539]

Choice

For questions 1–10, choose from the following:

a. wind-dispersed fruits   b. fruits dispersed by animals   c. water-dispersed fruits

1. ___ heavy wax coats [p.538]

2. ___ coconut palms [p.538]

3. ___ seed coats assaulted by digestive enzymes to assist in releasing embryos [p.538]

4. ___ maples [p.538]

5. ___ orchids [p.538]

6. ___ air sacs [p.538]

7. ___ hooks, spines, hairs, and sticky surfaces [p.538]

8. ___ cacao [p.538]

9. ___ dandelions [p.538]

10. ___ cockleburs, bur clover, and bedstraw [p.538]

Short Answer

11. Explain why, although it might not appear so, the following statement may actually reflect an adaptation for reproductive success in terms of Darwinian evolutionary theory: "Although giant saguaro cactus plants produce many flowers on each plant, perhaps less than 50 percent of these flowers set fruit with viable seeds." [p.539]

_____

_____

_____

_____

_____

31.7. ASEXUAL REPRODUCTION OF FLOWERING PLANTS [pp.540–541]

Selected Words: *Populus tremuloides* [p.540], *Larrea divaricata* [p.540], *Daucus carota* [p.540]

Boldfaced, Page-Referenced Terms

[p.540] vegetative growth _____

[p.540] parthenogenesis _____

[p.540] tissue culture propagation _____

Matching

Match the following asexual reproductive modes of flowering plants.

1. ___corm [p.540]
2. ___bulb [p.540]
3. ___parthenogenesis [p.540]
4. ___runner [p.540]
5. ___vegetative propagation on modified stems [p.540]
6. ___rhizome [p.540]
7. ___tuber [p.540]
8. ___tissue culture propagation (induced propagation) [p.540]
9. ___vegetative growth [p.540]

A. In a general sense, new plants develop from tissues or organs that drop or separate from parent plants
B. New shoots arise from axillary buds (enlarged tips of slender underground rhizomes)
C. New plants arise from cells in parent plant that were not irreversibly differentiated; a laboratory technique
D. New plants arise at nodes of underground horizontal stem
E. New plant arises from axillary bud on short, thick vertical underground stem
F. New plants arise at nodes on an aboveground horizontal stem
G. New plant arises from an axillary bud on a short underground stem
H. Involves asexual reproduction using runners, rhizomes, corms, tubers, and bulbs
I. Embryo develops without nucleus or cellular fusion

Matching

Choose the most appropriate example for each modified stem.

10. ___bulb [p.540]

11. ___rhizome [p.540]

12. ___tuber [p.540]

13. ___runner [p.540]

14. ___corm [p.540]

A. Potato
B. Strawberry
C. Gladiolus
D. Onion, lily
E. Bermuda grass

Self-Quiz

___ 1. The joint evolution of flowers and their pollinators is known as _____. [p.530]
   a. adaptation
   b. coevolution
   c. joint evolution
   d. covert evolution

___ 2. A stamen is _____. [p.533]
   a. composed of a stigma
   b. the mature male gametophyte
   c. the site where microspores are produced
   d. part of the vegetative phase of an angiosperm

___ 3. The portion of the carpel that contains an ovule is the _____. [p.533]
   a. stigma
   b. anther
   c. style
   d. ovary

___ 4. The phase in the life cycle of plants that gives rise to spores is known as the _____. [p.533]
   a. gametophyte
   b. embryo
   c. sporophyte
   d. seed

___ 5. A gametophyte is _____. [p.532]
   a. a gamete-producing plant
   b. haploid
   c. both a and b
   d. the plant produced by the fusion of gametes

___ 6. A characteristic of a seed is that it _____. [p.536]
   a. contains an embryo sporophyte
   b. represents an arrested growth stage
   c. is covered by hardened and thickened integuments
   d. all of these

___ 7. An immature fruit is a(n) _____ and an immature seed is a(n) _____. [p.536]
   a. ovary; megaspore
   b. ovary; ovule
   c. megaspore; ovule
   d. ovule; ovary

___ 8. In flowering plants, one sperm nucleus fuses with that of an egg, and a zygote forms that develops into an embryo. Another sperm fuses with _____. [p.535]
   a. a primary endosperm cell to produce three cells, each with one nucleus
   b. a primary endosperm cell to produce one cell with one triploid nucleus
   c. both nuclei of the endosperm mother cell, forming a primary endosperm cell with a single triploid nucleus
   d. one of the smaller megaspores to produce what will eventually become the seed coat

___ 9. "Simple, aggregate, multiple, and accessory" refer to types of _____. [p.537]
   a. carpels
   b. seeds
   c. fruits
   d. ovaries

_10. "When a leaf falls or is torn away from a jade plant, a new plant can develop from the leaf, from meristematic tissue." This statement refers to _____. [p.540]
   a. parthenogenesis
   b. runners
   c. tissue culture propagation
   d. vegetative propagation

Chapter Objectives/Review Questions

1. _____ refers to two or more species jointly evolving as an outcome of close ecological interactions. [p.530]
2. Describe the role of a pollinator. [p.530]
3. Distinguish between sporophytes and gametophytes. [p.532]
4. Identify the various parts of a typical flower and state their functions. [pp.532–533]
5. Walled microspores form in pollen sacs and develop into _____ _____. [p.533]
6. Distinguish between a flower that is _perfect_ and one that is _imperfect_. [p.533]
7. Describe the condition called allergic rhinitis. [p.533]
8. Relate the sequence of events and structures involved that give rise to microspores and megaspores. [p.534]
9. _____ is the transfer of pollen grains to a receptive stigma. [p.534]
10. What structures represent the male gametophyte and female gametophyte in flowering plants? List the contents of each. [pp.534–535]
11. The endosperm mother cell in the embryo sac is composed of two haploid _____. [p.534]
12. Describe the double fertilization that occurs uniquely in the flowering plant life cycle. [p.535]
13. After pollination and double fertilization, an _____ and nutritive tissue form in the _____, which becomes a seed. [p.535]
14. How is endosperm formed? What is the function of endosperm? [p.534]
15. Describe the formation of the embryo sporophyte; describe the origin and formation of seeds and fruits. [p.536]
16. Review the general types of fruits produced by flowering plants. [p.536]
17. Seeds and fruits are structurally adapted for _____ by air currents, water currents, and many kinds of animals. [p.538]
18. Define the term _vegetative growth_. [p.540]
19. Distinguish among parthenogenesis, vegetative propagation, and tissue culture propagation; cite an example for each. [p.540]
20. List representative plant examples of a runner, a rhizome, a corm, a tuber, and a bulb. [p.540]

Integrating and Applying Key Concepts

In terms of botanical morphology, a flower is interpreted as "a reproductive shoot bearing organs." After studying floral organs in this chapter and leaf structure in text Chapter 29, can you think of any comparable structural evidence that might have led botanists to arrive at this conclusion?

32

PLANT GROWTH AND DEVELOPMENT

Interactive Exercises

Foolish Seedlings and Gorgeous Grapes [pp.544–545]

32.1. PATTERNS OF EARLY GROWTH AND DEVELOPMENT—AN OVERVIEW
   [pp.546–547]

Selected Words: bukane [p.544], *Gibberella fujikuroi* [p.544], *Eschscholzia californica* [p.544], *Vitis* [p.545], *Phaseolus vulgaris* [p.546], *Zea mays* [p.546]

Boldfaced, Page-Referenced Terms

[p.544] gibberellin _____

_____

[p.544] hormones _____

_____

[p.546] germination _____

_____

[p.546] imbibition _____

_____

Fill-in-the-Blanks

Before or after seed dispersal from the parent plant, the growth of the (1) _____ [p.546] idles. For seeds, (2) _____ [p.546] is the resumption of growth by an immature stage in the life cycle after a period of arrested development. Germination depends on (3) _____ [p.546] factors, such as temperature, moisture, and oxygen level in soil, and the number of seasonal daylight hours available. By a process known as (4) _____ [p.546], water molecules move into the seed. As more water moves in, the seed swells and its coat (5) _____ [p.546].

Once the seed coat splits, more oxygen reaches the embryo, and (6) _____ [p.546] respiration moves into high gear. The embryo's (7) _____ [p.546] cells begin to divide rapidly. In general, the (8) _____ [p.546] meristem is the first to be activated. Its meristematic descendants divide, elongate, and give rise to the (9) _____ _____ [p.546]. When this structure breaks through the seed coat, (10) _____ [p.546] is over.

For both monocots and dicots, the patterns of germination, growth, and development that unfold have a (11) _____ [p.546] basis; they are dictated by the plant's (12) _____ [p.546]. All cells in the new plant arise from the same cell, the (13) _____ [p.546]. Thus all cells inherit the same (14) _____ [p.546]. Unequal (15) _____ [p.546] divisions between daughter cells lead to differences in their (16) _____ [p.547] equipment and output. Activities in daughter cells start to vary as a result of (17) _____ [p.547] gene expression. As an example, genes governing the synthesis of growth-stimulating (18) _____ [p.547] are activated in some cells but not others. (19) _____ [p.547] among genes, hormones, and the environment govern how an individual plant grows and develops.

Labeling

Identify each numbered part of the accompanying illustration.

20. _____ [p.547] 26. _____ [p.547] 32. _____ [p.547]

21. _____ [p.547] 27. _____ [p.547] 33. _____ [p.547]

22. _____ [p.547] 28. _____ [p.547] 34. _____ [p.547]

23. _____ [p.547] 29. _____ [p.547] 35. _____ [p.547]

24. _____ [p.547] 30. _____ [p.547] 36. _____ [p.547]

25. _____ [p.547] 31. _____ [p.547] 37. _____ [p.547]

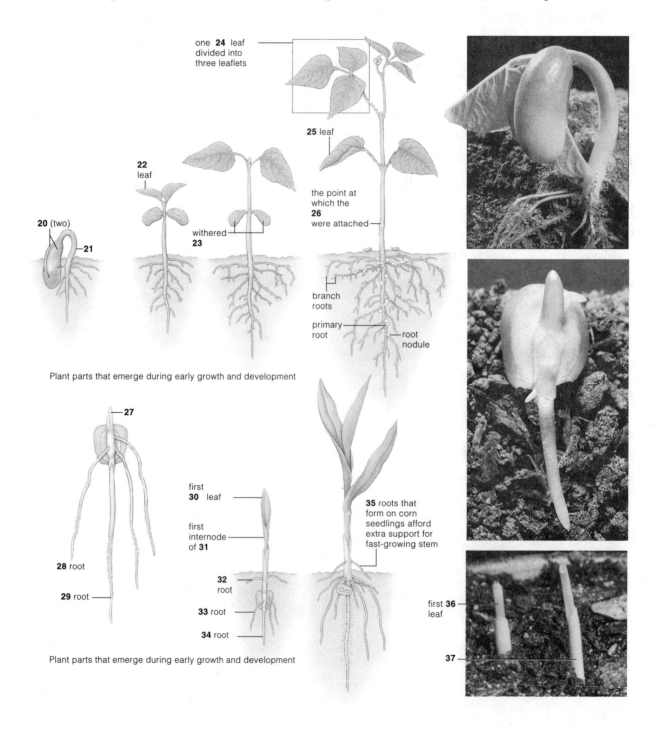

one **24** leaf divided into three leaflets

25 leaf

the point at which the **26** were attached

22 leaf

withered **23**

20 (two)

21

branch roots

primary root

root nodule

Plant parts that emerge during early growth and development

27

first **30** leaf

first internode of **31**

35 roots that form on corn seedlings afford extra support for fast-growing stem

28 root

29 root

32 root

33 root

34 root

first **36** leaf

37

Plant parts that emerge during early growth and development

32.2. HORMONAL EFFECTS ON PLANT GROWTH AND DEVELOPMENT [pp.548–549]

Selected Words: quantitative terms [p.548], qualitative terms [p.548], Agent Orange [p.549]

Boldfaced, Page-Referenced Terms

[p.548] growth _____

_____

[p.548] development _____

_____

[p.548] auxins _____

_____

[p.548] cytokinins _____

_____

[p.548] ethylene _____

_____

[p.548] abscisic acid _____

_____

[p.548] coleoptile _____

_____

[p.548] herbicides _____

_____

[p.549] apical dominance _____

_____

Choice

For questions 1–14, choose from the following:

        a. auxins   b. gibberellins   c. cytokinins   d. abscisic acid   e. ethylene

1. ___ Ancient Chinese burned incense to hurry fruit ripening. [p.549]

2. ___ Natural and synthetic versions are used to prolong the shelf life of cut flowers, lettuces, mushrooms, and other vegetables. [p.549]

3. ___ Orchardists spray trees with IAA to thin overcrowded seedlings in the spring. [p.548]

4. ___ Inhibits cell growth, induces bud dormancy, and prevents seeds from germinating prematurely; causes stomata to close when a plant is water stressed [p.549]

5. ___ IAA, the most pervasive naturally occurring compound of its type [p.548]

6. ___ Exposed to oranges and other citrus fruits to brighten their color of their rind before being displayed in the market [p.549]

7. ___ Influences stem lengthening; influences plant responses to gravity and light and promotes coleoptile lengthening [p.548]

8. ___ Promotes stem lengthening; helps buds and seeds end dormancy and resume growth in the spring [p.548]

9. ___ Used to prevent premature fruit drop—all fruit can be picked at the same time [p.548]

10. ___ Stimulate cell division [p.549]

11. ___ Used by food distributors to ripen green tomatoes and other fruit after it is shipped to grocery stores [p.549]

12. ___ Most abundant in root and shoot meristems and in the tissues of maturing fruit [p.549]

13. ___ Induces various aging responses, including fruit ripening and leaf drop [p.549]

14. ___ Some synthetic forms serve as herbicides. [p.548]

Matching

Choose the most appropriate answer for each term.

15. ___2,4-D [p.548]

16. ___hormone [p.548]

17. ___apical dominance [p.549]

18. ___herbicide [p.548]

19. ___growth [p.548]

20. ___promote stem lengthening [pp.548–549]

21. ___IAA [p.548]

22. ___development [p.548]

23. ___2,4,5–T [p.548]

24. ___coleoptile [p.548]

25. ___target cell [p.548]

A. Mixed with 2,4-D to produce Agent Orange for defoliation in Vietnam
B. The emergence of specialized, morphologically different body parts; measured in qualitative terms
C. Gibberellins
D. Hormonal effect that blocks lateral bud growth
E. Synthetic auxin used as an herbicide to selectively kill broadleaf weeds that compete vigorously with cereal plants for nutrients
F. An increase in the number, size, and volume of cells; measured in quantitative terms
G. A signaling molecule released from one cell that travels to target cells and stimulates or inhibits gene activity
H. The most important naturally occurring auxin
I. Any synthetic auxin compound used to kill some plants but not others
J. A thin sheath around the primary shoot of grass seedlings, such as corn plants

32.3. ADJUSTMENTS IN THE RATE AND DIRECTION OF GROWTH [pp.550–551]

Selected Words: trope [p.550], auxein [p.551], thigma [p.551]

Boldfaced, Page-Referenced Terms

[p.550] plant tropism _____

_____

[p.550] gravitropism _____

_____

[p.550] statoliths _____

_____

[p.550] phototropism _____

_____

[p.551] thigmotropism _____

_____

Choice

For questions 1–10, choose from the following:

        a. phototropism   b. gravitropism   c. thigmotropism   d. mechanical stress

1. ___ More intense sunlight on one side of a plant—stems curve toward the light [pp.550–551]
2. ___ Vines climbing around a fencepost as they grow upward [p.551]
3. ___ Plants grown outdoors have shorter stems than plants grown in a greenhouse. [p.551]
4. ___ A root turned on its side will curve downward. [p.550]
5. ___ Briefly shaking a plant daily inhibits the growth of the entire plant. [p.551]
6. ___ A potted seedling turned on its side—the growing stem curves upward [p.550]
7. ___ Leaves turn until their flat surfaces face light. [pp.550–551]
8. ___ Flavoprotein may be a central component. [p.551]
9. ___ Tendrils are sometimes involved. [p.551]
10. ___ Statoliths are generally involved. [p.550]

32.4. BIOLOGICAL CLOCKS AND THEIR EFFECTS [pp.552–553]

Selected Words: Pfr [p.552], Pr [p.552], *circadian* [p.552]

Boldfaced, Page-Referenced Terms

[p.552] biological clocks _____

_____

[p.552] phytochrome _____

_____

[p.552] circadian rhythm _____

_____

[p.552] photoperiodism _____

_____

[p.552] short-day plants _____

_____

[p.552] long-day plants _____

_____

[p.552] day-neutral plants _____

_____

Matching

Choose the most appropriate answer for each term.

1. ___photoperiodism [p.552]
2. ___Pr [p.552]
3. ___day-neutral plants [p.552]
4. ___phytochrome activation [p.552]
5. ___"long-day" plants [p.552]
6. ___circadian rhythms [p.552]
7. ___Pfr [p.552]
8. ___biological clocks [p.552]
9. ___rhythmic leaf movements [p.552]
10. ___phytochrome [p.552]

A. Biological activities that recur in cycles of twenty-four hours or so
B. Flower in spring when daylength exceeds a critical value
C. Internal time-measuring mechanisms with roles in adjusting daily activities
D. Any biological response to a change in the relative length of daylight and darkness in the twenty-four-hour cycle; active Pfr may be an alarm button for this process
E. A blue-green pigment that absorbs red or far-red wavelengths, with different results
F. An example of a circadian rhythm
G. Active form of phytochrome
H. Flower when mature enough to do so
I. Inactive form of phytochrome
J. May induce plant cells to take up free calcium ions or induce certain plant cell organelles to release them

Labeling

Identify each numbered part of the accompanying illustration.

11. _____ [p.552]
12. _____ [p.552]
13. _____ [p.552]
14. _____ [p.552]
15. _____ [p.552]

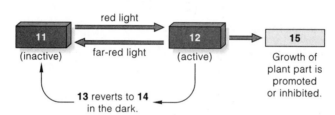

Fill-in-the-Blanks

(16) _____ [p.552] is any biological response to change in the relative length of daylight and darkness in the cycle of twenty-four hours. (17) _____ [p.552], phytochrome's active form, may be a switching mechanism for the process; it may trigger the synthesis of specific (18) _____ [p.552] in specific types of plant cells.

"Long-day plants" flower in spring when daylength becomes (19) [choose one] ❑ shorter ❑ longer [p.552] than some critical value. "Short-day plants" flower in late summer or early autumn when daylength becomes (20) [choose one] ❑ shorter, ❑ longer [p.552] than some critical value. "Day-neutral plants" flower whenever they become (21) _____ [p.552] enough to do so without regard to daylength. Spinach is a (22) _____ - _____ [p.552] plant because it will not flower and produce seeds unless it is exposed to fourteen hours of light every day for two weeks. Cocklebur is termed a (23) _____ - _____ [p.553] plant because it flowers after a single night that is longer than 8 1/2 hours.

32.5. LIFE CYCLES END, AND TURN AGAIN [pp.554–555]

32.6. *Commentary:* THE RISE AND FALL OF A GIANT [p.556]

Selected Words: vernalis [p.555], *Secale cereale* [p.555]

Boldfaced, Page-Referenced Terms

[p.554] abscission _____

_____

[p.554] senescence _____

_____

[p.554] dormancy _____

_____

[p.555] vernalization _____

_____

Choice

For questions 1–13, choose from the following:

  a. senescence   b. abscission   c. vernalization   d. entering dormancy   e. breaking dormancy

1. ___ Dropping of leaves or other parts from a plant [p.554]
2. ___ A process at work between fall and spring; temperatures become milder, and rain and nutrients become available again. [p.555]
3. ___ Strong cues are short days, long, cold nights, and dry, nitrogen-deficient soil [p.554]
4. ___ The process proceeds at tissues in the base of leaves, flowers, fruits, or other plant parts; a special zone is involved. [p.554]
5. ___ The sum total of processes leading to the death of plant parts or the whole plant. [p.554]
6. ___ A recurring cue for this process is a decrease in daylength. [p.554]
7. ___ When a plant stops growing under conditions that appear quite suitable for growth [p.554]
8. ___ Unless buds of some biennials and perennials are exposed to low winter temperatures, flowers will not form on their stems when spring rolls around. [p.555]
9. ___ The forming of ethylene in cells near the breakpoints may trigger the process. [p.554]
10. ___ Stopping nutrient drain to reproductive parts blocks this process. [p.554]
11. ___ Keeping germinating seeds of winter rye at near-freezing temperature to induce flowering the following summer [p.554]
12. ___ Many perennial and biennial plants start to shut down growth as autumn approaches and days grow shorter. [p.554]
13. ___ Can be postponed when gardeners remove flower buds from plants to maintain vegetative growth [p.554]

Self-Quiz

___ 1. Promoting fruit ripening and abscission of leaves, flowers, and fruits is a function ascribed to _____. [p.549]
   a. gibberellins
   b. ethylene
   c. abscisic acid
   d. auxins

___ 2. Auxins _____. [p.548]
   a. cause flowering
   b. promote stomatal closure
   c. promote cell division
   d. promote cell elongation in coleoptiles and stems

___ 3. _____ is demonstrated by a germinating seed whose first root always curves down while the stem always curves up. [p.550]
   a. Phototropism
   b. Photoperiodism
   c. Gravitropism
   d. Thigmotropism

___ 4. Light of _____ wavelengths is the main stimulus for phototropism. [p.551]
   a. blue
   b. yellow
   c. red
   d. green

___ 5. Plants whose leaves are open during the day but fold them at night are exhibiting a _____. [p.552]
   a. growth movement
   b. circadian rhythm
   c. biological clock
   d. both b and c are correct

___ 6. 2,4-D, a potent dicot weed killer, is a synthetic _____. [p.549]
   a. auxin
   b. gibberellin
   c. cytokinin
   d. phytochrome

___ 7. All the processes that lead to the death of a plant or any of its organs are called _____. [p.554]
   a. dormancy
   b. vernalization
   c. abscission
   d. senescence

___ 8. Phytochrome is converted to an active form, _____, at sunrise and reverts to an inactive form, _____, at sunset, at night, or in the shade. [p.552]
   a. Pr; Pfr
   b. Pfr; Pfr
   c. Pr; Pr
   d. Pfr; Pr

___ 9. Which of the following is not promoted by the active form of phytochrome? [pp.552,554]
   a. Entering dormancy
   b. Flowering
   c. Rhythmic leaf movements
   d. Stem elongation

___ 10. When a perennial or biennial plant stops growing under conditions suitable for growth, it has entered a state of _____. [p.554]
   a. senescence
   b. vernalization
   c. dormancy
   d. abscission

Chapter Objectives/Review Questions

1. _____ are signaling molecules. [p.544]
2. _____ is the process by which an immature stage in the life cycle resumes growth after a period of arrested development. [p.546]
3. Define imbibition and describe its role in germination. [p.546]
4. List the environmental factors that influence germination. [p.546]
5. The primary _____ breaks through the seed coat first. [p.546]
6. The basic patterns of growth and development are heritable, dictated by the plant's _____. [p.546]
7. Compare and contrast the major features of the growth and development of a monocot plant and a dicot plant. [p.546]
8. Plant _____ involves cell divisions and cell enlargements; plant _____ requires cell differentiation, as brought about by selective gene expression. [p.547]
9. Explain why plant growth is measured in quantitative terms and plant development, in qualitative terms. [p.548]
10. Describe the general role of plant hormones. [p.548]
11. _____ affect the lengthening of stems and responses to gravity and light; they also make coleoptiles grow longer. [p.548]
12. _____ promote stem lengthening and help buds and seeds break dormancy to resume growth in the spring. [p.548]
13. Certain synthetic auxins are used as _____, compounds that kill some plant species but not others. [pp.548–549]
14. _____ promote cell division and leaf expansion and retard leaf aging. [p.549]
15. _____ _____ promotes stomatal closure and promotes bud and seed dormancy. [p.549]
16. _____ promotes ripening of fruit and abscission of leaves, flowers, and fruits. [p.549]
17. Describe the form of growth inhibition known as apical dominance. [p.549]
18. Define phototropism, gravitropism, and thigmotropism, and give examples of each. [pp.550–551]
19. Explain how statoliths form the basis for gravity-sensing mechanisms. [p.550]
20. Plants make the strongest phototropic response to light of _____ wavelengths; _____ is a yellow pigment molecule that absorbs blue wavelengths. [p.551]
21. Provide two examples of how mechanical stress can affect plants. [p.551]
22. Plants have internal time-measuring mechanisms called biological _____. [p.552]
23. The alarm button for some biological clocks in plants is the blue-green pigment molecule known as _____. [p.552]
24. What are circadian rhythms? Give an example. [p.552]
25. Phytochrome is converted to an active form, _____, at sunrise, when red wavelengths dominate the sky. It reverts to an inactive form, _____, at sunset, at night, or even in shade, where far-red wavelengths predominate. [p.552]
26. _____ is a biological response to a change in the relative length of daylight and darkness in a twenty-four-hour cycle. [p.552]
27. _____ serves as a switching mechanism in the biological clock governing flowering responses. [p.552]
28. Describe the photoperiodic responses of "long-day," "short-day," and "day-neutral" plants. [p.552]
29. _____ is the dropping of leaves, flowers, fruits, or other plant parts. [p.554]
30. Describe the events that signal plant senescence. [p.554]
31. List environmental cues that send a plant into dormancy. [p.554]
32. The low-temperature stimulation of flowering is called _____. [p.555]
33. What conditions are instrumental in the dormancy-breaking process? [p.555]

Integrating and Applying Key Concepts

An oak tree has grown up in the middle of a forest. A lumber company has just cut down all the surrounding trees except for a narrow strip of woods that includes the oak. How will the oak be likely to respond as it adjusts to its changed environment? To what new stresses will it be exposed? Which hormones will most probably be involved in the adjustment?

33

TISSUES, ORGAN SYSTEMS, AND HOMEOSTASIS

Interactive Exercises

Meerkats, Humans, It's All the Same [pp.560–561]

33.1. EPITHELIAL TISSUE [pp.562–563]

Selected Words: anatomy [p.560], *physiology* [p.560], *Suricata suricatta* [p.561], *simple* epithelium [p.562], *stratified* epithelium [p.562], *Dendrobates* [p.563]

Boldfaced, Page-Referenced Terms

[p.560] internal environment _____

_____

[p.560] homeostasis _____

_____

[p.560] tissue _____

_____

[p.560] organ _____

_____

[p.560] organ system _____

_____

[p.561] division of labor _____

_____

[p.562] epithelium _____

_____

[p.562] tight junctions _____

_____

[p.562] adhering junctions _____

_____

[p.562] gap junctions _____

_____

[p.563] gland cells _____

_____

[p.563] exocrine glands _____

_____

[p.563] endocrine glands _____

_____

Matching

Choose the most appropriate answer for each term.

1. ___ internal environment [p.560]
2. ___ anatomy [p.560]
3. ___ physiology [p.560]
4. ___ homeostasis [p.560]
5. ___ tissue [p.560]
6. ___ organ [p.560]
7. ___ organ system [p.560]
8. ___ division of labor [p.561]

A. Consists of two or more organs that are interacting physically, chemically, or both in a common task
B. How the body functions
C. Consists of interstitial fluid (tissue fluids) and blood that bathes the living cells of any complex animal
D. Consists of different tissues that are organized in specific proportions and patterns
E. Cells, tissues, organs, and organ systems, split up the work in ways that contribute to the survival of the animal as a whole
F. How the animal body is structurally put together
G. An interactive group of cells and intercellular substances that take part in one or more particular tasks
H. With respect to the animal body, refers to stable operating conditions in the internal environment

Fill-in-the-Blanks

(9) _____ tissue has a free surface, which faces either a body fluid or the outside environment [p.562].

(10) _____ _____ has a single layer of cells and functions as a lining for body cavities, ducts, and tubes [p.562]. (11) _____ _____ has two or more layers and typically functions in protection, as it does in the skin [p.562]. (12) _____ [p.562] junctions are strands of proteins that help stop substances from leaking across a tissue. (13) _____ [p.562] junctions cement cells together. (14) _____ [p.562] junctions help communicate by promoting the rapid transfer of ions and small molecules among them.

(15) _____ [p.562] junctions in the epithelium of your stomach help prevent a condition called

(16) _____ [p.562] ulcer. (17) _____ [p.563] glands secrete mucus, saliva, earwax, milk, oil, digestive enzymes, and other cell products. These products are usually released onto a free (18) _____ [p.563] surface through ducts or tubes. (19) _____ [p.563] glands lack ducts; their products are (20) _____ [p.563], which are secreted directly into the fluid bathing the gland. Typically, the (21) _____ [p.563] picks up the hormone molecules and distributes them to target cells elsewhere in the body.

33.2. CONNECTIVE TISSUE [pp.564–565]

33.3. MUSCLE TISSUE [p.566]

33.4. NERVOUS TISSUE [p.567]

33.5. *Focus on Science:* FRONTIERS IN TISSUE RESEARCH [p.567]

Selected Words: fibroblasts [p.564], *plasma* [p.565], *contract* [p.566], *striated* [p.566], *laboratory-grown epidermis* [p.567], *designer organs* [p.567], type I *diabetes mellitus* [p.567]

Boldfaced, Page-Referenced Terms

[p.564] loose connective tissue _____

_____

[p.564] dense, irregular connective tissue _____

_____

[p.564] dense, regular connective tissue _____

_____

[p.564] cartilage _____

_____

[p.565] bone tissue _____

_____

[p.565] adipose tissue _____

_____

[p.565] blood _____

_____

[p.566] skeletal muscle tissue _____

_____

[p.566] smooth muscle tissue _____

_____

[p.566] cardiac muscle tissue _____

_____

[p.567] nervous tissue _____

_____

[p.567] neuroglia _____

_____

[p.567] neurons _____

_____

Choice

For questions 1–10, choose from the following types of connective tissue proper:

a. loose   b. dense, irregular   c. dense, regular

1. ___ Contains fibers, mostly collagen-containing ones, and a few fibroblasts. [p.564]
2. ___ Rows of fibroblasts often intervene between the bundles. [p.564]
3. ___ Has its fibers and cells loosely arranged in a semifluid ground substance. [p.564]
4. ___ Has parallel bundles of many collagen fibers and resists being torn apart. [p.564]
5. ___ Forms protective capsules around organs that do not stretch much. [p.564]
6. ___ Often serves as a support framework for epithelium. [p.564]
7. ___ Found in tendons, which attach skeletal muscle to bones. [p.564]
8. ___ Besides fibroblasts, it contains infection-fighting white blood cells. [p.564]
9. ___ Found in elastic ligaments, which attach bones to each other. [p.564]
10. ___ It is also present in the deeper part of skin. [p.564]

Complete the Table

11. After reading each description, supply the name of the specialized connective tissue.

Specialized Connective Tissue	Description
a. [p.565]	Chockfull of large fat cells; stores excess carbohydrates and proteins; richly supplied with blood
b. [pp.564–565]	Intercellular material, solid yet pliable, resists compression; structural models for vertebrate embryo bones; maintains shape of nose, outer ear, and other body parts; cushions joints
c. [p.565]	Derived mainly from connective tissue, has transport functions; circulating within plasma are a great many red blood cells, white blood cells, and platelets
d. [p.565]	The weight-bearing tissue of vertebrate skeletons, which support or protect softer tissues and organs; mineral-hardened with calcium-salt laden collagen fibers and ground substance; interact with skeletal muscles attached to them

Dichotomous Choice

Circle one of two possible answers given between parentheses in each statement.

12. Contractile cells of (skeletal/smooth) muscle tissue taper at both ends. [p.566]
13. The contractile walls of the heart are composed of (striated/cardiac) muscle tissue. [p.566]
14. Walls of the stomach and intestine contain (smooth/skeletal) muscle tissue. [p.566]
15. The only muscle tissue attached to bones is (skeletal/smooth). [p.566]
16. (Smooth/Skeletal) muscle cells are bundled together in parallel. [p.566]
17. "Involuntary" muscle action is associated with (smooth/skeletal) muscle tissue. [p.566]
18. The term *striated* means (bundled/striped). [p.566]
19. (Smooth/Skeletal) muscle tissue has a sheath of tough connective tissue enclosing several bundles of muscle cells. [p.566]
20. The function of smooth muscle tissue is to (pump blood/move internal organs). [p.566]
21. Cell junctions fuse together the plasma membranes of (smooth/cardiac) muscle cells. [p.566]
22. (Muscle/Nervous) tissue exerts the greatest control over the body's responsiveness to changing conditions. [p.567]
23. Excitable cells are the (neuroglia/neurons). [p.567]
24. (Neuroglia/Muscle) cells protect and structurally and metabolically support the neurons. [p.567]
25. When a (neuron/muscle cell) is suitably stimulated, an electrical "message" travels over its plasma membrane that may result in stimulation of other cells of the same type or of other types. [p.567]
26. Different types of (neuroglia/neurons) detect specific stimuli, integrate information, and issue or relay commands for response. [p.567]
27. The lives of people with type I diabetes mellitus might, in the future, be made more normal with (a designer organ/a sheet of laboratory-grown epidermis). [p.567]

Labeling and Matching

Label each of the following illustrations with one of the following terms: *connective, epithelial, muscle,* or *nervous tissue.* Complete the exercise by writing *all* appropriate letters and numbers from each of the following groups in the parentheses after each label.

28. _____ (_____) [p.564]

29. _____ (_____) [p.562]

30. _____ (_____) [p.566]

31. _____ (_____) [p.566]

32. _____ (_____) [p.564]

33. _____ (_____) [p.565]

34. _____ (_____) [p.562]

35. _____ (_____) [p.565]

36. _____ (_____) [p.567]

37. _____ (_____) [p.566]

38. _____ (_____) [p.562]

39. _____ (_____) [p.565]

A. Adipose
B. Bone
C. Cardiac
D. Dense, regular
E. Loose
F. Simple columnar
G. Simple cuboidal
H. Simple squamous
I. Smooth
J. Skeletal
K. Blood
L. Neurons
1. Absorption
2. Communication by means of electrical signals
3. Energy reserve
4. Contraction for voluntary movements
5. Diffusion
6. Padding
7. Contract to propel substances along internal passageways; not striated
8. Attaches muscle to bone and bone to bone
9. In vertebrates, provides the strongest internal framework of the organism
10. Elasticity
11. Secretion
12. Pumps circulatory fluid; striated
13. Insulation
14. Transport of nutrients and waste products to and from body cells

28.

29.

30.

31.

32.

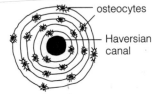
— osteocytes
— Haversian canal

33.

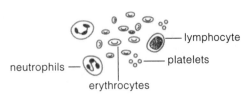

34.

neutrophils —
lymphocyte
platelets
erythrocytes

35.

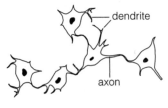
— dendrite
axon

36.

37.

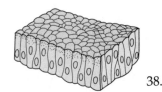
38.

39.

33.6. ORGAN SYSTEMS [pp.568–569]

Selected Words: *midsagittal* plane [p.568], *dorsal* [p.568], *ventral* [p.568], frontal plane [p.569], transverse plane [p.569], *anterior* [p.569], *posterior* [p.569], superior [p.569], inferior [p.569], distal [p.569], proximal [p.569], *germ* cells [p.569], *somatic* [p.569]

Boldfaced, Page-Referenced Terms

[p.569] ectoderm _____

_____

[p.569] mesoderm _____

_____

[p.569] endoderm _____

_____

Fill-in-the-Blanks

The brain is housed in the (1) _____ [p.568] cavity. The (2) _____ [p.568] cavity contains the spinal cord and the beginnings of spinal nerves. The heart and lungs are found within the (3) _____ [p.568] cavity. The stomach, spleen, liver, gallbladder, pancreas, small intestine, most of the large intestine, the kidneys, and ureters lie inside the (4) _____ [p.568] cavity. The (5) _____ [p.568] cavity contains the urinary bladder, sigmoid colon, rectum, and reproductive organs.

Complete the Table

6. Supply the name of the primary tissue of the embryo that does the job indicated by becoming specialized in particular ways. [All from p.569]

Primary Tissue	Functions
a.	Forms internal skeleton and muscle, circulatory, reproductive, and urinary systems
b.	Forms inner lining of gut and linings of major organs formed from the embryonic gut
c.	Forms outer layer of skin and the tissues of the nervous system

Labeling

Identify each numbered part of the accompanying illustration that reviews the directional terms and planes of symmetry for the human body.

7. _____ [p.569]

8. _____ [p.569]

9. _____ [p.569]

10. _____ [p.569]

11. _____ [p.569]

12. _____ [p.569]

13. _____ [p.569]

14. _____ [p.569]

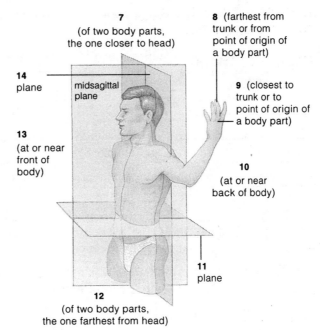

7
(of two body parts, the one closer to head)

8 (farthest from trunk or from point of origin of a body part)

9 (closest to trunk or to point of origin of a body part)

14 plane

midsagittal plane

13 (at or near front of body)

10 (at or near back of body)

11 plane

12 (of two body parts, the one farthest from head)

Labeling and Matching

Label each organ system described in the blank. Complete the exercise by matching and entering the proper letter of the following illustration in the parentheses after each label.

15. _____ (   ) Rapidly transport many materials to and from cells; help stabilize internal pH and temperature. [p.568]

16. _____ (   ) Rapidly deliver oxygen to the tissue fluid that bathes all living cells; remove carbon dioxide wastes of cells; help regulate pH. [p.569]

17. _____ (   ) Maintain the volume and composition of internal environment; excrete excess fluid and blood-borne wastes. [p.569]

18. _____ (   ) Support and protect body parts; provide muscle attachment sites; produce red blood cells; store calcium, phosphorus. [p.568]

19. _____ (   ) Hormonally control body function; work with nervous system to integrate short-term and long-term activities. [p.568]

20. _____ (   ) *Female:* produce eggs; after fertilization, afford a protected, nutritive environment for the development of new individual. *Male:* produce and transfer sperm to the female. Hormones of both systems also influence other organ systems. [p.569]

21. _____ (   ) Ingest food and water; mechanically, chemically break down food, and absorb small molecules into internal environment; eliminate food residues. [p.569]

22. _____ (   ) Move body and its internal parts; maintain posture; generate heat (by increases in metabolic activity). [p.568]

23. _____ (   ) Detect both external and internal stimuli; control and coordinate responses to stimuli; integrate all organ system activities. [p.568]

24. _____ (   ) Protect body from injury, dehydration, and some pathogens; control its temperature; excrete some wastes; receive some external stimuli [p.568]

25. _____ (   ) Collect and return some tissue fluid to the bloodstream; defend the body against infection and tissue damage [p.569]

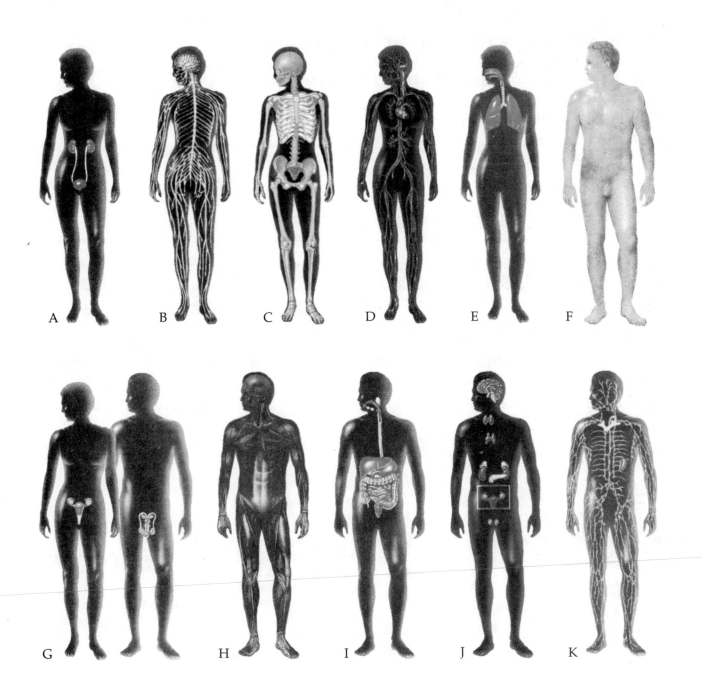

A

B

C

D

E

F

G

H

I

J

K

33.7. HOMEOSTASIS AND SYSTEMS CONTROL [pp.570–571]

Selected Words: interstitial [p.570], plasma [p.570], intensify [p.571]

Boldfaced, Page-Referenced Terms

[p.570] extracellular fluid _____

_____

[p.570] sensory receptors _____

_____

[p.570] stimulus _____

_____

[p.570] integrator _____

_____

[p.570] effectors _____

_____

[p.570] negative feedback mechanism _____

_____

[p.571] positive feedback mechanism _____

_____

Matching

Choose the most appropriate answer for each term.

1. ___extracellular fluid [p.570]
2. ___interstitial fluid [p.570]
3. ___plasma [p.570]
4. ___sensory receptors [p.570]
5. ___stimulus [p.570]
6. ___integrator [p.570]
7. ___effectors [p.570]
8. ___negative feedback mechanism [p.570]
9. ___positive feedback mechanism [p.571]

A. An activity alters a condition in the internal environment, and this triggers a response that reverses the altered condition
B. The fluid portion of blood
C. An example is the brain, a central command post where different bits of information are pulled together in the selection of a response
D. The fluid not inside cells
E. A specific change in the environment
F. Examples are muscles and glands
G. Fluid that occupies spaces between cells and tissues
H. Sets in motion a chain of events that intensify a change from an original condition—and after a limited time, the intensification reverses the change
I. Cells or cell parts that can detect a stimulus

Self-Quiz

___ 1. Which of the following is not included in connective tissues? [pp.564–565]
   a. Bone
   b. Blood
   c. Cartilage
   d. Skeletal muscle

___ 2. Gland cells are contained in _____ tissues. [p.563]
   a. muscular
   b. epithelial
   c. connective
   d. nervous

___ 3. Blood is considered to be a(n) _____ tissue. [p.565]
   a. epithelial
   b. muscular
   c. connective
   d. none of these

___ 4. _____ are abundant in tissues of the heart and stomach where they promote diffusion of ions and small molecules from cell to cell. [p.563]
   a. Adhesion junctions
   b. Filter junctions
   c. Gap junctions
   d. Tight junctions

___ 5. Muscle that is not striped and is involuntary is _____. [p.566]
   a. cardiac
   b. skeletal
   c. striated
   d. smooth

___ 6. A(n) _____ is a group of cells and intercellular substances, all interacting in one or more tasks. [p.560]
   a. organ
   b. organ system
   c. tissue
   d. cuticle

___ 7. A graduate student in developmental biology accidentally stabbed a fish embryo. Later the embryo developed into a creature that could not move and had no supportive or circulatory systems. Which embryonic tissue had suffered the damage? [p.569]

   a. ectoderm
   b. endoderm
   c. mesoderm
   d. protoderm

___ 8. A tissue whose cells are striated and fused at the ends by cell junctions so that the cells contract as a unit is called _____ tissue. [p.566]
   a. smooth muscle
   b. dense fibrous connective
   c. supportive connective
   d. cardiac muscle

___ 9. The secretion of tears, milk, sweat, and oil are functions of _____ tissues. [p.563]
   a. epithelial
   b. loose connective
   c. lymphoid
   d. nervous

___10. Memory, decision making, and issuing commands to effectors are functions of _____ tissue. [p.567]
   a. connective
   b. epithelial
   c. muscle
   d. nervous

___11. An animal that feels overheated from the sun moves to an environment that tends to cool its body. This is an example of _____. [p.570]
   a. intensifying an original condition
   b. positive feedback mechanism
   c. positive phototropic response
   d. negative feedback mechanism

___12. Which group is arranged correctly from smallest structure to largest, reading left to right? [p.566]
   a. muscle cells, muscle bundle, muscle
   b. muscle cells, muscle, muscle bundle
   c. muscle bundle, muscle cells, muscle
   d. none of the above

Matching

Choose the most appropriate answer for each term.

13. ___circulatory system [p.568]
14. ___digestive system [p.569]
15. ___endocrine system [p.568]
16. ___integumentary system [p.568]
17. ___muscular system [p.568]
18. ___nervous system [p.568]
19. ___reproductive system [p.569]
20. ___respiratory system [p.569]
21. ___skeletal system [p.568]
22. ___urinary system [p.569]

A. Picks up nutrients absorbed from gut and transports them to cells throughout body
B. Helps cells use nutrients by supplying them with oxygen and relieving them of CO_2 wastes
C. Helps maintain the volume and composition of body fluids that bathe the body's cells
D. Provides basic framework for the animal and supports other organs of the body
E. Uses chemical messengers to control and guide body functions
F. Produces younger, temporarily smaller versions of the animal
G. Breaks down larger food molecules into smaller nutrient molecules that can be absorbed by body fluids and transported to body cells
H. Consists of contractile parts that move the body through the environment and propel substances about in the animal
I. Serves as an electrochemical communications system in the animal's body
J. In the meerkat, serves as a heat catcher in the morning and protective insulation at night

Chapter Objectives/Review Questions

1. Explain how the meerkat maintains a rather constant internal environment in spite of changing external conditions. [p.560]
2. Cells are the basic units of life; in a multicellular animal, like cells are grouped into a(n) _____, and these are organized in specific proportions and patterns that compose a(n) _____. [p.560]
3. Explain how, if each cell can perform all its basic activities, organ systems contribute to cell survival. [pp.560–561,568]
4. _____ has a free surface, which faces either a body fluid or the outside environment. [p.562]
5. Distinguish simple epithelium from stratified epithelium. [p.562]
6. Name and describe three kinds of cell junctions that occur in epithelia and other tissues. [p.562]
7. Name and describe the various types of epithelial tissues as well as their location and general functions. [pp.562–563]
8. Define the term *gland*. [p.563]
9. _____ glands usually secrete their products onto a free epithelial surface through ducts or tubes; cite examples of their products. [p.563]
10. _____ glands lack ducts; their products are _____, which are secreted directly into the fluid bathing the gland. [p.563]
11. Distinguish between loose connective, dense, irregular, and dense, regular, connective tissues on the basis of their structures and functions. [p.564]
12. Cartilage, bone, adipose tissue, and blood are known as the specialized connective tissues; describe their structures and various functions. [pp.564–565]
13. Distinguish among skeletal, smooth, and cardiac muscle tissues in terms of location, structure, and function. [p.566]
14. Muscle tissues contain specialized cells that can _____. [p.566]
15. Of all tissues, _____ tissue exerts the greatest control over the body's responsiveness to changing conditions. [p.567]
16. _____ are excitable cells, the communication units of most nervous systems. [p.567]
17. Discuss the implications of lab-grown epidermis, designer organs, and research to put together packages of cells capable of producing specific life-saving substances that are absent in patients who suffer from genetic disorders or chronic diseases. [p.567]

18. List each of the eleven principal organ systems in humans, and list the main task of each. [pp.568–569]
19. List the major cavities in the human body and the organs they house. [p.568]
20. Name the directional terms and planes of symmetry used for description of the human body. [p.569]
21. _____ gives rise to the skin's outer layer and tissues of the nervous system; _____ gives rise to muscles, bones, and most of the circulatory, reproductive, and urinary systems; _____ gives rise to the lining of the digestive tract and to organs derived from it. [p.569]
22. Describe the ways by which extracellular fluid helps cells survive. [p.570]
23. Distinguish between interstitial fluid and plasma. [p.570]
24. Describe the relationships among receptors, integrators, and effectors in a negative feedback system. [p.570]
25. Prepare a diagram that illustrates the components necessary for negative feedback at the organ level. [p.570]
26. Explain the mechanisms involved in a positive feedback mechanism. [p.571]
27. Tell why negative and positive feedback mechanisms are viewed as homeostatic controls. [pp.570–571]

Interpreting and Applying Key Concepts

Explain why, of all places in the body, marrow is located on the interior of long bones. Explain why your bones are remodeled after you reach maturity. Why does your body not keep the same mature skeleton throughout life?

34

INFORMATION FLOW AND THE NEURON

Interactive Exercises

TORNADO! [pp.574–575]

34.1. NEURONS—THE COMMUNICATION SPECIALISTS [pp.576–577]
34.2. A CLOSER LOOK AT ACTION POTENTIALS [pp.578–579]

Selected Words: input zones [p.576], *conducting* zone [p.576], *trigger* zone [p.576], *output* zones [p.576], voltage difference [p.576], concentration gradient [p.577], *graded* signal [p.578], *local* signal [p.578], *all-or-nothing* event [p.578], charge reversal [p.578], *Loligo* "giant" axons [p.579], self-propagating [p.579]

Boldfaced, Page-Referenced Terms

[p.575] neurons _____

_____

[p.575] sensory neuron _____

_____

[p.575] stimulus _____

_____

[p.575] interneurons _____

_____

[p.575] motor neuron _____

_____

[p.575] neuroglia _____

_____

[p.576] dendrites _____

_____

[p.576] axon _____

_____

[p.576] resting membrane potential _____

_____

[p.576] action potential _____

_____

[p.577] sodium–potassium pumps _____

_____

[p.578] threshold level _____

_____

[p.578] positive feedback _____

_____

Fill-in-the-Blanks

Nerve cells that conduct messages are called (1) _____ [p.575]. (2) _____ [p.575] cells, which support and nurture the activities of neurons, make up more than half the volume of the nervous system. (3) _____ [p.575] neurons are receptors for environmental stimuli, (4) _____ [p.575] connect different neurons in the central nervous system, and (5) _____ [p.575] neurons are linked with muscles or glands. All neurons have a (6) _____ _____ [p.576] that contains the nucleus and the metabolic means to carry out protein synthesis. (7) _____ [p.576] are short, slender extensions of (6), and together these two neuronal parts are the neurons' "input zone" for receiving (8) _____ [p.576]. The (9) _____ [p.576] is a single long, cylindrical extension of the (10) _____ _____ [p.576]; in motor neurons, the (11) _____ [p.576] has finely branched (12) _____ [p.576] that terminate on muscle or gland cells and are "output zones," where messages are sent on to other cells.

A neuron at rest establishes unequal electric charges across its plasma membrane, and a (13) _____ _____ [p.576] is maintained. Another name for (13) is the (14) _____ _____ _____ [p.576]; it represents a tendency for activity to happen along the membrane. An electrical gradient also exists across the neuronal membrane; compared with the outside, the inside of a neuron at rest has an overall (15) _____ [p.576] charge. For many neurons in most animals, the difference in charge across the neuronal membrane is about 70 (16) _____ [p.576]. Weak disturbances of the neuronal membrane might set off only slight changes across a small patch, but strong disturbances can cause a(n) (17) _____ _____ [p.576], which is an abrupt, short-lived reversal in the polarity of charge across the plasma membrane of the neuron. How is the resting membrane potential established, and what restores it between action potentials? The concentrations of (18) _____ [p.576] ions (K^+), sodium ions (19) (___$^+$) [p.576], and other charged substances are not the same on the inside and outside of the neuronal membrane.

(20) _____ [p.577] proteins that span the membrane affect the diffusion of specific types of ions across it. (21) _____ [p.577] proteins that span the membrane pump sodium and potassium ions against their concentration gradients across it by using energy stored in ATP. A neuronal membrane has many more positively charged (22) _____ [p.577] ions inside than out and many more positively charged (23) _____ [p.577] ions outside than inside. There are about (24) _____ [p.577] times more potassium ions on the cytoplasmic side as outside, and there are about (25) _____ [p.577] times more sodium ions outside as inside. These ions can cross the membrane only by traveling along passages through (26) _____ [p.577] proteins. Some channel proteins leak ions through them all the time; others have (27) _____ [p.577] that open only when stimulated. Transport proteins called (28) _____-_____ _____ [p.577] counter the leakage of ions across the neuronal membrane and maintain the resting membrane potential.

Labeling

Identify the parts of the following illustration of a neuron.

29. _____ _____ [p.576]

30. _____ _____ [p.577]

31. _____ [p.577]

32. _____-_____ _____ [p.577]

33. _____ _____ [p.577]

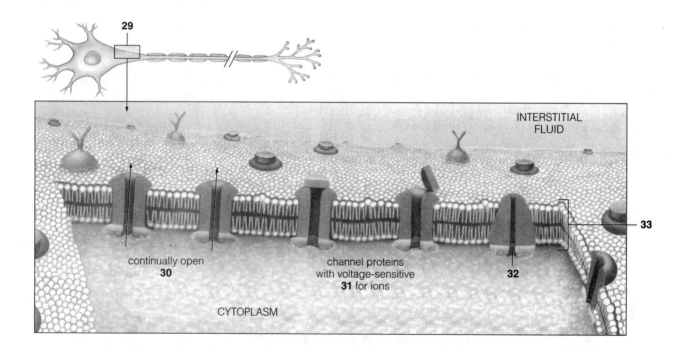

Fill-in-the-Blanks

In all neurons, stimulation at an input zone produces (34) _____ [p.578] signals that do not spread very far (half a millimeter or less). (35) _____ [p.578] means that signals can vary in magnitude—small or large—depending on the intensity and (36) _____ [p.578] of the stimulus. When stimulation is intense or prolonged, graded signals can spread into an adjacent (37) _____ _____ [p.578] of the membrane—the site where action potentials can be initiated. For a fraction of a second, the cytoplasmic side of a bit of membrane becomes positive with respect to the outside. The (38) _____ [p.578] that travels along the neuronal membrane is nothing more than short-lived changes in the membrane potential.

Once an action potential has been achieved, it is an (39) _____-_____-_____ [p.578] event; its amplitude will not change even if the strength of the stimulus changes. The minimum change in membrane potential needed to achieve an action potential is the (40) _____ [p.578] value. Each action potential is followed by a time of insensitivity to stimulation.

Labeling

Identify the aspects of the action potential waveform illustrated at the right. [All from p.579]

41. _____ _____

42. _____

43. _____ _____ _____

44. _____

45. _____

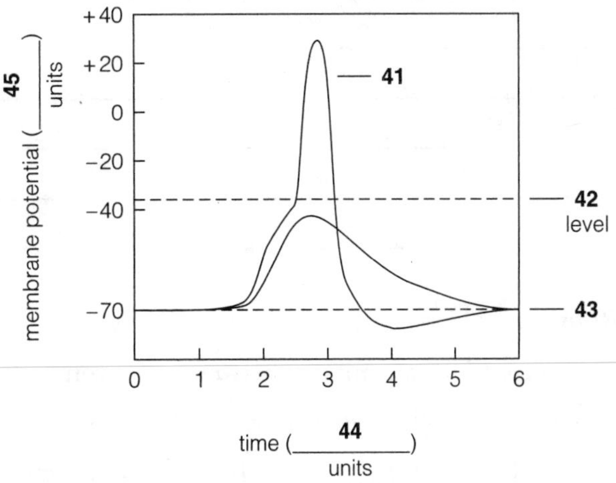

34.3. CHEMICAL SYNAPSES [pp.580–581]
34.4. PATHS OF INFORMATION FLOW [pp.582–583]
34.5. *Focus of Health:* SKEWED INFORMATION FLOW [p.584]

Selected Words: *presynaptic cell* [p.580], *postsynaptic cell* [p.580], *excitatory* effect [p.580], *inhibitory* effect [p.580], serotonin [p.580], norepinephrine [p.580], dopamine [p.580], GABA (gamma aminobutyric acid) [p.580], substance P [p.580], endorphins [p.580], *depolarizing* effect [p.581], *hyperpolarizing* effect [p.581], summation [p.581], *divergent, convergent,* and *reverberating* circuits [p.582], multiple sclerosis [p.582], *Clostridium botulinum* [p.584], botulism [p.584], *Clostridium tetani* [p.584], *tetanus* [p.584]

Boldfaced, Page-Referenced Terms

[p.580] neurotransmitters _____

[p.580] chemical synapses _____

_____

[p.580] acetylcholine (ACh) _____

_____

[p.580] neuromodulators _____

_____

[p.581] EPSPs (excitatory postsynaptic potentials) _____

_____

[p.581] IPSPs (inhibitory postsynaptic potentials) _____

_____

[p.581] synaptic integration _____

_____

[p.582] nerves _____

_____

[p.582] myelin sheath _____

_____

[p.582] Schwann cells _____

_____

[p.582] reflexes _____

_____

Fill-in-the-Blanks

The junction specialized for transmission between a neuron and another cell is called a (1) _____
_____ [p.580]. Usually, the signal being sent to the receiving cell is carried by chemical messengers
called (2) _____ [p.580]. (3) _____ [p.580] is an example of this type of chemical messenger that
diffuses across the synaptic cleft, combines with protein receptor molecules on the muscle cell membrane,
and soon thereafter is rapidly broken down by enzymes. (4) _____ [p.580] acts on brain cells that
govern sleeping, sensory perception, temperature regulation, and emotional states. GABA (gamma
aminobutyric acid) is the most common (5) _____ [p.580] signal in the brain. (6) _____ [p.580] are
neuromodulators that inhibit perceptions of pain and may have roles in memory and learning, emotional
depression, and sexual behavior. At an (7) _____ [p.581] synapse, the membrane potential is driven
toward the threshold value and increases the likelihood that an action potential will occur. At an
(8) _____ [p.581] synapse, the membrane potential is driven away from the threshold value, and the
receiving neuron is less likely to achieve an action potential. A specific neurotransmitter can have either
excitatory or inhibitory effects depending on which type of protein channel it opens up in the (9) _____
[p.581] membrane.

(10) _____ _____ [p.581] at the cellular level is the moment-by-moment tallying of all excitatory and inhibitory signals acting on a neuron. Incoming information is (11) _____ [p.581] by cell bodies, and the charge differences across the membranes are either enhanced or inhibited. An (12) _____ [p.581] postsynaptic potential (EPSP) brings the membrane closer to threshold and has a depolarizing effect. An inhibitory postsynaptic potential (IPSP) drives the membrane away from threshold and either has a (13) _____ [p.581] effect or maintains the membrane at its resting level.

Some narrow-diameter neurons are wrapped in lipid-rich (14) _____ [p.582] produced by specialized neuroglial cells called Schwann cells; each of these is separated from the next by a(n) (15) _____ _____ [p.582]—a small gap where the axon is exposed to extracellular fluid. An action potential jumps from one node to the next in line and in the largest myelinated axon, signals travel (16) _____ [p.582] meters per second.

Labeling

Label the parts of the nerve shown in the accompanying illustrations. [All from p.582]

17. _____
18. _____ _____
19. _____ _____
20. _____
21. _____ _____
22. _____ _____

a nerve fasicle (many **20** bundled in connective tissue)

outer wrapping of the nerve

Labeling

Label the parts of neurons and types of neurons in the illustration. [All from p.583]

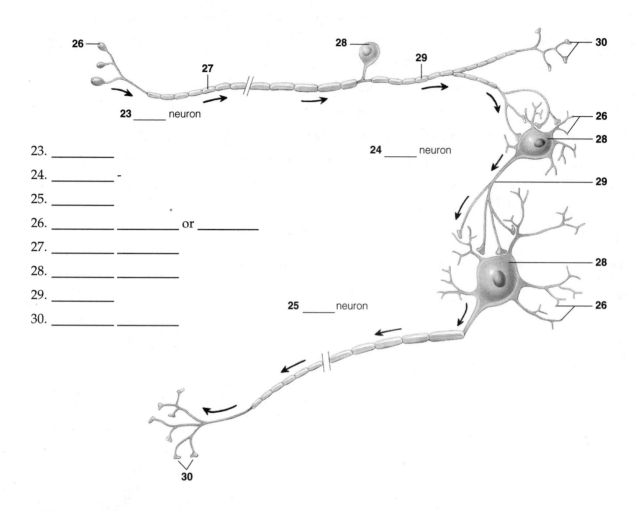

23. _____
24. _____ -
25. _____
26. _____ _____ or _____
27. _____ _____
28. _____ _____
29. _____
30. _____ _____

Fill-in-the-Blanks

A (31) _____ [p.582] is an involuntary sequence of events elicited by a stimulus. During a (32) _____ _____ [p.583], a muscle contracts involuntarily whenever conditions cause a stretch in length; many of these help you maintain an upright posture despite small shifts in balance. Located within skeletal muscles are length-sensitive organs called (33) _____ _____ [p.583], which generate action potentials when stretched beyond a critical point; these potentials are conducted rapidly to the (34) _____ _____ [p.583], where they are communicated to motor neurons leading right back to the muscle that was stretched. When action potentials reach the end of a motor neuron, they cause (35) _____ [p.580] to be released that serve as chemical signals to adjacent muscle cells. Muscles (36) _____ [p.580] in response to the signals.

Imbalances can occur at chemical synapses; a neurotoxin produced by *Clostridium tetani* blocks the release of inhibitory (37) _____ [p.584] on motor neurons, which may cause tetanus—a prolonged, spastic paralysis that can lead to death.

Matching

Match the following choices with the correct number in the diagram. Two of the numbers match with two lettered choices. [All from p.583]

38. ___
39. ___
40. ___
41. ___
42. ___
43. ___
44. ___
45. ___
46. ___

A. Response
B. Action potentials generated in motor neuron and propagated along its axon toward muscle
C. Motor neuron synapses with muscle cells
D. Muscle cells contract
E. Local signals in receptor endings of sensory neuron
F. Muscle spindle stretches
G. Action potentials generated in all muscle cells innervated by motor neuron
H. Stimulus
I. Axon endings synapse with motor neuron
J. Spinal cord
K. Action potential propagated along sensory neuron toward spinal cord

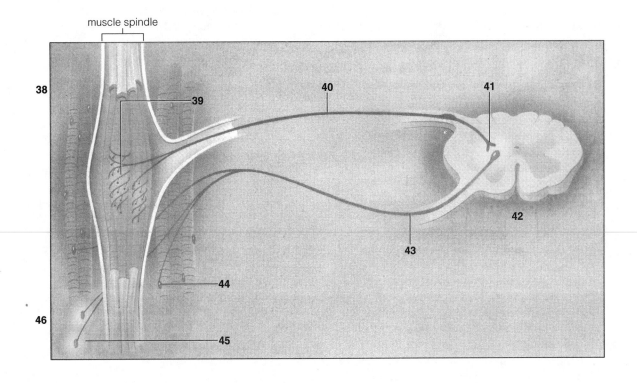

Self-Quiz

___ 1. Which of the following is *not* true of an action potential? [p.579]
   a. It is a short-range message that can vary in size.
   b. It is an all-or-none brief reversal in membrane potential.
   c. It doesn't decay with distance.
   d. It is self-propagating.

___ 2. The conducting zone of a neuron is the _____. [p.576]
   a. axon
   b. axonal terminals
   c. cell body
   d. dendrite

___ 3. The output zone of a neuron is the _____. [p.576]
   a. axon
   b. axonal endings
   c. cell body
   d. dendrite

___ 4. _____ carry out the specified responses to the stimuli. [p.575]
   a. Neuromodulators
   b. Neuroglia
   c. Muscles or glands
   d. Schwann cells

___ 5. An action potential is brought about by _____. [p.576]
   a. a sudden membrane impermeability
   b. the movement of negatively charged proteins through the neuronal membrane
   c. the movement of lipoproteins to the outer membrane
   d. a local change in membrane permeability caused by a greater-than-threshold stimulus

___ 6. The resting membrane potential _____. [p.576]
   a. exists as long as a voltage difference sufficient to do work exists across a membrane
   b. occurs because there are more potassium ions outside the neuronal membrane than there are inside
   c. occurs because of the unique distribution of receptor proteins located on the dendrite exterior
   d. is brought about by a local change in membrane permeability caused by a greater-than-threshold stimulus

___ 7. The phrase "all or none" used in conjunction with discussion about an action potential means that _____. [p.578]
   a. a resting membrane potential has been received by the cell
   b. an impulse does not diminish or dissipate as it travels away from the trigger zone
   c. the membrane either achieves total equilibrium or remains as far from equilibrium as possible
   d. propagation along the neuron is much faster than in other neurons

___ 8. Endorphins _____. [p.580]
   a. are neuromodulators
   b. block perceptions of pain
   c. may play a role in causing emotional depression
   d. are involved in all of the above roles

___ 9. An action potential passes from neuron to neuron across a synaptic cleft by _____. [p.580]
   a. myelin bridges
   b. the resting membrane potential
   c. neurotransmitter substances
   d. neuromodulator substances

___10. _____ are responsible for integration in the nervous system. [p.575]
   a. Interneurons
   b. Schwann cells
   c. Motor neurons
   d. Sensory neurons

Chapter Objectives/Review Questions

1. Draw a neuron and label it according to its three general zones, its specific structures, and the specific function(s) of each structure. [p.576]
2. Define resting membrane potential; explain what establishes it and how it is used by the cell neuron. [p.576]
3. Define action potential by stating its three main characteristics. [p.576]
4. Define sodium–potassium pump and state how it helps maintain the resting membrane potential. [pp.576–577]
5. Describe the distribution of the invisible array of large proteins, ions, and other molecules in a neuron, both at rest and as a neuron experiences a change in potential. [pp.576–579]
6. Explain the chemical basis of the action potential. Look at Figure 34.6 in your text and determine which part of the curve represents the following:
   a. the point at which the stimulus was applied;
   b. the events prior to achievement of the threshold value;
   c. the opening of the ion gates and the diffusing of the ions;
   d. the change from net negative charge inside the neuron to net positive charge and back again to net negative charge; and
   e. the active transport of sodium ions out of and potassium ions into the neuron. [pp.576–579]
7. Explain how graded signals differ from action potentials. [p.578]
8. Define period of insensitivity and state what causes it. [p.579]
9. Understand how a nerve impulse is received by a neuron, conducted along a neuron, and transmitted across a synapse to a neighboring neuron, muscle, or gland. [pp.580–583]
10. Outline some of the ways by which information flow is regulated and integrated in the human body. [pp.580–584]
11. Distinguish the way excitatory synapses function from the way inhibitory synapses function. [p.581]
12. Define Schwann cell, unsheathed nodes, and myelin sheath and explain how each helps narrow-diameter neurons conduct nerve impulses quickly. [p.582]
13. Explain what a reflex is by drawing and labeling a diagram and telling how it functions. [pp.582–583]
14. Explain what the stretch reflex is and tell how it helps an animal survive. [p.583]

Integrating and Applying Key Concepts

What do you think might happen to human behavior if inhibitory postsynaptic potentials did not exist and if the threshold stimulus necessary to provoke an EPSP were much higher?

35

INTEGRATION AND CONTROL: NERVOUS SYSTEMS

Interactive Exercises

Why Crack the System? [pp.586–587]

35.1. INVERTEBRATE NERVOUS SYSTEMS [pp.588–589]

Selected Words: *drug* [p.586], *dealer* [p.586], *crack* [p.586], radial symmetry [p.588], *bilateral* symmetry [p.588], plexuses [p.589]

Boldfaced, Page-Referenced Terms

[p.588] nervous system _____

_____

[p.588] nerve net _____

_____

[p.588] reflex pathways _____

_____

[p.588] nerve _____

_____

[p.589] ganglia (singular, ganglion) _____

_____

Fill-in-the-Blanks

All animals except sponges have some type of (1) _____ [p.588] system in which nerve cells, such as (2) _____ [p.588], are oriented in signal-conducting and information-processing pathways. At the minimum, the cells making up the communication lines receive information about changing conditions outside and inside the (3) _____ [p.588], then elicit suitable responses from muscle and gland cells.

The first animals evolved in the (4) _____ [p.588]. It is in this environment that we still find animals with the simplest (5) _____ [p.588] systems. They are sea anemones, jellyfishes, and other cnidarians. These invertebrates display (6) _____ [p.588] symmetry.

Such animals have a (7) _____ [p.588] net, a loose mesh of nerve cells intimately associated with epithelial tissue. The nerve cells interact with (8) _____ [p.588] cells and contractile cells along reflex pathways in the same epithelial tissue. In (9) _____ [p.588] pathways, sensory stimulation triggers simple, stereotyped movements.

The nerve net itself extends through the animal's (10) _____ [p.588], but information flow through it is not highly focused. It simply commands the body wall to slowly contract and expand or move tentacles through the water.

Flatworms are the simplest animals having a (11) _____ [p.588] nervous system. There are equivalent body parts on the left and right sides of the body's (12) _____ [p.588] plane. The ladderlike nervous system of the flatworm has (13) (number) _____ [p.588] cordlike nerves running longitudinally through the body, with many side branches. A (14) _____ [p.588] is like a cable in which sensory axons, motor axons, or both are bundled together in a sheath of connective tissue.

Also, in the head end of some flatworms are two (15) _____ [p.589], each of which is a local cluster of nerve cell bodies that acts as a local integrating center. Did (16) _____ [p.589] nervous systems evolve from nerve nets? Maybe. In nearly all animals more complex than flatworms, we find local nerve nets or (17) _____ [p.589], such as the one in your intestinal wall. Chance mutations in ancient planulas may have favored a concentration of (18) _____ [p.589] cells in the leading end, not the trailing end. This allowed for more rapid, effective responses to varied stimuli. Natural (19) _____ [p.589] must have favored a concentration of sensory cells at the body's leading end. (20) _____ [p.589] and bilateral symmetry may have started this way.

35.2. VERTEBRATE NERVOUS SYSTEMS—AN OVERVIEW [pp.590–591]

Selected Words: afferent [p.591], efferent [p.591]

Boldfaced, Page-Referenced Terms

[p.590] neural tube _____

_____

[p.591] central nervous system _____

_____

[p.591] peripheral nervous system _____

_____

[p.591] white matter _____

_____

[p.591] gray matter _____

_____

[p.591] neuroglia _____

_____

Complete the Table

1. A comparison of the brains of some existing vertebrates (Fig. 35.4, p.590) suggests an evolutionary trend toward an expanded, more complex brain. Complete the following table by entering the words *forebrain, midbrain,* and *hindbrain* in the correct blanks to understand how the anterior end of the dorsal, hollow nerve cord expanded into functionally distinct regions, which also increased in complexity in certain lineages.

Brain Region	Basic Functions
a. [p.590]	Coordinates reflex responses to sight, sounds
b. [p.590]	Receives, integrates sensory information from nose, eyes, and ears; in land-dwelling vertebrates, contains the highest integrating centers
c. [p.590]	Provides reflex control of respiration, blood circulation, other basic tasks; in complex vertebrates, coordination of sensory input, motor dexterity, and possibly mental dexterity

Matching

Choose the most appropriate answer for each term.

2. ___neural tube [p.590]
3. ___central nervous system [p.591]
4. ___peripheral nervous system [p.591]
5. ___tracts [p.591]
6. ___white matter tracts [p.591]
7. ___gray matter tracts [p.591]
8. ___neuroglial cells [p.591]
9. ___afferent [p.591]
10. ___efferent [p.591]

A. Nerves carrying motor output away from the central nervous system to muscles and glands
B. The nerve cord that persists in all vertebrate embryos
C. Remember, protect, or structurally and functionally support neurons
D. The spinal cord and brain
E. Nerves carrying sensory input to the central nervous system
F. Consists of unmyelinated axons, dendrites, and nerve cell bodies and neuroglial cells
G. Consists mainly of nerves that thread through the rest of the body and carry signals into and out of the central nervous system
H. The communication lines inside the brain and spinal cord
I. Contains axons with glistening white myelin sheaths and specializes in rapid signal transmission

35.3. THE MAJOR EXPRESSWAYS [pp.592–593]

Selected Words: *spinal* nerves [p.592], *cranial* nerves [p.592], rebound effect [p.593], *meningitis* [p.593]

Boldfaced, Page-Referenced Terms

[p.592] somatic nerves _____

_____

[p.592] autonomic nerves _____

_____

[p.592] parasympathetic nerves _____

[p.593] sympathetic nerves _____

[p.593] fight–flight response _____

[p.593] spinal cord _____

[p.593] meninges _____

_____

Labeling

Label each numbered part of the accompanying illustration.

1. _____ [p.592]
2. _____ [p.592]
3. _____ [p.592]
4. _____ [p.592]
5. _____ _____ [p.592]

6. _____ [p.592]
7. _____ [p.592]
8. _____ [p.592]
9. _____ [p.592]

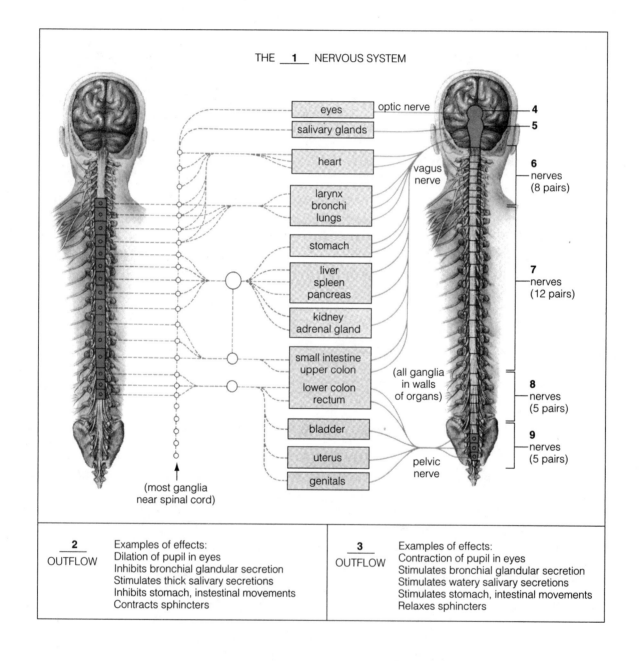

THE __1__ NERVOUS SYSTEM

eyes — optic nerve — 4
salivary glands — 5
heart
larynx bronchi lungs
stomach
liver spleen pancreas
kidney adrenal gland
small intestine upper colon
lower colon rectum
bladder
uterus
genitals

vagus nerve

(all ganglia in walls of organs)

pelvic nerve

6 — nerves (8 pairs)
7 — nerves (12 pairs)
8 — nerves (5 pairs)
9 — nerves (5 pairs)

(most ganglia near spinal cord)

__2__ OUTFLOW

Examples of effects:
Dilation of pupil in eyes
Inhibits bronchial glandular secretion
Stimulates thick salivary secretions
Inhibits stomach, instestinal movements
Contracts sphincters

__3__ OUTFLOW

Examples of effects:
Contraction of pupil in eyes
Stimulates bronchial glandular secretion
Stimulates watery salivary secretions
Stimulates stomach, intestinal movements
Relaxes sphincters

Labeling

Identify the numbered parts of the accompanying illustrations.

10. _____ _____ [p.593]

11. _____ [p.593]

12. _____ [p.593]

13. _____ [p.593]

14. _____ _____ [p.593]

15. _____ [p.593]

16. _____ _____ [p.593]

17. _____ _____ [p.593]

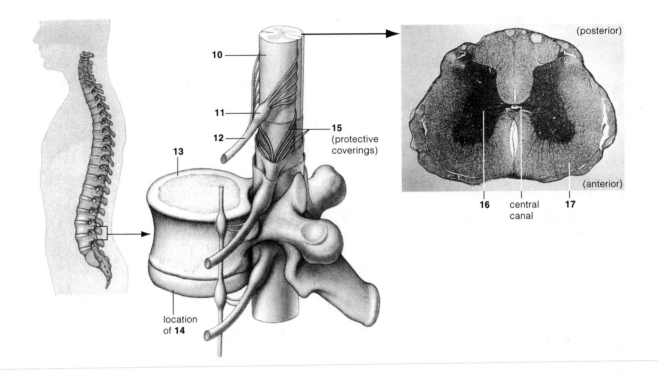

Choice

For questions 18–34, choose from the following:

      a. peripheral—somatic nerves   b. peripheral—autonomic sympathetic nerves
      c. peripheral—autonomic parasympathetic nerves   d. spinal cord nerves

18. ___ Dominate when the body is not receiving much outside stimulation [p.592]

19. ___ Carry signals about moving your head, trunk, and limbs [p.592]

20. ___ Dominate in times of sharpened awareness [p.593]

21. ___ Can be attacked by meningitis [p.593]

22. ___ Sensory axons inside these nerves deliver information from receptors in the skin, skeletal muscles, and tendons to the central nervous system [p.592]

23. ___ Tend to slow down the body overall and divert energy to basic "housekeeping" tasks, such as digestion [pp.592–593]

24. ___ The meninges, three tough, tubelike coverings, are part of the structure [p.593]

25. ___ A vital expressway for signals between the peripheral nervous system and the brain [p.593]

26. ___ The fight–flight response [p.593]

27. ___ Threads through a canal formed by bones of the vertebral column [p.593]

28. ___ Commands one's heart to beat faster [p.593]

29. ___ Tend to shelve housekeeping tasks [p.593]

30. ___ Their motor axons deliver commands from the brain and spinal cord to the body's skeletal muscles [p.592]

31. ___ Signals cause the release of epinephrine [p.593]

32. ___ Some of its interneurons exert direct control over certain reflex pathways [p.593]

33. ___ Commands your heart to beat a little slower [p.593]

34. ___ Gray matter that plays an important role in controlling reflexes for limb movement and organ activity [p.593]

35.4. FUNCTIONAL DIVISIONS OF THE VERTEBRATE BRAIN [pp.594–595]

Boldfaced, Page-Referenced Terms

[p.594] brain _____

_____

[p.594] brain stem _____

_____

[p.594] medulla oblongata _____

_____

[p.594] cerebellum _____

_____

[p.594] pons _____

_____

[p.594] tectum _____

_____

[p.595] cerebrum _____

_____

[p.595] thalamus _____

_____

[p.595] hypothalamus _____

_____

[p.595] reticular formation _____

_____

[p.595] cerebrospinal fluid _____

_____

[p.595] blood–brain barrier _____

_____

Fill-in-the-Blanks

The forebrain, midbrain, and hindbrain of vertebrates form three successive portions of the (1) _____ [p.594] tube. The nervous tissue that evolved first in all three regions is called the (2) _____ [p.594] stem. This area is still identifiable in the (3) _____ [p.594] brain, and it still contains many simple, basic reflex centers. Over evolutionary time, expanded layers of (4) _____ [p.594] matter developed from the brain stem. The more recent additions have been correlated with an increasing reliance on three major (5) _____ [p.594] organs: the nose, ears, and eyes. The forebrain and possibly parts of the (6) _____ [p.594] (part of the hindbrain) contain the newest additions of (7) _____ [p.594] matter.

The brain is not tidily subdivided into three main regions as it might seem. An evolutionarily ancient mesh of (8) _____ [p.595] still extends from the uppermost part of the spinal cord, on through the brain stem, and on into higher integrative centers of the cerebral cortex. This major network is the (9) _____ [p.595] formation. It persists as a low-level pathway to motor centers of the medulla oblongata and spinal cord. It can also activate centers in the (10) _____ [p.595] cortex and thereby help govern the activities of the nervous system as a whole.

The hollow (11) _____ [p.595] tube that first develops in vertebrate embryos persists in the adults, as a continuous system of fluid-filled cavities and canals. Within this system is the (12) _____ fluid [p.595], a clear extracellular fluid that cushions the tissues of the brain and spinal cord from sudden, jarring movements.

A mechanism called the (13) _____–_____ [p.595] barrier protects the brain and spinal cord by exerting some control over which solutes enter the cerebrospinal fluid. No other portion of extracellular fluid has solute concentrations maintained within such (14) _____ [p.595] limits. The barrier operates at the (15) _____ [p.595] membrane of certain cells.

Matching

Choose the most appropriate answer for each term.

16. ____medulla oblongata [p.594]

17. ____cerebellum [p.594]

18. ____pons [p.594]

19. ____tectum [p.594]

20. ____cerebrum [p.595]

21. ____thalamus [p.595]

22. ____hypothalamus [p.595]

23. ____olfactory lobes [pp.594–595]

A. Midbrain: roof of the midbrain, a more recent layering of gray matter; in mammals this structure became a reflex center that quickly relays sensory signals to those higher integrating centers

B. Hindbrain: integrates sensory input from the eyes, ears, and muscle spindles with motor signals from the forebrain; helps control motor dexterity and more recent expansions may be crucial in language and some other forms of mental dexterity

C. Forebrain: dealing mainly with odors from predators, prey, and potential mates

D. Hindbrain: contains reflex centers for vital tasks, such as respiration and blood circulation; also coordinates motor responses with certain complex reflexes, such as coughing; influences other brain regions, helping you sleep or wake up

E. Forebrain: a pair of outgrowths from the brain stem where olfactory input and responses to it became integrated; these outgrowths expanded greatly, especially during and after the vertebrate invasion of land

F. Hindbrain: bands of many axons extend from both sides of the cerebellum to this area; a major traffic center for information passing between the cerebellum and the higher integrating centers of the forebrain

G. Forebrain: evolved into the premier center for homeostatic control over the internal environment; became central to behaviors related to internal organ activities, such as thirst, hunger, and sex, and to emotional expression, such as sweating with fear

H. Forebrain: evolved as coordinating center for sensory input and as a relay station for signals to the cerebrum

35.5. A CLOSER LOOK AT THE HUMAN CEREBRUM [pp.596–597]

35.6. *Focus on Science:* SPERRY'S SPLIT-BRAIN EXPERIMENTS [p.598]

Selected Words: neuroglial cells [p.596], *motor* areas [p.596], *sensory* areas [p.596], *association* areas [p.596], Broca's area [p.596], *epilepsy* [p.598]

Boldfaced, Page-Referenced Terms

[p.596] cerebral hemispheres _____

_____

[p.596] cerebral cortex _____

_____

[p.597] limbic system _____

_____

Labeling

Identify each numbered part of the accompanying illustration.

1. _____ [p.596]
2. _____ [p.596]
3. _____ [p.596]
4. _____ [p.596]
5. _____ _____ [p.596]

6. _____ [p.596]
7. _____ [p.596]
8. _____ [p.596]
9. _____ [p.596]
10. _____ _____ [p.596]

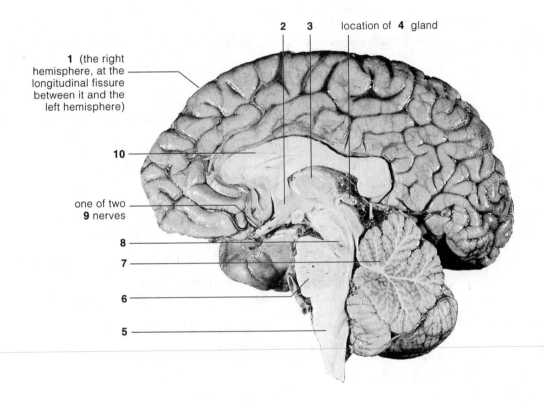

2   3   location of **4** gland

1 (the right
hemisphere, at the
longitudinal fissure
between it and the
left hemisphere)

10

one of two
9 nerves

8

7

6

5

Complete the Table

11. Complete the following table by identifying in the left column the area of the brain whose functions are described in the right column. Choose from cerebral cortex, occipital lobe; cerebral cortex, temporal lobe; cerebral cortex, parietal lobe; cerebral cortex, frontal lobe; left cerebral hemisphere; right cerebral hemisphere; cerebral cortex; limbic system; and corpus callosum.

Brain Area	Functions
a. [p.597]	Located inside the cerebral hemispheres; governs emotions and has roles in memory; distantly related to olfactory lobes and still deals with the sense of smell
b. [p.596]	Perception of sounds and of odors arises in primary cortical areas located here
c. [p.596]	Deals more with visual–spatial relationships, music, and other creative enterprises
d. [p.596]	Primary motor cortex controls coordinated movements of skeletal muscles; thumb, finger, and tongue muscles get much of the area's attention; Broca's area and the frontal eye field are located here
e. [p.596]	A transverse band of nerve tracts; carries signals back and forth between the hemispheres and coordinates their functioning
f. [p.596]	Located at the rear of this lobe; primary visual cortex, which receives sensory inputs from the eyes
g. [p.596]	Deals mainly with speech, analytical skills, and mathematics; dominates the right hemisphere in most people
h. [p.596]	The body is spatially mapped out in the primary somatosensory cortex; this area is the main receiving center for sensory input from the skin and joints; also deals with taste perception
i. [p.597]	A thin outer layer of gray matter on the left and right cerebral hemispheres

Short Answer

12. In an effort to relieve the frequent seizures of severe epilepsy, neural surgeon Roger Sperry cut the neural bridge of the corpus callosum of several of these patients. The seizures did subside in frequency and intensity. Summarize the subsequent findings of Sperry regarding the function of the corpus callosum. [p.598]

_____

_____

_____

_____

_____

35.7. MEMORY [p.599]

Selected Words: *short-term* storage [p.599], *long-term* storage [p.599], *facts* [p.599], *skills* [p.599], *amnesia* [p.599], *Parkinson's disease* [p.599], *Alzheimer's disease* [p.599]

Boldfaced, Page-Referenced Terms

[p.599] memory _____

_____

Labeling

Identify each numbered part of the accompanying illustration.

1. _____ [p.599]
2. _____ [p.599]
3. _____ [p.599]
4. _____ [p.599]
5. _____ [p.599]
6. _____ [p.599]
7. _____ _____ [p.599]
8. _____ _____ [p.599]
9. _____ and _____ [p.599]
10. _____ _____ [p.599]
11. _____ _____ [p.599]

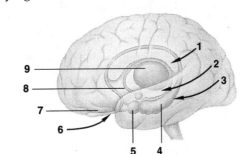

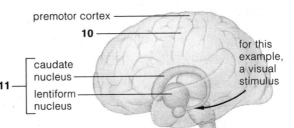

Fill-in-the-Blanks

(12) _____ [p.599] is the capacity of an individual's brain to store and retrieve information about past sensory experience. (13) _____ [p.599], and adaptive modifications of our behavior would be impossible without it. Information is stored in stages. (14) _____-_____ [p.599] storage is a stage of neural excitation that lasts a few seconds to a few hours; it is limited to a few bits of sensory information. In (15) _____-_____ [p.599] storage, seemingly unlimited amounts of information get tucked away more or less permanently.

Only some (16) _____ [p.599] input is chosen for transfer to brain structures involved in short-term memory. If information is (17) _____ [p.599], it is forgotten; otherwise, it is consolidated with banks of information in long-term storage structures.

The human brain processes facts separately from (18) _____ [p.599]. Explicit bits of information are soon forgotten or filed away in (19) _____-_____ [p.599] storage, along with the circumstance in which they were learned. (20) _____ [p.599] are gained by practicing specific motor activities and are best recalled by actually performing the motor activity involved.

Separate memory circuits handle different kinds of (21) _____ [p.599]. A circuit leading to fact memory starts with inputs at the sensory cortex that flow to the (22) _____ [p.599] and (23) _____ [p.599], structures in the limbic system. The (24) _____ [p.599] is the gatekeeper and connects the sensory cortex with parts of the thalamus and hypothalamus that govern emotional states. The (25) _____ [p.599] mediates learning and spatial relations. Information flows to the prefrontal cortex, where multiple banks of (26) _____ [p.599] memories are retrieved and used to stimulate or inhibit other parts of the brain. New input also flows to (27) _____ [p.599] ganglia, which send it back to the cortex in a feedback loop that reinforces the input until it can be consolidated in (28) _____-_____ [p.599] storage.

(29) _____ [p.599] memory also begins at the (30) _____ [p.599] cortex, but this circuit routes sensory input to the (31) _____ _____ [p.599], which promotes motor responses. The circuit for motor skills extends to the (32) _____ [p.599], the brain region that coordinates motor activity.

(33) _____ [p.599] is a loss of memory, the severity of which depends on whether the hippocampus, amygdala, or both are damaged, as by a severe head blow; this does not affect capacity to learn new (34) _____ [p.599]. By contrast, basal ganglia are destroyed and learning ability is lost during (35) _____ [p.599] disease, yet skill memory is retained. With a usual onset during later life, (36) _____ [p.599] disease is linked to structural changes in the cerebral cortex and (37) _____ [p.599]. Affected people often can remember long-standing (38) _____ [p.599] but they have difficulty remembering what has just happened to them. In time they become confused, depressed, and unable to complete a train of thought.

35.8. STATES OF CONSCIOUSNESS [p.600]

35.9. *Focus on Health:* DRUGGING THE BRAIN [pp.600–601]

35.10. *Focus on Science:* THE NOT-QUITE-COMPLETE TEEN BRAIN [pp.602–603]

Selected Words: *alpha rhythm* [p.600], *slow-wave sleep* pattern [p.600], *REM sleep* [p.600], rapid eye movements [p.600], *EEG arousal* [p.600], serotonin [p.600], *caffeine* [p.600], *nicotine* [p.600], *cocaine* [p.600], *amphetamines* [p.601], *crank* [p.601], *alcohol* [p.601], *cirrhosis* [p.601], analgesics [p.601], *codeine* [p.601], *heroin* [p.601], *LSD* [p.601], *marijuana* [p.601], *Cannabis* [p.601], *MRI* [p.602]

Boldfaced, Page-Referenced Terms

[p.600] consciousness _____

_____

[p.600] drug addiction _____

_____

Matching

Choose the most appropriate answer for each term.

1. ___consciousness [p.600]
2. ___EEGs [p.600]
3. ___alpha rhythm [p.600]
4. ___slow-wave sleep [p.600]
5. ___REM sleep [p.600]
6. ___EEG arousal [p.600]

A. Briefly punctuates slow-wave sleep; accompanied by REM, irregular breathing, faster heartbeat, and twitching fingers; experience vivid dreams
B. The prominent EEG wave pattern for someone who is relaxed, with eyes closed
C. The spectrum includes sleeping and aroused states, during which neural chattering shows up as wavelike patterns in EEGs
D. A transition that occurs when conscious effort is made to focus on external stimuli or even on one's own thoughts
E. Electrical recordings of the frequency and strength of membrane potentials at the surface of the brain
F. A pattern that dominates when sensory input is low and the mind is more or less idling; subjects seemed to be mulling over recent, ordinary events

Choice

For questions 7–15, choose from the following:

    a. stimulants   b. depressants, hypnotics   c. analgesics   d. psychedelics, hallucinogens

7. ___ amphetamines [p.601]
8. ___ caffeine [p.600]
9. ___ cocaine [pp.600–601]
10. ___ codeine [p.601]
11. ___ ethyl alcohol [p.601]
12. ___ heroin [p.601]
13. ___ lysergic acid diethylamide (LSD) [p.601]
14. ___ marijuana [p.601]
15. ___ nicotine [p.600]

Self-Quiz

___ 1. All nerves that lead away from the central nervous system are _____. [p.591]
    a. efferent nerves
    b. sensory nerves
    c. afferent nerves
    d. spinal nerves
    e. peripheral nerves

___ 2. _____ nerves dominate when the body is not receiving much outside stimulation. [p.592]
    a. Ganglia
    b. Pacemaker
    c. Sympathetic
    d. Parasympathetic
    e. All of the above

___ 3. What humans comprehend, communicate, remember, and voluntarily act on arises in the _____. [p.596]
   a. medulla
   b. thalamus
   c. hypothalamus
   d. cerebellum
   e. cerebral cortex

___ 4. The _____ are the protective coverings of the brain and spinal cord. [p.593]
   a. ventricles
   b. meninges
   c. tectums
   d. olfactory bulbs
   e. pineal glands

___ 5. The _____ persists as a low-level pathway to motor centers of the medulla oblongata and spinal cord; it also can activate centers in the cerebral cortex and thereby govern the activities of the nervous system as a whole. [p.595]
   a. medulla
   b. pons
   c. thalamus
   d. hypothalamus
   e. reticular formation

___ 6. The left hemisphere of the brain is responsible for _____. [p.596]
   a. music
   b. artistic ability and spatial relationships
   c. speech, analytical skills, and mathematics
   d. abstract abilities

___ 7. The part of the brain that controls the basic responses necessary to maintain life processes (respiration, blood circulation) is _____. [p.594]
   a. the cerebral cortex
   b. the cerebellum
   c. the corpus callosum
   d. the medulla

___ 8. To produce a split-brain individual, an operation would be required to sever the _____. [p.598]
   a. pons
   b. fissure of Rolando
   c. hypothalamus
   d. reticular formation
   e. corpus callosum

___ 9. The _____ integrates sensory input from the eyes, ears, and muscle spindles with motor signal from the forebrain; it also helps control motor dexterity. [p.594]
   a. cerebrum
   b. pons
   c. cerebellum
   d. hypothalamus
   e. thalamus

___10. The _____ evolved as a coordinating center for sensory input and as a relay station for signals to the cerebrum. [p.595]
   a. medulla
   b. pons
   c. reticular formation
   d. hypothalamus
   e. thalamus

Chapter Objectives/Review Questions

1. Describe a "nerve net." [p.588]
2. Fully explain how the shift from radial to bilateral symmetry within invertebrate animals influenced the complexity of nervous systems. [pp.588–589]
3. The nerve cord that persists in all vertebrate embryos is called the _____ _____. [p.590]
4. Define and contrast the vertebrate central and peripheral nervous systems. [p.591]
5. The communication lines inside the brain and spinal cord are called _____, not nerves. [p.591]
6. Compare the structures of the spinal cord and brain with respect to white matter and gray matter. [p.591]
7. _____ cells protect or structurally and functionally support neurons. [p.591]
8. In terms of sensory input and motor output, _____ means "to bring to," and _____ means "to carry outward." [p.591]
9. Distinguish between somatic nerves and autonomic nerves. [p.592]
10. Explain how parasympathetic nerve activity balances sympathetic nerve activity. List activities of the sympathetic and parasympathetic nerves in regulating pupil diameter, rate of heartbeat, activities of the gut, and elimination of urine. [pp.592–593]

11. Describe the basic structural and functional organization of the spinal cord. In your answer, distinguish spinal cord from vertebral column. [p.593]
12. List the parts of the brain found in the hindbrain, midbrain, and forebrain, and tell the basic functions of each. [p.594]
13. The _____ formation is an ancient mesh of interneurons that still persists as a low-level pathway to motor centers of the medulla oblongata and spinal cord. [p.595]
14. Describe the function of cerebrospinal fluid. [p.595]
15. In terms of structure and function, explain how the mechanism called the blood–brain barrier protects the brain and spinal cord. [p.595]
16. Describe how the cerebral hemispheres are structurally and functionally related to the other parts of the forebrain. [pp.596–597]
17. The _____ cerebral hemisphere affords most of the control over speech, mathematics, and analytical skills; the _____ hemisphere affords most control over spatial abilities, music, and other creative enterprises. [p.596]
18. The cerebral cortex is functionally divided into _____ areas (control of voluntary motor activity), _____ areas (perception of the meaning of sensations), and _____ areas (information integration that precedes conscious action). [p.596]
19. The _____ system, which is located inside the cerebral hemispheres, governs emotions and has roles in memory. [p.597]
20. State what the results of the "split-brain" experiments suggest about the functioning of the cerebral hemispheres. [p.598]
21. Distinguish between short-term information storage and long-term storage. [p.595]
22. The human brain processes _____ separately from _____. [p.599]
23. Define amnesia, Parkinson's disease, and Alzheimer's disease; list the characteristics of each. [p.599]
24. Explain what an electroencephalogram is and what EEGs can tell us about the levels of conscious experience. Describe three typical EEG patterns and tell which level of consciousness each characterizes. [p.600]
25. List the major classes of psychoactive drugs and provide an example of each class. [pp.600–601]
26. Cite evidence gathered by researchers showing that the teenage brain is not completely developed and may account for stereotypic behavior through these ages. [pp.602–603]

Integrating and Applying Key Concepts

Suppose that anger is eventually determined to be caused by excessive amounts of specific transmitter substances in the brains of angry people. Also suppose that an inexpensive antidote to anger that neutralizes these anger-producing transmitter substances is readily available. Can violent murderers now argue that they have been wrongfully punished because they were victimized by their brain's transmitter substances and could not have acted in any other way? Suppose an antidote is prescribed to curb violent tempers in an easily angered person. Suppose also that the person forgets to take the pill and subsequently murders a family member. Can the murderer still claim to be victimized by transmitter substances?

36

SENSORY RECEPTION

Interactive Exercises

Different Strokes for Different Folks [pp.606–607]

36.1. SENSORY RECEPTORS AND PATHWAYS—AN OVERVIEW [pp.608–609]

Selected Words: "ultrasounds" [p.606], *compound* sensations [p.608], sensory receptors [p.608], stimulus [p.608], summation [p.608], baroreceptor [p.608], *amplitude* [p.609], *frequency* [p.609]

Boldfaced, Page-Referenced Terms

[p.606] echolocation _____

_____

[p.608] sensory systems _____

_____

[p.608] sensation _____

_____

[p.608] perception _____

_____

[p.608] mechanoreceptors _____

_____

[p.608] thermoreceptors _____

_____

[p.608] pain receptors (nociceptors)_____

_____

[p.608] chemoreceptors _____

_____

[p.608] osmoreceptors _____

_____

[p.608] photoreceptors_____

_____

[p.609] sensory adaptation _____

_____

[p.609] somatic sensations _____

_____

[p.609] special senses _____

_____

Fill-in-the-Blanks

A sensory system consists of sensory receptors for specific stimuli, (1) _____ _____ [p.608] that conduct information from those receptors to the brain, and (2) _____ _____ [p.608] where information is evaluated. A(n) (3) _____ [p.608] is conscious awareness of change in internal or external conditions; this is not to be confused with (4) _____ [p.608], which is an understanding of what sensation means. The specialized peripheral endings of sensory neurons that detect specific kinds of stimuli are (5) _____ _____ [p.608]. A (6) _____ [p.608] is any form of energy that activates a specific type of sensory receptor. (7) _____ [p.608] detect the chemical energy of specific substances dissolved in the fluid surrounding them; (8) _____ [p.608] detect mechanical energy associated with changes in pressure, position, or acceleration; (9) _____ [p.608] detect the energy of visible and ultraviolet wavelengths of light; and (10) _____ [p.608] detect radiant energy associated with temperature changes.

Every type of sensation is caused by (11) _____ _____ [p.608] arriving from particular nerve pathways activating specific neurons in the (12) _____ [p.608]. Besides sensing the kind of stimulus, the brain also interprets variations in (13) _____ _____ [p.609]. Interpretation is based on the (14) _____ [p.609] of action potentials propagated along single axons and the (15) _____ [p.609] of axons carrying action potentials from a given tissue.

Sometimes the (16) _____ [p.609] of action potentials decreases or stops even when a stimulus is being maintained at constant strength; such a decrease is known as (17) _____ _____ [p.609].

Some mechanoreceptors only signal a (18) _____ [p.608] in a stimulus; if a stimulus is constant, but deserves no response, these receptors will quit responding. But there are other types of receptors that adapt slowly or not at all. (19) _____ [p.609] receptors that continually inform the brain about the changes in (20) _____ [p.609] of particular muscles help maintain balance and posture.

Sensory receptors that are present at more than one body location generally inform the brain about (21) _____ _____ [p.609], that is, how different parts of the body feel at any particular time. Special senses are restricted to specific locations, such as inside the eyes or ears.

Matching

Select the best match for each of the following items. [All from p.608]

22. ___ Is associated with vision

23. ___ Is associated with pain

24. ___ Detects odors

25. ___ Detects sounds

26. ___ Detects CO_2 concentration in the blood

27. ___ Detects environmental temperature

28. ___ Detects internal body temperature

29. ___ Detects touch

30. ___ Rods and cones

31. ___ Hair cells in the ear's organ of Corti

32. ___ Pacinian corpuscles in the skin

33. ___ Olfactory receptors in nose

34. ___ Any stimulus that causes tissue damage

35. ___ Is associated with the movement of fluid in the inner ear

A. chemoreceptors
B. mechanoreceptors
C. nociceptors
D. photoreceptors
E. thermoreceptors

36.2. SOMATIC SENSATIONS [pp.610–611]
36.3. SENSES OF TASTE AND SMELL [p.612]

Selected Words: Meissner's corpuscle [p.610], the bulb of Krause [p.610], Ruffini endings [p.610], Pacinian corpuscle [p.610], *somatic* pain [p.610], *visceral* pain [p.610], bradykinins [p.610], histamine [p.610], prostaglandins [p.610], substance P [p.611], *hyperalgesia* [p.611], *referred pain* [p.611], *phantom pain* [p.611], *chemical* senses [p.612], taste buds [p.612], olfactory bulbs [p.612], vomeronasal organ [p.612]

Boldfaced, Page-Referenced Terms

[p.610] somatosensory cortex _____

_____

[p.610] free nerve endings _____

_____

[p.610] encapsulated receptors _____

_____

[p.610] pain _____

_____

[p.612] taste receptors _____

_____

[p.612] olfactory receptors _____

_____

[p.612] pheromones _____

_____

Fill-in-the-Blanks

Signals from receptors in the skin and joints travel to the primary (1) _____ _____ [p.610], which is a strip little more than an inch wide running from the top of the (2) _____ [p.610, Fig. 36.4] to just above the (3) _____ [p.610, Fig. 36.4] on the surface of each (4) _____ _____ [p.610]. The largest portion of the somatosensory cortex is A (see accompanying figure), which receives signals coming from the (5) _____ [p.610]. The second largest region is C, which receives signals coming from the (6) _____ [p.610].

The somatic sensations (awareness of (7) _____ [p.610], pressure, heat, (8) _____ [p.610], and pain) start with receptor endings that are embedded in (9) _____ [p.610] and other tissues at the body's surfaces, in (10) _____ [p.610] muscles, and in the walls of internal organs. All skin (11) _____ [p.610] are easily deformed by pressure on the skin's surface; these make you aware of touch, vibrations, and pressure. (12) _____ [p.610] nerve endings serve as "heat" receptors, and their firing of action potentials increases with increases in temperature. (13) _____ [p.610] is the perception of injury to some body region; the perception begins when (14) _____ [recall p.608], which include free nerve endings, send signals via the thalamus to the parietal lobe of the brain, where they are interpreted. When the (15) _____ _____ [p.610] pass a certain threshold, the signals generated are translated into sensations of pain. The brain sometimes gets confused and may associate perceived pain with a tissue some distance from the damaged area; this phenomenon is called (16) _____ _____ [p.611]. Usually the nerve pathways to both the injured and the mistaken area pass through the same segment of spinal cord. (17) _____ [p.610] in skeletal muscle, joints, tendons, ligaments, and (18) _____ [p.610] are responsible for awareness of the body's position in space and of limb movements.

Labeling and Matching

Identify each indicated part of the accompanying illustration. Complete the exercise by entering the appropriate letter in the parentheses that follow the labels. [All from p.611]

19. _____ _____ _____ (   )

20. _____ _____ (   )

21. _____ _____ (   )

22. _____

23. _____

24. _____ _____ (   )

A. React continually to ongoing stimuli
B. Contribute to sensations of rapid vibrations
C. Involved in sensing heat, light pressure, and pain
D. Stimulated by slower vibrations

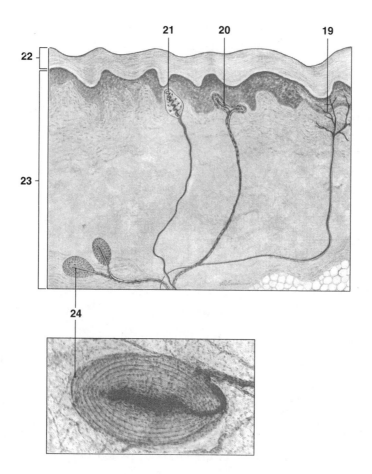

Fill-in-the-Blanks

(25) _____ [p.612] receptors detect molecules that become dissolved in fluid next to some body surface. Receptors are the modified dendrites of (26) _____ [p.612] neurons. In the case of taste, these receptors, when located on animal tongues, are often part of sensory organs, (27) _____ _____ [p.612], which are enclosed by circular papillae. Animals smell substances by means of (28) _____ [p.612] receptors, such as the ones in your (29) _____ [p.612]; humans have about (30) _____ [p.612] million of these in a nose. Sensory nerve pathways lead from the nasal cavity to the region of the brain where odors are identified and associated with their sources—the (31) _____ [p.612] bulb and nerve tract. (32) _____

[use logic] receptors sampling odors from food in the mouth are important for our sense of taste. When we suffer from the common cold and have a "runny" nose, odor molecules from food have difficulty reaching the olfactory receptors and contributing their information signals; our senses of taste and (33) _____ [p.612] are both dulled.

36.4. SENSE OF BALANCE [p.613]
36.5. SENSE OF HEARING [pp.614–615]

Selected Words: *equilibrium* position [p.613], cristae [p.613], *dynamic* equilibrium [p.613], "semicircular" canals [p.613], cupola [p.613], vestibular nerve [p.613], *static* equilibrium [p.613], *motion sickness* [p.613], *amplitude* [p.614], intensity [p.614], *frequency* [p.614], *pitch* [p.614], pinna [p.614], auditory canal [p.614], eardrum [p.614], hammer, anvil, stirrup [p.614], oval window [p.614], organ of Corti [p.615], basilar membrane [p.615], tectorial membrane [p.615], auditory nerve [p.615]

Boldfaced, Page-Referenced Terms

[p.613] inner ears _____

_____

[p.613] vestibular apparatus _____

_____

[p.613] hair cells _____

_____

[p.613] otoliths ("ear stones") _____

_____

[p.614] hearing _____

[p.614] middle ear _____

_____

[p.614] external ear _____

_____

[p.614] cochlea _____

_____

[p.615] acoustical receptors _____

_____

Fill-in-the-Blanks

The (1) _____ [p.614] (perceived loudness) of sound depends on the height of the sound wave. The (2) _____ [p.614] (perceived pitch) of sound depends on how fast the wave changes occur. The faster the vibrations, the (3) [choose one] ❑ higher, ❑ lower [p.614] the sound. Hair cells are (4) [choose one] ❑ nociceptors, ❑ mechanoreceptors, ❑ thermoreceptors [p.613] that detect vibrations. The hammer, anvil, and stirrup are located in the (5) [choose one] ❑ inner, ❑ middle [p.614] ear. The (6) _____ [p.614] is a

coiled tube that resembles a snail shell and contains the (7) _____ _____ _____ [pp.614–615]—
the organ that changes vibrations into electrochemical impulses. Structures that detect rotational
acceleration in humans are (8) _____ _____ [p.613].

Labeling

Identify each indicated part of the accompanying illustrations.

9. _____ _____ [p.614]

10. _____ [p.614]

11. _____ _____ [p.614]

12. _____ _____ [p.614]

13. _____ _____ [pp.614–615]

14. _____ _____ [p.615]

15. _____ _____ [p.615]

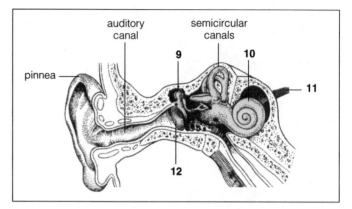

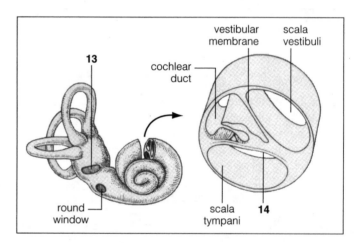

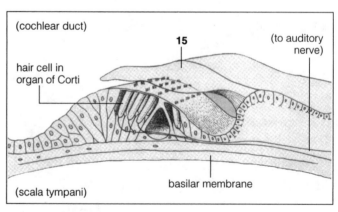

36.6. SENSE OF VISION [pp.616–617]

36.7. STRUCTURE AND FUNCTION OF VERTEBRATE EYES [pp.618–619]

36.8. *Focus on Science:* **DISORDERS OF THE HUMAN EYE** [pp.620–621]

36.9. CASE STUDY: FROM SIGNALING TO VISUAL PERCEPTION [pp.622–623]

Selected Words: sclera [p.618], choroid [p.618], ciliary muscle [p.618], iris [p.618], retina [p.617], pupil [p.618], fovea [p.618], optic nerve [p.618], aqueous humor [p.618], vitreous body [p.618], optic disk [p.618], fiberlike ligaments [p.619], rhodopsin [p.620], red-green color blindness [p.622] , astigmatism [p.622], nearsightedness [p.622], farsightedness [p.622], cataracts [p.623], glaucoma [p.623]

Boldfaced, Page-Referenced Terms

[p.616] vision _____

_____

[p.616] eyes _____

_____

[p.616] ocellus (singular, ocelli)_____

_____

[p.616] visual field _____

_____

[p.616] lens _____

_____

[p.616] cornea _____

_____

[p.617] compound eye _____

_____

[p.617] camera eyes _____

_____

[p.619] visual accommodation _____

_____

[p.620] rod cells _____

_____

[p.620] cone cells _____

_____

Fill-in-the-Blanks

Light is a stream of (1) _____ [p.620]—discrete energy packets. (2) _____ [p.616] is a process in which photons are absorbed by pigment molecules and photon energy is transformed into the electrochemical energy of a nerve signal. (3) _____ [p.616] requires precise light focusing onto a layer of photoreceptive cells that are dense enough to sample details of the light stimulus, followed by image formation in the brain. (4) _____ [p.616] are simple clusters of photosensitive cells, usually arranged in a cuplike depression in the epidermis. (5) _____ [p.616] are well-developed photoreceptor organs that allow at least some degree of image formation. The (6) _____ [p.616] is a transparent cover of the lens area, and the (7) _____ [p.617] consists of tissue containing densely packed photoreceptors. Compound eyes contain several thousand photosensitive units known as (8) _____ [p.617]. In the vertebrate eye, lens adjustments assure that the (9) _____ _____ [p.619] for a specific group of light rays lands on the retina. (10) _____ _____ [p.619] refers to the lens adjustments that bring about precise focusing onto the retina. (11) _____ [p.623] people focus light from nearby objects posterior to the retina. (12) _____ [p.621] cells are concerned with daytime vision and, usually, color perception. A (13) _____ [p.621] is a funnel-shaped pit on the retina that provides the greatest visual acuity.

Labeling

Identify each indicated part of the accompanying illustration. [All from p.618]

14. _____ _____
15. _____
16. _____
17. _____
18. _____ _____
19. _____ _____
20. _____
21. _____
22. _____ _____
23. _____ _____
24. _____

Self-Quiz

___ 1. According to the mosaic theory, _____. [p.617]
   a. the basement membrane's pigment molecules prevent the scattering of light
   b. light falling on the inner area of an "on-center" field activates firing of the cells
   c. hair cells in the semicircular canals cooperate to detect rotational acceleration
   d. each ommatidium detects information about only one small region of the visual field; many ommatidia contribute "bits" to the total image
   e. all of the above

___ 2. The principal place in the human ear where sound waves are amplified is _____. [p.614]
   a. the pinna
   b. the ear canal
   c. the middle ear
   d. the organ of Corti
   e. none of the above

___ 3. The place where vibrations are translated into patterns of nerve impulses is _____. [p.615]
   a. the pinna
   b. the ear canal
   c. the middle ear
   d. the organ of Corti
   e. none of the above

For questions 4–8, choose from the following answers:
   a. fovea [p.621]
   b. cornea [p.618]
   c. iris [pp.618–619]
   d. retina [p.620]
   e. sclera [p.618]

___ 4. The white protective fibrous tissue of the eye is the _____.

___ 5. Rods and cones are located in the _____.

___ 6. The highest concentration of cones is in the _____.

___ 7. The adjustable ring of contractile and connective tissues that controls the amount of light entering the eye is the _____.

___ 8. The outer transparent protective covering of part of the eyeball is the _____.

___ 9. Visual accommodation involves the ability to _____. [p.619]
   a. change the sensitivity of the rods and cones by means of transmitters
   b. change the width of the lens by relaxing or contracting certain muscles
   c. change the curvature of the cornea
   d. adapt to large changes in light intensity
   e. all of the above

___ 10. Nearsightedness is caused by _____. [p.622]
   a. eye structure that focuses an image in front of the retina
   b. uneven curvature of the lens
   c. eye structure that focuses an image posterior to the retina
   d. uneven curvature of the cornea
   e. none of the above

Chapter Objectives/Review Questions

1. Define and distinguish among chemoreceptors, mechanoreceptors, photoreceptors, and thermoreceptors. Name at least one example of each type that appears in an animal. [p.608]
2. Distinguish the types of stimuli detected by tactile and stretch receptors from those detected by hearing and equilibrium receptors. [pp.608,610,613–614]
3. Explain how a taste bud works, and distinguish the types of stimuli it detects from those detected by touch or stretch receptors. [pp.610,612]
4. Explain how the three semicircular canals of the human ear detect changes of position and acceleration in a variety of directions. [p.613]
5. Follow a sound wave from pinna to organ of Corti; mention the name of each structure it passes and state where the sound wave is amplified and where the pattern of pressure waves is translated into electrochemical impulses. [pp.614–615]
6. State how low- and high-amplitude sounds affect the organ of Corti. [pp.614–615]
7. State how low- and high-pitch sounds affect the organ of Corti. [pp.614–615]
8. Explain what a visual system is, and list four of the five aspects of a visual stimulus that are detected by different components of a visual system. [p.616]
9. Contrast the structure of compound eyes with the structures of invertebrate eyespots and of the human eye. [pp.616–618]
10. Define nearsightedness and farsightedness and relate each to eyeball structure. [pp.622–623]
11. Describe how the human eye perceives color, and black-and-white. [pp.620–621]
12. Explain the general principles that affect how light is detected by photoreceptors and changed into electrochemical messages. [pp.620–621]
13. Indicate the causes of the following disorders of the human eye: (a) red-green color blindness, (b) astigmatism, (c) cataracts, (d) glaucoma, and (e) retinal detachment. [pp.622–623]

Integrating and Applying Key Concepts

How might human behavior be changed if human eyes were compound eyes composed of ommatidia and if humans perceived only vibrations—as fish do—rather than sounds?

37

ENDOCRINE CONTROL

Interactive Exercises

Hormone Jamboree [pp.626–627]

37.1. THE ENDOCRINE SYSTEM [pp.628–629]

Selected Terms: vomeronasal organ [p.628]

Boldfaced, Page-Referenced Terms

[p.628] hormones _____

_____

[p.628] neurotransmitters _____

_____

[p.628] local signaling molecules _____

_____

[p.628] pheromones _____

_____

[p.629] endocrine system _____

_____

Matching

Choose the most appropriate answer for each term.

1. ___ hormones [p.628]
2. ___ neurotransmitters [p.628]
3. ___ vomeronasal organ [p.628]
4. ___ target cells [p.628]
5. ___ local signaling molecules [p.628]
6. ___ pheromones [p.628]

A. Signaling molecules released from axon endings of neurons that act swiftly on target cells
B. A pheromone detector discovered in humans
C. Released by many types of body cells and alter conditions within localized regions of tissues
D. Nearly odorless secretions of particular exocrine gland; common signaling molecules that act on cells of other animals of the same species and help integrate social behavior
E. Secretions from endocrine glands, endocrine cells, and some neurons that the bloodstream distributes to nonadjacent target cells
F. Cells that have receptors for any given type of signaling molecule

Complete the Table

7. Complete the following table by identifying the numbered components of the endocrine system shown in the illustration on the facing page, as well as the hormones produced by each.

Gland Name	Number	Hormones Produced
[p.629] a. Hypothalamus		
[p.629] b. Pituitary, anterior lobe		
[p.629] c. Pituitary, posterior lobe		
[p.629] d. Adrenal glands (cortex)		
[p.629] e. Adrenal glands (medulla)		
[p.629] f. Ovaries (two)		
[p.629] g. Testes (two)		
[p.629] h. Pineal		
[p.629] i. Thyroid		
[p.629] j. Parathyroids (four)		
[p.629] k. Thymus		
[p.629] l. Pancreatic islets		

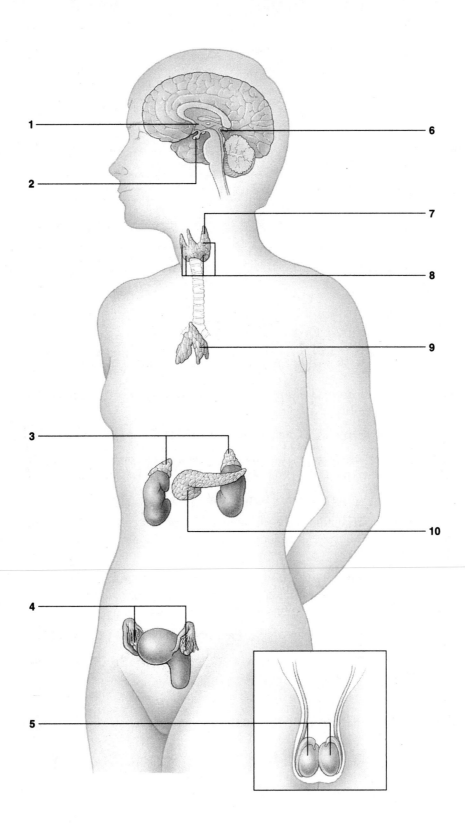

1

2

3

4

5

6

7

8

9

10

37.2. SIGNALING MECHANISMS [pp.630–631]

Selected Words: *testicular feminization syndrome* [p.630]

Boldfaced, Page-Referenced Terms

[p.630] steroid hormones _____

_____

[p.631] peptide hormones _____

_____

[p.631] second messenger _____

_____

Choice

For questions 1–10, choose from the following:

a. steroid hormones   b. peptide hormones

1. ___ Lipid-soluble molecules derived from cholesterol; can diffuse directly across the lipid bilayer of a target cell's plasma membrane [p.630]

2. ___ Various peptides, polypeptides, and glycoproteins [p.631]

3. ___ One example involves testosterone, defective receptors, and a condition called testicular feminization syndrome. [p.630]

4. ___ Hormones that often require assistance from second messengers [p.631]

5. ___ Hormones that bind to receptors at the plasma membrane of a cell; the receptor then activates specific membrane-bound enzyme systems, which in turn initiate reactions leading to the cellular response [p.631]

6. ___ Lipid-soluble molecules that move through the target cell's plasma membrane to the nucleus, where it binds to some type of protein receptor. The hormone–receptor complex moves into the nucleus and interacts with specific DNA regions to stimulate or inhibit transcription of mRNA. [p.630]

7. ___ Water-soluble signaling molecules that may incorporate anywhere from 3 to 180 amino acids [p.631]

8. ___ Involves molecules such as cyclic AMP that activates many enzymes in cytoplasm that, in turn, cause alteration in some cell activity [p.631]

9. ___ Glucagon is an example. [p.631]

10. ___ Cyclic AMP relays a signal into the cell's interior to activate protein kinase A. [p.631]

37.3. THE HYPOTHALAMUS AND PITUITARY GLAND [pp.632–633]

Selected Words: *posterior* lobe [p.632], *anterior* lobe [p.632], *second* capillary bed [p.632]

Boldfaced, Page-Referenced Terms

[p.632] hypothalamus _____

_____

[p.632] pituitary gland _____

_____

[p.633] releasers _____

_____

[p.633] inhibitors _____

_____

Choice-Match

Label each hormone listed below with an "A" if it is secreted by the anterior lobe of the pituitary, a "P" if it is released from the posterior pituitary, or an "I" if it is released from intermediate tissue. Complete the exercise by entering the letter of the corresponding action in the parentheses after each label.

1. __( ) ACTH [p.632]
2. __( ) ADH [p.632]
3. __( ) FSH [p.632]
4. __( ) STH (GH) [p.632]
5. __( ) LH [p.632]
6. __( ) MSH [p.632]
7. __( ) OCT [p.632]
8. __( ) PRL [p.632]
9. __( ) TSH [p.632]

A. Stimulates egg and sperm formation in ovaries and testes
B. Targets pigmented cells in skin and other surface coverings; induces color changes in response to external stimuli and affects some behaviors
C. Stimulates and sustains milk production in mammary glands
D. Stimulates progesterone secretion, ovulation, and corpus luteum formation in females; promotes testosterone secretion and sperm release in males
E. Induces uterine contractions and milk movement into secretory ducts of the mammary glands
F. Stimulates release of thyroid hormones from the thyroid gland
G. Acts on the kidneys to conserve water required in control of extracellular fluid volume
H. Stimulates release of adrenal steroid hormones from the adrenal cortex
I. Promotes growth in young; induces protein synthesis and cell division; roles in adult glucose and protein metabolism

Dichotomous Choice

Circle one of two possible answers given between parentheses in each statement.

10. The (hypothalamus/pituitary gland) region of the brain monitors internal organs and activities related to their functioning, such as eating and sexual behavior; it also secretes some hormones. [p.632]
11. The (posterior/anterior) lobe of the pituitary stores and secretes two hormones, ADH and OCT, that are produced by the hypothalamus. [p.632]
12. The (posterior/anterior) lobe of the pituitary produces and secretes its own hormones that govern the release of hormones from other endocrine glands. [p.632]
13. Humans lack the (posterior/intermediate) lobe of the pituitary gland, one that is possessed by many other vertebrates. [p.632]
14. Most hypothalamic hormones acting in the anterior pituitary lobe are (releasers/inhibitors) and cause target cells there to secrete hormones of their own. [p.633]
15. Some hypothalamic hormones slow secretion from their targets in the anterior pituitary; these are classed as (releasers/inhibitors). [p.632]

37.4. EXAMPLES OF ABNORMAL PITUITARY OUTPUT [p.634]

Selected Words: gigantism [p.634], pituitary dwarfism [p.634], acromegaly [p.634], diabetes insipidus [p.634]

Complete the Table

1. Complete the following table to summarize examples of abnormal pituitary output.

Condition	Hormone/Abnormality	Characteristics
[p.634] a.	Excessive somatotropin produced during childhood	Affected adults are proportionally similar to a normal person but larger
[p.634] b.	Insufficient somatotropin produced during childhood	Affected adults are proportionally similar to a normal person but much smaller
[p.634] c.	Diminished ADH secretion by a damaged posterior pituitary lobe	Large volumes of dilute urine are secreted, causing life-threatening dehydration
[p.634] d.	Excessive somatotropin output during adulthood when long bones can no longer lengthen	Abnormal thickening of bone, cartilage, and other connective tissues in the hands, feet, and jaws

37.5. SOURCES AND EFFECTS OF OTHER HORMONES [p.635]

Complete the Table

1. Complete the following table by matching the gland/organ and the hormone(s) produced by it to the descriptions of hormone action. Refer to text Table 37.3, p.635.

Gland/Organ	Hormones
A. adrenal cortex	a. thyroxine and triiodothyronine
B. adrenal medulla	b. glucagon
C. thyroid	c. PTH
D. parathyroids	d. androgens (includes testosterone)
E. testes	e. somatostatin
F. ovaries	f. thymosins
G. pancreas (alpha cells)	g. glucocorticoids
H. pancreas (beta cells)	h. estrogens, general (includes progesterone)
I. pancreas (delta cells)	i. epinephrine
J. thymus	j. melatonin
K. pineal	k. insulin
	l. progesterone
	m. mineralocorticoids (including aldosterone)
	n. calcitonin
	o. norepinephrine

Gland/Organ	Hormone	Hormone Action
[p.635] a.		Elevates calcium levels in blood
[p.635] b.		Influences carbohydrate metabolism of insulin-secreting cells
[p.635] c.		Required in egg maturation and release, and in preparation of uterine lining for pregnancy and its maintenance in pregnancy; influences growth, development, and genital development; maintains sexual traits
[p.635] d.		Promote protein breakdown and conversion to glucose in most cells
[p.635] e.		Lowers blood sugar level in muscle and adipose tissue
[p.635] f.		In general, required in sperm formation, genital development and maintenance of sexual traits, influences growth and development
[p.635] g.		In most cells, regulates metabolism, plays roles in growth and development
[p.635] h.		In the gonads, influences daily biorhythms and influences gonad development and reproductive cycles
[p.635] i.		In liver, muscle, and adipose tissues; raises blood sugar level, fatty acids; increases heart rate and force of contraction
[p.635] j.		Targets lymphocytes, has roles in immunity
[p.635] k.		Raises blood sugar level
[p.635] l.		Prepares and maintains uterine lining for pregnancy; stimulates breast development
[p.635] m.		In kidneys, promotes sodium reabsorption and control of salt–water balance
[p.635] n.		In smooth muscle cells of blood vessels, promotes constriction or dilation of blood vessels
[p.635] o.		In bone, lowers calcium levels in blood

37.6. FEEDBACK CONTROL OF HORMONAL SECRETIONS [pp.636–637]

Selected Words: *inhibit* [p.636], *stimulate* [p.636], *fight–flight response* [p.636], *goiter* [p.637], *hypothyroidism* [p.637], *hyperthyroidism* [p.637], *primary* reproductive organs [p.637]

Boldfaced, Page-Referenced Terms

[p.636] negative feedback _____

_____

[p.636] positive feedback _____

_____

[p.636] adrenal cortex _____

_____

[p.636] adrenal medulla _____

_____

[p.636] thyroid gland _____

_____

[p.637] gonads _____

_____

Matching

Choose the most appropriate answer for each term.

1. ___adrenal medulla [p.636]
2. ___ACTH [p.636]
3. ___nervous system [p.636]
4. ___glucocorticoids [p.636]
5. ___CRH [p.636]
6. ___cortisol [p.636]
7. ___fight–flight response [p.636]
8. ___cortisol-like drugs [p.636]
9. ___adrenal cortex [p.636]
10. ___feedback mechanisms [p.636]

A. Results from actions of epinephrine and norepinephrine in times of excitement or stress
B. Outer portion of each adrenal gland; some of its cells secrete hormones such as glucocorticoids
C. Negative ultimately inhibits hormone secretion, positive stimulates further hormone secretion
D. Initiates a stress response during chronic stress, injury, or illness; cortisol helps to suppress inflammation
E. Inner portion of the adrenal gland; neurons located here release epinephrine and norepinephrine
F. Stimulates the adrenal cortex to secrete cortisol; this helps raise the level of glucose by preventing muscle cells from taking up more blood glucose
G. Used to counter asthma and other chronic inflammatory disorders
H. Help maintain blood glucose concentration and help suppress inflammatory responses
I. Secreted by the hypothalamus in response to falling glucose blood levels; stimulates the anterior pituitary to secrete ACTH
J. Secreted by the adrenal cortex; blocks the uptake and use of blood glucose by muscle cells; also stimulates liver cells to form glucose from amino acids

Fill-in-the-Blanks

Thyroxine and triiodothyronine are the main hormones secreted by the human (11) _____ gland [p.636]. They are critical for normal development of many tissues, and they control overall (12) _____ [pp.636–637] rates in humans and other warm-blooded animals. The synthesis of thyroid hormones requires (13) _____ [p.637], which is obtained from food. When iodine is absorbed from the gut, iodine is converted to (14) _____ [p.637]. In the absence of that form of the element, blood levels of thyroid hormones decrease. The anterior pituitary responds by secreting (15) _____ [p.637]. When thyroid hormones cannot be synthesized, the feedback signal continues—and so does TSH secretion. THS secretion continues and overstimulates the thyroid gland and causes (16) _____ [p.637], an enlargement of the thyroid gland. (17) _____ [p.637] results from insufficient blood-level concentrations of thyroid hormones. Such (18) _____ [p.637] adults are often overweight, sluggish, dry-skinned, intolerant of cold, and sometimes confused and depressed. (19) _____ [p.637] results from excess concentrations of thyroid hormones. Affected adults show an increased heart rate, heat intolerance, elevated blood pressure, profuse sweating, and weight loss even when caloric intake increases. Affected individuals typically are nervous and agitated, and have trouble sleeping. The primary reproductive organs are known as (20) _____ [p.637]. These organs produce and secrete sex (21) _____ [p.637] essential to reproduction. These organs are known as the (22) _____ [p.637] in human males and (23) _____ [p.637] in females. Testes secrete (24) _____ [p.637]. Ovaries secrete estrogens and progesterone. All these hormones influence (25) _____ sexual traits [p.637]. Both types of organs produce (26) _____, or sex cells [p.637].

37.7. RESPONSES TO LOCAL CHEMICAL CHANGES [pp.638–639]

Selected Words: rickets [p.638], exocrine cells [p.638], endocrine cells [p.638], alpha cells [p.638], glucagon [p.638], beta cells [p.638], insulin [p.638], delta cells [p.638], diabetes mellitus [p.638], type 1 diabetes [p.639], type 2 diabetes [p.639]

Boldfaced, Page-Referenced Terms

[p.638] parathyroid glands _____

_____

[p.638] pancreatic islet _____

_____

Fill-in-the-Blanks

Humans have four (1) _____ [p.638] glands positioned next to the posterior (or back) of the human thyroid; they secrete (2) _____ [p.638] in response to a low (3) _____ [p.638] level in blood. This hormone induces living bone cells to secrete enzymes that digest bone tissue and thereby release (4) _____ [p.638] and other minerals to interstitial fluid, then to the blood. It enhances calcium (5) _____ [p.518] from the filtrate flowing from the nephrons of the kidneys. PTH induces some kidney cells to secrete enzymes that act on blood-borne precursors of the active form of vitamin (6) _____ [p.638], a hormone. This hormone stimulates (7) _____ [p.638] cells to increase calcium absorption from the gut lumen. In a child with vitamin D deficiency, too little calcium and phosphorus are absorbed, and so rapidly growing bones develop improperly. The resulting bone disorder is called (8) _____ [p.638], characterized by bowed legs, a malformed pelvis, and in many cases a malformed skull and rib cage.

Complete the Table

9. Complete the following table to summarize function of the pancreatic islets.

Pancreatic Islet Cells	Hormone Secreted	Hormone Action
[p.638] a. Alpha cells		
[p.638] b. Beta cells		
[p.638] c. Delta cells		

Dichotomous Choice

Circle one of two possible answers given between parentheses in each statement.

10. Insulin deficiency can lead to diabetes mellitus, a disorder in which the glucose level (rises/decreases) in the blood, then in the urine. [p.638]
11. In a person with diabetes mellitus, urination becomes (reduced/excessive), so the body's water–solute balance becomes disrupted; people become abnormally dehydrated and thirsty. [pp.638–639]
12. Lacking a steady glucose supply, body cells of people with diabetes mellitus begin breaking down their own fats and proteins for (energy/water). [p.639]
13. Weight loss occurs and (amino acids/ketones) accumulate in blood and urine; this promotes excessive water loss with a life-threatening disruption of brain function. [p.639]
14. After a meal, blood glucose rises; pancreatic beta cells secrete (glucagon/insulin); targets use glucose or store it as glycogen. [p.639]
15. Blood glucose levels decrease between meals. (Glucagon/Insulin) is secreted by stimulated pancreas alpha cells; targets convert glycogen back to glucose, which then enters the blood. [p.639]
16. In (type 1 diabetes/type 2 diabetes) the body mistakenly mounts an autoimmune response against its own insulin-secreting beta cells and destroys them. [p.639]
17. Juvenile-onset diabetes is also known as (type 1 diabetes/type 2 diabetes); these patients survive with insulin injections. [p.639]
18. In (type 1 diabetes/type 2 diabetes), insulin levels are close to or above normal, but target cells fail to respond to insulin. [p.639]
19. (Type 1 diabetes/Type 2 diabetes) usually is manifested during middle age and is less dramatically dangerous than the other type; beta cells produce less insulin as a person ages. [p.639]

37.8. HORMONES AND THE ENVIRONMENT [pp.640–641]

Selected Words: melatonin [p.640], "jet lag" [p.640], seasonal affective disorder [SAD] [p.640], "winter blues" [p.640], *Xenopus laevis* [p.640]

Boldfaced, Page-Referenced Terms

[p.640] pineal gland _____

_____

[p.640] biological clock _____

_____

[p.640] puberty _____

_____

[p.641] molting _____

_____

[p.641] ecdysone _____

_____

Choice

For questions 1–10, choose from the following:

a. melatonin   b. ecdysone

1. ___ The hormone that controls molting [p.641]
2. ___ Hormone secreted by the pineal gland [p.640]
3. ___ High blood levels of this hormone in winter (long nights) suppress sexual activity in hamsters. [p.640]
4. ___ Winter blues [p.640]
5. ___ Chemical interactions that cause an old cuticle to detach from the epidermis and muscles [p.641]
6. ___ Triggers puberty [p.640]
7. ___ Suppresses growth of a bird's gonads in fall and winter [p.640]
8. ___ Jet lag [p.640]
9. ___ Waking up at sunrise [p.640]
10. ___ Produced and stored in molting glands by insects and crustaceans [p.641]

Short Answer

11. What gland and its secretions are suspected to be involved in recently observed developmental deformities in frogs? Why? [pp.640–641] _____

_____

_____

_____

_____

_____

Self-Quiz

___ 1. The _____ region of the forebrain monitors internal organs, influences certain forms of behavior, and secretes some hormones. [p.632]
   a. hypothalamus
   b. pancreas
   c. thyroid
   d. pituitary
   e. thalamus

___ 2. Neurons of the _____ produce ADH and oxytocin that are stored within axon endings of the _____. [p.632]
   a. anterior pituitary, posterior pituitary
   b. adrenal cortex, adrenal medulla
   c. posterior pituitary, hypothalamus
   d. posterior pituitary, thyroid
   e. hypothalamus, posterior pituitary

___ 3. If you were lost in the desert and had no fresh water to drink, the level of _____ in your blood would increase as a means to conserve water. [p.632]
   a. insulin
   b. corticotropin
   c. oxytocin
   d. antidiuretic hormone
   e. salt

Choice

For questions 4–6, choose from the following answers:

a. estrogen   b. PTH   c. FSH   d. somatotropin   e. prolactin

___ 4. _____ stimulates bone cells to release calcium and phosphate and the kidneys to conserve it. [p.638]

___ 5. _____ stimulates and sustains milk production in mammary glands. [p.632]

___ 6. _____ is the hormone associated with pituitary dwarfism, gigantism, and acromegaly. [p.634]

For questions 7–9, choose from the following answers:

a. adrenal medulla   b. adrenal cortex   c. thyroid   d. anterior pituitary   e. posterior pituitary

___ 7. The _____ produces glucocorticoids that help increase the level of glucose in blood. [p.636]

___ 8. The gland that is most closely associated with emergency situations is the _____. [p.636]

___ 9. The overall metabolic rates of warm-blooded animals, including humans, depend on hormones secreted by the _____ gland. [pp.636–637]

For question 10, choose from the following answers:

a. parathyroid hormone   b. aldosterone   c. calcitonin   d. mineralocorticoids
e. none of the above

___ 10. If all sources of calcium were eliminated from your diet, your body would secrete more _____ in an effort to release calcium stored in your body and send it to the tissues that require it. [p.638]

Matching

Choose the most appropriate answer for each term.

11. ___ADH [p.632]

12. ___ACTH and TSH [p.636]

13. ___FSH and LH [pp.632–633]

14. ___GnRH [p.633]

15. ___STH [pp.632,634]

16. ___glucocorticoids [p.636]

17. ___cortisol [p.636]

18. ___epinephrine and norepinephrine [p.636]

19. ___thyrosine and triiodothyronine [p.636]

20. ___estrogens and progesterone [pp.629,635]

21. ___PTH [p.638]

22. ___glucagon [p.638]

23. ___testosterone [pp.629,635]

24. ___insulin [p.638]

25. ___pheromones [p.628]

26. ___somatostatin [p.633]

27. ___melatonin [p.640]

28. ___ecdysone [p.641]

A. In times of excitement or stress, these adrenal medulla hormones help adjust blood circulation and fat and carbohydrate metabolism

B. Hormones secreted by the ovaries; influence secondary sexual traits

C. Hormone secreted by beta pancreatic cells; lowers blood glucose level

D. Abnormal amount of this anterior pituitary lobe hormone has different effects on human growth during childhood and adulthood

E. Anterior pituitary lobe hormones that orchestrate secretions from the adrenal gland and thyroid gland, respectively

F. Adrenal cortex hormones that help increase the level of glucose in blood

G. Hormone secreted by alpha pancreatic cells; raises the blood glucose level

H. The hormone that largely controls molting in insects and crustaceans

I. Major thyroid hormones having widespread effects such as controlling the overall metabolic rates of warm-blooded animals

J. A glucocorticoid that comes into play when the body is under stress, as when the blood glucose level declines below a set point

K. Hypothalamic hormone that brings about secretion of FSH and LH, which are gonadotropins

L. Hormone secreted by the testes; influence secondary sexual traits

M. Hormone secreted by the pineal gland; influences the growth and development of gonads

N. Antidiuretic hormone produced by the posterior pituitary lobe; promotes water reabsorption when the body must conserve water

O. Hormone secreted by delta pancreatic cells; helps control digestion and absorption of nutrients; can also block secretion of insulin and glucagon

P. Nearly odorless hormonelike secretions of certain exocrine glands; they diffuse through water or air to cellular targets outside the animal body

Q. Anterior pituitary lobe hormones that act through the gonads to influence gamete formation and secretion of the sex hormones required in sexual reproduction

R. Hormone secreted by the parathyroid glands in response to low blood calcium levels

Chapter Objectives/Review Questions

1. Hormones, neurotransmitters, local signaling molecules, and pheromones are all known as _____ molecules that carry out integration. [p.628]
2. _____ are the secretory products of endocrine glands, endocrine cells, and some neurons. [p.628]
3. Define the terms *neurotransmitters, local signaling molecules,* and *pheromones*. [p.628]
4. Collectively, the body's sources of hormones came to be called the _____ system. [p.629]
5. Locate and name the components of the human endocrine systems on a diagram such as text Figure 37.2b. [p.629]
6. Target cells have _____ receptors that hormones and other signaling molecules interact with and have diverse effects on physiological processes. [p.630]
7. Contrast the proposed mechanisms of hormonal action on target cell activities by (a) steroid hormones and (b) peptide hormones that are proteins or are derived from proteins. [pp.630–631]
8. The _____ and the pituitary gland interact closely as a major neural-endocrine control center. [p.632]
9. Explain how, even though the anterior and posterior lobes of the pituitary are compounded as one gland, the tissues of each part differ in character. [pp.632–633]
10. Identify the hormones released from the posterior lobe of the pituitary, and state their target tissues. [p.632]
11. Identify the hormones produced by the anterior lobe of the pituitary, and tell which target tissues or organs each acts on. [pp.632–633]
12. Most hypothalamic hormones acting in the anterior lobe are _____; they cause target cells to secrete hormones of their own but others are _____ that slow secretion from their targets. [p.633]
13. Pituitary dwarfism, gigantism, and acromegaly are all associated with abnormal secretion of _____ by the pituitary gland. [p.634]
14. One cause of _____ _____ is damage to the pituitary's posterior lobe and the diminished secretion or lack of ADH. [p.634]
15. Outline the major human hormone sources, their secretions, main targets, and primary actions as shown on text Table 37.3. [p.635]
16. With _____ feedback, an increase or decrease in the concentration of a secreted hormone triggers events that inhibit further secretion. [p.636]
17. With _____ feedback, an increase in the concentration of a secreted hormone triggers events that stimulate further secretion. [p.636]
18. The adrenal _____ secretes glucocorticoids. [p.636]
19. The _____ system initiates the stress response. [p.636]
20. Describe the role of cortisol in a stress response. [p.636]
21. The adrenal _____ contains neurons that secrete epinephrine and norepinephrine. [p.636]
22. List the features of the fight–flight response. [p.636]
23. Over time, a sustained response is made to the low blood level of TSH produced by the thyroid gland; this causes an enlargement known as a form of _____. [p.637]
24. Describe the characteristics of hypothyroidism and hyperthyroidism. [p.637]
25. The _____ are primary reproductive organs that produce and secrete hormones with essential roles in reproduction. [p.637]
26. Name the glands that secrete PTH, and state the function of this hormone. [p.638]
27. Describe an ailment called rickets, and state its cause. [p.638]
28. Give two examples that illustrate the effects of local signaling molecules. [p.638]
29. Name the hormones secreted by alpha, beta, and delta pancreatic cells; list the effect of each. [p.638]
30. Describe the symptoms of diabetes mellitus, and distinguish between type 1 and type 2 diabetes. [pp.638–639]
31. The pineal gland secretes the hormone _____; relate two examples of the action of this hormone. [p.640]
32. Explain the cause of "jet lag" and "winter blues." [p.640]
33. Discuss possible causes of frog deformities, a relatively recent observation in nature. [pp.640–641]
34. The invertebrate hormone _____ is related to the control of the phenomenon known as _____, which occurs among crustaceans and insects. [p.641]

Integrating and Applying Key Concepts

Suppose you suddenly quadruple your already high daily consumption of calcium. State which body organs would be affected, and tell how they would be affected. Name two hormones whose levels would most probably be affected, and tell whether your body's production of them would increase or decrease. Suppose you continue this high rate of calcium consumption for ten years. Predict which organs would be subject to the most stress as a result.

38

PROTECTION, SUPPORT, AND MOVEMENT

Interactive Exercises

Of Men, Women, and Polar Huskies [pp.644–645]

38.1. INTEGUMENTARY SYSTEM [pp.646–647]

38.2. A LOOK AT HUMAN SKIN [p.648]

38.3. *Focus on Health:* SUNLIGHT AND SKIN [p.649]

Selected Words: chitin [p.646], chromatophores [p.647], melanin [p.647], keratin [p.647], vitamin D [p.647], *stratified* epithelium [p.648], carotene [p.648], collagen [p.648], follicles [p.648], *cold sweats* [p.648], *acne* [p.648], "split ends" [p.648], *hirsutism* [p.648], *aging* [p.649], elastin [p.649], *cold sores* [p.649]

Boldfaced, Page-Referenced Terms

[p.646] integument _____

_____

[p.646] cuticle _____

_____

[p.646] skin _____

_____

[p.646] epidermis _____

_____

[p.646] dermis _____

_____

[p.648] keratinocytes _____

_____

[p.648] melanocytes _____

_____

[p.648] sweat glands _____

_____

[p.648] oil glands _____

_____

[p.648] hair _____

_____

[p.649] Langerhans cells _____

_____

[p.649] Granstein cells _____

_____

Fill-in-the-Blanks

Human skin is an organ system that consists of two layers: the outermost (1) _____ [p.646], which contains mostly dead cells, and the (2) _____ [p.646], which contains hair follicles, nerves, tiny muscles associated with the hairs, and various types of glands. The (3) _____ [p.646] layer, with its loose connective tissue and store of fat in (4) _____ [recall Chapter 37] tissue, lies beneath the skin.

Labeling

Label the numbered parts of the following illustration. [All from p.646]

5. _____
6. _____ _____
7. _____ _____
8. _____ _____
9. _____ _____
10. _____ _____
11. _____ _____
12. _____
13. _____
14. _____

Choice

Match the following proteins (or protein derivatives) with the particular ability that each substance lends to the skin. For questions 15–21, choose from these letters: [All from p.648]

a. collagen   b. elastin   c. keratin   d. melanin   e. hemoglobin

15. ___ Fibers that run through the dermis and lend a flexible but substantive structure to it

16. ___ Protects against loss of moisture

17. ___ Protects against ultraviolet radiation and sunburn

18. ___ Helps skin to be stretchable, yet return to its previous shape

19. ___ Beaks, hooves, hair, fingernails, and claws contain a lot of this

20. ___ Helps ward off bacterial attack by making the skin surface rather impermeable

21. ___ Located in red blood cells; binds with O_2

Fill-in-the-Blanks

The sun's ultraviolet wavelengths stimulate (22) _____ [p.648] production in melanocytes.

(23) _____ [p.648] glands lubricate and soften hair and the skin; (24) _____ [p.648] is a skin inflammation occurring after bacteria have successfully infected the ducts leading from these glands.

(25) _____ [p.648], excessive hairiness, may result when the body produces abnormal amounts of testosterone.

Cholecalciferol, otherwise known as (26) _____ _____ [p.649], is a steroidlike compound that helps the human body absorb (27) _____ [p.649] from food. (26) is produced by special cells in the skin on exposure to (28) _____ [p.649] from a precursor molecule that is related to (29) _____ [p.649]. (26) is released into the bloodstream and it travels to absorptive cells in the lining of the (30) _____ [p.649], where it acts.

Sunlight also damages two types of cells that mobilize the body's (31) _____ [p.649] system. (32) _____ [p.649] cells are phagocytes that are produced in the bone marrow but that later reside in the skin, where they envelope and digest viruses and bacteria; they then signal other parts of the immune system to join the battle and round up the foreign invaders. Exposure to excess ultraviolet radiation can damage (32) cells and render the skin susceptible to viral infections such as (33) _____ _____ [p.649]. (34) _____ [p.649] cells issue suppressor signals that dampen immune responses and keep them under control.

38.4. TYPES OF SKELETONS [pp.650–651]
38.5. CHARACTERISTICS OF BONE [pp.652–653]
38.6. HUMAN SKELETAL SYSTEM [pp.654–655]

Selected Words: longitudinal muscles [p.650], radial muscles [p.650], hydraulic pressure [p.650], cuticle [p.650], cartilage [p.651], *compact* bone [p.652], *spongy* bone [p.652], Haversian system [p.652], *osteoporosis* [p.653], calcitonin [p.653], parathyroid hormone (PTH) [p.653], *appendicular* and *axial* portions [p.654], pelvic and pectoral girdles [p.654], *herniated disks* [p.654], *fibrous, cartilaginous* and *synovial* joints [p.654], *strain* [p.654], *sprain* [p.654], *osteoarthritis* [p.654], *rheumatoid arthritis* [p.654], vertebral column, cranium, clavicle, sternum, scapula, radius, carpal bones, femur, tibia, tarsal bones, metatarsals [p.655]

Boldfaced, Page-Referenced Terms

[p.650] hydrostatic skeleton _____

_____

[p.650] exoskeleton _____

_____

[p.650] endoskeleton _____

_____

[p.652] bones _____

_____

[p.652] red marrow _____

_____

[p.652] yellow marrow _____

_____

[p.652] osteoblasts _____

_____

[p.652] osteocytes _____

_____

[p.653] osteoclasts _____

_____

[p.654] vertebrae _____

_____

[p.654] intervertebral disks _____

_____

[p.654] joints _____

_____

[p.654] ligaments _____

_____

Fill-in-the-Blanks

All motor systems are based on muscle cells that are able to (1) _____ [p.650] and (2) _____ [p.650], and on the presence of a medium against which the (3) _____ [p.650] force can be applied. Longitudinal and radial muscle layers work as in (4) _____ [p.650] to each other; the action of one motor element opposes the action of another. A membrane filled with fluid resists compression and can act as a (5) _____ [p.650] skeleton. Arthropods have muscles attached to an (6) [choose one] ❏ endoskeleton, ❏ exoskeleton [p.650]. (7) _____ [p.652] are hard compartments that enclose and protect the brain, spinal cord, heart, lungs, and other vital organs of vertebrates. Bones support and anchor (8) _____ [p.652] and soft organs, such as eyes. (9) _____ [p.652] systems are found in the long bones of mammals and contain living bone cells that receive their nutrients from the blood. (10) _____ _____ [p.652] is a major site of blood cell formation. Bone tissue serves as a "bank" for (11) _____ [p.652], (12) _____ [p.652], and other mineral ions; depending on metabolic needs, the body deposits ions into and withdraws ions from this "bank."

Bones develop from (13) _____ [p.653] secreting material inside the shaft and on the surface of the cartilage model. Bone can also give ions back to interstitial fluid as bone cells dissolve out component minerals and remodel bone in response to (14) _____ [p.653] signals, lack of exercise, and calcium deficiencies in the diet. Extreme decreases in bone density result in (15) _____ [p.653], particularly among older women.

The (16) _____ [p.654] portion of the human skeleton includes the skull, vertebral column, ribs, and breastbone; the (17) _____ [p.654] portion includes the pectoral and pelvic girdles and the forelimbs and hindlimbs, when they exist, in vertebrates.

(18) _____ [p.654] joints are freely movable and are lubricated by a fluid secreted into the capsule of dense connective tissue that surrounds the bones of the joint. Bones are often tipped with (19) _____ [p.654]; as a person ages, the cartilage at (20) _____ [p.654] joints may simply wear away, a condition called (21) _____ [p.654]. By contrast, in (22) _____ _____ [p.654], the synovial membrane becomes inflamed, cartilage degenerates, and bone becomes deposited in the joint.

Labeling

Identify each indicated part of the accompanying illustrations. [All from p.652]

23. _____ _____

24. _____ _____ _____

25. _____ _____

26. _____ _____

27. _____ _____ _____

28. _____ _____

29. _____ _____

30. _____ _____

31. _____

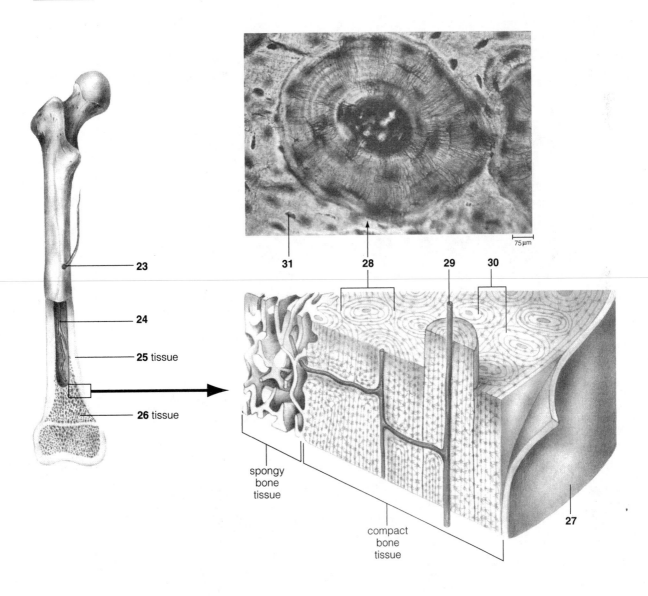

23

24

25 tissue

26 tissue

31   28   29   30

75 μm

spongy
bone
tissue

compact
bone
tissue

27

32. _____

33. _____

34. _____

35. _____

36. _____

37. _____ _____

38. _____

39. _____

40. _____ _____

41. _____

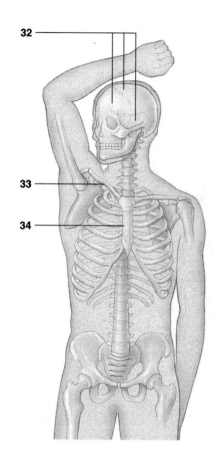

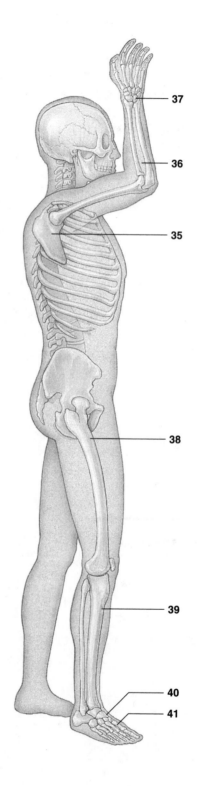

38.7. SKELETAL-MUSCULAR SYSTEMS [pp.656–657]

38.8. A CLOSER LOOK AT MUSCLES [pp.658–659]

38.9. ENERGY FOR CONTRACTION [p.660]

38.10. CONTROL OF CONTRACTION [pp.660–661]

38.11. PROPERTIES OF WHOLE MUSCLES [p.662]

38.12. *Focus on Science:* **PEBBLES, FEATHERS, AND MUSCLE MANIA** [p.663]

Selected Words: lever system [p.656], *skeletal* muscle [p.656], triceps brachii [p.657], pectoralis major [p.657], external oblique [p.657], rectus abdominis [p.657], quadriceps femoris [p.657], tibialis anterior [p.657], gastrocnemius [p.657], biceps femoris [p.657], gluteus maximus [p.657], latissimus dorsi [p.657], trapezius [p.657], deltoid [p.657], biceps brachii [p.657], myofibrils [p.658], Z bands [p.658], *thin* filaments [p.658], *thick* filaments [p.659], T tubules [p.660], *isometric, isotonic* and *lengthening* contractions [p.662], *muscle fatigue* [p.662], *aerobic exercise* [p.662], strength training [p.662], 'roid rage [p.663]

Boldfaced, Page-Referenced Terms

[p.656] skeletal muscles _____

_____

[p.656] tendons _____

_____

[p.658] sarcomeres _____

_____

[p.658] actin _____

_____

[p.659] myosin _____

_____

[p.659] sliding-filament model _____

_____

[p.659] cross-bridge formation _____

_____

[p.660] creatine phosphate _____

_____

[p.660] oxygen debt _____

_____

[p.661] action potential _____

_____

[p.661] sarcoplasmic reticulum _____

_____

[p.662] muscle tension _____

_____

[p.662] motor unit _____

_____

[p.662] muscle twitch _____

_____

[p.662] tetanus _____

_____

[p.662] exercise _____

_____

[p.663] anabolic steroids _____

_____

Sequence

Arrange in order of decreasing size. [All from p.658]

1. ___   A. Muscle fiber (muscle cell)

2. ___   B. Myosin filament

3. ___   C. Muscle bundle

4. ___   D. Muscle

5. ___   E. Myofibril

6. ___   F. Actin filament

Labeling

Identify each indicated part of the accompanying illustration.

7. _____ _____ [p.657]

8. _____ _____ [p.657]

9. _____ _____ [p.657]

10. _____ _____ [p.657]

11. _____ _____ [p.657]

12. _____ _____ [p.657]

13. _____ [p.657]

14. _____ [p.657]

15. _____ [p.657]

16. _____ _____ [p.657]

17. _____ _____ [p.656]

18. _____ _____ [p.656]

19. _____ _____ [p.656]

20. _____ _____ [p.656]

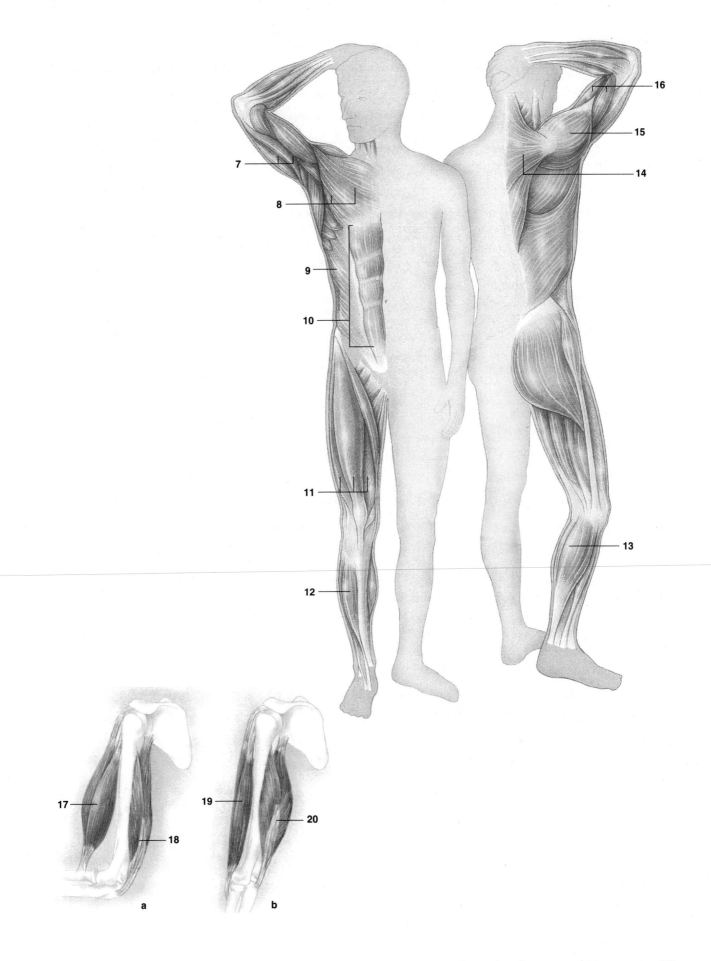

Fill-in-the-Blanks

There are three types of muscle tissue: (21) _____ [p.656], which is striated, involuntary, and located in the heart; smooth, which is involuntary, not striated, and mostly located in the walls of internal organs in vertebrates; and (22) _____ [p.656], which is striated, largely voluntary, and generally attached to (23) _____ [p.656] or cartilage by means of (24) _____ [p.656]. (25) _____ [p.656] muscle interacts with the skeleton to bring positional changes of body parts and to move the animal through its environment.

Together, the skeleton and its attached muscles are like a system of levers in which rigid rods, (26) _____ [p.656], move about at fixed points, called (27) _____ [p.656]. Most attachments are close to joints, so a muscle has to shorten only a small distance to produce a large movement of some body part.

When the (28) _____ _____ [pp.656–657] contracts, the elbow joint bends (flexes). As it relaxes and as its partner, the (29) _____ _____ [pp.656–657], contracts, the forelimb extends and straightens.

Muscle cells have three properties in common: contractility, elasticity, and (30) _____ [p.661]. Like a (31) _____ [p.661], a muscle cell is "excited" when a wave of electrical disturbance, a (an) (32) _____ _____ [p.661], travels along its cell membrane. Neurons that send signals to muscles are (33) _____ [p.660] neurons.

Each muscle cell contains (34) _____ [p.658]: threadlike structures packed together in parallel array. Every (34) is functionally divided into (35) _____ [p.658], which appear to be striped and are arranged one after another along its length. Each myofibril contains (36) _____ [p.659] filaments, which have cross-bridges, and (37) _____ [p.658] filaments, which are thin and lack cross-bridges. According to the (38) _____-_____ [p.659] model, (39) _____ [p.659] filaments physically slide along actin filaments and pull them toward the center of a (40) _____ [p.659] during a contraction. The energy that drives the forming and breaking of the cross-bridges comes immediately from (41) _____ [p.659], which obtained its phosphate group from (42) _____ _____ [p.660], which is stored in muscle cells.

(43) _____ [p.659] of an entire muscle is brought about by the combined decreases in length of the individual sarcomeres that make up the myofibrils of the muscle cells.

The parallel orientation of muscle's component parts directs the force of contraction toward a (44) _____ [p.659] that must be pulled in some direction.

When excited by incoming signals from motor neurons, muscle cell membranes cause (45) _____ [p.661] ions to be released from the (46) _____ _____ [p.661]. These ions remove all obstacles that might interfere with myosin heads binding to the sites along (47) _____ [p.661] filaments. When muscle cells relax/rest, calcium ions are sent back to the sarcoplasmic reticulum by (48) _____ _____ [p.661].

By controlling the (49) _____ _____ [p.661] that reach the (50) _____ _____ [p.661] in the first place, the nervous system controls muscle contraction by controlling calcium ion levels in muscle tissue.

Complete the Table

For each item in the following table, state its specific role in muscle contraction.

Aerobic respiration [p.660]	51.
ATP [p.660]	52.
Calcium ions [p.661]	53.
Creatine phosphate [p.660]	54.
Glycogen [p.660]	55.
Glycolysis alone [p.660]	56.
Motor neuron [p.660]	57.
Myosin heads [p.659]	58.
Sarcomere [p.658]	59.
Sarcoplasmic reticulum [p.661]	60.

Fill-in-the-Blanks

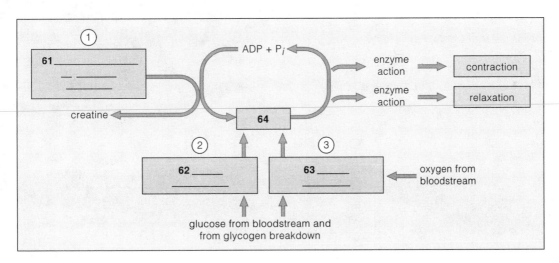

Fill-in-the-Blanks

The larger the (65) _____ [p.662] of a muscle, the greater its strength. A (66) _____ [p.662] neuron and the muscle cells under its control are called a (67) _____ _____ [p.662]. A (68) _____ _____ [p.662] is a response in which a muscle contracts briefly when reacting to a single, brief stimulus and then relaxes. The strength of the muscular contraction depends on how far the (69) _____ [p.662] response has proceeded by the time another signal arrives. (70) _____ [p.662] is the state of contraction in which a motor unit that is being stimulated repeatedly is maintained. Aerobic exercise is not intense, but is long in duration; such exercise increases the number of (71) _____ [p.662] in both fast and slow muscle cells and builds more capillaries to bring blood to the area. Strength training involves intense, short-duration exercise (weight lifting), which causes fast-acting muscle cells to form more (72) _____ [p.662] and more glycolytic enzymes.

Labeling

Label the numbered parts of the accompanying illustrations. [All from p.658]

73. _____ bundle

74. _____ _____

75. _____

76. _____

77. _____ _____

78. _____ _____

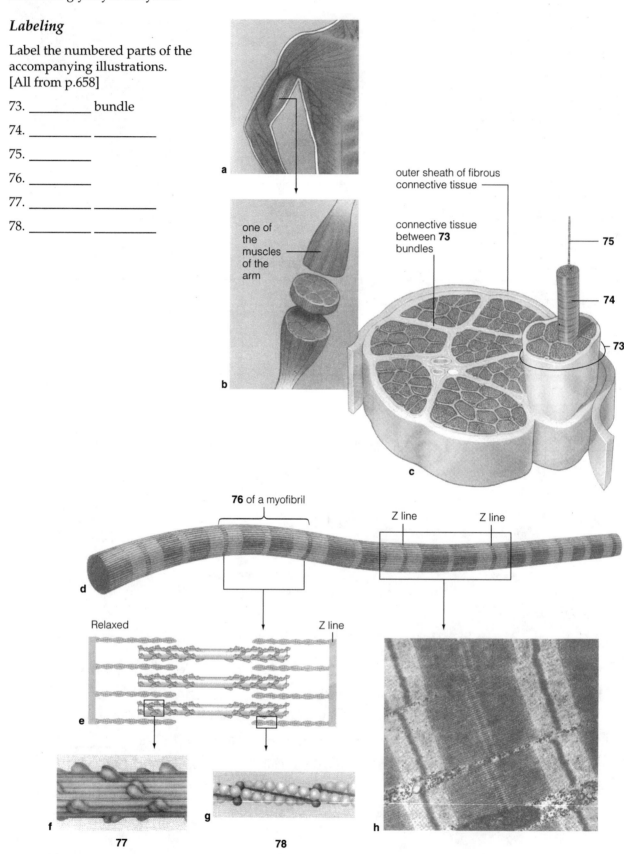

Labeling

Identify the numbered parts of the accompanying illustrations.

79. _____
  _____
  _____
  _____ [p.660]

80. _____ _____
  _____ [p.660]

81. _____
  _____
  _____
  _____ [p.660]

82. _____ _____
  _____ [p.660]

83. _____
  _____ [p.662]

84. _____ _____ [p.662]

85. _____ _____ [p.662]

86. _____ (_____)[p.662]

87. _____ _____
  _____ [p.662]

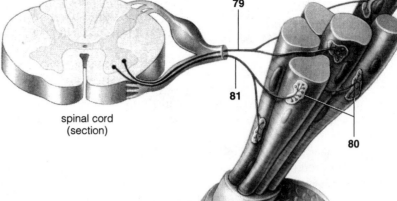

spinal cord
(section)

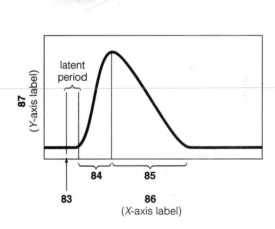

Fill-in-the-Blanks

(88) _____ _____ [p.663] are synthetic hormones that mimic testosterone in building greater
(89) _____ [p.663] mass in both men and women. It is illegal for competitive athletes to use them
because of unfair advantage and because of the side effects. In men, (90) _____ [p.663], baldness,
shrinking (91) _____ [p.663], and infertility are the first signs of damage. In women, (92) _____
[p.663] hair becomes more noticeable, (93) _____ _____ [p.663] become irregular, breasts may
shrink, and the (94) _____ [p.663] may become grossly enlarged. Aside from these physical side effects,
some men experience uncontrollable (95) _____ [p.663], delusions, and wildly manic behavior.

Self-Quiz

For questions 1–7, choose from the following answers:

    a. bone [p.652]
    b. cartilage [p.652]
    c. epidermis [pp.646,648]
    d. dermis [p.646]
    e. hypodermis [p.646]

___ 1. Fat cells in adipose tissue are most likely to be located in this.

___ 2. Keratinized, flattened cells are most likely to be located in this.

___ 3. Melanin in melanocytes is most likely to be here.

___ 4. Smooth muscles attached to hairs are probably here.

___ 5. This makes up the original "model" of the skeletal framework.

___ 6. The receiving ends of sensory receptors are most likely here.

___ 7. This serves as a "bank" for withdrawing and depositing calcium and phosphate ions.

For questions 8–11, choose from the following answers:

    a. ligaments [p.654]
    b. osteoblasts [p.653]
    c. osteoclasts [p.653]
    d. red marrow [p.652]
    e. tendons [p.656]

___ 8. _____ secrete bone-dissolving enzymes.

___ 9. Major site of blood cell formation.

___10. _____ remove Ca^{++} and $PO_4^{\equiv}$ ions from blood and build bone.

___11. _____ attach muscles to bone.

For questions 12–16, choose from the following answers:

    a. an action potential [p.661]
    b. cross-bridge formation [p.659]
    c. the sliding-filament model [p.659]
    d. tension [p.662]
    e. tetanus [p.662]

___12. _____ is a mechanical force that causes muscle cells to shorten if it is not exceeded by opposing forces.

___13. A wave of electrical disturbance that moves along a neuron or muscle cell in response to a threshold stimulus.

___14. _____ is a large contraction caused by repeated stimulation of motor units that are not allowed to relax.

___15. _____ is assisted by calcium ions and ATP.

___16. _____ explains how myosin filaments move to the centers of sarcomeres and back.

For questions 17–20, choose from the following answers:

    a. actin [p.658]
    b. myofibril [p.658]
    c. myosin [p.659]
    d. sarcomere [p.658]
    e. sarcoplasmic reticulum [p.661]

___17. A(n) _____ contains many repetitive units of muscle contraction.

___18. The repetitive unit of muscle contraction.

___19. Thin filaments that depend on calcium ions to clear their binding sites so that they can attach to parts of thick filaments.

___20. _____ stores calcium ions and releases them in response to an action potential.

Chapter Objectives/Review Questions

1. Name the four functions of human skin. [pp.645,647]
2. Describe the two-layered structure of human skin, and identify the items located in each layer. [p.646]
3. Describe several ways in which sunlight affects human skin. [p.649]
4. Compare invertebrate and vertebrate motor systems in terms of skeletal and muscular components and their interactions. [pp.650–651]
5. Explain the various roles of osteoblasts, mature bone cells, cartilage models, and long bones in the development of human bones. [pp.652–653]
6. Identify human bones by name and location. [pp.654–655]
7. Refer to Figure 38.19 of your main text and indicate (a) a muscle used in sit-ups, (b) another used in dorsally flexing and inverting the foot, and (c) another used in flexing the elbow joint. [p.657]
8. Explain in detail the structure of muscles, from the molecular level to the organ systems level. Then explain how biochemical events occur in muscle contractions and how antagonistic muscle action refines movements. [pp.658–659]
9. Describe the fine structure of a muscle fiber; use terms such as *myofibril*, *sarcomere*, *motor unit*, *actin*, and *myosin*. [pp.658–659]
10. List, in sequence, the biochemical and fine structural events that occur during the contraction of a skeletal muscle fiber, and explain how the fiber relaxes. [pp.658–661]
11. Distinguish twitch contractions from tetanic contractions. [p.662]
12. Describe the effects of anabolic steroids on men and women. [p.663]

Integrating and Applying Key Concepts

If humans had an exoskeleton rather than an endoskeleton, would they move differently from the way they do now? Name any advantages or disadvantages that having an exoskeleton instead of an endoskeleton would present in human locomotion.

39

CIRCULATION

Interactive Exercises

Heartworks [pp.666–667]

39.1. CIRCULATORY SYSTEMS—AN OVERVIEW [pp.668–669]

Selected Words: electrocardiograms (ECGs) [p.666], "internal environment" [p.668], *closed* circulatory system [p.668], *open* circulatory system [p.668], blood flow *velocity* [p.668], blood flow *rate* [p.668], gills [p.669], lungs [p.669], one circuit [p.669], double circuit [p.669]

Boldfaced, Page-Referenced Terms

[p.668] circulatory system _____

_____

[p.668] interstitial fluid _____

_____

[p.668] blood _____

_____

[p.668] heart _____

_____

[p.668] capillary beds _____

_____

[p.668] capillaries _____

_____

[p.669] pulmonary circuit _____

_____

[p.669] systemic circuit _____

_____

[p.669] lymphatic system _____

_____

Fill in the Blanks

Cells survive by taking in from their surroundings what they need, substances called (1) _____ [p.667], and giving back to their surroundings materials that they don't need: (2) _____ [p.667]. In most animals, substances move rapidly to and from living cells by way of a (3) _____ [p.668] circulatory system. (4) _____ [p.668], a fluid connective tissue within the (5) _____ [p.668] and blood vessels, is the transport medium.

Most of the cells of animals are bathed in a(n) (6) _____ _____ [p.667]; blood is constantly delivering nutrients and removing wastes from that fluid. The (7) _____ [p.668] generates the pressure that keeps blood flowing. Blood flows (8) [choose one] ❒ rapidly ❒ slowly [p.668] through large-diameter vessels to and from the heart, but where the exchange of nutrients and wastes occurs, in the (9) _____ [p.668] beds, the blood is divided up into vast numbers of smaller-diameter vessels with tremendous surface area that enables the exchange to occur by diffusion. An elaborate network of drainage vessels attracts excess interstitial fluid and reclaimable (10) _____ [p.669] and returns them to the circulatory system. This network is part of the (11) _____ [p.669] system, which also helps clean disease agents out of the blood.

Nutrients are absorbed into the blood from the (12) _____ [p.667] and (13) _____ [p.667] systems. Carbon dioxide is given to the (14) _____ [p.667] system for elimination, and excess water, solutes, and wastes are eliminated by the (15) _____ [p.667] system.

Label the numbered parts in the following illustrations. [All from p.668]

16. _____

17. _____ _____

18. _____

Short Answer

Describe the kind of circulatory system in:

19. Creature A. _____

20. Creature B. _____

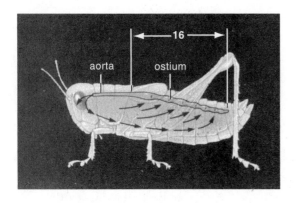

A

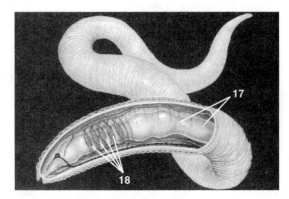

B

39.2. CHARACTERISTICS OF BLOOD [pp.670–671]

39.3. *Focus on Health:* BLOOD DISORDERS [p.672]

39.4. BLOOD TRANSFUSION AND TYPING [pp.672–673]

Selected Words: plasma proteins [p.670], erythrocytes [p.670], leukocytes [p.671], neutrophils, eosinophils, basophils, monocytes, and lymphocytes [p.671], macrophages [p.671], megakaryocytes [p.671], *hemorrhagic, chronic, hemolytic, iron deficiency,* and B_{12} *deficiency* anemias [p.672], *sickle-cell* anemia [p.672], *thalassemias* [p.672], *polycythemias* [p.672], "blood doping" [p.672], *infectious mononucleosis* [p.672], *leukemias* [p.672], *"self"* markers [p.672], *erythroblastosis fetalis* [p.673]

Boldfaced, Page-Referenced Terms

[p.670] plasma _____

_____

[p.670] red blood cells _____

_____

[p.670] white blood cells _____

_____

[p.670] platelets _____

_____

[p.671] stem cells _____

_____

[p.671] cell count _____

_____

[p.672] anemias _____

_____

[p.672] blood transfusions _____

_____

[p.672] agglutination _____

_____

[p.673] ABO blood typing _____

_____

[p.673] Rh blood typing _____

_____

Complete the Table

1. Complete the following table, which describes the components of blood. [All from p.670]

Components	Relative Amounts	Functions
Plasma portion (50%–60% of total volume):		
Water	91%–92% of plasma volume	Solvent
a. (albumin, globulins, fibrinogen, etc.)	7%–8%	Defense, clotting, lipid transport, roles in extracellular fluid volume, and so forth
Ions, sugars, lipids, amino acids, hormones, vitamins, dissolved gases	1%–2%	Roles in extracellular fluid volume, pH, and so on
Cellular portion (40%–50% of total volume):		
b.	4,800,000–5,400,000 per microliter	O_2, CO_2 transport
White blood cells:		
c.	3,000–6,750	Phagocytosis
d.	1,000–2,700	Immunity
Monocytes (macrophages)	150–720	Phagocytosis
Eosinophils	100–360	Roles in inflammatory response, immunity
Basophils	25–90	Roles in inflammatory response, anticlotting
e.	250,000–300,000	Roles in clotting

Fill-in-the-Blanks

Blood is a highly specialized fluid (2) _____ [p.670] tissue that helps stabilize internal (3) _____ [p.670] and equalize internal temperature throughout an animal's body. Oxygen binds with the (4) _____ [p.671] atom in a hemoglobin molecule. The red blood (5) _____ _____ [p.671] in males is about 5.4 million cells per microliter of blood; in females, it is 4.8 million per microliter. The plasma portion constitutes approximately (6) _____ [p.670] to _____ [p.670] percent of the total blood volume. Erythrocytes are produced in the (7) _____ _____ _____ [p.671]. (8) _____ _____ [p.671] are immature cells not yet fully differentiated. (9) _____ [p.671] and monocytes are highly mobile and phagocytic; they chemically detect, ingest, and destroy bacteria, foreign matter, and dead cells. (10) _____ [p.671] (thrombocytes) are membrane-bound cell fragments that aid in forming blood clots by releasing substances that initiate the process.

In humans, red blood cells lack their (11) _____ [p.671], but they contain enough resources to sustain them for about (12) _____ [p.671] months. Platelets also have no (13) _____ [p.671], but they last a maximum of (14) _____ [p.671] days in the human bloodstream.

If you are blood type (15) _____ [p.673], you have no antibodies against A or B markers in your plasma.

(16) _____ [pp.672–673] is a response in which antibodies act against "foreign" cells bearing specific markers and cause them to clump together.

Labeling and Matching

Identify the numbered cell types in the following illustration. Complete the exercise by matching and entering the letter of the appropriate function in the parentheses after the given cell types. A letter may be used more than once. [Blanks are from p.671; parentheses are from p.670]

17. _____
18. _____ _____ ( )
19. _____ ( )
20. _____ ( )
21. _____ ( )
22. _____ ( )
23. _____
24. _____ ( )

A. Phagocytosis
B. Plays a role in the inflammatory response
C. Plays a role in clotting
D. Immunity
E. O_2, CO_2 transport
F. Play a role in producing blood cells

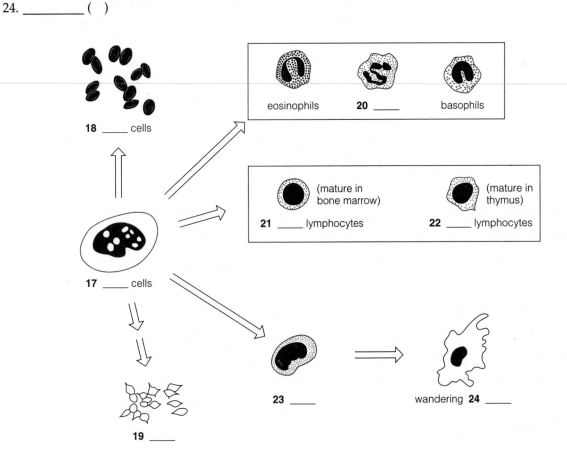

18 _____ cells

eosinophils   20 _____   basophils

(mature in bone marrow)
21 _____ lymphocytes

(mature in thymus)
22 _____ lymphocytes

17 _____ cells

19 _____

23 _____

wandering 24 _____

Matching

[All from p.672 except 27]

25. ___B₁₂ deficiency
26. ___chronic anemias
27. ___erythroblastosis fetalis [p.673]
28. ___hemolytic anemias
29. ___hemorrhagic anemias
30. ___infectious mononucleosis
31. ___leukemias
32. ___polycythemias
33. ___sickle-cell anemia
34. ___thalassemias

A. A potential hazard for strict vegetarians and alcoholics
B. A category of cancers that suppress or impair white blood cell formation in bone marrow
C. Disorder caused by sluggish blood flow caused by far too many red blood cells; "blood doping" and some bone marrow cancers
D. Caused by mixing Rh⁺ and Rh⁻ blood types; RhoGam can prevent this
E. Abnormal forms of hemoglobin caused by a gene mutation
F. Results from a sudden blood loss as from a severe wound
G. Disorders caused by specific infectious bacteria and parasites as they replicate inside red blood cells and then lyse them
H. An Epstein–Barr virus causes this highly contagious disease, which results from too many monocytes and lymphocytes
I. These disorders result from ongoing but slight blood loss; hemorrhoids, a bleeding ulcer, or monthly blood loss by premenopausal women could be the cause

39.5. HUMAN CARDIOVASCULAR SYSTEM [pp.674–675]

39.6. THE HEART IS A LONELY PUMPER [pp.676–677]

Selected Words: "cardiovascular" [p.674], *pulmonary* circuit [p.674], *systemic* circuit [p.674], pericardium [p.676], myocardium [p.676], coronary circulation [p.676], endothelium [p.676], atrium, atria [p.676], ventricle [p.676], *one-way* valves [p.676], systole [p.676], diastole [p.676], striated [p.677], SA node [p.677], AV node [p.677]

Boldfaced, Page-Referenced Terms

[p.674] arteries _____

_____

[p.674] arterioles _____

_____

[p.674] venules _____

_____

[p.674] veins _____

_____

[p.674] aorta _____

_____

[p.676] cardiac cycle _____

_____

[p.677] cardiac conduction system _____

_____

[p.677] cardiac pacemaker _____

_____

Fill-in-the-Blanks

The heart is a pumping station for two major blood transport routes: the (1) _____ [p.674] circulation to and from the lungs, and the (2) _____ [p.674] circulation to and from the rest of the body. In the pulmonary circuit, the heart pumps (3) _____ [p.674]-poor blood to the lungs through the pulmonary arteries; then the (4) _____ [p.675]-enriched blood flows back to the (5) _____ [p.674] through the pulmonary veins.

In an amphibian, reptile, bird, or mammal heart, blood first enters into chambers called (6) _____ [p.676]; blood is pumped from the heart out of the thicker-walled chambers called (7) _____ [p.676]. Each contraction period is called (8) _____ [p.676]; each relaxation period is (9) _____ [p.676]. During a cardiac cycle, contraction of the (10) _____ [p.677] is the driving force for blood circulation; (11) _____ [p.677] contraction helps fill the ventricles.

Fill-in-the-Blanks

Look at Figure 39.11 in your text and use a red pen to redden all tubes in this illustration (except the pulmonary artery) that are indicated by "aorta" or "artery." Memorize the names, then fill in all the blanks. Also redden the parts of the heart that contain oxygen-rich blood. [All from p.675]

12. _____
VEINS
(from brain, neck, head tissues)

13. _____
_____
(from neck, shoulder, arms—veins of upper body)

14. _____
VEINS
(from lungs to heart)

15. _____
VEIN
(from liver to inferior vena cava)

16. _____
VEIN
(from kidneys back to heart)

17. _____
_____
(receives blood from all veins below the diaphragm)

18. _____
VEINS
(carry blood from pelvic organs and lower abdominal wall)

27. _____
ARTERIES

26. _____
ARTERIES
(to lungs)

25. _____
ARTERIES
(to cardiac muscle)

24. _____
ARTERY
(to arm, hand)

23. _____
ARTERY
(to kidney)

22. _____
AORTA
(to digestive tract, pelvic organs)

21. _____
ARTERIES
(to pelvic organs, lower abdominal wall)

20. _____
ARTERY
(to thigh and inner knees)

19. _____
VEIN
(from thigh and inner knee)

Labeling

Identify each indicated part of the accompanying illustrations. [All from p.676]

28. _____

29. _____ _____ _____

30. _____ _____ _____

31. _____ _____

32. _____ _____ _____

33. _____ _____ _____

34. _____ _____ _____

35. _____ _____ _____

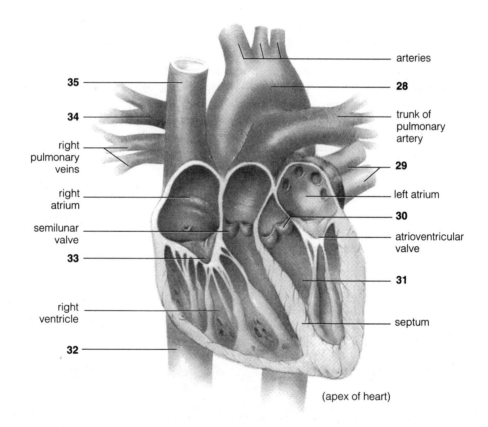

arteries

35

34

right pulmonary veins

right atrium

semilunar valve

33

right ventricle

32

28

trunk of pulmonary artery

29

left atrium

30

atrioventricular valve

31

septum

(apex of heart)

39.7. BLOOD PRESSURE IN THE CARDIOVASCULAR SYSTEM [pp.678–679]

39.8. FROM CAPILLARY BEDS BACK TO THE HEART [pp.680–681]

39.9. *Focus on Health*: CARDIOVASCULAR DISORDERS [pp.682–683]

Selected Words: systolic and *diastolic* pressure [p.678], *pulse pressure* [p.678], epinephrine [p.679], angiotensin [p.679], mean arterial pressure [p.679], carotid arteries [p.679], aortic arch [p.679], medulla oblongata [p.679], *edema* [p.680], *elephantiasis* [p.680], hypertension [p.682], *atherosclerosis* [p.682], *heart attack* [p.682], *strokes* [p.682], *arteriosclerosis* [p.682], *atherosclerotic plaque* [p.683], *thrombus* [p.683], *embolus* [p.683], *angina pectoris* [p.683], *stress electrocardiograms* [p.683], *angiography* [p.683], *coronary bypass surgery* [p.683], *laser angioplasty* [p.683], *balloon angioplasty* [p.683], *arrhythmias* [p.683], *bradycardia* [p.683], *tachycardia* [p.683], *atrial fibrillation* [p.683], *ventricular fibrillation* [p.683]

Boldfaced, Page-Referenced Terms

[p.678] blood pressure _____

_____

[p.679] vasodilation _____

_____

[p.679] vasoconstriction _____

_____

[p.679] baroreceptor reflex _____

_____

[p.680] ultrafiltration _____

_____

[p.680] reabsorption _____

_____

[p.683] LDLs, low-density lipoproteins _____

_____

[p.683] HDLs, high-density lipoproteins _____

_____

Fill-in-the-Blanks

Blood pressure is normally high in the (1) _____ [p.678] immediately after leaving the heart, but then it drops as the fluid passes along the circuit through different kinds of blood vessels. (2) _____ [p.678] are pressure reservoirs that keep blood flowing smoothly away from the heart while the (3) _____ [p.678] are relaxing. Energy in the form of (4) _____ [p.678] is lost as it overcomes (5) _____ [p.678] to the flow of blood. Arterial walls are thick, muscular, and (6) _____ [p.678] and have large diameters. Arteries present [choose one] (7) ❏ much ❏ little [p.678] resistance to blood flow, so pressure [choose one] (8) ❏ drops a lot ❏ does not drop much [p.678] in the arterial portion of the systemic and pulmonary circuits.

The greatest drop in pressure occurs at [choose one] (9) ❏ capillaries ❏ arterioles ❏ veins [p.679]. With this slowdown, blood flow can now be allotted in different amounts to different regions of the body in response to signals from the (10) _____ [p.679] system and endocrine system or even changes in local chemical conditions. (11) _____ [p.678] are control points where adjustments can be made in the volume of blood flow to be delivered to different capillary beds. They offer great resistance to flow, so there is a major drop in (12) _____ [p.679] in these tubes.

When a person is resting, blood pressure is influenced most by reflex centers in the (13) _____ _____ [p.679]. When the resting level of blood pressure increases, reflex centers command the heart to [choose one] (14) ❏ beat more slowly ❏ beat faster [p.679], and command smooth muscle cells in arteriole walls to [choose one] (15) ❏ contract less forcefully ❏ relax [p.679], which results in [choose one] (16) ❏ vasodilation ❏ vasoconstriction [p.679].

A(n) (17) _____ [p.680] is a blood vessel with such a small diameter that red blood cells must flow through it single file; its wall consists of no more than a single layer of (18) _____ [p.680] cells resting on a basement membrane. In each (19) _____ _____ [p.680], small molecules move between the bloodstream and the (20) _____ [p.680] fluid.

Capillary beds are (21) _____ [p.680] zones where substances are exchanged between blood and interstitial fluid. Capillaries have the thinnest walls across which (22) _____ _____ [p.680] drives fluid forcing molecules of water and solutes to move in the same direction. (23) _____ [pp.680–681] serve as large-diameter, low-resistance transport tubes back to the heart. Leading into (23) are (24) _____ [p.680]. (25) _____ [pp.680–681] serve as temporary reservoirs for blood volume; (26) _____ [p.680] are narrower in diameter, some solutes diffuse across their somewhat thinner walls, and some control over capillary pressure also is exerted at these vessels. Veins contain (27) _____ [p.680] that prevent blood from flowing backward; they contain [choose one] (28) ❑ 20–30 ❑ 35–45 ❑ 50–60 [p.681] percent of the total blood volume.

Labeling

Identify each indicated part of the accompanying illustrations. [All from p.678]

29. _____

30. _____

31. _____

32. _____

33. _____ _____

34. _____

a. 29

outer coat 33 basement membrane endothelium

34

b. 30

outer coat smooth muscle between elastic layers basement membrane endothelium

c. 31

outer coat smooth muscle rings over elastic layer basement membrane endothelium

d. 32

basement membrane endothelium

True–False

If the statement is true, write a T in the blank. If false, make it true by changing the underlined word.

_____35. The pulse <u>rate</u> is the difference between the systolic and the diastolic pressure readings. [p.678]

_____36. Because the total volume of blood remains constant in the human body, blood pressure must <u>also</u> <u>remain</u> <u>constant</u> throughout the circuit. [p.678]

Fill-in-the-Blanks

One cause of (37) _____ [p.682] is the rupture of one or more blood vessels in the brain. A(n)

(38) _____ _____ [p.683] blocks a coronary artery. (39) _____ [p.683] is a term for a formation

that can include cholesterol, calcium salts, and fibrous tissue. It is not healthful to have a high concentration

of (40) _____ [p.683]-density lipoproteins in the bloodstream. A clot that stays in place is a

(41) _____ [p.683], but a clot that travels in the bloodstream is an embolus.

39.10. HEMOSTASIS [p.684]
39.11. LYMPHATIC SYSTEM [pp.684–685]

Selected Words: lymph vascular system [p.684], lymph capillaries [p.684], lymph vessels [p.684], "valves" [p.685], *lymphoid* organs and tissues [p.685], tonsils [p.685], red pulp [p.685], white pulp [p.685]

Boldfaced, Page-Referenced Terms

[p.684] hemostasis _____

_____

[p.685] lymph nodes _____

_____

[p.685] spleen _____

_____

[p.685] thymus gland _____

_____

Fill-in-the-Blanks

Bleeding is stopped by several mechanisms that are referred to as (1) _____ [p.684]; the mechanisms

include blood vessel spasm, (2) _____ _____ _____ [p.684], and blood (3) _____ [p.684].

Once the platelets reach a damaged vessel, through chemical recognition they adhere to exposed

(4) _____ [p.684] fibers in damaged vessel walls. Reactions cause rod-shaped proteins to assemble into

long (5) _____ [p.684] fibers. These trap blood cells and components of plasma. Under normal

conditions, a clot eventually forms at the damaged site.

Choice

Choose from these possibilities for numbers 6–10.

a. bone marrow   b. lymph nodes   c. lymph vascular system   d. spleen   e. thymus gland

6. ___ A huge reservoir of red blood cells and a filter of pathogens and used-up blood cells from the blood; contains red pulp. [p.685]

7. ___ Delivers water and plasma proteins from capillary beds to the blood circulation, delivers fats from the small intestine to the blood, and delivers pathogens, foreign cells, and material and cellular debris to the organized disposal centers. [pp.684–685]

8. ___ Immature T lymphocytes become mature here, and hormones are produced here. [p.685]

9. ___ Contain white blood cells that destroy invading bacteria and viruses as they are filtered from the lymph. [p.685]

10. ___ Lymphocytes, which are a kind of white blood cell, are produced here before they go on to other parts of the lymphatic system. [p.685; recall p.671]

Identification/Fill-in-the-Blanks

11. _____ [p.685]

12. _____ gland [p.685]

13. _____ duct [p.685]

14. _____ [p.685]

15. _____ _____ [p.685]

16. organized arrays of _____ [p.684]

17. _____ _____ [p.685]

18. _____ [p.684] vessels reclaim fluid lost from the bloodstream, purify the blood of microorganisms

19. and transport _____ [p.684] from the

20. _____ _____ [p.684] to the bloodstream.

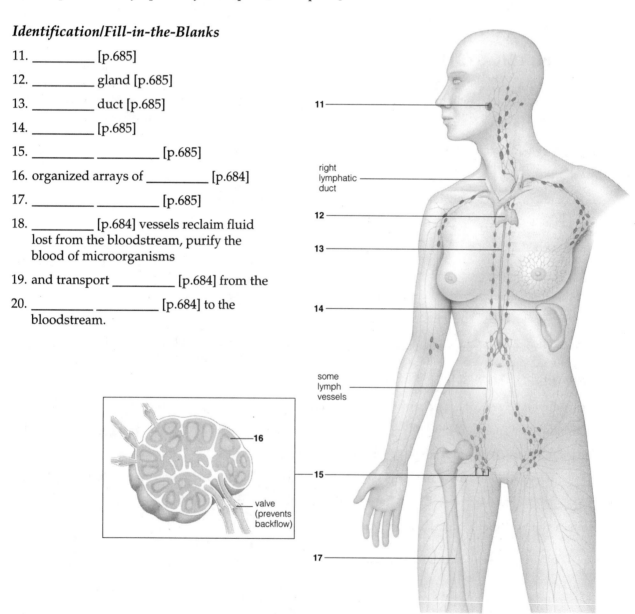

11 ——

right lymphatic duct ——

12 ——

13 ——

14 ——

some lymph vessels ——

16

15 ——

valve (prevents backflow)

17 ——

Self-Quiz

Multiple Choice

___ 1. Most of the oxygen in human blood is transported by _____. [pp.670–671]
   a. plasma
   b. serum
   c. platelets
   d. hemoglobin
   e. leukocytes

___ 2. Of all the different kinds of white blood cells, two classes of _____ are the ones that respond to specific invaders and confer immunity to a variety of disorders. [p.671]
   a. basophils
   b. eosinophils
   c. monocytes
   d. neutrophils
   e. lymphocytes

___ 3. Open circulatory systems generally lack _____. [p.668]
   a. a heart
   b. arterioles
   c. capillaries
   d. veins
   e. arteries

___ 4. Red blood cells originate in the _____. [p.671; recall p.652]
   a. liver
   b. spleen
   c. yellow bone marrow
   d. thymus gland
   e. red bone marrow

___ 5. Hemoglobin contains _____. [p.671]
   a. copper
   b. magnesium
   c. sodium
   d. calcium
   e. iron

___ 6. The pacemaker of the human heart is the _____. [p.677]
   a. sinoatrial node
   b. semilunar valve
   c. inferior vena cava
   d. superior vena cava
   e. atrioventricular node

___ 7. During systole, _____. [p.676]
   a. oxygen-rich blood is pumped to the lungs
   b. the heart muscle tissues contract
   c. the atrioventricular valves suddenly open
   d. oxygen-poor blood from all parts of the human body, except the lungs, flows toward the right atrium
   e. none of the above

___ 8. _____ are reservoirs of blood pressure in which resistance to flow is low. [p.678]
   a. Arteries
   b. Arterioles
   c. Capillaries
   d. Venules
   e. Veins

___ 9. Begin with a red blood cell located in the superior vena cava and travel with it in proper sequence as it goes through the following structures. Which will be *last* in sequence? [p.676]
   a. aorta
   b. left atrium
   c. pulmonary artery
   d. right atrium
   e. right ventricle

___ 10. The lymphatic system is the principal avenue in the human body for transporting _____. [p.684]
   a. fats
   b. wastes
   c. carbon dioxide
   d. amino acids
   e. interstitial fluids

Matching

11. ___agglutination [pp.672–673]

12. ___angiogram, angiography [p.683]

13. ___atherosclerosis [pp.682–683]

14. ___carotid arteries [p.679]

15. ___coronary arteries [p.675]

16. ___edema [p.680]

17. ___embolism, embolus [p.683]

18. ___erythroblastosis fetalis [p.673]

19. ___hypertension [p.682]

20. ___inferior vena cava [p.675]

21. ___jugular veins [p.675]

22. ___renal arteries [p.675]

23. ___stroke [p.682]

24. ___tachycardia [p.683]

25. ___thrombosis, thrombus [p.683]

A. Receive(s) blood from the brain, tissues of the head and neck

B. The heart's own blood supplier(s)

C. Excessively fast rate of heart beat; occurs during heavy exercising

D. Damage to brain caused by damaged blood vessels

E. High blood pressure

F. A blood clot that is on the move from one place to another

G. Diagnostic technique that uses opaque dyes and x-rays to discover blocked blood vessels

H. Deliver(s) blood to the head, neck, brain

I. The clumping of red blood cells, or of antibodies with antigens

J. Delivers blood to kidneys, where its composition and volume are adjusted

K. A blood clot that is lodged in a blood vessel and is blocking it

L. Receive(s) blood from all veins below the diaphragm

M. Progressive thickening of the arterial wall and narrowing of the arterial lumen (space)

N. Blood cell destruction caused by mismatched Rh blood types (Rh⁻ mother and Rh⁺ fetus)

O. Accumulation of excess fluid in interstitial spaces; extreme in elephantiasis

---

Chapter Objectives/Review Questions

1. Distinguish between open and closed circulatory systems. [pp.668–669]
2. Describe the composition of human blood, using percentages of volume. [p.670]
3. Distinguish the five types of leukocytes from each other in terms of structure and functions. [pp.670–671]
4. State where erythrocytes, leukocytes, and platelets are produced. [p.671]
5. Describe how blood is typed for the ABO blood group and for the Rh factor. [pp.672–673]
6. Trace the path of blood in the human body. Begin with the aorta and name all major components of the circulatory system through which the blood passes before it returns to the aorta. [pp.674–675]
7. Explain what causes a heart to beat. Then describe how the rate of heartbeat can be slowed down or speeded up. [pp.676–677]
8. Describe how the structures of arteries, capillaries, and veins differ. [p.678]
9. Explain what causes high pressure and low pressure in the human circulatory system. Then show where major drops in blood pressure occur in humans. [pp.678–679]
10. List the factors that cause blood to leave the heart and the factors that cooperate to return blood to the heart. [pp.678–682]
11. Explain how veins and venules can act as reservoirs of blood volume. [pp.680–681]
12. Describe how hypertension develops, how it is detected, and whether it can be corrected. [p.682]
13. Distinguish a stroke from a coronary artery blockage, occlusion. [pp.682–683]
14. State the significance of high- and low-density lipoproteins to cardiovascular disorders. [p.683]
15. List in sequence the events that occur in the formation of a blood clot. [p.684]
16. Describe the composition and function of the lymphatic system. [pp.684–685]

Integrating and Applying Key Concepts

You observe that some people appear as though fluid had accumulated in their lower legs and feet. Their lower extremities resemble those of elephants. You inquire about what is wrong and are told that the condition is caused by the bite of a mosquito that is active at night. Construct a testable hypothesis that would explain (1) why the fluid was not being returned to the torso, as normal, and (2) what the mosquito did to its victims.

40

IMMUNITY

Interactive Exercises

Russian Roulette, Immunological Style [pp.688–689]

40.1. THREE LINES OF DEFENSE [p.690]

40.2. COMPLEMENT PROTEINS [p.691]

40.3. INFLAMMATION [pp.692–693]

Selected Words: "immune" [p.688], Jenner [p.688], *smallpox* [p.688], Pasteur [p.689], Koch [p.689], *anthrax* [p.689], *Bacillus anthracis* [p.689], athlete's foot [p.690], *Lactobacillus* [p.690], *nonspecific* and *specific* responses [p.690], complement "coat" [p.691], edema [p.693], *chemotaxins* [p.693], *interleukins* [p.693], *lactoferrin* [p.693], *endogenous pyrogen* [p.693], "set point" [p.693], *interleukin-1* [p.693], thromboplastin [p.693], fibrin [p.693]

Boldfaced, Page-Referenced Terms

[p.689] vaccination _____

[p.690] pathogens _____

[p.690] lysozyme _____

[p.691] complement system _____

[p.691] lysis _____

[p.692] neutrophils _____

[p.692] eosinophils _____

[p.692] basophils _____

[p.692] macrophages _____

[p.692] acute inflammation _____

[p.692] mast cells _____

[p.692] histamine _____

[p.693] fever _____

Fill-in-the-Blanks

The best way to deal with damaging foreign agents is to prevent their entry. Several barriers prevent pathogens from crossing the boundaries of your body. Intact skin and (1) _____ [p.690] membranes are effective barriers. (2) _____ [p.690] is an enzyme that destroys the cell wall of many bacteria. (3) _____ [p.690] fluid destroys many food-borne pathogens in the gut. Normal (4) _____ [p.690] residents of the skin, gut, and vagina outcompete pathogens for resources and help keep their numbers under control.

Even simple aquatic invertebrates defend themselves with (5) _____ [p.690] cells and antimicrobial substances including lysozymes that cause bacteria to burst open. In most animals, when a sharp object cuts through the skin and foreign microbes enter, some plasma proteins come to the rescue and seal the wound with a (6) _____ [p.690] mechanism. (7) _____ [p.690] white blood cells engulf bacteria soon thereafter.

Also, vertebrates have about twenty plasma proteins, collectively referred to as the (8) _____ _____ [pp.691–692] that are activated one after another in a "cascade" of reactions to help destroy invading microorganisms. When the complement system is activated, circulating basophils and mast cells in tissues release (9) _____ [p.692], which increases the permeability of (10) _____ [p.692] and makes them "leaky," so fluid seeps out and causes the inflamed area to become swollen and warm.

Vertebrates have (11) _____ [p.690], general defenses such as those just mentioned that isolate and destroy threatening agents identified as "nonself." (11) defenses do not identify *which specific* foreign agent needs to be destroyed; they only identify that there is *some* foreign agent that needs to be destroyed. If foreign agents penetrate the boundaries of your body, (11) defenses constitute a fast way to exterminate them. If (11) defenses fail to control invading pathogens, (12) _____ [p.690] defenses come to the rescue, in which well-coordinated armies of (13) _____ [p.690] blood cells make highly focused counterattacks against pathogens whose specific identity has been identified. These armies and the organs of our body that house them constitute our (14) _____ [p.690] system.

Complete the Table

Complete the following table by providing the specific functions carried out by each of the four different kinds of white blood cells listed. [All from pp.690,692]

Type of Cells	Functions
Basophils	15.
Eosinophils	16.
Neutrophils	17.
Macrophages	18.

40.4. THE IMMUNE SYSTEM [pp.694–695]
40.5. LYMPHOCYTE BATTLEGROUNDS [p.696]
40.6. CELL-MEDIATED RESPONSES [pp. 696–697]
40.7. ANTIBODY-MEDIATED RESPONSES [pp.698–699]

Selected Words: *immunological specificity* [p.694], *immunological memory* [p.694], *self* markers [p.694], *nonself* markers [p.694], *effector* cells [p.694], *memory* cells [p.694], *differentiation* [p.694], "touch killing" [p.694], *processed antigen* [p.694], *antibody-mediated* responses [p.695], *primary* and *secondary* immune responses [p.695], cell-mediated responses [p.696], thymus gland [p.696], clone [p.696], "double signal" [p.696], *perforins* [p.696], *IgM, IgD, IgG, IgA, IgE* [All from p.699]

Boldfaced, Page-Referenced Terms

[p.694] B lymphocytes (=B cells) _____

_____

[p.694] T lymphocytes (=Tcells) _____

_____

[p.694] immune system _____

_____

[p.694] antigen _____

_____

[p.694] MHC markers _____

_____

[p.694] antigen–MHC complexes _____

_____

[p.694] antigen-presenting cell _____

_____

[p.694] helper T cells _____

_____

[p.694] cytotoxic T cells _____

_____

[p.695] antibodies _____

_____

[p.696] TCRs, T-cell receptors _____

_____

[p.696] apoptosis _____

_____

[p.697] natural killer cells (NK cells) _____

_____

[p.697] immunoglobulins, Igs _____

_____

Fill-in-the-Blanks

If the (1) _____ [p.690] defenses (fast-acting white blood cells such as neutrophils, eosinophils, and basophils; slower-acting macrophages; and the plasma proteins involved in clotting and complement) fail to repel the microbial invaders, then the body calls on its (2) _____ _____ [p.694], which identifies *specific* targets to kill and *remembers* the identities of its targets. Your own unique (3) _____ _____ [p.694] patterns identify your cells as "self" cells. Any other surface pattern is, by definition, (4) _____ [p.694], and doesn't belong in your body.

The principal actors of the immune system are (5) _____ [p.694] descended from stem cells (consult Fig. 39.7) in the bone marrow that have two different strategies of action to deal with their different kinds of enemies. (6) _____ _____ [p.695] clones secrete antibodies and act principally against the extracellular (ones that stay outside the body cells) enemies that are pathogens in blood or on the cell surfaces of body tissues. (7) _____ _____ [p.696] clones descend from lymphocytes that matured in the (8) _____ [p.696] gland, where they acquired specific markers on their cell surfaces; they defend principally against intracellular pathogens such as (9) _____ [pp.695–696], and against any cells that are perceived as abnormal or foreign, such as (10) _____ [p.695] cells and cells of organ transplants.

Each kind of cell, virus, or substance bears unique molecular configurations (patterns) that give it a unique (11) _____ [p.694]. A(n) (12) _____ [p.694] is any molecular configuration that causes the formation of lymphocyte armies. Any cell that processes and displays (12) together with a suitable MHC molecule is known as an (13) _____-_____ [p.694] cell that can activate lymphocytes to undergo rapid cell divisions. Lymphocyte subpopulations that fight and destroy enemies are known as (14) _____ [p.694] cells; among these are (15) _____ _____ [p.694] cells that promote the formation of large populations of effector cells and memory cells, and (16) _____ _____ [p.694] cells that eliminate infected body cells and tumor cells by a "touch killing." Together with other "killers," they execute the (17) _____-_____ [pp.694,697] immune responses. By contrast, (6) produce Y-shaped antigen-binding receptor molecules called (18) _____ [p.695]. (6) and (18) carry out the (19) _____-_____ [p.695] immune responses.

Other lymphocyte subpopulations, (20) _____ [p.694] cells, enter a resting phase, but they "remember" the specific agent that was conquered and will undertake a larger, more rapid response if it shows up again.

Labeling

Identify each numbered part in the accompanying illustration. [All from p.695]

21. _____ - _____ _____

22. _____ _____ _____ _____

23. _____ _____ _____

24. _____

25. _____ _____ _____

26. _____ _____ _____

27. _____

28. _____

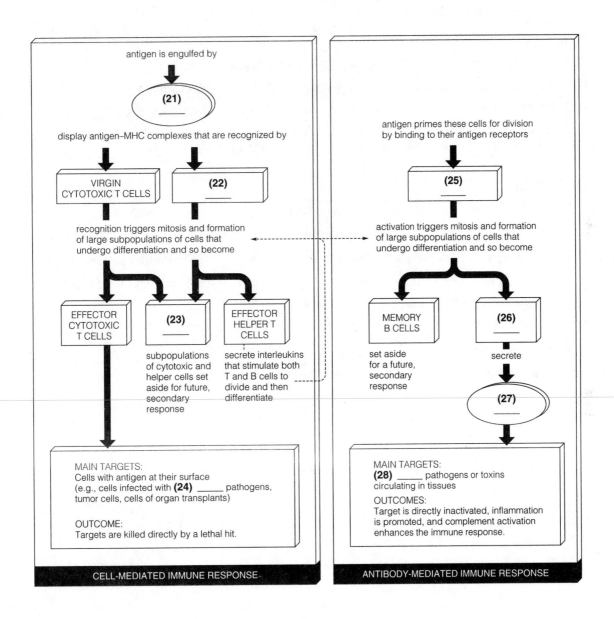

Choice

Match each of the white blood cell types with the appropriate function.

a. effector cytotoxic T cells   b. effector helper T cells   c. macrophages   d. memory cells
e. effector B cells

29. ___ Lymphocytes that directly destroy by "touch-killing" tumor cells, cells of organ transplants, or body cells already infected by certain viruses [pp.695,697]

30. ___ A portion of B and T cell populations that were set aside as a result of a first encounter, now circulate freely and respond rapidly to any later attacks by the same type of invader [p.695, recall p.694]

31. ___ Lymphocytes and their progeny that produce antibodies [pp.695,698]

32. ___ Lymphocytes that secrete interleukins; stimulate the rapid division of B cells and cytotoxic T cells [p.695]

33. ___ "Big eater" white blood cells that develop from monocytes, engulf anything perceived as foreign, and alert helper T cells to the presence of specific foreign agents [recall p.692]

Fill-in-the-Blanks

In addition to effector B cells and cytotoxic T cells (executioner lymphocytes that mature in the thymus),

(34) _____ _____ [p.697] cells mature in other lymphoid tissues and search out any cell that is

either coated with complement proteins or antibodies, *or* bears any foreign molecular pattern. When

cytotoxic T cells find foreign cells, they secrete (35) _____ [pp.696–697] and other toxic substances to

touch-kill and induce cell death by apoptosis.

Labeling

Identify each numbered part in the accompanying illustration.

36. _____ _____ [p.697]

37. _____ _____ [p.697]

38. _____ _____ [p.697]

39. _____ - _____ [p.697]

40. _____ _____ cell [p.698]

41. _____ [p.698]

42. _____ _____ [p.698]

43. _____ _____ [p.698]

Below:
Example of a(n) (36) _____ _____-mediated immune response to a bacterial invasion. [p.697]

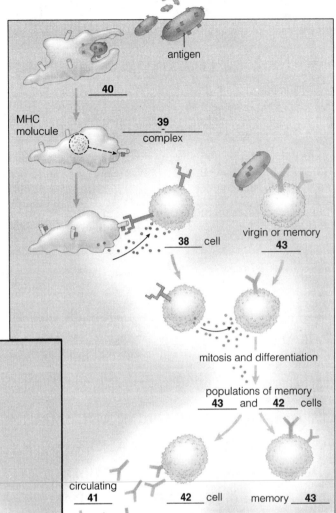

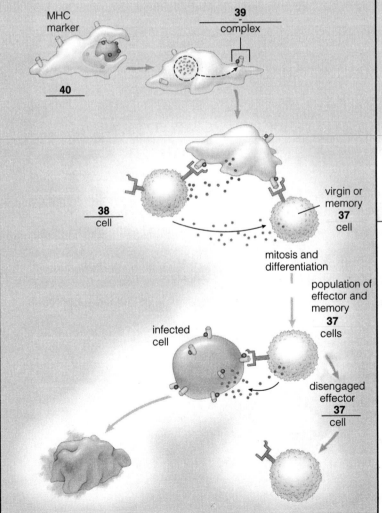

Above:
Example of a(n) (41) _____ [p.698]-mediated immune response to a bacterial invation.

Note: If the same number is used more than once, that is because the label is the same in all such situations. If you have labeled it correctly on the lefthand illustration, then its identification should also be correct on the other.

Fill-in-the-Blanks

The (44) _____ _____ _____ [p.695] is the route taken during a first-time contact with an antigen; the secondary immune response is more rapid because patrolling battalions of (45) _____ _____ [p.695] are in the bloodstream on the lookout for enemies they have conquered before. When these meet up with recognizable (46) _____ [p.695], they divide at once, producing large clones of B or T cells within two to three days.

Complete the Table

Immunoglobulin Type	Function
47. [p.699]	Enter mucus-coated surfaces of the respiratory, digestive, and reproductive tracts where they neutralize pathogens
48. [p.699]	Triggers inflammation when parasitic worms attack the body; play a role in allergic responses
49. [p.699]	Activate complement proteins; neutralize many toxins; long-lasting; can cross placenta and protect developing fetus; also present in colostrum from mammary glands
50. [p.699]	First to be secreted during immune responses; after binding to antigen, trigger complement cascade; also tag invaders and bind them in clumps for later phagocytosis

40.8. *Focus on Health:* CANCER AND IMMUNOTHERAPY [p.699]

40.9. IMMUNE SPECIFICITY AND MEMORY [pp.700–701]

40.10. DEFENSES ENHANCED, MISDIRECTED, OR COMPROMISED [pp.702–703]

40.11. *Focus on Health:* AIDS—THE IMMUNE SYSTEM COMPROMISED [pp.704–705]

Selected Words: carcinomas [p.699], sarcomas [p.699], leukemia [p.699], *immunotherapy* [p.699], C. Milstein and G. Kohler [p.699], *monoclonal antibodies* [p.699], "plantibodies" [p.699], *LAK* cells [p.699], *therapeutic vaccines* [p.699], *dendritic* cells [p.699], melanoma [p.699], *variable* regions [p.700], V segments and J segments [p.700], M. Burnet [p.701], *clonal selection* hypothesis [p.701], *active* immunization [p.702], booster injection [p.702], *passive* immunization [p.702], *asthma* [p.702], *hay fever* [p.702], *anaphylactic shock* [p.702], antihistamine [p.702], *Grave's disorder* [p.703], *myasthenia gravis* [p.703], *rheumatoid arthritis* [p.703], *severe combined immunodeficiencies* (SCIDs) [p.703], AIDS [p.704], *Pneumocystis carinii* [p.704], protease inhibitors, AZT, and ddi [p.705]

Boldfaced, Page-Referenced Terms

[p.702] immunization _____

_____

[p.702] vaccine _____

_____

[p.702] allergens _____

_____

[p.702] allergy _____

_____

[p.703] autoimmune response _____

_____

Fill-in-the-Blanks

Various strategems have been developed that enhance immunological defenses against tumors as well as against certain pathogens; collectively they are referred to as (1) _____ [p.699].

The term (2) _____ [p.699] refers to cells that have lost control over cell division. Milstein and Kohler developed a means of producing large amounts of (3) _____ _____ [p.699], which are produced by clones of proliferating (4) _____ [p.699] cells: Mouse B cells fused with tumor cells that divide nonstop. A different therapy involves researchers extracting (5) _____ [p.699] from tumors, activating them by exposing them to a (6) _____ [p.699], and injecting these *LAK* cells back into the patient, where they actively kill tumor cells.

Antibodies are plasma proteins that are part of the (7) _____ [p.699] group of proteins. Some of these circulate in blood; others are present in other body fluids or bound to B cells. (8) _____ [p.698] are Y-shaped proteins that lock onto specific foreign targets and thereby tag them for destruction by phagocytes or by activating the complement system. While B or T cells are maturing, different regions of the genes that code for antigen receptors are shuffled at random into one of millions of possible combinations; this process of (9) _____ [p.700] produces a gene that codes for just one of millions of possible antigen receptors.

Labeling

Identify each numbered part in the accompanying illustration and its legend. [All from p.701]

10. _____

11. _____

12. _____ _____

13. _____ _____

14. _____ _____

Clonal selection of a (12) _____ _____ cell, the descendants of which produced the specific (10) _____ that can combine with a specific antigen.

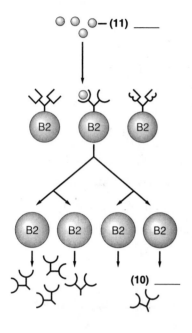

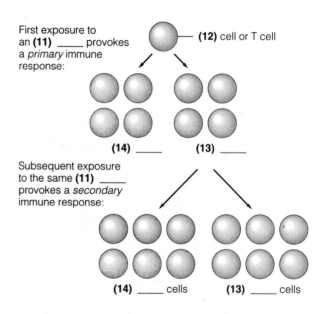

Fill-in-the-Blanks

The (15) _____ _____ _____ [p.701] explains how an individual has immunological memory, which refers to the body's capacity to make a secondary immune response to any subsequent encounter with the same type of (16) _____ [p.701] that provoked the primary response.

Deliberately provoking the production of memory lymphocytes is known as (17) _____ [p.702]. In a(n) (18) _____ [p.702] immunization, a vaccine containing antigens is injected into the body or taken orally. The first one elicits a (19) _____ _____ _____ [p.702], and the second one (a booster shot) elicits a secondary immune response, which causes the body to produce more antibodies and (20) _____ _____ [p.702] to provide long-lasting protection.

Complete the Table

Indicate with a check (√) the recommended age(s) of vaccination. [All from p.702]

Age Vaccination Is Administered

Disease	(a) At birth	(b) At 2 mos.	(c) 1–4 mos.	(d) 4 mos.	(e) 6 mos.	(f) 6–18 mos.	(g) 12–15 mos.	(h) 12–18 mos.	(i) 4–6 yrs.	(j) 11–12 yrs.
21. Diphtheria										
22. Hemophilus influenzae										
23. Hepatitis B										
24. Measles										
25. Mumps										
26. Polio										
27. Rubella										
28. Tetanus										
29. Whooping cough (pertussis)										

Fill-in-the-Blanks

(30) _____ [p.702] is an altered secondary response to a normally harmless substance that may actually cause injury to tissues. (31) _____ _____ [p.703] is a disorder in which the body mobilizes its forces against certain of its own tissues. (32) _____ _____ [p.703] is an example of this kind of disorder in which antibodies tag acetylcholine receptors on skeletal muscle cells and cause progressive weakness. (33) _____ _____ [p.703] is a similar kind of disorder in which skeletal joints are chronically inflamed.

AIDS is a constellation of disorders that follow infection by the (34) _____ _____ _____ [p.704]. In the United States, transmission has occurred most often among intravenous drug abusers who share needles and among (35) _____ _____ [p.704]. HIV is a (36) _____ [p.704]; its genetic material is RNA rather than DNA, and it has several copies of an enzyme (37) _____ _____ [p.704], which uses the viral RNA as a template for making DNA, which is then inserted into a host chromosome.

HIV is transmitted when (38) _____ _____ [p.704] of an infected person enter another person's tissues. The virus cripples the immune system by attacking (39) _____ _____ [p.704] cells and antigen-presenting (40) _____ [p.704]. In sub-Saharan Africa, 1997 reported AIDS cases numbered (41) _____ [p.704] (number) people. Worldwide, an estimated (42) _____ [p.704] (number) are infected.

Self-Quiz

Multiple Choice

___ 1. All the body's phagocytes are derived from stem cells in the _____. [p.692]
   a. spleen
   b. liver
   c. thymus
   d. bone marrow
   e. thyroid

___ 2. The plasma proteins that are activated when they contact a bacterial cell are collectively known as the _____ system. [p.691]
   a. shield
   b. complement
   c. IgG
   d. MHC
   e. HIV

___ 3. _____ are divided into two groups: T cells and B cells. [p.694]
   a. Macrophages
   b. Lymphocytes
   c. Platelets
   d. Complement cells
   e. Cancer cells

___ 4. _____ produce and secrete antibodies that set up bacterial invaders for subsequent destruction by macrophages. [p.698]
   a. B cells
   b. Phagocytes
   c. T cells
   d. Bacteriophages
   e. Thymus cells

___ 5. Antibodies are shaped like the letter _____. [p.698]
   a. Y
   b. W
   c. Z
   d. H
   e. E

___ 6. The markers for every cell in the human body are referred to by the letters _____. [p.694]
   a. HIV
   b. MBC
   c. RNA
   d. DNA
   e. MHC

___ 7. Effector B cells _____. [pp.698–699]
   a. fight against extracellular pathogens and toxins circulating in tissues
   b. develop from antigen-presenting cells
   c. manufacture and secrete antibodies
   d. do not divide and form clones
   e. all of the above

___ 8. Clones of B or T cells are _____. [p.694]
   a. being produced continually
   b. sometimes known as memory cells if they keep circulating in the bloodstream
   c. only produced when their surface proteins recognize other specific proteins previously encountered
   d. produced and mature in the bone marrow
   e. both (b) and (c)

___ 9. Whenever the body is reexposed to a specific sensitizing agent, IgE antibodies cause _____. [pp.699,702]
   a. prostaglandins and histamine to be produced
   b. clonal cells to be produced
   c. histamine to be released
   d. the immune response to be suppressed
   e. none of the above

___10. The clonal selection hypothesis explains _____. [p.701]
   a. how self cells are distinguished from nonself cells
   b. how B cells differ from T cells
   c. how so many different kinds of antigen-specific receptors can be produced by lymphocytes
   d. how memory cells are set aside from effector cells
   e. how antigens differ from antibodies

Matching

Choose the most appropriate description for each term.

11. ____allergy [p.702]

12. ____antibody [p.698]

13. ____antigen [p.694]

14. ____macrophage [p.692]

15. ____clone [p.696]

16. ____complement [p.691]

17. ____histamine [p.692]

18. ____MHC marker [p.694]

19. ____effector B cell [p.699]

20. ____T cell [p.696]

A. Begins its development in bone marrow, but matures in the thymus gland
B. Cells that have directly or indirectly descended from the same parent cell
C. A potent chemical that causes blood vessels to dilate and let protein pass through the vessel walls
D. Y-shaped immunoglobulin
E. A nonself marker
F. The progeny of turned-on B cells
G. A group of about twenty proteins that participate in the inflammatory response
H. An altered secondary immune response to a substance that is normally harmless to other people
I. The basis for self-recognition at the cell surface
J. Principal perpetrator of phagocytosis

Chapter Objectives/Review Questions

1. Describe typical external barriers that organisms present to invading organisms. [p.690]
2. List and discuss four nonspecific defense responses that serve to exclude microbes from the body. [pp.690–692]
3. List the three general types of cells that form the basis of the vertebrate immune system. [p.690]
4. Explain how the complement system is related to an inflammatory response. [pp.691–692]
5. Describe the sequence of events that occur during inflammatory responses. [pp.692–693]
6. Explain why the immune system of mammals usually does not attack "self" tissues. Understand how vertebrates (especially mammals) recognize and discriminate between self and nonself tissues. [p.694]
7. Distinguish the roles of T cells from the roles of B cells. [pp.694–695]
8. Explain what is meant by *primary immune response*, as contrasted with *secondary immune response*. [p.695]
9. Distinguish between the antibody-mediated response pattern and the cell-mediated response pattern. [pp.696–699]
10. Explain what monoclonal antibodies are, and tell how they are currently being used in passive immunization and cancer treatment. [pp.699,702]
11. Describe how recognition proteins and antibodies are made. State how they are used in immunity. [pp.700–701]
12. Describe the clonal selection hypothesis, and tell what it helps to explain. [p.701]
13. Describe two ways that people can be immunized against specific diseases. [p.702]
14. Distinguish allergy from autoimmune disorder. [pp.702–703]
15. Describe some examples of immune failures, and identify as specifically as you can which weapons in the immunity arsenal failed in each case. [pp.703–705]
16. Describe how AIDS specifically interferes with the human immune system. [pp.704–705]

Integrating and Applying Key Concepts

Suppose you wanted to get rid of forty-seven warts that you have on your hands by treating them with monoclonal antibodies. Outline the steps you would have to take.

41

RESPIRATION

Interactive Exercises

Conquering Chomolungma [pp.708–709]

41.1. THE NATURE OF RESPIRATION [p.710]

41.2. INVERTEBRATE RESPIRATION [p.711]

41.3. VERTEBRATE RESPIRATION [pp.712–713]

Selected Words: hypoxia [p.708], hyperventilate [p.708], homeostasis [p.709], surface-to-volume ratio [p.710], ventilation [p.710], transport pigments [p.710], book lungs [p.711], *external* gills [p.712], *internal* gills [p.712], internal respiratory surfaces [p.712]

Boldfaced, Page-Referenced Terms

[p.709] respiration _____

_____

[p.709] respiratory systems _____

_____

[p.710] pressure gradients _____

[p.710] partial pressure _____

_____

[p.710] respiratory surface _____

_____

[p.710] Fick's law_____

_____

[p.710] hemoglobin _____

_____

[p.711] integumentary exchange _____

_____

[p.711] gills _____

_____

[p.711] tracheal respiration _____

_____

[p.712] countercurrent flow _____

_____

[p.712] lungs _____

_____

[p.713] vocal cords _____

_____

[p.713] glottis _____

_____

Fill-in-the-Blanks

A(n) (1) _____ [p.711] is an outfolded, thin, moist membrane endowed with blood vessels; it may (as in fish) be protected by bony covering, or (as in aquatic insects) it may be naked. Gas transfer is enhanced by (2) _____ _____ [p.712], in which water flows past the bloodstream in the opposite direction. Insects have (3) _____ [p.711] (chitin-lined air tubes leading from the body surface to the interior). At sea level, atmospheric pressure is approximately 760 mm Hg, and oxygen represents about (4) _____ [p.710] percent of the total volume of air.

The energy to drive animal activities comes mainly from (5) _____ _____ [p.709], which uses (6) _____ [p.709] and produces (7) _____ _____ [p.709] wastes. In a process called (8) _____ [p.709], animals move (6) into their internal environment and give up (7) to the external environment.

All respiratory systems make use of the tendency of any gas to diffuse down its (9) _____ _____ [p.710]. Such a (9) exists between (6) in the atmosphere [(10) _____ [p.710] pressure] and the metabolically active cells in body tissues, where (6) is used rapidly; pressure is (11) ❏ highest ❏ lowest [p.709] here. Another (9) exists between (7) in body tissues where (12) [choose one] ❑ high ❑ low [p.709] pressure exists and the atmosphere, with its (13) ❑ higher ❑ lower [p.709] amount of (7).

The more extensive the (14) _____ _____ [p.710] of a respiratory surface membrane and the larger the differences in (15) _____ _____ [p.710] across it, the faster a gas diffuses across the membrane. (16) _____ [p.710] is an important transport pigment, each molecule of which can bind loosely with as many as four O_2 molecules in the lungs.

A(n) (17) _____ [p.712] is an internal respiratory surface in the shape of a cavity or sac. In all lungs, (18) _____ [p.713] carry gas to and from one side of the respiratory surface, and (19) _____ [pp.710,713] in blood vessels carries gas to and from the other side.

Labeling

Identify the numbered parts of the accompanying illustration, which shows the respiratory system of many fishes. [All from p.712]

20. _____

21. _____ _____

22. _____-_____ _____

23. _____-_____ _____

24. _____

25. _____

water in

20 water out

gill arch

21

22

23

direction of **24** flow (gray arrow) and **25** flow (black arrow)

Fill-in-the-Blanks

Oxygen is said to exert a (26) _____ _____ [p.710] of 760/21 or 160 mm Hg. (27) _____ [pp.709–710] alone moves oxygen from the alveoli into the bloodstream, and it is enough to move (28) _____ _____ [pp.709–710] in the reverse direction. (29) _____ [p.708] is the medical name for oxygen deficiency; it is characterized by faster breathing, faster heart rate, and significant ion imbalances in cerebrospinal fluid at altitudes of 8,000 feet above sea level.

41.4. HUMAN RESPIRATORY SYSTEM [pp.714–715]

41.5. BREATHING—CYCLIC REVERSALS IN AIR PRESSURE GRADIENTS [pp.716–717]

Selected Words: laryngitis [p.715], pleural membrane [p.715], thoracic cavity [p.715], *pleurisy* [p.715], *respiratory* bronchioles [p.715], "bronchial tree" [p.715], vocal cords [p.715], glottis [p.715], *inhalation* [p.716], *exhalation* [p.716], *atmospheric* pressure [p.716], *intrapulmonary* pressure [p.716], *intrapleural* pressure [p.716]

Boldfaced, Page-Referenced Terms

[p.714] alveolus, alveoli _____

_____

[p.715] pharynx _____

_____

[p.715] larynx _____

_____

[p.715] epiglottis _____

_____

[p.715] trachea _____

_____

[p.715] bronchus (pl., bronchi) _____

_____

[p.715] diaphragm _____

_____

[p.715] bronchioles _____

_____

[p.716] respiratory cycles _____

_____

[p.717] vital capacity _____

_____

[p.717] tidal volume _____

_____

Fill-in-the-Blanks

During inhalation, the (1) _____ [p.717] moves downward and flattens, and the (2) _____ _____ [p.717] moves outward and upward; when these things happen, the chest cavity volume (3) [choose one] ❑ increases, ❑ decreases [p.717], and the internal pressure (4) [choose one] ❑ rises, ❑ drops, ❑ stays [p.717] the same. Every time you take a breath, you are (5) _____ [p.717] the respiratory surfaces of your lungs. The (6) _____ _____ [pp.714–715] surrounds each lung. In succession, air passes through the nasal cavities, pharynx, past the epiglottis, through the (7) _____ [p.715] (the space between the true vocal cords), and into the (8) _____ [p.715], into the trachea, and then to the (9) _____ [p.715], (10) _____ [p.715], and alveolar ducts. Exchange of gases occurs across the epithelium of the (11) _____ [p.715].

Labeling

Identify each indicated part of the accompanying illustration. [All from p.714]

12. _____ _____
13. _____
14. _____
15. _____
16. _____ _____
17. _____
18. _____
19. _____
20. _____ _____
21. _____ _____
22. _____ _____
23. _____
24. _____ (singular) _____ (plural)
25. _____

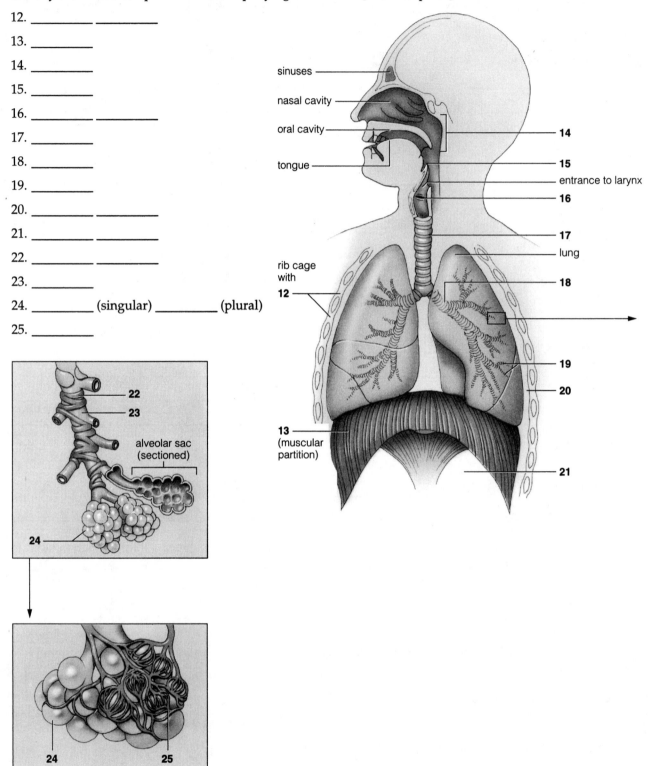

sinuses

nasal cavity

oral cavity

tongue

14

15

entrance to larynx

16

17

lung

18

rib cage with

12

19

20

13 (muscular partition)

21

22

23

alveolar sac (sectioned)

24

24   **25**

Complete the Table

26. Complete the following table with the structures that carry out the functions listed.

Structure	Function
a. [pp.714–715]	Thin-walled sacs where O_2 diffuses into body fluids and CO_2 diffuses out
b. [p.715]	Increasingly branched airways that connect the trachea and alveoli
c. [p.715]	Muscle sheet that contracts during inhalation
d. [p.715]	Airway where breathing is blocked while swallowing and where sound is produced
e. [p.715]	Airway that enhances speech sounds; connects nasal cavity with larynx

41.6. GAS EXCHANGE AND TRANSPORT [pp.718–719]

41.7. *Focus on Health:* WHEN THE LUNGS BREAK DOWN [pp.720–721]

41.8. RESPIRATION IN UNUSUAL ENVIRONMENTS [pp.722–723]

41.9. *Focus on Science:* RESPIRATION IN LEATHERBACK SEA TURTLES [pp.724–725]

Selected Words: bicarbonate (HCO_3^-) [p.718], buffer [p.719], breathing *rhythm* [p.719], breathing *magnitude* [p.719], medulla oblongata [p.719], pons [p.719], aortic bodies [p.719], carotid bodies [p.719], partial pressure gradients [p.719], *apnea* [p.719], *sudden infant death syndrome* (SIDS) [p.719], *bronchitis* [p.720], *emphysema* [p.720], *secondhand smoke* [p.720], "smoker's cough" [p.721], lung cancer [p.721], *pot* [p.721], *carbon monoxide poisoning* [p.722], myoglobin [p.723], plexuses [p.723], nitrogen narcosis [p.723], "the bends" [p.723], decompression sickness [p.723]

Boldfaced, Page-Referenced Terms

[p.718] heme groups _____

_____

[p.718] oxyhemoglobin (HbO_2) _____

_____

[p.718] carbaminohemoglobin (HbO_2) _____

_____

[p.718] carbonic anhydrase _____

_____

[p.722] acclimatization _____

_____

[p.722] erythropoietin _____

_____

Fill-in-the-Blanks

An inward-directed gradient for (1) _____ [p.718] is maintained because inhalations continually replenish (1) and cells continually use it during the last stage of the (2) _____ _____ [recall p.709] process. An outward-directed gradient for (3) _____ _____ [p.718] is maintained because cells continually produce (3) and exhalation removes it. Each (4) _____ [p.718] molecule is constructed of four polypeptide chains compactly bound to four iron-containing, nitrogenous heme groups; the iron atom of each heme group binds reversibly with (1). Of all inhaled (1), 98.5 percent is bound to the heme groups of (4).

About 60 percent of the carbon dioxide in the blood is transported as (5) _____ [p.718]. Without (6) _____ [p.718], the plasma would be able to carry only about 2 percent of the oxygen that whole blood carries. When oxygen-rich blood reaches a(n) (7) _____ [p.718] tissue capillary bed, oxygen diffuses outward, and carbon dioxide moves from tissues into the capillaries. When the (8) _____ _____ [p.719] of carbon dioxide is lower in the alveoli than in the neighboring blood capillaries, carbonic acid dissociates to form water and carbon dioxide. Chemoreceptors in the brain monitor and coordinate signals coming in from arterial walls, from blood vessels, and from other brain regions. One respiratory center, the (9) _____ _____ [p.719], regulates contractions of the diaphragm and intercostal muscles associated with inhalation and exhalation.

(10) _____ [p.720] is the distension of lungs and the loss of gas exchange efficiency such that running, walking, and even exhaling are painful experiences. At least 90 percent of all (11) _____ _____ [p.721] deaths are the result of cigarette smoking; only about 10 percent of afflicted individuals will survive.

When a diver ascends, (12) _____ [p.723] tends to move out of the tissues and into the bloodstream. If the ascent is too rapid, many bubbles of (12) collect at the (13) _____ [p.723], hence the common name, "the bends," for what is otherwise known as (14) _____ [p.723] sickness.

Self-Quiz

Matching

1. ___bronchioles [p.715]
2. ___bronchitis [p.720]
3. ___carbonic anhydrase [p.718]
4. ___emphysema [p.720]
5. ___Fick's law [p.710]
6. ___glottis [p.715]
7. ___hypoxia [p.708]
8. ___intercostal muscles [pp.714,716]
9. ___larynx [p.715]
10. ___oxyhemoglobin [p.718]
11. ___pharynx [p.715]
12. ___pleurisy [p.715]
13. ___tidal volume [p.717]
14. ___ventilation [p.710]
15. ___vital capacity [p.717]

A. Membrane that encloses human lung becomes inflamed and swollen; painful breathing generally results
B. HbO_2
C. The amount of air inhaled and exhaled during normal breathing of a human at rest; generally about 500 ml
D. Throat passageway that connects to *both* the respiratory tract below *and* the digestive tract
E. One gas will diffuse faster than another if its surface area is more extensive and its partial pressure is greater than that of the other gas.
F. Inflammation of the two principal passageways that lead air into the human lungs
G. Contract when air is leaving the lungs; relax when lungs are filling with air
H. The opening into the "voicebox"
I. Finer and finer branchings that lead to alveoli
J. Maximum volume of air that can move out of your lungs after a single, maximal inhalation
K. An enzyme that increases the rate of production of H_2CO_3 from CO_2 and H_2O
L. Lungs have become distended and inelastic so that walking, running and even exhaling are difficult.
M. Where sound is produced by vocal cords
N. Movements that keep air or water moving across a respiratory surface
O. Too little oxygen is being distributed in the body's tissues.

Multiple Choice

___16. Most forms of life depend on _____ to obtain oxygen and eliminate carbon dioxide. [pp.709–710]
a. active transport
b. bulk flow
c. diffusion
d. osmosis
e. muscular contractions

___17. _____ is the most abundant gas in Earth's atmosphere. [p.710]
a. Water vapor
b. Oxygen
c. Carbon dioxide
d. Hydrogen
e. Nitrogen

___18. With respect to respiratory systems, countercurrent flow is a mechanism that explains how _____. [p.712]
a. oxygen uptake by blood capillaries in the lamellae of fish gills occurs
b. ventilation occurs
c. intrapleural pressure is established
d. sounds originating in the vocal cords of the larynx are formed
e. all of the above

___19. _____ have the most efficient respiratory system. [p.713]
a. Amphibians
b. Reptiles
c. Birds
d. Mammals
e. Humans

___20. During inhalation, _____. [p.716]
    a. the pressure in the thoracic cavity (intrapleural pressure) is less than the pressure within the lungs (intrapulmonary pressure)
    b. the pressure in the chest cavity (intrapleural pressure) is greater than the pressure within the lungs (intrapulmonary pressure)
    c. the diaphragm moves upward and becomes more curved
    d. the thoracic cavity volume decreases
    e. all of the above

___21. Oxygen moves from alveoli to the bloodstream _____. [p.718]
    a. whenever the concentration of oxygen is greater in alveoli than in the blood
    b. by means of active transport
    c. by using the assistance of carbaminohemoglobin
    d. principally due to the activity of carbonic anhydrase in the red blood cells
    e. by all of the above

___22. Immediately before reaching the alveoli, air passes through the _____. [p.715]
    a. bronchioles
    b. glottis
    c. larynx
    d. pharynx
    e. trachea

___23. Hemoglobin _____. [p.718]
    a. releases oxygen more readily in tissues with high rates of cellular respiration
    b. tends to release oxygen in places where the temperature is lower
    c. tends to hold on to oxygen when the pH of the blood drops
    d. tends to give up oxygen in regions where partial pressure of oxygen exceeds that in the lungs
    e. all of the above

___24. Oxyhemoglobin releases O_2 when _____. [p.718]
    a. carbon dioxide concentrations are high
    b. body temperature is lowered
    c. pH values are high
    d. CO_2 concentrations are low
    e. all of the above occur

___25. Nonsmokers live an average of _____ longer than people in their mid-twenties who smoke two packs of cigarettes each day. [p.721]
    a. 6 months
    b. 1–2 years
    c. 3–5 years
    d. 7–9 years
    e. over 12 years

Chapter Objectives/Review Questions

1. Understand how the human respiratory system is related to the circulatory system, to cellular respiration, and to the nervous system. [pp.709,718–719]
2. Understand the behavior of gases and the types of respiratory surfaces that participate in gas exchange. [pp.710–713]
3. Describe how incoming oxygen is distributed to the tissues of insects, and contrast this process with the process that occurs in mammals. [pp.711,713]
4. Define *countercurrent flow* and explain how it works. State where such a mechanism is found. [pp.712–713]
5. List all the principal parts of the human respiratory system, and explain how each structure contributes to transporting oxygen from the external world to the bloodstream. [pp.714–715]
6. Describe the relationship of the human lung to the pleural sac and to the thoracic cavity. [pp.716–717]
7. Describe what happens to carbon dioxide when it dissolves in water under conditions normally present in the human body. [p.718]
8. List the factor(s) that cause(s) oxygen to diffuse from the bloodstream into the tissues far from the lungs. Then list the factors that cause carbon dioxide to diffuse into the bloodstream from the same tissues. [pp.718–719]
9. Explain why oxygen diffuses from alveolar air spaces, through interstitial fluid, and across capillary epithelium. Then explain why carbon dioxide diffuses in the reverse direction. [pp.718–719]
10. List the structures involved in detecting carbon dioxide levels in the blood and in regulating the rate of breathing. Name the location of each structure. [p.719]

11. Distinguish bronchitis from emphysema. Then explain how lung cancer differs from emphysema. [pp.720–721]
12. List some of the ways that respiratory systems are adapted to unusual environments. [pp.722–723]

Integrating and Applying Key Concepts

Consider the amphibians—animals that generally have aquatic larval forms (tadpoles) and terrestrial adults. Outline the respiratory changes that you think might occur as an aquatic tadpole metamorphoses into a land-going juvenile.

42

DIGESTION AND HUMAN NUTRITION

Interactive Exercises

Lose It—And It Finds Its Way Back [pp.728–729]

42.1. THE NATURE OF DIGESTIVE SYSTEMS [pp.730–731]

42.2. OVERVIEW OF THE HUMAN DIGESTIVE SYSTEM [p.732]

42.3. INTO THE MOUTH, DOWN THE TUBE [p.733]

Selected Words:

anorexia nervosa [p.728], *bulimia* [p.728], *food-gathering* region [p.730], *food-processing* region [p.730], *crop* [p.730], gizzard [p.730], "chewing cud" [p.731], *lumen* [p.732], *caries* [p.733], *gingivitis* [p.733], *periodontal disease* [p.733], oral cavity [p.733], mouth [p.733]

Boldfaced, Page-Referenced Terms

[p.729] nutrition _____

_____

[p.729] digestive system _____

_____

[p.730] incomplete digestive system _____

_____

[p.730] complete digestive system _____

_____

[p.730] mechanical processing and motility _____

_____

[p.730] secretion _____

_____

[p.730] digestion _____

_____

[p.730] absorption _____

_____

[p.730] elimination _____

_____

[p.731] ruminants _____

_____

[p.732] gut _____

_____

[p.733] tooth _____

_____

[p.733] tongue _____

_____

[p.733] saliva _____

_____

[p.733] pharynx _____

_____

[p.733] esophagus _____

_____

[p.733] sphincter _____

_____

Fill-in-the-Blanks

(1) _____ [p.729] is a large concept that encompasses processes by which food is ingested, digested, absorbed, and later converted to the body's own (2) _____ [p.729], lipids, proteins, and nucleic acids. A digestive system is some form of body cavity or tube in which food is reduced first to (3) _____ [p.729] and then to small (4) _____ [p.729]. Digested nutrients are then (5) _____ [p.729] into the internal environment. The (6) _____ [p.729] system distributes nutrients to cells throughout the body. The (7) _____ [p.729] system supplies oxygen to the cells so that they can oxidize the carbon atoms of food molecules, thereby changing them to the waste product (8) _____ _____ [p.729], which is eliminated by the same system. And if excess water, salts, and wastes accumulate in the blood, the (9) _____ [p.729] system and skin will maintain the volume and composition of blood and other body fluids. A(n) (10) _____ [p.730] digestive system has only one opening, two-way traffic, and a highly branched gut cavity that serves both digestive and (11) _____ [p.730] functions. A(n) (12) _____ [p.730] digestive system has a tube or cavity with regional specializations and a(n) (13) _____ [p.730] at each end. (14) _____ [p.730] involves the muscular movement of the gut wall, but (15) _____ [p.730] is the release into the lumen of enzyme fluids and other substances required to carry out digestive functions. The pronghorn antelope feeds on plant material that breaks down slowly. Antelopes are (16) _____ [p.731]: hoofed mammals that have multiple stomach chambers; nutrients are released slowly when the animal rests. In comparison with human molars, the antelope molar has a much higher (17) _____ [p.731]; its teeth wear down rapidly because their plant diet is mixed with abrasive bits of dirt.

The human digestive system is a tube, 21–30 feet long in an adult, that has regions specialized for different aspects of digestion and absorption; they are, in order, the mouth, pharynx, esophagus, (18) _____ [p.732], (19) _____ _____ [p.732], large intestine, rectum, and (20) _____ [p.732]. Various (21) _____ [p.732] structures secrete enzymes and other substances that are also essential to the breakdown and absorption of nutrients; these include the salivary glands, liver, gallbladder, and (22) _____ [p.732].

Saliva contains an enzyme (23) _____ _____ [p.733] that breaks down starch. Contractions force the larynx against a cartilaginous flap called the (24) _____ [p.733], which closes off the trachea. The (25) _____ [p.733] is a muscular tube that propels food to the stomach.

Labeling

Identify each numbered structure in the accompanying illustration. [All from p.732]

26. _____ _____
27. _____
28. _____
29. _____
30. _____
31. _____ _____
32. _____ _____
33. _____
34. _____
35. _____
36. _____ _____

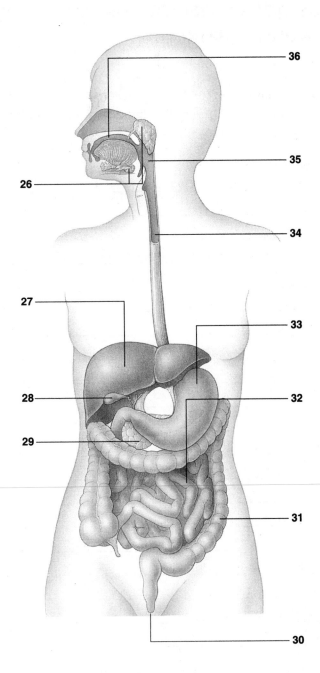

42.4. DIGESTION IN THE STOMACH AND SMALL INTESTINE [pp.734–735]

42.5. ABSORPTION IN THE SMALL INTESTINE [pp.736–737]

42.6. DISPOSITION OF ABSORBED ORGANIC COMPOUNDS [p.738]

42.7. THE LARGE INTESTINE [p.739]

Selected Words: heartburn [p.734], *peptic ulcer* [p.734], *Helicobacter pylori* [p.734], peristalsis [p.734], duodenum, jejunum, and ileum [p.734], trypsin and chymotrypsin [p.735], "emulsion" [p.735], secretin [p.735], CCK (cholecystokinin) [p.735], GIP (glucose insulinotropic peptide) [p.735], mucosa [p.735], gastrin [p.739], constipation [p.739], appendicitis [p.739], colon cancer [p.739]

Boldfaced, Page-Referenced Terms

[p.734] stomach _____

_____

[p.734] gastric fluid _____

_____

[p.734] chyme _____

_____

[p.734] pancreas _____

_____

[p.734] liver _____

_____

[p.734] gallbladder _____

_____

[p.735] bile _____

_____

[p.735] emulsification _____

_____

[p.736] villi (singular, villus) _____

_____

[p.736] microvilli (singular, microvillus) _____

_____

[p.737] segmentation _____

_____

[p.737] micelle formation _____

_____

[p.739] colon _____

_____

[p.739] bulk _____

_____

[p.739] appendix _____

_____

Complete the Table

1. Complete the following table by naming the organs described.

Organ	Main Functions
a. [recall p.733]	Mechanically breaks down food, mixes it with saliva
b. [recall p.733]	Moisten food; start polysaccharide breakdown; buffer acidic foods in mouth
c. [p.734]	Stores, mixes, dissolves food; kills many microorganisms; starts protein breakdown; empties in a controlled way
d. [p.736]	Digests and absorbs most nutrients
e. [p.734]	Produces enzymes that break down all major food molecules; produces buffers against hydrochloric acid from stomach
f. [p.735]	Secretes bile for fat emulsification; secretes bicarbonate, which buffers hydrochloric acid from stomach
g. [p.735]	Stores, concentrates bile from liver
h. [p.739]	Stores, concentrates undigested matter by absorbing water and salts
i. [p.739]	Controls elimination of undigested and unabsorbed residues

Fill-in-the-Blanks

Carbohydrates include sugars and (2) _____ [recall pp.733–734], the name commonly given to polysaccharides. Rice, cereal, pasta, bread, and white potatoes are composed of many polysaccharide molecules that are too large to be absorbed into the internal environment. If these foods are chewed thoroughly, (3) _____ _____ [p.734] in the mouth digests them to the (4) _____ [p.734] (double sugar) level. Since little carbohydrate digestion occurs in the (5) _____ [p.734], if you gulped down your food, starch digestion would again begin in the (6) _____ _____ [p.734] where (7) _____ _____ [p.734] produced by the pancreas would do what should have been done in the mouth. Digestion of disaccharides to monosaccharides (simple sugars) also occurs in the (8) _____ _____ [p.734]. The enzymes responsible are (9) _____ [p.734] with names such as *sucrase, lactase,* and *maltase.*

Proteins are digested to protein fragments, beginning in the (10) _____ [p.734] by (11) _____ [p.734] secreted by the lining of the stomach. Protein fragments are subsequently digested to smaller protein fragments in the (12) _____ _____ [p.734] by enzymes known as trypsin and chymotrypsin produced by the (13) _____ [p.734]. Eventually the smaller protein fragments are digested to (14) _____ _____ [p.734] by means of carboxypeptidase produced by the pancreas and by aminopeptidase produced by glands in the intestinal lining.

Another name for fat is *triglycerides*. (15) _____ [p.734], produced by the pancreas but acting in the (16) _____ _____ [p.735] breaks down one triglyceride molecule into three (17) _____ _____ [pp.734–735] molecules and one glycerol molecule. (18) _____ [p.735], which was made by the liver, stored in the (19) _____ [p.735] and which does not contain digestive enzymes, emulsifies the fat droplet (converts it into small droplets coated with bile salts) thereby increasing the surface area of the substrate on which (20) _____ [p.735] can act.

Nutrients are also mostly digested and absorbed in the (21) _____ _____ [p.736]. (18) is made by the liver, is stored in the gallbladder, and works in the (22) _____ _____ [p.734]. (23) _____ [p.734] is an example of an enzyme that is made by the pancreas but works in the small intestine to convert protein fragments to amino acids. (24) _____ _____ [p.734] are made in the pancreas, but convert DNA and RNA into nucleotides in the small intestine. Any alternating progression of contracting and relaxing muscle movements along the length of a tube is known as (25) _____ [p.734].

True–False

If the statement in true, write a T in the blank. If false, make it true by changing the underlined word.

_____ 26. Amylase digests starch, lipase digests lipids, and <u>peptidases</u> break peptide bonds in protein fragments, yielding amino acids. [p.734]

_____ 27. ATP is the end product of <u>digestion</u>. [Recall Chapter 8, electron transport phosphorylation]

_____ 28. The appendix has no known <u>digestive</u> functions. [p.739]

_____ 29. Water and <u>sodium</u> ions are absorbed into the bloodstream from the lumen of the large intestine. [p.739]

_____ 30. Fatty acids and monoglycerides recombine into fats inside epithelial cells lining the <u>colon</u>. [p.737]

Short Answer

To answer the following questions, consult Figure 42.12 and pp. 738; 740–741.

31. What is the pool of amino acids used for in the human body? _____

_____

32. Which breakdown products result from carbohydrate and fat digestion? _____

_____

33. Monosaccharides, free fatty acids, and monoglycerides all have three uses; identify them. _____

_____

42.8. HUMAN NUTRITIONAL REQUIREMENTS [pp.740–741]

42.9. VITAMINS AND MINERALS [pp.742–743]

42.10. *Focus on Science:* WEIGHTY QUESTIONS, TANTALIZING ANSWERS [pp.744–745]

Selected Words: the zone diet [p.740], *sucrose polyester* [p.741], *complete* proteins [p.741], *incomplete* proteins [p.741], net protein utilization (NPU) [p.741], vitamin K, vitamins C and E, beta-carotene precursor [p.743], free radicals [p.743], body mass index (BMI) [p.744], *caloric intake* [p.744], *energy output* [p.744], *identical twins* [p.745], "set point" [p.745]

Boldfaced, Page-Referenced Terms

[p.740] kilocalories _____

_____

[p.740] food pyramids _____

_____

[p.741] essential fatty acids _____

_____

[p.741] essential amino acids _____

_____

[p.742] vitamins _____

_____

[p.742] minerals _____

_____

[p.744] obesity _____

_____

[p.745] *ob* gene _____

_____

[p.745] leptin _____

_____

Complete the Table

1. Complete the following table by determining how many kilocalories the people described should take in daily, given the stated exercise level, in order to *maintain* their weight [p.744].

Height	Age	Sex	Level of Physical Activity	Present Weight (lbs.)	Number of Kilocalories/Day
5′6″	25	Female	Moderately active	138	a.
5′10″	18	Male	Very active	145	b.
5′8″	53	Female	Not very active	143	c.

Fill-in-the-Blanks

If your body mass index (BMI) is 27 or higher, your risk for developing type 2 diabetes, heart disease, hypertension, breast cancer, colon cancer, gout, gallstones, and/or osteoarthritis increases dramatically. You can calculate your BMI as follows: Multiply your weight in pounds times (2) _____ [p.744], and divide the result by your height in inches squared. (3) _____ _____ [p.740] are the body's main sources of energy; they should make up (4) _____ to _____ [p.740] percent of the human daily caloric intake. (5) _____ [p.741] and cholesterol are components of animal cell membranes.

Fat deposits are used primarily as (6) _____ _____ [p.741], but they also cushion many organs and provide insulation. Lipids should constitute less than (7) _____ [p.741] percent of the human diet. One teaspoon a day of polyunsaturated oil supplies all (8) _____ _____ _____ [p.741] that the body cannot synthesize. (9) _____ [p.741] are digested to twenty common amino acids, of which eight are (10) _____ [p.741], cannot be synthesized, and must be supplied by the diet. Animal proteins such as (11) _____ [p.741] and (12) _____ [p.741] contain high amounts of essential amino acids (that is, they are complete). (13) _____ [p.742] are organic substances needed in small amounts in order to build enzymes or help them catalyze metabolic reactions. (14) _____ [p.742] are inorganic substances needed for a variety of uses.

Review Problems

You are a nineteen-year old male, very sedentary (TV, sleep, and computers), 6 feet tall, medium frame, and you weigh 195 lbs.

15. Use Figure 42.16 to calculate the number of calories required to sustain your desired weight. Are you underweight, overweight, or just right? [p.744] _____

_____

_____

_____

16. How many calories are you allowed to ingest every day? [p.744] _____

_____

_____

_____

Complete the Table

Use the new, improved food pyramid to construct a one-day diet that would eventually allow you to reach that weight if you ate a similar diet every day. Place your choices in the following table as a diet for the person described above. [p.740]

How many servings from each group below are you allowed to have daily? [see legend for Figure 42.14]	What, specifically, could you choose to eat? [What are your preferences?]
17. complex carbohydrates	17b.
18. fruits	18b.
19. vegetables	19b.
20. dairy group	20b.
21. assorted proteins	21b.
22. The "sin" group at the top	22b.

Fill-in-the-Blanks [All from p.740]

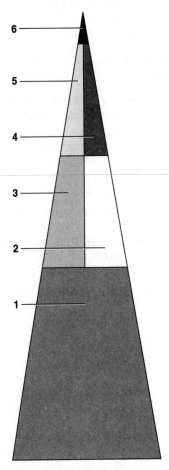

Use the (23) _____ _____ diagram at the left, as revised in 1992, to devise a well-balanced diet for yourself. Group 1, the trapezoidal base, represents the group of complex (24) _____, which includes rice, pasta, cereal, and (25) _____. From this group, (26) (choose 1) ❑ 0, ❑ 2–3, ❑ 2–4, ❑ 3–5, ❑ 6–11 servings from this group every day are needed to supply energy and fiber. Group 2 represents the (27) _____ group. Use the choices in (26) to indicate the number of servings (28) _____ that are needed from this group each day. Group 3 is the (29) _____ group, from which (30) _____ servings are needed each day. Choices include mango, oranges, and (31) _____, cantaloupe, pineapple, or 1 cup of fresh (32) _____. Group 4 includes foods that are a source of nitrogen: nuts, poultry, fish, legumes, and (33) _____. From this group, the body's (34) _____ and nucleic acids are constructed. (35) _____ servings are required every day because the human body cannot synthesize eight of the twenty essential (36) _____ _____ that are used to construct proteins, and must get them in their food supplies. The foods in Group 5, the (37) _____, yogurt and cheese group, supply calcium, vitamins A, D, B$_2$, B$_{12}$. You need (38) _____ servings every day. The foods in Group 6 provide extra calories but few vitamins and minerals; (39) _____ servings are needed every day.

Self-Quiz

Multiple Choice

___ 1. The process that moves nutrients into the blood or lymph is _____. [p.730]
   a. ingestion
   b. absorption
   c. assimilation
   d. digestion
   e. none of the above

___ 2. The enzymatic digestion of proteins begins in the _____. [p.734]
   a. mouth
   b. stomach
   c. liver
   d. pancreas
   e. small intestine

___ 3. The enzymatic digestion of starches begins in the _____. [p.734]
   a. mouth
   b. stomach
   c. liver
   d. pancreas
   e. small intestine

___ 4. The greatest amount of absorption of digested nutrients occurs in the _____. [p.736]
   a. stomach
   b. pancreas
   c. liver
   d. colon
   e. small intestine

___ 5. Glucose moves through the membranes of the small intestine mainly by _____. [p.737]
   a. peristalsis
   b. osmosis
   c. diffusion
   d. active transport
   e. bulk flow

___ 6. Which of the following is *not* found in bile? [p.735]
   a. lecithin
   b. salts
   c. digestive enzymes
   d. cholesterol
   e. pigments

___ 7. The average American consumes approximately _____ pounds of sugar per year. [p.740]
   a. 25
   b. 50
   c. 75
   d. 100
   e. 135

___ 8. Of the following, _____ has (have) the highest net protein utilization. [Fig.42.15a]
   a. beans
   b. eggs
   c. fish
   d. meat
   e. bread

___ 9. One hour after a meal, the blood richest in nutrients would be in the _____. [Recall Chapter 39, pp.674–675]
   a. abdominal aorta
   b. hepatic portal vein
   c. hepatic vein
   d. pulmonary artery
   e. vena cava

___10. The element needed by humans for blood clotting, nerve impulse transmission, and bone and tooth formation is _____. [p.743]
   a. magnesium
   b. iron
   c. calcium
   d iodine
   e. zinc

Matching

Match the best lettered item with its correct numbered item at the left.

11. ___anorexia nervosa [p.728]

12. ___vitamins C and E [p.742]

13. ___bulimia [p.728]

14. ___complex carbohydrates [p.740]

15. ___essential amino acids [p.741]

16. ___essential fatty acids [p.741]

17. ___mineral [p.743]

18. ___rickets [p.742]

19. ___scurvy [p.742]

20. ___vitamin [p.742]

A. Linoleic acid is one example
B. Phenylalanine, lysine, and methionine are three of eight
C. Combine with free radicals; counteract their destructive effects on DNA and cell membranes
D. Obsessive dieting + skewed perception of body weight
E. Vitamin C deficiency
F. Vitamin D deficiency in young children
G. Organic substance that helps enzymes to do their jobs; required in small amounts for good health.
H. Feasting followed by vomiting or taking laxatives
I. Inorganic substance required for good health
J. Long chains of simple sugars; in pasta and white potatoes

Chapter Objectives/Review Questions

1. Distinguish between incomplete and complete digestive systems and tell which is characterized by (a) specialized regions, (b) two-way traffic, and (c) discontinuous feeding. [pp.730–731]
2. Define and distinguish among motility, secretion, digestion, and absorption. [p.730]
3. List all parts (in order) of the human digestive system through which food actually passes. Then list the auxiliary organs that contribute one or more substances to the digestive process. [p.732]
4. Explain how, during digestion, food is mechanically broken down. Then explain how it is chemically broken down. [p.733]
5. Tell which foods undergo digestion in each of the following parts of the human digestive system and state what the food is broken into: mouth, stomach, small intestine, large intestine. [p.734]
6. List the enzyme(s) that act in (a) the oral cavity, (b) the stomach, and (c) the small intestine. Then tell where each enzyme was originally made. [p.734]
7. Describe how the digestion and absorption of fats differ from the digestion and absorption of carbohydrates and proteins. [pp.733–737]
8. Explain how the human body manages to meet the energy and nutritional needs of the various body parts even though the person may be feasting sometimes and fasting at other times. [pp.734–735,738]
9. Describe the cross-sectional structure of the small intestine, and explain how its structure is related to its function. [pp.735–737]
10. List the items that leave the digestive system, and enter the circulatory system during the process of absorption. [pp.736–738]
11. State which processes occur in the colon (large intestine). [p.739]
12. Reproduce from memory the Food Pyramid Diagram as revised in 1992. Identify each of the six components, list the numerical range of servings permitted from each group and also list some of the choices available. [p.740]
13. Compare the contributions of carbohydrates, proteins, and fats to human nutrition with the contributions of vitamins and minerals. [pp.740–743]
14. Construct an ideal diet for yourself for one 24-hour period. Calculate the number of calories necessary to maintain your weight (see p. 744) and then use the Food Pyramid (p. 740) to choose exactly what to eat and how much. [pp.740,744]
15. Summarize current ideas for promoting health by eating properly. [pp.740–744]
16. Summarize the daily nutritional requirements of a 25-year-old man who weighs 135 pounds, works at a desk job, and exercises very little. State what he needs in energy, carbohydrates, proteins, and lipids, and name at least six vitamins and six minerals that he needs to include in his diet every day. [pp.740–744]
17. Distinguish vitamins from minerals, and state what is meant by net protein utilization. [pp.741–743]
18. Name five minerals that are important in human nutrition, and state the specific role of each. [p.743]

Integrating and Applying Key Concepts

Suppose you could not eat solid food for two weeks and you had only water to drink. List in correct sequential order the measures your body would take to try to preserve your life. Mention the command signals that are given as one after another critical point is reached, and tell which parts of the body are the first and the last to make up for the deficit.

43

THE INTERNAL ENVIRONMENT

Interactive Exercises

Tale of the Desert Rat [pp.748–749]

43.1. URINARY SYSTEM OF MAMMALS [pp.750–751]

Selected Words: "metabolic water" [p.748], *solutes* [p.749], ammonia [p.750], uric acid [p.750], *glomerular* capillaries [p.751], *peritubular* capillaries [p.751], renal pelvis [p.751], renal cortex [p.751], renal medulla [p.751]

Boldfaced, Page-Referenced Terms

[p.749] interstitial fluid _____

_____

[p.749] blood _____

_____

[p.749] extracellular fluid _____

_____

[p.750] urinary excretion _____

_____

[p.750] urea _____

_____

[p.751] urinary system _____

_____

[p.751] kidneys _____

_____

[p.751] urine _____

_____

[p.751] ureter _____

_____

[p.751] urinary bladder _____

_____

[p.751] urethra _____

_____

[p.751] nephrons _____

_____

[p.751] Bowman's capsule _____

_____

[p.751] glomerulus _____

_____

[p.751] proximal tubule _____

_____

[p.751] loop of Henle _____

_____

[p.751] distal tubule _____

_____

[p.751] collecting duct _____

_____

Fill-in-the-Blanks

The body gains water by absorbing water from the slurry in the lumen of the small intestine and from (1) _____ [p.750] during condensation reactions. The mammalian body loses water mostly by excretion of (2) _____ [p.750], evaporation through the skin and (3) _____ _____ [p.750], elimination of feces from the gut, and (4) _____ [p.748] as the body is cooled. Cellular secretions and wastes, including carbon dioxide, enter (5) _____ _____ [p.750], then the blood.

The body gains solutes by absorption of substances from the gut, by the secretion of hormones and other substances, and by (6) _____ [p.750], which produces CO_2 and other waste products of degradative reactions. Besides CO_2, there are several major metabolic wastes that must be eliminated: (7) _____ [p.750], formed by amino groups being split from amino acids; (8) _____ [p.750], which is produced in the liver during reactions that link two ammonia molecules to CO_2 and release a molecule of water; and (9) _____ _____ [p.750], which is formed in reactions that break down nucleic acids.

Labeling

Identify each indicated part of the accompanying illustrations.

10. _____ [p.750]

11. _____ [p.750]

12. _____ _____ [p.750]

13. _____ [p.750]

14. _____ [p.751]

15. _____ [p.751]

16. _____ [p.751]

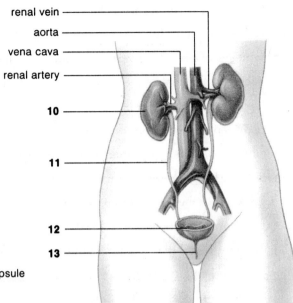

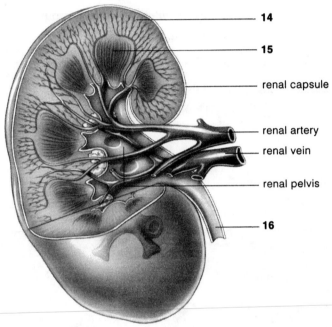

Fill-in-the-Blanks

In mammals, urine formation occurs in a pair of (17) _____ [p.751]. Each contains about a million tubelike blood-filtering units called (18) _____ [p.751]. The function of (17) depends on intimate links between the (18) and the (19) _____ [p.751]. Blood containing excess water and solutes enters each kidney by means of a renal artery, which subdivides into smaller blood vessels called (20) _____ [p.751]. Each (20) leads into a set of capillaries inside a (21) _____ _____ [p.751], the wall of which forms a cup around a cluster of (22) _____ [p.751] capillaries. Most of the water and (23) _____ [p.751] are driven by blood pressure from the blood-filtering unit, the (24) _____ [p.751], across membranous barriers into the lumen of the (21) and on into the (25) _____ _____ [p.751]; the remainder of the blood, now containing relatively little water, few solutes, and many large proteins, continues onward, back out of the (21) and into a second set of (26) _____ [p.751] capillaries, which surround the slender urine-forming tubule and reabsorb most of the water and needed solutes. From the (25), what was filtered out of the blood (the filtrate) continues around a hairpin-shaped (27) _____ _____ _____ [p.751] and up into the (28) _____ _____ [p.751]. It ends up in a (29) _____ _____ [p.751] that leads to the tube, called the (30) _____ [p.751], that conducts urine from the kidney to the urinary bladder for storage.

Every mammal has a (31) _____ [p.751] system that filters water and (32) _____ [p.751] from the blood. Then it reclaims both in amounts necessary to maintain (33) _____ [p.751] fluid, and it eliminates the rest in the form of urine.

Labeling

Identify each indicated part of the accompanying illustration. [All from p.751]

34. _____ _____
35. _____ _____
36. _____ _____
37. _____ _____
38. _____ _____
39. _____ _____
40. _____ _____
    _____

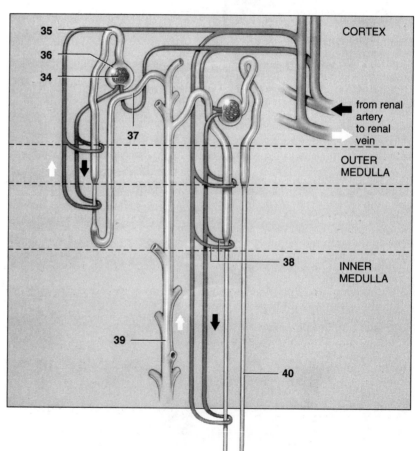

43.2. URINE FORMATION [pp.752–753]

43.3. *Focus on Health:* WHEN KIDNEYS BREAK DOWN [p.754]

43.4. THE ACID–BASE BALANCE [p.754]

43.5. ON FISH, FROGS, AND KANGAROO RATS [p.755]

Selected Words: hypothalamus [p.753], renin [p.753], adrenal cortex [p.753], *uremic toxicity* [p.754], *glomerulonephritis* [p.754], *kidney stones* [p.754], *kidney dialysis machine* [p.754], *"dialysis"* [p.754], *hemodialysis* [p.754], *peritoneal dialysis* [p.754], *metabolic acidosis* [p.754], *bicarbonate–carbon dioxide* buffer system [p.754]

Boldfaced, Page-Referenced Terms

[p.752] filtration _____

_____

[p.752] tubular reabsorption _____

_____

[p.752] tubular secretion _____

_____

[p.753] ADH, antidiuretic hormone _____

_____

[p.753] aldosterone _____

_____

[p.753] angiotensin II _____

_____

[p.753] thirst center _____

_____

[p.754] renal failure _____

_____

[p.754] acid–base balance _____

_____

Fill-in-the-Blanks

Three processes, (1) _____ [p.752], (2) _____ _____ [p.752], and (3) _____ _____ [p.752], cooperate to produce urine in nephrons. (1) occurs by blood pressure forcing most water and solutes (except proteins) out through the walls of the (4) _____ [p.752] capillaries. (2) occurs along the (5) _____'s [p.752] tubular regions as appropriate amounts of water and solutes move into neighboring (6) _____ [p.752] capillaries. (3) occurs at the tubule wall but in the opposite direction to (2); to pump any substance against its natural gradient in an active transport process, ATP is required.

Approximately two-thirds of the filtrate's water move into the peritubular capillaries from the (7) _____ _____ [p.753], as do some of the sodium and other ions. As the filtrate moves along the loop of Henle, water from the filtrate is attracted into (6) by (8) _____ [p.753] before the turn; after the turn, the wall of the loop is impermeable to water, but active transport pumps (9) _____ [p.753] out. Filtrate arriving at the distal tubule is (10) _____ [p.753] and is ready for hormone-induced adjustments.

Two hormones, ADH and (11) _____ [p.753], adjust the reabsorption of water and (12) _____ [p.753] along the distal tubules and collecting ducts. An increase in the secretion of aldosterone causes (13) [check one] ❏ more, ❏ less [p.753] sodium to be excreted in the urine. Increased secretion of (14) _____ [p.753] enhances water reabsorption at distal tubules and collecting ducts when the body must conserve water. When excess water must be excreted, ADH secretion is (15) [check one] ❏ stimulated, ❏ inhibited [p.753]. (16) _____ _____ [p.754] form when uric acid, calcium salts, and other wastes settle out of the urine and collect in the renal pelvis; if they interfere with urine flow, they must be removed to prevent (17) _____ _____ [p.754], from which almost 13 million people in the United States suffer. (18) _____ [p.754] is a procedure that connects a machine to a blood vessel and pumps the patient's blood through tubes of cellophane-like material so that wastes from the blood diffuse across the membrane into a surrounding warm saline bath.

True–False

If the underlined term is false, write the correct term in the blank.

_____ 19. When the body rids itself of excess water, urine becomes more <u>dilute</u>. [p.753]

_____ 20. Water reabsorption into capillaries is achieved by osmosis and <u>active transport</u>. [p.753]

Fill-in-the-Blanks

The (21) _____ [p.754] control the acid–base balance of body fluids by controlling the levels of dissolved ions, especially (22) _____ [p.754] ions. The extracellular pH of humans must be maintained between 7.37 and (23) _____ [p.754]. (24) [check one] ❏ Acids ❏ Bases [p.754] lower the pH and (25) [check one] ❏ acids ❏ bases [p.754] raise it. If you drink a gallon of orange juice, which is acidic, the pH would be (26) [check one] ❏ raised ❏ lowered [recall Chapter 2; logic], but the effect is minimized when excess (27) _____ [p.754] ions are neutralized by (28) _____ [p.754] ions in the bicarbonate-carbon dioxide buffer system. Only the (29) _____ [p.754] system eliminates excess H^+ and restores buffers. Desert-dwelling kangaroo rats have very long (30) _____ _____ _____ [p.755] so that nearly all (31) _____ [p.755] that reaches their very long collecting ducts is reabsorbed. In freshwater, bony fishes and amphibians tend to gain (32) _____ [p.755] and lose (33) _____ [p.755]; they produce (34) [check one] ❏ very dilute ❏ very concentrated urine [p.755].

43.6. MAINTAINING THE BODY'S CORE TEMPERATURE [pp.756–757]
43.7. TEMPERATURE REGULATION IN MAMMALS [pp.758–759]

Selected Words: hypothalamus [p.756], peripheral thermoreceptors [p.758], *brown* adipose tissue [p.758], *hypothermia* [p.758], *frostbite* [p.759], "panting" [p.759], *hyperthermia* [p.759], *fever* [p.759]

Boldfaced, Page-Referenced Terms

[p.756] core temperature _____

_____

[p.756] radiation _____

_____

[p.756] conduction _____

_____

[p.756] convection _____

_____

[p.757] evaporation _____

_____

[p.757] ectotherms _____

_____

[p.757] endotherms _____

_____

[p.757] heterotherms _____

_____

[p.758] peripheral vasoconstriction _____

_____

[p.758] pilomotor response _____

_____

[p.758] shivering response _____

_____

[p.758] nonshivering heat production _____

_____

[p.759] peripheral vasodilation _____

_____

[p.759] evaporative heat loss _____

_____

Fill-in-the-Blanks

In the brain of mammals, the (1) _____ [p.758] is the seat of temperature control. Thermoreceptors located deep in the body are called (2) _____ [p.756] thermoreceptors. The (3) _____ _____ [p.758] contains smooth muscles that erect hairs or feathers and create an insulative layer of still air that helps prevent heat loss. (4) _____ _____ [p.758] is a response to cold stress in which the bloodstream's convective delivery of heat to the body's surface is reduced. A drop in core body temperature below tolerance levels is referred to as (5) _____ [p.758].

True–False

If the underlined term makes the statement false, write the correct term in the blank.

_____6. Jackrabbits are <u>endotherms</u>. [p.757]

_____7. When the core temperature of the human body is about <u>86°F</u>, consciousness is lost and heart muscle action becomes irregular. [p.758]

_____8. When the core temperature of the human body reaches <u>77°F</u>, ventricular fibrillation sets in and death soon follows. [p.758]

Matching

Choose the most appropriate answer to match with the following terms.

9. ___conduction [p.756]

10. ___convection [p.756]

11. ___ectotherm [p.757]

12. ___endotherm [p.757]

13. ___evaporation [p.757]

14. ___heterotherm [p.757]

15. ___radiation [p.756]

A. Body temperature determined more by heat exchange with the environment than by metabolic heat

B. Heat transfer by air or water heat-bearing currents away from or toward a body

C. Body temperature determined largely by metabolic activity and by precise controls over heat produced and heat lost

D. Direct transfer of heat energy between two objects in direct contact with each other

E. The emission of energy in the form of infrared or other wavelengths that are converted to heat by the absorbing body

F. Body temperature fluctuating at some times and heat balance controlled at other times

G. In changing from the liquid state to the gaseous state, the energy required is supplied by the heat content of the liquid

Self-Quiz

___ 1. An entire subunit of a kidney that purifies blood and restores solute and water balance is called a _____. [p.751]
   a. glomerulus
   b. loop of Henle
   c. nephron
   d. ureter
   e. none of the above

___ 2. The last portion of the excretory system passed by urine before it is eliminated from the body is the _____. [p.751]
   a. renal pelvis
   b. bladder
   c. ureter
   d. collecting ducts
   e. urethra

___ 3. Filtration of the blood in the kidney takes place in the _____. [p.752]
   a. loop of Henle
   b. proximal tubule
   c. distal tubule
   d. Bowman's capsule
   e. all of the above

___ 4. _____ primarily controls the concentration of water in urine. [p.753]
   a. Aldosterone
   b. Antidiuretic hormone
   c. Epinephrine
   d. Glucagon
   e. Renin

___ 5. _____ primarily controls the concentration of sodium in urine. [p.753]
a. Insulin
b. Glucagon
c. Antidiuretic hormone
d. Aldosterone
e. Epinephrine

___ 6. Hormonal control over the reabsorption or tubular secretion of sodium primarily affects _____. [p.753]
a. Bowman's capsules
b. distal tubules and collecting ducts
c. proximal tubules
d. the urinary bladder
e. loops of Henle

___ 7. During reabsorption, sodium ions cross the proximal tubule walls into the interstitial fluid principally by means of _____. [p.753]
a. osmosis
b. countercurrent multiplication
c. bulk flow
d. active transport
e. all of the above

___ 8. In humans, the thirst center is located in the _____. [p.753]
a. adrenal cortex
b. thymus
c. heart
d. adrenal medulla
e. hypothalamus

___ 9. The longer the _____, the greater an animal's capacity to conserve water and to concentrate solutes to be excreted in the urine. [p.755]
a. loop of Henle
b. proximal tubule
c. ureter
d. Bowman's capsule
e. collecting tubule

___10. Normally, the extracellular pH of the human body must be maintained between _____ and _____; only the urinary system eliminates excess _____ and restores _____. [p.754]
a. 6.45–7.30; NH_4^+; urea
b. 7.37–7.43; H^+; buffers
c. 7.50–7.85; H^+; glucose
d. 7.90–8.30; NH_4^+; urea
e. 8.15–8.35; OH^-; glucose

___11. Which of the following is/are not involved in driving the exchanges of heat that maintain the body's core temperature? [p.752, 756–757]
a. conduction
b. convection
c. evaporation
d. filtration
e. radiation

Chapter Objectives/Review Questions

1. List some of the factors that can change the composition and volume of body fluids. [pp.748–750]
2. List three soluble by-products of animal metabolism that various kinds of animals excrete in their urine. [p.750]
3. List successively the parts of the human urinary system that constitute the path of urine formation and excretion. [pp.750–751]
4. Locate the processes of filtration, reabsorption, and tubular secretion along a nephron, and tell what makes each process happen. [pp.752–753]
5. State explicitly how the hypothalamus, adrenal cortex, and distal tubules and collecting ducts of the nephrons are interrelated in regulating water and solute levels in body fluids. [p.753]
6. List two kidney disorders and explain what can be done if kidneys become too diseased to work properly. [p.754]
7. Describe the role of the kidney in maintaining the pH of the extracellular fluids between 7.37 and 7.43. [p.754]
8. List the ways in which ectotherms are disadvantaged by not being able to maintain a particular body temperature. Describe the things ectotherms can do to lessen their vulnerability. [pp.756–757]
9. Distinguish between ectotherms and endotherms, and give two examples of each group. [p.757]

10. Explain how endotherms maintain their body temperature when environmental temperatures fall. [pp.757–759]
11. Define the term *hypothermia* and state the situations in which a human might experience the disorder. [pp.758–759]
12. Explain how endotherms maintain their body temperature when environmental temperatures rise 3 to 4 degrees Fahrenheit above standard body temperature. [p.759]

Integrating and Applying Key Concepts

The hemodialysis machine used in hospitals is expensive and time consuming. So far, artificial kidneys capable of allowing people who have nonfunctional kidneys to purify their blood by themselves, without having to go to a hospital or clinic, have not been developed. Which aspects of the hemodialysis procedure do you think have presented the most problems in development of a method of home self-care? If you had an unlimited budget and were appointed head of a team to develop such a procedure and its instrumentation, what strategy would you pursue?

Crossword Puzzle—Urinary Systems

Across

1. Connects urinary bladder to exterior. [p.751]
4. _____ enhances sodium reabsorption, and is produced by the adrenal cortex. [p.753]
6. Extracellular _____ bathes the body's cells. [p.749]
7. _____ acid is the least toxic nitrogenous waste and is constructed from nucleic acid breakdown. [p.750]
8. Bowman's _____ encloses the glomerulus. [p.751]
10. _____ is an enzyme secreted by kidney cells that detaches part of a protein circulating in the blood so that the new protein can be made into a hormone that acts on the adrenal cortex. [p.753]
11. A cluster of capillaries enclosed by a Bowman's capsule that filters blood. [pp.751–752]
13. Connects a kidney to the urinary bladder. [p.751]
14. An organ that adjusts the volume and composition of blood and helps maintain the composition of extracellular fluid. [p.751]
16. A small tube. [p.751]
17. Relating to kidney function. [p.754]
19. _____ of Henle. [p.751]
20. Nitrogenous waste formed from an amino group. [p.750]
21. Nitrogenous waste formed in the liver, and relatively harmless. [p.750]
22. Water and solutes move out of the nephron tubule, then into adjacent capillaries. [p.752]
24. The urinary _____ includes three organs and three larger waste-containing tubes. [p.751]
25. More than a million of these units are packed inside each fist-sized kidney. [p.751]
26. The elimination of fluid wastes. [p.750]
27. The _____ tubule is the part of the nephron that is farthest from the Bowman's capsule. [p.751]

Down

1. Product of kidneys. [p.751]
2. _____ fluid bathes the body's cells. [p.749]
3. Loop of _____. [p.751]
4. _____ glands are perched on top of the kidneys. [pp.750,753]
5. _____ moves excess H⁺ and a few other substances by active transport from the capillaries into the cells of the nephron wall and then into the urine. [p.752]
6. Water and small-molecule solutes are forced from the blood into the Bowman's capsule. [p.752]
9. Urinary _____ stores urine. [pp.750–751]
12. The middle region of the kidney. [p.751]
13. _____ excretion is a process that dumps excess mineral ions and metabolic wastes. [p.752]
15. The _____ tubule is nearest to its Bowman's capsule. [p.751]
18. The outer region of kidney, adrenal gland, or brain. [p.751]
23. The "water conservation" hormone produced by the posterior pituitary. [p.753]

44

PRINCIPLES OF REPRODUCTION AND DEVELOPMENT

Interactive Exercises

From Frog to Frog and Other Mysteries [pp.762–763]

44.1. THE BEGINNING: REPRODUCTIVE MODES [pp.764–765]

44.2. STAGES OF DEVELOPMENT—AN OVERVIEW [pp.766–767]

Selected Words: *reproductive timing* [p.764], *oviparous* [p.764], hermaphrodite [p.764], *ovoviviparous* [p.764], *viviparous* [p.764], internal fertilization [p.765]

Boldfaced, Page-Referenced Terms

[p.762] zygote _____

_____

[p.764] sexual reproduction _____

_____

[p.764] asexual reproduction _____

_____

[p.765] yolk _____

_____

[p.766] embryo _____

_____

[p.766] gamete formation _____

_____

[p.766] fertilization _____

_____

[p.766] cleavage _____

_____

[p.766] blastomeres _____

_____

[p.766] gastrulation _____

_____

[p.766] ectoderm _____

_____

[p.766] endoderm _____

_____

[p.766] mesoderm _____

_____

[p.766] organ formation _____

_____

[p.766] growth and tissue specialization _____

_____

Fill-in-the-Blanks

New sponges budding from parent sponges and a flatworm dividing into two flatworms represent examples of (1) _____ [p.764] reproduction. This type of reproduction is useful when gene-encoded traits are strongly adapted to a limited set of (2) _____ [p.764] conditions. Separation into male and female sexes requires special reproductive structures, control mechanisms, and behaviors; this cost is offset by a selective advantage: (3) _____ [p.764] in traits among the offspring. Males and females of the same species must synchronize their (4) _____ _____ [p.764] to ensure that their gametes mature and find each other at the correct time for fertilization to occur. (5) _____ [p.764] animals are nourished by maternal tissues, not yolk only, until the time of birth.

Copperheads are (6) _____ [p.764]: Fertilization is internal; the fertilized eggs develop inside the mother's body without additional nourishment, and the young are born live. Birds are (7) _____ [p.764]: Eggs with large yolk reserves are released from and develop outside the mother's body.

(8) _____ _____ [p.766] is considered the first stage of animal development. Rich stores of substances such as yolk become assembled in localized regions of the (9) _____ [p.765] cytoplasm. When sperm and egg unite and their DNA mingles and is reorganized, the process is referred to as (10) _____ [p.766]. At the end of fertilization, a(n) (11) _____ [p.766] is formed. (12) _____ [p.766] includes the repeated mitotic divisions of a zygote that segregate the egg cytoplasm into a cluster of cells known as (13) _____ [p.766]; the entire cluster is known as a blastula. (14) _____ [p.766] is the process that arranges cells into three primary tissue layers. Ectoderm eventually will give rise to skin epidermis and the (15) _____ [p.766] system; endoderm forms the inner lining of the (16) _____ [p.766] and associated digestive glands. Mesoderm forms the circulatory system, the (17) _____ [p.766], the muscles and connective tissues. Each stage of (18) _____ [p.766] development builds on structures that were formed during the stage preceding it. (19) _____ [p.767] cannot proceed properly unless each stage is successfully completed before the next begins.

True–False

If false, explain why.

_____20. Yolk is a substance rich in carbohydrates that nourishes embryonic stages. [p.755]

_____21. Most animals reproduce sexually. [p.764]

_____22. Sexual reproduction is less advantageous in predictable environments; asexual reproduction is more advantageous in predictable environments. [p.764]

_____23. The eggs of leopard frogs (*Rana pipiens*) are fertilized externally in the water. [p.762]

_____24. Gastrulation precedes organ formation. [p.766]

Sequence

Arrange the following events in correct chronological sequence. Write the letter of the first step next to 25, the letter of the second step next to 26, and so on. [All from p.766]

25. ___   A. Gastrulation

26. ___   B. Fertilization

27. ___   C. Cleavage

28. ___   D. Growth, tissue specialization

29. ___   E. Organ formation

30. ___   F. Gamete formation

Complete the Table

31. Complete the following table by entering the correct germ layer (ectoderm, mesoderm, or endoderm) that forms the tissues and organs listed. [All from p.766]

Tissues/Organs	Germ Layer (= primary tissue layer)
Muscle, circulatory organs	a.
Nervous tissues	b.
Inner lining of the gut	c.
Circulatory organs (blood vessels, heart)	d.
Outer layer of the integument	e.
Reproductive and excretory organs	f.
Organs derived from the gut	g.
Most of the skeleton	h.
Connective tissues of the gut and integument	i.

44.3. EARLY MARCHING ORDERS [pp.768–769]

44.4. HOW DO SPECIALIZED TISSUES AND ORGANS FORM? [pp.770–771]

Selected Words: "maternal messages" [p.768], cleavage furrow [p.768], vegetal pole [p.769], animal pole [p.769], morula [p.769], "inner cell mass" [p.769], *identical twins* [p.769], neural tube [p.770], gastrula [p.770]

Boldfaced, Page-Referenced Terms

[p.768] oocyte _____

_____

[p.768] sperm _____

_____

[p.768] gray crescent _____

_____

[p.769] cytoplasmic localization _____

_____

[p.769] blastula _____

_____

[p.769] blastocyst _____

_____

[p.770] neural tube _____

_____

[p.770] cell differentiation _____

_____

[p.771] morphogenesis _____

_____

[p.771] apoptosis _____

_____

True–False

If 1, 2, or 3 is false, explain why. If 4 or 5 is false because of the underlined term, write the correct term in the blank.

_____ 1. During gastrulation, maternal controls over gene activity are activated and begin the process of differentiation in each cell's nucleus. [p.770]

_____ 2. In complex eukaryotes, development until gastrulation is governed by DNA in the nucleus of the zygote. [p.768]

_____ 3. In a developing chick embryo, the heart begins to beat at some time between 30 and 36 hours after fertilization. [p.770]

_____ 4. Sperm penetration into the cytoplasm of the egg brings about specific structural changes and chemical reactions. [p.768]

_____ 5. Body parts become folded, tubes become hollowed out, and eyelids, lips, noses, and ears all become slit or perforated by apoptosis. [p.771]

Fill-in-the-Blanks

In amphibian eggs, sperm penetration on one side of an egg causes pigment granules on the opposite side of the egg to flow toward the point of (6) _____ _____ [p.768]. A lightly pigmented area near the frog egg's midsection called the (7) _____ _____ [p.768] results. It is a visible marker of the site where the (8) _____ _____ [p.768] will be established and where gastrulation will begin.

The third stage of animal development, (9) _____ [recall pp.766,768], is characterized by the subdividing and compartmentalizing of the zygote; no growth occurs at this stage, and usually a hollow ball of cells, the (10) _____ [p.769], is formed. Much of the information that determines how structures will be spatially organized in the embryo begins with the distribution of (11) _____ _____ [pp.768–769] in the oocyte. These consist of mRNAs, regionally distributed enzymes and other proteins, yolk, and other factors. As cleavage membranes divide up the cytoplasm, various cytoplasmic determinants become localized in different daughter cells; this process, called (12) _____ _____ [p.769], helps seal the developmental fate of the descendants of those cells.

The fourth stage, (13) _____ [recall pp.766,770], is concerned with the formation of ectoderm, mesoderm, and endoderm, the (14) _____ [p.770] tissue layers of the embryo; at the end of this stage, the (15) _____ [p.770] is formed, and resembles an early embryo. Cleavage of a frog's zygote is complete, because there is so little yolk that cleavage membranes can subdivide the entire cytoplasmic mass; in the chick, however, there is so much yolk that cleavage membranes cannot subdivide the entire mass.

Cleavage is therefore said to be incomplete, and the chick grows from a primitive streak on the surface of the (16) _____ [p.769] mass into a chick embryo complete with wing and leg buds and beating heart during the first (17) _____ [p.770] days.

Through (18) _____ _____ [p.770], a single fertilized egg gives rise to an assortment of different types of specialized cells; these differentiated cells have the same number and same kinds of (19) _____ [p.770] because they are all descended by mitosis from the same zygote. However, from gastrulation onward, certain genes are (20) _____ [p.770] in some cells but not in others.

In every vertebrate, an imaginary straight line connecting the head to the tail end defines where a (21) _____ _____ [p.770], the forerunner of the brain and spinal cord, will form. All organs of the adult begin formation by two principal processes: (22) _____ _____ [p.770], in which a cell selectively activates specific genes (and not others) and synthesizes some proteins not found in other cell types, lays the groundwork for (23) _____ [p.771], which is a program of orderly changes in an embryo's size, shape, and proportions that results in tissues becoming specialized to function in a specific way and form the early stages of organs.

(24) _____ [p.771] (from root words that mean "giving rise to shape") creates the specialized tissues and early organs characteristic of that species.

During (25) _____ _____ _____ [p.771], cells send out pseudopods and use them to move them along prescribed routes; forerunners of neurons interconnect this way as a nervous system is forming. (26) _____ [p.771] cues tell the cells when to stop migration. As (27) _____ [p.771] lengthen and rings of (28) _____ [p.771] in cells constrict, sheets of cells expand and fold inward and outward, as in neural tube formation. Sometimes, as in the formation of the human hand, specific cells in the early form of the hand die on schedule according to cues in their genetic programs; this form of programmed cell death, called (29) _____ [p.771], rids the developing embryo of cells not needed in the next developmental phases.

Sometimes entire organs (such as testes in human males) change position in the developing organism, but the inward or outward folding of (30) _____ _____ [p.771] is seen more often.

44.5. EXPERIMENTAL EVIDENCE OF CELL INTERACTIONS [pp.772–773]

44.6. *Focus on Science:* TO KNOW A FLY [pp.774–775]

44.7. AN UNDERLYING UNITY IN ANIMAL DEVELOPMENT [pp.776–777]

44.8. FROM THE EMBRYO ONWARD [p.777]

44.9. WHY DO ANIMALS AGE? [p.778]

44.10. *Commentary:* DEATH IN THE OPEN [p.779]

Selected Words: superficial cleavage pattern [p.774], *maternal effect* genes [p.775], *gap* genes [p.775], *pair-rule* genes [p.775], *segment polarity* genes [p.775], *homeotic* genes [p.775], *physical, architectural* and *phyletic* constraints [p.776], *larvae* [p.777], *incomplete* metamorphosis [p.777], *complete* metamorphosis [p.777], ecdysone [p.777], juvenile hormone [p.777], telomerase [p.778]

Boldfaced, Page-Referenced Terms

[p.772] embryonic induction _____

_____

[p.772] pattern formation _____

_____

[p.772] AER _____

_____

[p.773] morphogens _____

_____

[p.775] fate map _____

_____

[p.776] theory of pattern formation _____

_____

[p.776] homeotic genes _____

_____

[p.777] juvenile _____

_____

[p.777] metamorphosis _____

_____

[p.777] molting _____

_____

[p.778] aging _____

_____

[p.778] telomeres _____

_____

Fill-in-the-Blanks

(1) _____ _____ [p.772] consists of several processes that transform a gastrula (with its three germ layers) into an embryo with established developmental axes and organ rudiments (undeveloped lumps of tissue) in place and recognizable.

As the embryo develops, one group of cells may produce a substance (say, a growth factor) that diffuses to another group of cells and turns on protein synthesis in those cells. Such interaction among embryonic cells is called (2) _____ _____ [p.772].

Spemann demonstrated that the process known as (3) _____ _____ [p.773] occurs in salamander embryos, where one body part differentiates because of signals it receives from an adjacent body part. (4) _____ [p.773] are slowly degradable proteins that form concentration gradients as they diffuse from an inducing tissue into adjoining tissues.

In many animals, the embryo develops into a motile, independent (5) _____ [p.777], which extends the food supply and range of the population. A larva necessarily must undergo (6) _____ [p.777] in order to become a juvenile. Fruitflies show (7) _____ [p.777] metamorphosis. Normal cells cap the ends of chromosomes with repetitive DNA sequences called (8) _____ [p.778]; a bit of each is lost during each nuclear division, and when only a nub remains, cells stop dividing and will eventually die.

Matching

In *Drosophila*, specific types of genes act as master organizers that establish the location, shapes, and sizes of various body parts in the animal during pattern formation. Match the correct gene with its function. [All from p.775]

9. ___ gap genes

10. ___ homeotic genes

11. ___ maternal effect genes

12. ___ pair-rule genes

13. ___ segment polarity genes

A. Divide the embryo into segment-size units

B. Produces substances that accumulate in bands that each correspond to two body segments

C. Map out broad regions of the body; different concentrations of these switch on pair-rule genes

D. Directly and collectively govern the developmental fate of each body segment

E. Transcribe specific mRNAs and indirectly cause specific regulatory proteins to be translated; these products become localized in different parts of the egg cytoplasm and are activated in the zygote, where they activate or suppress gap genes.

True–False

If the underlined term makes the statement false, write the correct term in the blank.

_____ 14. The aging and death of a cell may be coded in large part in its DNA; <u>external</u> signals activate those DNA messages and tell the cell that it is time to die. [p.778]

_____ 15. A process of predictable cellular deterioration is built into the life cycle of all organisms that consist of <u>differentiated cells that show considerable specialization</u>. [p.778]

Self-Quiz

___ 1. Animals such as birds lay eggs with large amounts of yolk; embryonic development happens within the egg covering outside the mother's body. Birds are said to be _____. [p.764]
   a. ovoviviparous
   b. viviparous
   c. oviparous
   d. parthenogenetic
   e. none of the above

___ 2. The process of cleavage most commonly produces a(n) _____. [pp.767,769]
   a. zygote
   b. blastula
   c. gastrula
   d. third germ layer
   e. organ

___ 3. Incomplete cleavage of cells at the yolk periphery is characteristic of the cleavage pattern of _____. [p.769]
a. frogs
b. sea urchins
c. chickens
d. humans
e. none of the preceding

___ 4. The formation of three germ (embryonic) tissue layers occurs during _____. [p.766]
a. gastrulation
b. cleavage
c. pattern formation
d. morphogenesis
e. neural plate formation

___ 5. The differentiation of a body part in response to signals from an adjacent body part is _____. [p.772]
a. contact inhibition
b. ooplasmic localization
c. embryonic induction
d. pattern formation
e. none of the above

___ 6. A homeotic mutation _____. [p.775]
a. may cause a leg to develop on the head where an antenna should grow
b. may affect pattern formation
c. affects morphogenesis
d. may alter the path of development
e. all of the above

___ 7. Shortly after fertilization, the zygote is subdivided into a multicelled embryo during a process known as _____. [p.766]
a. meiosis
b. parthenogenesis
c. embryonic induction
d. cleavage
e. invagination

___ 8. Muscles differentiate from _____ tissue. [p.766]
a. ectoderm
b. mesoderm
c. endoderm
d. parthenogenetic
e. yolky

___ 9. The gray crescent is _____. [p.768]
a. formed where the sperm penetrates the egg
b. part of only one blastomere after the first cleavage
c. the yolky region of the egg
d. where the first mitotic division begins
e. formed opposite from where the sperm enters the egg

___10. The nervous system differentiates from _____ tissue. [p.766]
a. ectoderm
b. mesoderm
c. endoderm
d. yolky
e. homeotic

Chapter Objectives/Review Questions

1. Explain how a spherical zygote becomes a multicellular adult with arms and legs. [pp.762,770–771]
2. Describe how asexual reproduction differs from sexual reproduction. Know the advantages and problems associated with having separate sexes. [pp.764–765]
3. Define the terms *oviparous, viviparous,* and *ovoviviparous*. For each of the three developmental strategies, cite an example of an animal that goes through it. [p.764]
4. Explain why evolutionary trends in many groups of organisms tend toward developing more complex, sexual strategies rather than retaining simpler, asexual strategies. [pp.764–765]
5. Explain how the amount of yolk in an ovum can influence an animal's cleavage pattern. [pp.765,769]
6. Name each of the three embryonic tissue layers and the organs formed from each. [p.766]
7. Describe early embryonic development, and distinguish among the following: gamete formation, fertilization, cleavage, gastrulation, organ formation, and tissue specialization. [p.766]
8. Compare the early stages of frog, sea urchin, and chick development (see Figures 44.4, 44.8, and 44.9) with respect to egg size and type of cleavage pattern (incomplete or complete). [pp.766–770]
9. Explain what causes polarity to occur during oocyte maturation in the mother, and state how polarity influences later development. [pp.768–769]
10. Define the term *gastrulation* and state what process begins at this stage that did not happen during cleavage. [pp.766,770]

11. Define the term *differentiation* and give two examples of cells in a multicellular organism that have undergone differentiation. [pp.770–771]
12. Define what is meant by *larva*. Distinguish metamorphosis from morphogenesis. [pp.771,777]
13. Distinguish complete from incomplete metamorphosis. [p.777]

Integrating and Applying Key Concepts

If embryonic induction did not occur in a human embryo, how would the eye region appear? What would happen to the forebrain and epidermis? If controlled cell death did not happen in a human embryo, how would its hands appear? Its face?

45

HUMAN REPRODUCTION
AND DEVELOPMENT

Interactive Exercises

The Journey Begins [pp.782–783]

45.1. REPRODUCTIVE SYSTEM OF HUMAN MALES [pp.784–785]
45.2. MALE REPRODUCTIVE FUNCTION [pp.786–787]

Selected Words: blastocyst [p.782], inner cell mass [p.782], blastula [p.782], gonads [p.783], accessory glands and ducts [p.783], scrotum [p.784], epididymis [p.784], vasa deferentia (singular, vas deferens) [p.784], ejaculatory ducts [p.784], urethra [p.784], penis [p.784], seminal vesicles [p.785], prostate gland [p.785], bulbourethral glands [p.785], *prostate cancer* [p.785], *testicular cancer* [p.785], PSA (prostate-specific antigen) test [p.785]

Boldfaced, Page-Referenced Terms

[p.783] testes (singular, testis) _____

_____

[p.783] ovaries (singular, ovary) _____

_____

[p.783] secondary sexual traits _____

_____

[p.784] puberty _____

_____

[p.784] seminiferous tubules _____

_____

[p.784] semen _____

_____

[p.786] Sertoli cells _____

_____

[p.786] Leydig cells _____

_____

[p.786] testosterone _____

_____

[p.787] LH, luteinizing hormone _____

_____

[p.787] FSH, follicle-stimulating hormone _____

_____

[p.787] GnRH _____

_____

Fill-in-the-Blanks

The numbered items on the illustrations that follow represent missing information; complete the numbered blanks in the narrative below to supply the missing information on the illustration. Some illustrated structures are numbered more than once to aid identification.

Within each testis and after repeated (1) _____ [p.786] divisions of undifferentiated diploid cells just inside the (2) _____ [pp.784,786] tubule walls, (3) _____ [p.786] occurs to form haploid, eventually mature (4) _____ [p.786]. Males produce sperm continuously from puberty onward. Sperm leaving a testis enter a long, coiled duct, the (5) _____ [p.784]; the sperm are stored in the last portion of this organ. When a male is sexually aroused, muscle contractions quickly propel the sperm through a thick-walled tube, the (6) _____ _____ [p.784], then to ejaculatory ducts and finally the (7) _____ [p.784], which opens at the tip of the penis. During the trip to the urethra, glandular secretions become mixed with the sperm to form semen. (8) _____ _____ [p.785] secrete fructose to nourish the sperm and prostaglandins to induce contractions in the female reproductive tract. (9) _____ _____ [p.785] secretions help neutralize vaginal acids. (10) _____ [p.785] glands secrete mucus to lubricate the penis, aid vaginal penetration, and improve sperm motility.

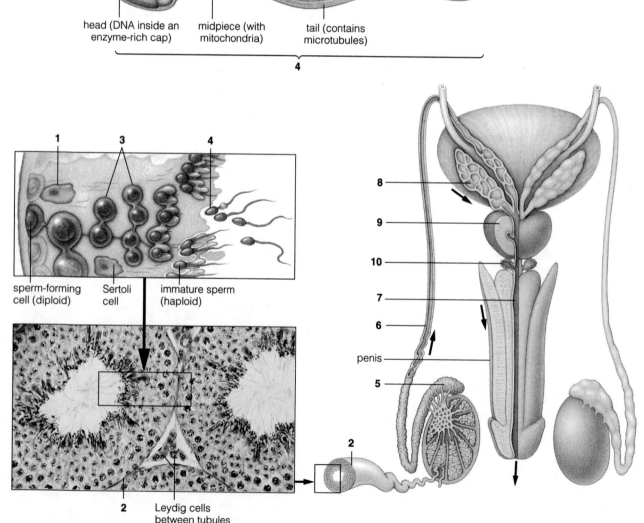

head (DNA inside an enzyme-rich cap)   midpiece (with mitochondria)   tail (contains microtubules)

4

sperm-forming cell (diploid)   Sertoli cell   immature sperm (haploid)

2   Leydig cells between tubules

penis

Dichotomous Choice

Circle one of two possible answers given between parentheses in each statement.

11. Testosterone is secreted by (Leydig/hypothalamus) cells. [p.786]
12. (Testosterone/FSH) governs the growth, form, and functions of the male reproductive tract. [p.786]
13. Sexual behavior, aggressive behavior, and secondary sexual traits are associated with (LH/testosterone). [p.786]
14. LH and FSH are secreted by the (anterior/posterior) lobe of the pituitary gland. [p.787]
15. The (testes/hypothalamus) governs sperm production by controlling interactions among testosterone, LH, and FSH. [p.787]
16. When blood levels of testosterone (increase/decrease), the hypothalamus stimulates the pituitary to release LH and FSH, which travel the bloodstream to the testes. [p.787]
17. Within the testes, (LH/FSH) acts on Leydig cells; they secrete testosterone, which enters the sperm-forming tubes. [p.787]
18. FSH enters the sperm-forming tubes and diffuses into (Sertoli/Leydig) cells to improve testosterone uptake. [p.787, Fig. 45.5]
19. When blood testosterone levels (increase/decrease) past a set point, negative feedback loops to the hypothalamus slow down testosterone secretion. [p.787, Fig. 45.5]

Labeling [All from p.787]

20. _____

21. _____ _____

22. _____ _____

23. _____ _____

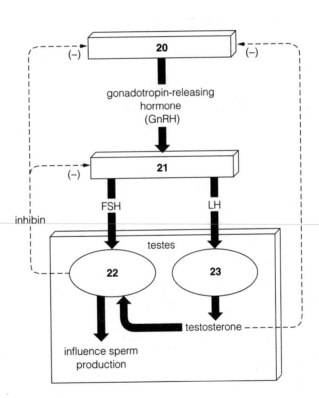

45.3. REPRODUCTIVE SYSTEM OF HUMAN FEMALES [pp.788–789]

45.4. FEMALE REPRODUCTIVE FUNCTION [pp.790–791]

45.5. VISUAL SUMMARY OF THE MENSTRUAL CYCLE [p.792]

Selected Words: oviducts [p.788], myometrium [p.788], cervix [p.788], vagina [p.788], labia majora [p.788], labia minora [p.788], clitoris [p.788], *estrous* cycle [p.788], menstruation [p.788], *menopause* [p.789], *endometriosis* [p.789], granulosa cells [p.790]

Boldfaced, Page-Referenced Terms

[p.788] oocytes _____

_____

[p.788] uterus _____

_____

[p.788] endometrium _____

_____

[p.788] menstrual cycle _____

_____

[p.788] follicular phase _____

_____

[p.788] ovulation _____

_____

[p.788] luteal phase _____

_____

[p.789] estrogens _____

_____

[p.789] progesterone _____

_____

[p.790] follicle _____

_____

[p.790] zona pellucida _____

_____

[p.790] secondary oocyte _____

_____

[p.790] polar bodies _____

_____

[p.791] corpus luteum _____

_____

Fill-in-the-Blanks

The numbered items on the illustrations that follow represent missing information; complete the numbered blanks in the narrative below to supply the missing information on the illustration. Some illustrated structures are numbered more than once to aid identification.

An immature egg (oocyte) is released from one (1) _____ [pp.788–789] of a pair. From each ovary, an (2) _____ [pp.788–789] forms a channel for transport of the immature egg to the (3) _____ [pp.788–789], a hollow, pear-shaped organ where the embryo grows and develops. The lower, narrowed part of the uterus is the (4) _____ [pp.788–789]. The uterus has a thick layer of smooth muscle, the (5) _____ [pp.788–789], lined inside with connective tissue, glands, and blood vessels; this lining is called the (6) _____ [pp.788–789]. The (7) _____ [pp.788–789], a muscular tube, extends from the cervix to the body surface; this tube receives sperm and functions as part of the birth canal. At the body surface are external genitals (vulva) that include organs for sexual stimulation. Outermost is a pair of fat-padded skin folds, the (8) _____ _____ [pp.788–789]. Those folds enclose a smaller pair of skin

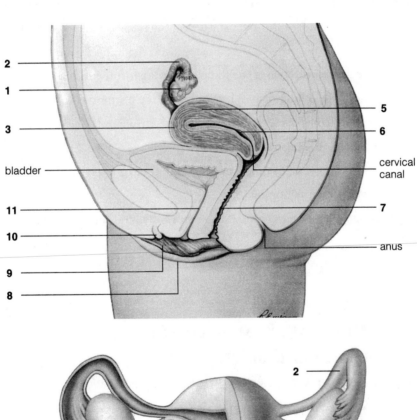

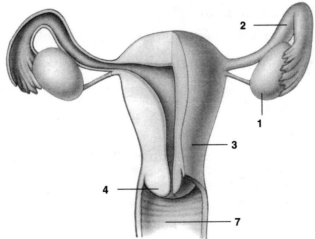

folds, the (9) _____ _____ [pp.788–789]. The smaller folds partly enclose the (10) _____ [pp.788–789], an organ sensitive to stimulation. The location of the opening to the (11) _____ [pp.788–789] is about midway between the clitoris and the vaginal opening. (12) _____ [p.790] occurs in the ovaries so that a normal female infant has about 2 million primary oocytes with the division process halted in the [check one] (13) ☐ I ☐ II [p.790] stage. By age seven, only about (14) _____ [p.790] remain. A primary oocyte surrounded by a nourishing layer of granulosa cells is called a(n) (15) _____ [p.790].

When a female enters puberty, the (16) _____ [p.790] secretes a hormone (GnRH) that makes the (17) _____ _____ [p.790] secrete the follicle-stimulating hormone (FSH) and luteinizing hormone (LH). These hormones are carried by the blood to all parts of the body, but each month, one (or more) follicles respond by growing and secreting the steroid hormones called (18) _____ [p.790]. The primary oocyte completes the (12) (13) division 8–10 hours before (19) _____ [p.790] occurs in response to a surge of (20) _____ [p.790] being produced by the (17) on day 12–14 of a twenty-eight-day cycle.

The first five days of the cycle are occupied by (21) _____ [p.792], the deterioration and expulsion of (22) _____ [p.791] tissues that line the uterus. During the next week, (18) stimulate new (22) tissues to be constructed.

Following ovulation, the granulosa cells of the follicle are transformed into a (23) _____ _____ [pp.791–792] by the midcycle surge of LH. The (23) secretes its key hormone (24) _____ [pp.791–792] and some estrogen. (24) prepares the reproductive tract for the arrival of the (25) _____ [p.791], which develops after an egg is fertilized. (24) also maintains the (26) _____ [p.791] during pregnancy. If a (25) does not burrow into the (26), the (23) self-destructs after approximately twelve days by secreting prostaglandins that disrupt its own functioning.

After this, (24) and estrogen levels in the blood decline rapidly, and the (26) tissues die and disintegrate; together with blood escaping from the (26) tissues, the (27) _____ _____ [p.791] begins and continues for three to six days.

45.6. PREGNANCY HAPPENS [p.793]

45.7. FORMATION OF THE EARLY EMBRYO [pp.794–795]

45.8. EMERGENCE OF THE VERTEBRATE BODY PLAN [p.796]

45.9. WHY IS THE PLACENTA SO IMPORTANT? [p.797]

45.10. EMERGENCE OF DISTINCTLY HUMAN FEATURES [pp.798–799]

45.11. *Focus on Health:* MOTHER AS PROVIDER, PROTECTOR, POTENTIAL THREAT [pp.800–801]

Selected Words: erection [p.793], ejaculation [p.793], *orgasm* [p.793], spongy tissue [p.793], glans penis [p.793], *embryonic* period [p.794], *fetal* period [p.794], *first, second, third* trimesters [p.794], *amniotic* cavity [p.795], *pregnancy tests* [p.795], "primitive streak" [p.796], pattern formation [p.796], embryonic inductions [p.796], neural tube [p.796], notochord [p.796], pharyngeal arches [p.796], umbilical cord [p.797], *teratogens* [p.800], rubella [p.801], thalidomide [p.801], anti-acne drugs [p.801], fetal alcohol syndrome (FAS) [p.801], secondhand smoke [p.801]

Boldfaced, Page-Referenced Terms

[p.793] coitus _____

_____

[p.793] ovum (plural, ova) _____

_____

[p.794] fetus _____

_____

[p.794] implantation _____

_____

[p.795] amnion _____

_____

[p.795] yolk sac _____

_____

[p.795] chorion _____

_____

[p.795] allantois _____

_____

[p.795] HCG, human chorionic gonadotropin _____

_____

[p.796] gastrulation _____

_____

[p.796] somites _____

_____

[p.797] placenta _____

_____

Fill-in-the-Blanks

Fertilization generally takes place in the (1) _____ [pp.793–794, Fig.45.10]; six or seven days after conception, (2) _____ [p.794] begins as the blastocyst sinks into the endometrium. Extensions from the chorion fuse with the endometrium of the uterus to form a (3) _____ [p.795], the organ of interchange between mother and fetus. By the end of the (4) _____ [p.798] week, all major organs have formed; the offspring is now referred to as a(n) (5) _____ [p.798].

Labeling

Identify each numbered part of the accompanying illustrations.

6. _____ _____ [p.795]

7. _____ _____ [p.795]

8. _____ _____ [p.795]

9. _____ _____ [p.795]

10. _____ _____ [pp.795,798]

11. _____ _____ _____ _____ [p.798]

12. _____ [p.795]

13. _____ [p.795]

14. _____ _____ [p.798]

15. _____ _____ [p.797]

16. _____ [p.798]

17. _____ _____ [p.798]

18. _____ [p.798]

19. _____ [p.799]

20. _____ [p.798]

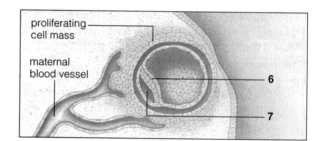

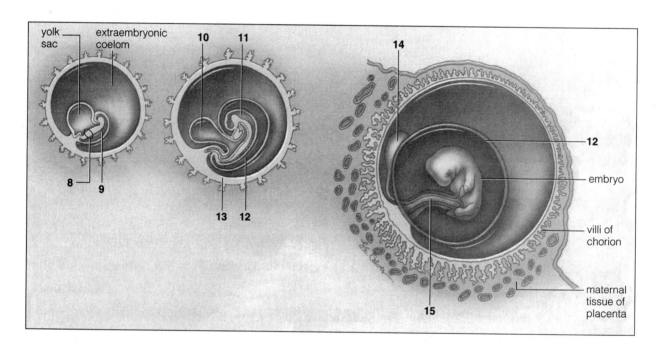

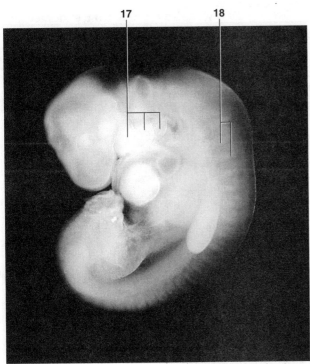

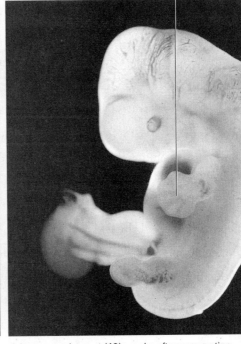

A human embryo at (**16**) weeks after conception.

A human embryo at (**19**) weeks after conception.

Matching

Link each of the four extraembryonic membranes with its function. [All from p.795]

21. ___allantois

22. ___amnion

23. ___chorion

24. ___yolk sac

A. Outermost membrane; will become part of the spongy, blood-engorged placenta
B. Directly encloses the embryo and cradles it in a buoyant protective fluid
C. In humans, some of this membrane becomes a site for blood cell formation, some will become the forerunner of gametes
D. In humans, this membrane serves in early blood formation and formation of the urinary bladder

Fill-in-the-Blanks

At-home human pregnancy tests use a treated "dip-stick" that changes color when (25) _____ _____ _____ [p.795] secreted by the blastocyst is present in the mother's urine. (25) also stimulates the (26) _____ _____ [p.795] to continue secreting progesterone and estrogen.

By the time a woman has missed her first menstrual period, generally during the third week after conception, cleavage is completed and (27) _____ [p.796] is well under way as the three primary tissue layers (ectoderm, mesoderm, and endoderm) are forming. On day 15, a faint band, the (28) _____ _____ [p.796], appears around a depression along the axis of the embryonic disk; it marks the beginning of (27) in vertebrate embryos. During days 18–23, morphogenetic processes cause neural folds to arch upward and merge to form the (29) _____ _____ [p.796]. (30) _____ [p.796] appear that will later form most of axial skeleton, skeletal muscles, and much of the dermal layer of the skin. By days 24–25 (during the fourth week), (31) _____ _____ [p.796] are visible that will contribute in forming the face, neck, mouth, larynx, pharynx, and nasal cavities.

During the third week after conception, the (32) _____ [p.797] was also forming from endometrial and extraembryonic membrane tissues; a(an) (33) _____ _____ [p.797] connects the developing embryo to it. Oxygen and (34) _____ [p.797] exit from maternal blood vessels into the embryo. Coming from the embryo are carbon dioxide and other (35) _____ [p.797] that will quickly be disposed of by the mother's lungs and kidneys.

By the end of the fourth week (36) _____ [p.798] form from limb buds and fingers and toes form from embryonic paddles. During the months that follow, organs become recognizable from the undeveloped lumps and bumps that preceded them. Growth of the (37) _____ [p.798] surpasses that of any other region of the body.

During the second trimester, the fetus is moving most of its facial (38) _____ [p.798] and is covered with hair. The eyes are sealed shut.

During the last trimester, the fetus gains weight, develops its lungs, immune system, and temperature regulation. By the ninth month, the rate of surviving birth has increased to (39) _____ [p.799] percent.

45.12. FROM BIRTH ONWARD [pp.802–803]

45.13. CONTROL OF HUMAN FERTILITY [pp.804–805]

45.14. *Focus on Health:* SEXUALLY TRANSMITTED DISEASES [pp.806–807]

45.15. *Focus on Bioethics:* TO SEEK OR END PREGNANCY [p.808]

Selected Words: *breech birth* [p.802], "afterbirth" [p.802], *afterbaby blues* [p.802], *breast cancer* [p.803], *prenatal* stage [p.803], *postnatal* stage [p.803], *abstinence* [p.804], *rhythm method* [p.804], *withdrawal* [p.804], *douching* [p.804], *vasectomy* [p.804], *tubal ligation* [p.804], *spermicidal foam* [p.805], *spermicidal jelly* [p.805], *IUDs* [p.805], *diaphragm* [p.805], *condoms* [p.805], *birth control pill* [p.805], *Depo-Provera* injection [p.805], *Norplant* [p.805], *morning-after pill* [p.805], *Preven* [p.805], *RU-486* [p.805], Type II *Herpes simplex* [p.806], *AIDS* [p.806], *gonorrhea* [p.806], *syphilis* [p.807], *chancre* [p.807], *pelvic inflammatory disease* (PID) [p.807], *Chlamydia trachomatis* [p.807], *NGU* (chlamydial nongonococcal urethritis) [p.807], *genital warts* [p.807]

Boldfaced, Page-Referenced Terms

[p.802] labor _____

_____

[p.802] lactation _____

_____

[p.803] prolactin _____

_____

[p.803] oxytocin _____

_____

[p.806] sexually transmitted diseases (STDs) _____

_____

[p.808] *in vitro* fertilization _____

_____

[p.808] abortion _____

_____

Fill-in-the-Blanks

On Earth each day, (1) _____ [p.804] more babies are born than people die. Each year in the United States, we still have about (2) _____ [p.804] teenage pregnancies and (3) _____ [p.804] abortions. The most effective method of preventing conception is complete (4) _____ [p.804]. (5) _____ [p.805] are about 85–93 percent reliable and help prevent venereal disease. A (6) _____ [p.805] is a flexible, dome-shaped disk, used with a spermicidal foam or jelly, that is placed over the cervix. In the United States, the most widely used contraceptive is the Pill—an oral contraceptive of synthetic (7) _____ [p.805] and (8) _____ [p.805] that suppress the release of FSH and LH from the (9) _____ [recall p.792] and thereby prevent the cyclic maturation and release of eggs. Two forms of surgical sterilization are vasectomy and (10) _____ _____ [p.804].

Matching

Match each of the following with *all* applicable diseases. Choose from:

> A. AIDS   B. chlamydial infection   C. genital herpes   D. gonorrhea
> E. pelvic inflammatory disease   F. syphilis

11. ___Can damage the brain and spinal cord in ways leading to various forms of insanity and paralysis [pp.806–807]

12. ___Has no cure [pp.806–807]

13. ___Can cause violent cramps, fever, vomiting, and sterility due to scarring and blocking of the oviducts [p.807]

14. ___Caused by a motile, corkscrew-shaped bacterium, *Treponema pallidum* [p.807]

15. ___Infected women typically have miscarriages, stillbirths, or sickly infants [p.807]

16. ___Caused by a bacterium with pili, *Neisseria gonorrhoeae* [p.807]

17. ___Caused by direct contact with the viral agent; about 25 million people in the United States infected by it [p.807]

18. ___Produces a chancre (localized ulcer) 1–8 weeks following infection [p.807]

19. ___Chronic infections by this can lead to cervical cancer [p.807]

20. ___Can lead to lesions in the eyes that cause blindness in babies born to mothers with this [p.807]

21. ___Acyclovir decreases the healing time and may also decrease the pain and viral shedding from the blisters [p.807]

22. ___Eight weeks following infection, a flattened, painless chancre harbors many motile bacteria [p.807]

23. ___May be cured by antibiotics but can be infected again [p.807]

24. ___Can be treated with tetracycline and sulfonamides [p.807]

25. ___Generally preventable by correct condom usage [pp.806–807]

Self-Quiz

For questions 1–5, choose from the following answers:
- a. AIDS
- b. Chlamydial infection
- c. Genital herpes
- d. Gonorrhea
- e. Syphilis

—— 1. _____ is a disease caused by a spherical bacterium (*Neisseria*) with pili; it is curable by prompt diagnosis and treatment. [pp.806–807]

—— 2. _____ is a disease caused by a spiral bacterium (*Treponema*) that produces a localized ulcer (a chancre). [p.807]

—— 3. _____ is an incurable disease caused by a retrovirus (an RNA-based virus). [p.806]

—— 4. _____ is a disease caused by an intracellular parasite that lives in the genital and urinary tracts; also causes NGU. [p.807]

—— 5. _____ is an extremely contagious viral infection (DNA-based) that causes sores on the facial area and reproductive tract; it is also incurable. [p.807]

For questions 6–8, choose from the following answers:
- a. blastocyst
- b. allantois
- c. yolk sac
- d. oviduct
- e. cervix

—— 6. The _____ lies between the uterus and the vagina. [p.788]

—— 7. The _____ is a pathway from the ovary to the uterus. [p.788]

—— 8. The _____ results from the process known as cleavage. [p.794]

For questions 9–12, choose from the following answers:
- a. Leydig cells
- b. seminiferous tubules
- c. vas deferens
- d. epididymis
- e. prostate

—— 9. The _____ connects a structure on the surface of the testis with the ejaculatory duct. [p.784]

——10. Testosterone is produced by the _____. [p.786]

——11. Meiosis occurs in the _____. [p.786]

——12. Sperm mature and become motile in the _____. [p.784]

Chapter Objectives/Review Questions

1. Distinguish between primary and secondary sexual traits and between gonads and accessory reproductive organs. [p.783]
2. Follow the path of a mature sperm from the seminiferous tubules to the urethral exit. List every structure encountered along the path and state the contribution to the nurture of the sperm. [p.784]
3. Describe how a man examines himself for testicular cancer. [p.785]
4. Diagram the structure of a sperm, label its components, and state the function of each. [pp.786–787]
5. List in order the stages that compose spermatogenesis. [pp.786–787]
6. Name the four hormones that directly or indirectly control male reproductive function. Diagram the negative feedback mechanisms that link the hypothalamus, anterior pituitary, and testes in controlling gonadal function. [pp.786–787]
7. Compare the function of the Leydig cells of the testis with the function of the ovarian follicle and the corpus luteum. [pp.786,790–791]
8. Distinguish the follicular phase of the menstrual cycle from the luteal phase and explain how the two cycles are synchronized by hormones from the anterior pituitary, hypothalamus, and ovaries. [pp.788–789]
9. Trace the path of a sperm from the urethral exit to the place where fertilization normally occurs. Mention in correct sequence all major structures of the female reproductive tract that are passed along the way, and state the principal function of each structure. [pp.788,793]
10. State which hormonal event brings about ovulation and which other hormonal events bring about the onset and finish of menstruation. [pp.790–792]
11. List the physiological factors that bring about erection of the penis during sexual stimulation and the factors that bring about ejaculation. [p.793]
12. List the similar events that occur in both male and female orgasm. [p.793]
13. Describe the events that occur during the first month of human development. State how much time cleavage and gastrulation require, when organogenesis begins, and what is involved in implantation and placenta formation. [pp.793–798]
14. State when the embryo begins to be referred to as a fetus and during which month fetuses born prematurely survive. [p.798]
15. Explain why the mother must be particularly careful of her diet, health habits, and life-style during the first trimester after fertilization (especially during the first six weeks). [pp.800–801]
16. Describe two different types of sterilization. [p.804]
17. Identify the factors that encourage and discourage methods of human birth control. [pp.804–805]
18. State which birth control methods help prevent venereal disease. [pp.804–805]
19. Identify the three most effective birth control methods used in the United States and the four least effective birth control methods. [p.805]
20. For each STD described in the Focus, state the causative organism and the symptoms of the disease. [pp.806–807]
21. State the physiological circumstances that would prompt a couple to try in vitro fertilization. [p.808]

Integrating and Applying Key Concepts

What rewards do you think a society should give potential parents who have at most two children during their lifetimes? In cases of rape and/or incest that cause a pregnancy, should society be able to dictate the fate of the victimized female? Should the man responsible be punished? In the absence of rewards or punishments, how can a society encourage women not to have abortions and yet ensure that the human birth rate does not continue to increase?

46

POPULATION ECOLOGY

Interactive Exercises

Tales of Nightmare Numbers [pp.812–813]

46.1. CHARACTERISTICS OF POPULATIONS [p.814]

46.2. *Focus on Science:* ELUSIVE HEADS TO COUNT [p.815]

46.3. POPULATION SIZE AND EXPONENTIAL GROWTH [pp.816–817]

Selected Words: *pre-reproductive, reproductive,* and *post-reproductive* ages [p.814], *habitat* [p.814], *crude* density [p.814], *Larrea* [p.814]

Boldfaced, Page-Referenced Terms

[p.813] ecology _____

_____

[p.814] demographics _____

_____

[p.814] population size _____

_____

[p.814] age structure _____

_____

[p.814] reproductive base _____

_____

[p.814] population density _____

_____

[p.814] population distribution _____

_____

[p.815] capture–recapture method _____

_____

[p.816] immigration _____

_____

[p.816] emigration _____

_____

[p.816] migrations _____

_____

[p.816] zero population growth _____

_____

[p.816] per capita _____

_____

[p.816] net reproduction per individual per unit time, or r _____

_____

[p.817] exponential growth _____

_____

[p.817] doubling time _____

_____

[p.817] biotic potential _____

_____

Matching

Choose the most appropriate answer for each term.

1. ___demographics [p.814]
2. ___population size [p.814]
3. ___population density [p.814]
4. ___habitat [p.814]
5. ___population distribution [p.814]
6. ___age structure [p.814]
7. ___reproductive base [p.814]
8. ___crude density [p.814]
9. ___pre-reproductive, reproductive, and post-reproductive [p.814]
10. ___clumped dispersion [p.814]
11. ___nearly uniform dispersion [p.814]
12. ___random dispersion [p.814]

A. Includes pre-reproductive and reproductive age categories
B. The general pattern in which the individuals of the population are dispersed through a specified area
C. When individuals of a population are more evenly spaced than they would be by chance alone
D. The number of individuals in some specified area or volume of a habitat
E. Occurs only when individuals of a population neither attract nor avoid one another when conditions are fairly uniform through the habitat, and when resources are available all the time
F. The number of individuals in each of several to many age categories
G. The measured number of individuals in a specified area
H. The number of individuals that contribute to a population's gene pool
I. The type of place where a species normally lives
J. Categories of a population's age structure
K. The vital statistics of a population
L. Individuals of a population form aggregations at specific habitat sites; most common dispersion pattern

Fill-in-the-Blanks

Population size increases as a result of births and (13) _____ [p.816]—the arrival of new residents from other populations of the species. Population size decreases as a result of deaths and (14) _____ [p.816], whereby individuals permanently move out of the population. Population size also changes on a predictable basis as a result of daily or seasonal events called (15) _____ [p.816]. However, this is a recurring round trip between two areas, so these transient effects need not be initially considered. If it is assumed that immigration is balancing emigration over time, the effects of both on population size may be ignored. This allows the definition of (16) _____ _____ [p.816] growth as an interval during which the number of births is balanced out by the number of deaths. The population size has no overall increase or decrease.

Births, deaths, and other variables that might affect population size can be measured in terms of (17) _____ _____ [p.816] rates, or rates per individual. *Capita* means (18) _____ [p.816], as in head counts. Visualize 5,000 mice living in a large cornfield. If the female mice collectively give birth to 2,000 mice per month, the birth rate would be (19) _____ [p.816] per mouse, per month. If 500 of the 5,000 die during the month, the death rate would be (20) _____ [p.816] per mouse per month.

If it is assumed the birth rate and death rate remain constant, both can be combined into a single variable—the net (21) _____ [p.816] per individual per unit (22) _____ [p.816], or *r* for short. For the mouse population in the example just given, the value of *r* is (23) _____ [p.816] per mouse per month. This example provides a method of representing population growth by the equation, (24) _____ [p.816].

If the monthly increases in the individuals in a population are plotted against time, one finds a graph line in the shape of a (25) _____ [p.817]. Then one knows that the (26) _____ [p.817] growth is being tracked. The length of time it takes for a population to double in size is its (27) _____ [p.817] time. When a population displays its (28) _____ _____ [p.817] in terms of growth, it is the maximum rate of increase per individual under ideal conditions.

Problems

For questions 29–33, consider the equation $G = rN$, where G = the population growth rate per unit time, r = the net population growth rate per individual per unit time, and N = the number of individuals in the population.

29. Assume that r remains constant at 0.2.

   a. As the value of G increases, what happens to the value of N? [pp.816–817]

   _____

   b. If the value of G decreases, what happens to the value of N? [pp.816–817]

   _____

   c. If the net reproduction per individual stays the same and the population grows faster, then what must happen to the number of individuals in the population? [pp.816–817]

   _____

30. If a society decides it is necessary to lower its value of N through reproductive means because supportive resources are dwindling, it must lower either its net reproduction per individual per unit or its [pp.816–817]

   _____

31. The equation $G = rN$ expresses a direct relationship between G and $r \times N$. If G remains constant and N increases, what must the value of r do? (In this situation, r varies inversely with N.) [pp.816–817] _____

   _____

32. Look at line (a) in the graph at the right. After seven hours have elapsed, approximately how many individuals are in the population? [pp.816–817]_____

   _____

33. Look at line (b) in the same graph.

   a. After 24 hours have elapsed, approximately how many individuals are in the population? [pp.816–817]

   _____

   b. After 28 hours have elapsed, approximately how many individuals are in the population? [pp.816–817]

   _____

46.4. LIMITS ON THE GROWTH OF POPULATIONS [pp.818–819]

46.5. LIFE HISTORY PATTERNS [pp.820–821]

46.6. *Focus on Science:* NATURAL SELECTION AND THE GUPPIES OF TRINIDAD
[pp.822–823]

Selected Words: sustainable supply of resources [p.818], *bubonic plague* [p.819], *pneumonic plague* [p.819], *Yersinia pestis* [p.819], *Type I* curves [p.821], *Type II* curves [p.821], *Type III* curves [p.821]

Boldfaced, Page-Referenced Terms

[p.818] limiting factor _____

_____

[p.818] carrying capacity _____

_____

[p.818] logistic growth _____

_____

[p.819] density-dependent control _____

_____

[p.819] density-independent factors _____

_____

[p.820] life history pattern _____

_____

[p.820] cohort _____

_____

[p.821] survivorship curve _____

_____

Fill-in-the-Blanks

If (1) _____ [p.818] factors (essential resources in short supply) act on a population, population growth tapers off. (2) _____ _____ [p.818] refers to the maximum number of individuals of a population that can be sustained indefinitely by the environment. S-shaped growth curves are characteristic of (3) _____ [p.818] population growth. The plot of (3) growth levels off once the (4) _____ _____ [p.818] is reached. In the equation $G = r_{max}N[(K - N)/K]$, as the value of N approaches the value of K, and K and r_{max} remain constant, the value of G (5) [check one] ❐ increases ❐ decreases ❐ cannot be determined by humans, even if they know algebra [pp.818–819]. As the value of r_{max} increases and G and N remain constant, the value of K (the carrying capacity) (6) [check one] ❐ increases ❐ decreases ❐ cannot be determined by humans, even if they know algebra [pp.818–819]. In an overcrowded population, predators, parasites, and disease agents serve as (7) _____-_____ [p.819] controls. When an event such as a freak summer snowstorm in the Colorado Rockies causes more deaths or fewer births in a butterfly population (with no regard to crowding or dispersion patterns), the controls are said to be (8) _____- _____ [p.819] factors. Food availability is a density- (9) _____ [p.819] factor that works to cut back population size when it approaches the environment's (10) _____ _____ [p.819]. Environmental disruptions such as forest fires and floods are density-independent (11) _____ [p.819] that may push a population above or below its tolerance range for a given variable.

The term (12) _____ _____ [p.820] pattern refers to the individuals of a species in a population that exhibit particular morphological, physiological, and behavioral traits that are adaptive to different conditions at different times in the life cycle. (13) Life and health _____ [p.820] companies use information summarized from life tables, which show trends in mortality and life expectancy. A(n) (14) _____ [p.820] is a group of individuals that is tracked from the time of birth until the last one dies. A(n) (15) _____ [p.820] table lists the completed data on a population's age-specific death schedule. Such tables can be converted to more cheery (16) "_____" schedules, which list the number of individuals that reach some specified age (x). (17) _____ [p.820] curves are the graph lines that emerge when ecologists plot the age-specific survival of a cohort in a particular habitat. Type (18) _____ [p.820] survivorship curves illustrate populations having low survivorship early in life. Organisms whose life cycles reflect high survivorship until fairly late in life with a later large increase in deaths have a type (19) _____ [p.820] survivorship curve. Organisms such as lizards, small mammals, and some songbirds are just as likely to be killed or die of disease at any age and reflect the type (20) _____ [p.820] survivorship curve. Provided people have access to good health care, human populations typically show a type (21) _____ [p.820] survivorship curve.

Matching

Choose the most appropriate answer(s) for each phrase. A letter may be used more than once, and a blank may contain more than one letter.

22. ___cohort example [p.820]

23. ___example(s) of density-independent controls [p.819]

24. ___example(s) of density-dependent controls [p.819]

25. ___selective effects of predation by killifish and pike cichlids [pp.822–823]

A. Drought, floods, earthquakes
B. Bubonic plague and pneumonic plague
C. Food availability
D. Adrenal enlargement in wild rabbits in response to crowding
E. Heavy applications of pesticides in your backyard
F. All the human babies born in New York City during 1987
G. Natural selection in the life history patterns of Trinidadian guppies
H. Freak snowstorms in the Rocky Mountains

46.7. HUMAN POPULATION GROWTH [pp.824–825]
46.8. CONTROL THROUGH FAMILY PLANNING [pp.826–827]
46.9. POPULATION GROWTH AND ECONOMIC DEVELOPMENT [pp.828–829]
46.10. SOCIAL IMPACT OF NO GROWTH [p.829]

Selected Words: *Homo habilis* [p.825], *Vibrio cholerae* [p.825], *baby-boomers* [p.826], *preindustrial stage* [p.828], *transitional* stage [p.828], *industrial stage* [p.828], *postindustrial* stage [p.828], *postpone* [p.829]

Boldfaced, Page-Referenced Terms

[p.826] total fertility rate _____

_____

[p.828] demographic transition model _____

_____

Graph Construction

1. Construct a graph of the following data in the space provided on the next page. [p.825]

Year	Estimated World Population
1650	500,000,000
1850	1,000,000,000
1930	2,000,000,000
1975	4,000,000,000
1986	5,000,000,000
1993	5,500,000,000
1995	5,700,000,000
1997	5,800,000,000
2050 (projected)	9,000,000,000+

a. Estimate the year that the world contained 3 billion humans. _____. (Consult your graph.)
b. Estimate the year that Earth will house 8 billion humans. _____. (Consult your graph.)
c. Do you expect Earth to house 8 billion humans within your lifetime? _____. (Consult your graph.)

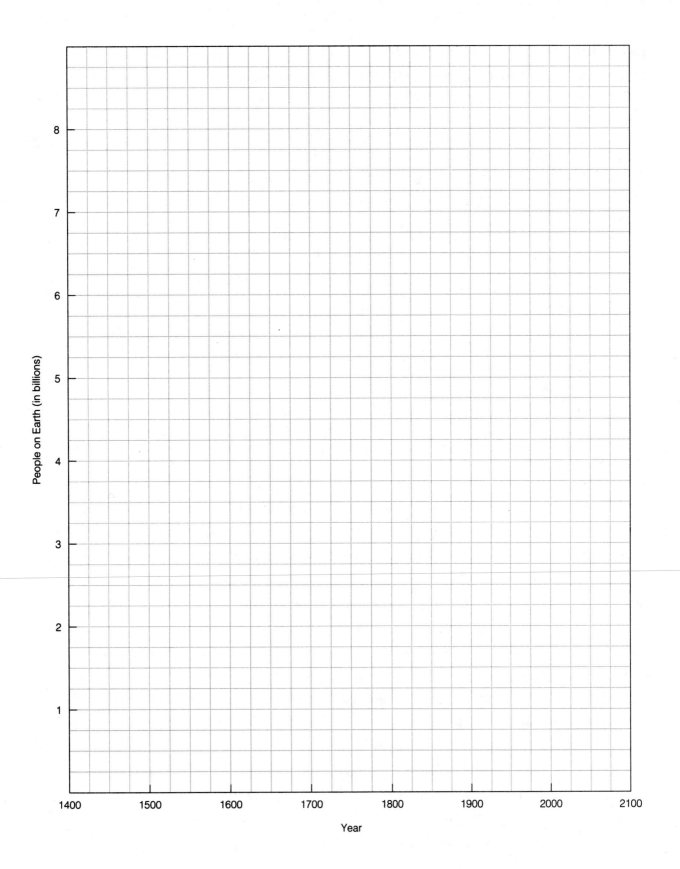

People on Earth (in billions)

Year

True–False

If the statement is true, write a T in the blank. If the statement is false, make it correct by changing the underlined word(s) and writing the correct word(s) in the answer blank.

_____2. In 1999, the human population surpassed 8 billion. [p.824]

_____3. Even if we could double our present food supply, death from starvation could still reach 20 million to 40 million people a year. [p.824]

_____4. Compared with the geographic spread of other organisms, it has taken the human population an exceptionally long period of time to expand into diverse new environments. [p.824]

_____5. Managing food supplies through agriculture has had the effect of increasing the carrying capacity for the human population. [p.824]

_____6. By bringing many disease agents under control and by tapping into concentrated existing forms of energy, humans used factors that had previously limited their population growth. [p.825]

_____7. The stupendously accelerated growth of the human population continues and can be sustained indefinitely. [p.825]

Fill-in-the-Blanks

If the annual current average population growth rate of 1.47 percent is maintained for the human population, there may be more than (8) _____ [p.826] billion people by the year 2050. Large efforts will have to be made to increase the amount of (9) _____ _____ [p.826] to sustain that many people. Such gross manipulation of resources is likely to intensify (10) _____ [p.826], which almost certainly will adversely affect our water supplies, the atmosphere, and productivity on land and in the seas. (11) _____ _____ [p.826] programs educate individuals about choosing how many children they will have, and when. Carefully developed and administered programs may bring about a long-term decline in (12) _____ [p.826] rates. To arrive at zero population growth, the average "replacement rate" would have to be slightly higher than (13) _____ [p.826] children per couple, for some (14) _____ [p.826] children die before reaching reproductive age. A more useful measure of global population trends is the total (15) _____ [p.826] rate. This is defined as the average number of children born to women during their reproductive years, as estimated on the basis of current age-specific birth rates. In 1997, the average rate was (16) _____ [p.826] children per woman, an impressive decline from 1950, when the rate was (17) _____ [p.826]. In the United States, a cohort of 78 million (18) _____-_____ [p.826] is being tracked since it formed in 1946 when soldiers returned home from World War II. Although the United States has a narrow reproductive base, more than (19) _____-_____ [p.826] of the world population falls in the broad pre-reproductive base. One way to slow growth is for women to bear children during their early (20) _____ [p.826], rather than in the mid-teens or early twenties. (21) _____ [p.827] is the country described to have the world's most extensive family planning program. Family planning programs on a global scale are designed to help (22) _____ [p.827] the size of the human population. Even if the human population reaches a level of zero growth, it will continue to grow for sixty years, for its (23) _____ [p.827] base already consists of a staggering number of individuals.

Choice

For questions 24–29, choose from the following age structure diagrams:

a. rapid growth   b. slow growth   c. zero growth   d. negative growth

24. ___ United States [p.827]

25. ___ China [p.827]

26. ___ Canada [p.827]

27. ___ Mexico [p.827]

28. ___ Australia [p.827]

29. ___ India [p.827]

Sequence

Arrange the following stages of the demographic transition model in correct chronological sequence. Write the letter of the first step next to 30, the letter of the second step next to 31, and so on.

30. ___   A. *Industrial stage:* population growth slows and industrialization is in full swing [p.828]

31. ___   B. *Preindustrial stage:* harsh living conditions, high birth rates and low death rates, slow population growth [p.828]

32. ___   C. *Postindustrial stage:* zero population growth is reached; birth rate falls below death rate, and population size slowly decreases [p.828]

33. ___   D. *Transitional stage:* industrialization begins, food production rises, and health care improves; death rates drop, birth rates remain high to give rapid population growth [p.828]

Fill-in-the-Blanks

Today, the United States, Canada, Australia, Japan, nations of the former Soviet Union, and most countries of Western Europe are in the (34) _____ stage [p.828]. Their growth rate is slowly (35) [check one] ❑ increasing ❑ decreasing [p.828]. In Germany, Bulgaria, Hungary, and some other countries, the death rates exceed the birth rates, and the populations are getting (36) _____ [p.828]. Mexico and other less developed countries are in the (37) _____ [p.828] stage. In many countries, population growth outpaces economic growth, death rates will increase, and they may be stuck in the (38) _____ stage [p.828]. Enormous disparities in (39) _____ [p.828] development are driving forces for immigration and emigration. Claiming that population growth affects economic health, many governments restrict (40) _____ [p.828].

India has a whopping (41) _____ [p.829] percent of the human population. By comparison, the United States has merely (42) _____ [p.829] percent. The highly industrialized United States produces (43) _____ [p.829] percent of all the world's goods and services. It uses (44) _____ [p.829] percent of the world's processed minerals and available, nonrenewable sources of energy. Its people consume (45) _____ [p.829] times more goods and services than the average person in India. The United States generates at least (46) _____ [p.829] percent of the global pollution and trash. By contrast, India produces about (47) _____ [p.829] percent of all goods and services and uses (48) _____ [p.829] percent of the available minerals and nonrenewable energy resources. India generates only about

(49) _____ [p.829] percent of the pollution and trash. Tyler Miller, Jr., estimated that it would take

(50) _____ [p.829] billion impoverished individuals in India to have as much environmental impact as

(51) _____ [p.829] million Americans.

For all species on our planet, the biological implications of extremely rapid (52) _____ [p.829] are

staggering. Yet so are the (53) _____ [p.829] implications of what will happen when and if the human

population declines to the point of zero population growth—and stays there. If the population ever does

reach and maintain zero growth over time, a larger proportion of the individuals will end up in the

(54) _____ [p.829] age brackets. To care for older, nonproductive citizens, fewer productive ones must

carry more and more of the (55) _____ [p.829] burden. Through our special abilities to undergo rapid

cultural evolution, we have (56) _____ [p.829] the action of most of the factors that limit growth. No

amount of (57) _____ [p.829] intervention can repeal the ultimate laws governing population growth, as

imposed by the (58) _____ _____ [p.829] of the environment.

Self-Quiz

___ 1. The number of individuals that contribute to a population's gene pool is _____. [p.814]
   a. the population density
   b. the population growth
   c. the population birth rate
   d. the population size

___ 2. The number of individuals in a given area or volume of a habitat is _____. [p.672]
   a. the population density
   b. the population growth
   c. the population birth rate
   d. the population size

___ 3. A population that is growing exponentially in the absence of limiting factors can be illustrated accurately by a(n) _____. [p.817]
   a. S-shaped curve
   b. J-shaped curve
   c. curve that terminates in a plateau phase
   d. tolerance curve

___ 4. How are the individuals in a population most often dispersed? [p.814]
   a. clumped
   b. very uniform
   c. nearly uniform
   d. random

___ 5. Assuming immigration is balancing emigration over time, _____ may be defined as an interval in which the number of births is balanced by the number of deaths. [p.816]
   a. the lack of a limiting factor
   b. exponential growth
   c. saturation
   d. zero population growth

___ 6. Assuming the birth and death rate remain constant, both can be combined into a single variable, r, or _____. [p.816]
   a. the per capita rate
   b. the minus migration factor
   c. exponential growth
   d. the net reproduction per individual per unit time

___ 7. _____ is a way to express the growth rate of a given population. [p.817]
   a. Doubling time
   b. Population density
   c. Population size
   d. Carrying capacity

8. Any population that is not restricted in some way will grow exponentially _____. [p.817]
   a. except in the case of bacteria
   b. irrespective of doubling time
   c. if the death rate is even slightly greater than the birth rate
   d. all of the above

9. The maximum number of individuals of a population (or species) that a given environment can sustain indefinitely defines _____. [p.818]
   a. the carrying capacity of the environment
   b. exponential growth
   c. the doubling time of a population
   d. density-independent factors

10. In natural communities, some feedback mechanisms operate whenever populations change in size; they are _____. [p.819]
    a. density-dependent factors
    b. density-independent factors
    c. always intrinsic to the individuals of the community
    d. always extrinsic to the individuals of the community

11. _____ are any essential resources that are in short supply for a population. [p.818]
    a. Density-independent factors
    b. Extrinsic factors
    c. Limiting factors
    d. Intrinsic factors

12. Which of the following is *not* characteristic of logistic growth? [p.818]
    a. S-shaped curve
    b. leveling off of growth as carrying capacity is reached
    c. unrestricted growth
    d. slow growth of a low-density population followed by rapid growth

13. The population growth rate (G) is equal to the _____ net population growth rate per individual (r) and number of individuals (N). [p.816]
    a. sum of
    b. product of
    c. doubling of
    d. difference between

14. $G = r_{max} N (K - N/K)$ represents _____. [p.818]
    a. exponential growth
    b. population density
    c. population size
    d. logistic growth

15. The beginning of industrialization, a rise in food production, improvement of health care, rising birth rates, and declining death rates describes the _____ stage of the demographic transition model. [p.828]
    a. preindustrial
    b. transitional
    c. industrial
    d. postindustrial

16. A group of individuals that is typically tracked from the time of birth until the last species dies is a _____. [p.820]
    a. cohort
    b. species
    c. type I curve group
    d. density-independent group

17. The survivorship curve typical of industrialized human populations is type _____. [p.821]
    a. I
    b. II
    c. III
    d. none of the above types

Chapter Objectives/Review Questions

1. Define the following terms: *ecology, demographics, population, habitat, population size, population density, population distribution, age structure,* and *reproductive base.* [pp.813–814]
2. In a population, the pre-reproductive ages and the reproductive ages together are counted as the _____ _____. [p.814]
3. List and describe the three patterns of dispersion illustrated by populations in a habitat. [p.814]
4. For mobile animals, ecologists sample population density by using a _____-_____ method. [p.815]
5. Distinguish immigration from emigration; define the term *migration.* [p.816]

6. Define zero population growth and describe how achieving it would affect population size. [p.816]
7. In the equation $G = rN$, as long as r holds constant, any population will show _____ growth. [p.817]
8. Calculate a population growth rate (G); use values for birth, death, and number of individuals (N) that seem appropriate. [p.816]
9. State how increasing the death rate of a population affects its doubling time. [p.817]
10. _____ _____ is the maximum rate of increase per individual under ideal conditions. [p.817]
11. The _____ rate of increase in population growth depends on the age at which each generation starts to reproduce, how often each individual reproduces, how many offspring are produced. [p.817]
12. List several examples of limiting factors, and explain how they influence population curves. [p.818]
13. Explain the phrase "exceeding the carrying capacity," as applied to a population of organisms in an environment. [p.818]
14. Explain the meaning of the logistic growth equation, and calculate values of G by using the logistic growth equation. Explain the meaning of r_{max} and K. [pp.818–819]
15. Contrast the conditions that promote J-shaped growth curves with those that promote S-shaped curves in populations. [pp.817–818]
16. Define the term *density-dependent growth* of populations; cite one example. [p.819]
17. Define the term *density-independent factors* and list two examples; indicate how such factors affect populations. [p.819]
18. Most species have a _____ _____ pattern in that its individuals exhibit particular morphological, physiological, and behavioral traits that are adaptive to different conditions at different times in the life cycle. [p.820]
19. Life insurance companies and ecologists track a _____, a group of individuals from the time of birth until the last one dies. [p.820]
20. Explain the significance and use of life tables; list and interpret the three survivorship curves. [p.820]
21. Explain how the construction of life tables and survivorship curves can be useful to humans in managing the distribution of scarce resources. [pp.820–821]
22. Guppy populations targeted by killifish tend to be larger, less streamlined, and more brightly colored, and guppy populations targeted by pike-cichlids tend to be smaller, more streamlined, and duller in color patterning. Other life history pattern differences exist between the two groups. After consideration of the research results obtained by Reznick and Endler, provide an explanation for these differences. [pp.822–823]
23. List three possible reasons why growth of the human population is out of control. [p.824]
24. Most governments are trying to lower birth rates by _____ _____ programs. [p.826]
25. Define the term *total fertility rate*. [p.826]
26. What is the significance of the cohort of 78 million baby-boomers? [p.826]
27. List and describe the four stages of the demographic transition model. [p.828]
28. Differences in population growth among countries correlate with levels of _____ development. [p.828]
29. Generally compare the implications of the resource consumptions of India and the United States. [p.829]
30. In the final analysis, no amount of _____ intervention can repeal the ultimate laws governing population growth, as imposed by the _____ _____ of the environment. [p.829]

Integrating and Applying Key Concepts

Assume that the world has reached zero population growth. The year is 2110, and there are 10.5 billion individuals of *Homo pollutans* on Earth. You have seen stories on the community television screen about how people used to live 120 years ago. List the ways that life has changed, and comment on the events that no longer happen because of the enormous human population.

47

SOCIAL INTERACTIONS

Interactive Exercises

Deck the Nest With Sprigs of Green Stuff [pp.832–833]

47.1. BEHAVIOR'S HERITABLE BASIS [pp.834–835]

47.2. LEARNED BEHAVIOR [p.836]

Selected Words: *Sturnus vulgaris* [p.832], *Ornithonyssus sylviarum* [p.833], *Daucus carota* [p.833], *intermediate response* [p.834]

Boldfaced, Page-Referenced Terms

[p.833] animal behavior _____

_____

[p.835] song system _____

_____

[p.835] instinctive behavior _____

_____

[p.835] sign stimuli _____

_____

[p.835] fixed action pattern _____

_____

[p.836] learned behavior _____

_____

[p.836] imprinting _____

_____

Matching

Choose the most appropriate answer for each term.

1. ___intermediate response [p.834]
2. ___song system [p.835]
3. ___instinctive behavior [p.835]
4. ___sign stimuli [p.835]
5. ___fixed action pattern [p.835]
6. ___learned behavior [p.836]
7. ___imprinting [p.836]

A. Animals process and integrate information gained from experiences, then use it to vary or change responses to stimuli
B. Well-defined environmental cues that trigger suitable responses
C. Term applied to genetically based behavioral reactions of hybrid offspring
D. Time-dependent form of learning; triggered by exposure to sign stimuli and usually occurring during sensitive periods of young animals
E. A behavior performed without having been learned by actual environmental experience
F. Consists of several brain structures that will govern activity of a vocal organ's muscles
G. A program of coordinated muscle activity that runs to completion independently of feedback from the environment

Dichotomous Choice

Circle one of two possible answers given between parentheses in each statement.

8. For garter snake populations living along the California coast, the food of choice is (the banana slug/ tadpoles and small fishes). [p.834]
9. In Stevan Arnold's experiments, newborn garter snakes that were offspring of coastal parents usually (ate/ignored) a chunk of slug as the first meal. [p.834]
10. Newborn garter snake offspring of (coastal/inland) parents ignored cotton swabs drenched in essence of slug and only rarely ate the slug meat. [p.834]
11. The differences in the behavioral eating responses of coastal and inland garter snakes (were/were not) learned. [p.834]
12. Hybrid garter snakes with coastal and inland parents exhibited a feeding response that indicated a(n) (environmental/genetic) basis for this behavior. [p.834]
13. In zebra finches and some other songbirds, singing behavior is an outcome of seasonal differences in the secretion of melatonin, a hormone secreted by the (gonads/pineal gland). [p.835]
14. In songbirds in spring, melatonin secretion is suppressed and gonads are released from hormonal suppression; they (increase/decrease) in size and step up their secretions of estrogen and testosterone. [p.835]
15. Even before a male bird hatches, a high (estrogen/testosterone) level stimulates development of a masculinized brain. [p.835]
16. Later, at the start of the breeding season, a male's enlarged gonads secrete even more (estrogen/ testosterone) that binds to cells and induces metabolic changes in the sound system to prepare the bird to sing when suitably stimulated. [p.835]
17. (Hormones/Genes) underlie animal-behavior-coordinated responses to stimuli. [p.835]

Complete the Table

18. Complete the following table with examples of instinctive and learned behavior.

Category	Examples
[p.835] a. Instinctive behavior	
[p.836] b. Learned behavior	

Matching

Choose the most appropriate answer for each category of learned behavior.

19. ___imprinting [p.836]

20. ___classical conditioning [p.836]

21. ___operant conditioning [p.836]

22. ___habituation [p.836]

23. ___spatial or latent learning [p.836]

24. ___insight learning [p.836]

A. Birds living in cities learn not to flee from humans or cars, which pose no threat to them.
B. Chimpanzees abruptly stack several boxes and use a stick to reach suspended bananas out of reach.
C. In response to a bell, dogs salivate even in the absence of food.
D. Bluejays store information about dozens or hundreds of places where they have stashed food.
E. Baby geese formed an attachment to Konrad Lorenz if separated from the mother shortly after hatching.
F. A toad learns to avoid stinging or bad-tasting insects after its first attempt to eat them.

47.3. THE ADAPTIVE VALUE OF BEHAVIOR [p.837]

Boldfaced, Page-Referenced Terms

[p.837] natural selection _____

_____

[p.837] reproductive success _____

_____

[p.837] adaptive behavior _____

_____

[p.837] social behavior _____

_____

[p.837] selfish behavior _____

_____

[p.837] altruistic behavior _____

_____

[p.837] territory _____

_____

Fill-in-the-Blanks

(1) _____ _____ [p.837] is the result of differences in reproductive success among individuals of a population that differ from one another in their heritable traits. Some versions of a trait may be better than others at helping the individual leave offspring. Thus alleles for those versions (2) _____ [p.837] in a population. Using the theory of (3) _____ [p.837] by natural selection as a point of departure, we should be able to develop and test possible explanations of why a behavior is (4) _____ [p.837]. We should be able to discern how some actions bestow (5) _____ [p.837] benefits that offset reproductive costs (disadvantages) associated with them. If a behavior is (6) _____ [p.837], it must promote the (7) _____ [p.837] production of offspring.

Complete the Table

8. Complete the following table of terms describing forms of individual adaptive behavior.

Behavior	Description
[p.837] a. Reproductive success	
[p.837] b. Adaptive behavior	
[p.837] c. Social behavior	
[p.837] d. Selfish behavior	
[p.837] e. Altruistic behavior	

47.4. COMMUNICATION SIGNALS [pp.838–839]

Selected Words: *signaling* pheromones [p.838], *priming* pheromones [p.838]

Boldfaced, Page-Referenced Terms

[p.838] communication signals _____

_____

[p.838] signaler _____

_____

[p.838] signal receivers _____

_____

[p.838] pheromones _____

_____

[p.838] composite signal _____

_____

[p.838] communication display _____

_____

[p.838] threat display _____

_____

[p.839] courtship displays _____

_____

[p.839] tactile displays _____

_____

[p.839] illegitimate receiver _____

_____

[p.839] illegitimate signalers _____

_____

Choice

For questions 1–12, choose from the following:

a. chemical signal, signaling pheromones
b. chemical signal, priming pheromones
c. acoustical signal
d. composite signal
e. communication display, social signal

f. communication display, threat display
g. communication display, courtship
h. communication display, tactile signal
i. illegitimate signaler
j. illegitimate receiver

1. ___ Assassin bugs hook dead termite bodies on their dorsal surfaces and acquire termite scent; this deception allows assassin bugs to hunt termite victims more easily. [p.839]

2. ___ Male songbirds sing to secure territory and attract a female. [p.838]

3. ___ The play bow of dogs and wolves [p.838]

4. ___ Bombykol molecules released by female silk moths serve as sex attractants. [p.838]

5. ___ A male bird might emit calls while bowing low, as if to peck the ground for food. [p.839]

6. ___ Termites act defensively when detecting scents from invading ants whose scent signals are meant to elicit cooperation from other ants. [p.839]

7. ___ A dominant male baboon's "yawn" that exposes large canines [pp.838–839]

8. ___ A volatile odor in the urine of certain male mice triggers and enhances estrus in female mice. [p.838]

9. ___ After finding a source of pollen or nectar, a foraging honeybee returns to its colony (a hive) and performs a complex dance. [p.839]

10. ___ Tungara frogs issue nighttime calls to females and rival males; the call is a "whine" followed by a "chuck." [p.838]

11. ___ A male firefly has a light-generating organ that emits a bright, flashing signal. [p.839]

12. ___ Ears laid back against the head of a zebra convey hostility, but ears pointing up convey its absence; a zebra with laid-back ears isn't too riled up when its mouth is open just a bit, but when the mouth is gaping, watch out. [p.838]

47.5. MATES, PARENTS, AND INDIVIDUAL REPRODUCTIVE SUCCESS [pp.840–841]

Selected Words: Harpobittacus apicalis [p.840], *Centrocercus urophasianus* [p.841]

Boldfaced, Page-Referenced Terms

[p.840] sexual selection _____

_____

[p.841] lek _____

_____

Complete the Table

1. Complete the following table to supply the common names of the animals that fit the text examples of sexual selection.

Animals	Descriptions of Sexual Selection
[pp.840–841] a.	Females select the males that offer them superior material goods; females permit mating only after they have eaten the "nuptial gift" for about five minutes.
[p.841] b.	Males congregate in a lek or communal display ground; each male stakes out a few square meters as his territory; females are attracted to the lek to observe male displays and usually select and mate with only one male.
[p.841] c.	Females of a species cluster in defendable groups at a time they are sexually receptive; males compete for access to the clusters; combative males are favored.
[p.841] d.	Extended parental care improves the likelihood that the current generation of offspring will survive; this behavior comes at a reproductive cost to the adults.

47.6. BENEFITS OF LIVING IN SOCIAL GROUPS [pp.842–843]

47.7. COSTS OF LIVING IN SOCIAL GROUPS [p.844]

47.8. EVOLUTION OF ALTRUISM [p.845]

47.9. *Focus on Science:* WHY SACRIFICE YOURSELF? [pp.846–847]

47.10. *Focus on Science:* ABOUT THOSE NAKED MOLE-RATS [p.848]

Selected Words: Ovibos moschatus [p.842], *Parus major* [p.842], caring for one's *relatives* [p.845], *indirect genetic contribution* [p.845], "self-sacrifice" genes [p.845], *Heterocephalus glaber* [p.848], *DNA fingerprinting* [p.848]

Boldfaced, Page-Referenced Terms

[p.842] selfish herd _____

_____

[p.843] dominance hierarchies _____

_____

[p.845] theory of indirect selection _____

_____

Matching

Choose the most appropriate answer for each term.

1. ___ cost–benefit approach [p.842]
2. ___ disadvantages to sociality [p.844]
3. ___ dominance hierarchy [p.843]
4. ___ cooperative predator avoidance [p.842]
5. ___ the selfish herd [pp.842–843]

A. Competition for resources, rapid depletion of food resources, cannibalism, and greater vulnerability to disease

B. A simple society brought together by reproductive self-interest; larger, more powerful male bluegills tend to claim the central locations

C. An attempt to identify the costs and benefits of sociality in terms of reproductive success of the individual

D. Adult musk oxen form a circle around their young while they face outward, and the "ring of horns" successfully deters the wolves; writhing, regurgitating reaction of Australian sawfly caterpillers to a disturbance

E. Some individuals of a baboon troop adopt a subordinate status with respect to the other members

Complete the Table

6. Complete the following table to supply examples for each listed category of the costs of living in social groups.

Costs of Living in Social Groups	Examples
[p.844] a. An increase in the competition for food	
[p.844] b. Encourages the spread of contagious diseases and parasites	
[p.844] c. Risks of being killed or exploited by others in the group	

Fill-in-the-Blanks

A(n) (7) _____ [p.845] animal that gives way to a dominant one is acting in its own self-interest. Such (8) _____ [p.845] animals are found among many vertebrate groups. A female wolf's (9) _____ [p.845] success increases when she monopolizes the benefits that males offer. Within a wolf pack, there is usually a(n) (10) _____ [p.845] breeding female and male. The other wolves are (11) _____ [p.845] sisters, aunts, brothers, and uncles. Their (12) _____ [p.845] behavior is to hunt and bring back food to members that remain in the den and guard the pups.

Altruistic behavior is most extreme in certain (13) _____ [p.845] societies. When a(n) (14) _____ [p.845] bee plunges its stinger into an invader of the hive, it commits suicide.

Altruistic individuals of a social group do not contribute their (15) _____ [p.845] to the next generation, but yet seem to perpetuate the genetic basis for their altruistic behavior over evolutionary time. According to William Hamilton's theory of (16) _____ _____ [p.845], those genes associated with caring for relatives—not one's direct descendants—tend to be favored in certain situations.

This form of altruism can be thought of as an extension of (17) _____ [p.845]. For example, if an uncle helps his niece survive long enough to reproduce, he has made an (18) _____ [p.845] genetic contribution to the next generation, as measured in terms of the genes that he and his relatives share. Altruism costs him; he may lose his own opportunities to (19) _____ [p.845].

Similarly, nonbreeding workers in insect societies indirectly promote their "self-sacrifice" genes through (20) _____ [p.845] behavior directed toward relatives. Thus, when a guard bee drives her stinger into a raccoon, she inevitably dies—but her siblings in the hive will perpetuate some of her (21) _____ [p.845]. Sterility and extreme self-sacrifice are rare among social groups of (22) _____ [p.845].

Unlike any other known vertebrate, the highly social naked mole-rat individuals live out their lives as (23) _____ [p.848] helpers in their social group. In each mole-rat clan, there is a single reproducing female, and she mates with one to three males. All other members of the clan care for the "queen" and "king" (or kings) and their offspring. A self-sacrificing naked mole-rat helps to (24) _____ [p.848] a very high proportion of the forms of genes that it carries. Studies using a technique called DNA fingerprinting suggest that the (25) _____ [p.848] of helpers and the helped among mole-rats may be as much as 90 percent identical.

47.11. AN EVOLUTIONARY VIEW OF HUMAN SOCIAL BEHAVIOR [p.849]

Selected Words: redirected adaptive behaviors [p.849]

Boldfaced, Page-Referenced Terms

[p.849] adoption _____

_____

Dichotomous Choice

Circle one of two possible answers given between parentheses in each statement.

1. Many people seem to believe that attempts to identify the adaptive value of a particular (animal/human) trait is an attempt to define its moral or social advantage. [p.849]
2. "Adaptive" refers to (a trait with moral value/a trait valuable in gene transmission). [p.849]
3. Cardinals feeding goldfish or Emperor penguins fighting to adopt orphans represent examples of (adopting by mistake/redirecting adaptive behaviors).
4. John Alcock suggests that husbands and wives who have lost an only child or who fail to produce children themselves should be especially prone to adopt (strangers/relatives). [p.849]
5. The human adoption process (can/cannot) be considered adaptive when indirect selection favors adults who direct parenting assistance to relatives. [p.849]
6. Joan Silk showed that in some traditional societies, people (will not/will) adopt related children far more often than nonrelated ones. [p.849]

7. In large, industrialized societies in which agencies and other means of adoption exist, individuals become parents of (related/nonrelated) children. [p.849]
8. It (is/is not) possible to test evolutionary hypotheses about the adaptive value of human behaviors. [p.849]
9. *Adaptive* behavior and *socially desirable* behavior (are not/are) separate issues. [p.849]
10. Strong parenting mechanisms evolved in the past, and it may be that their redirection toward a (relative/nonrelative) says more about human evolutionary history than it does about the transmission of one's genes. [p.849]

Self-Quiz

____ 1. The observable, coordinated responses that animals make to stimuli are what we call animal _____. [p.833]
   a. imprinting
   b. instincts
   c. behavior
   d. learning

____ 2. In _____, a particular behavior is performed without having been learned by actual experience in the environment. [p.835]
   a. natural selection
   b. altruistic behavior
   c. sexual selection
   d. instinctive behavior

____ 3. Newly hatched goslings follow any large moving objects to which they are exposed shortly after hatching; this is an example of _____. [p.836]
   a. homing behavior
   b. imprinting
   c. piloting
   d. migration

____ 4. A young toad flips its sticky-tipped tongue and captures a bumblebee that stings its tongue; in the future, the toad leaves bumblebees alone. This is _____. [p.836]
   a. instinctive behavior
   b. a fixed reaction pattern
   c. altruistic
   d. learned behavior

____ 5. Pavlov's dog experiments represent an example of _____. [p.836]
   a. classical conditioning
   b. latent learning
   c. selfish behavior
   d. habituation

____ 6. _____ provides an example of an illegitimate signaler. [p.839]
   a. A soldier termite killing an ant on cue
   b. An assassin bug with acquired termite odor
   c. A termite pheromone alarm signal
   d. The "yawn" of a dominant male baboon

____ 7. The claiming of the more protected central locations of the bluegill colony by the largest, most powerful males suggests _____. [pp.842–843]
   a. cooperative predator avoidance
   b. the selfish herd
   c. a huge parent cost
   d. self-sacrificing behavior

____ 8. A chemical odor in the urine of male mice triggers and enhances estrus in female mice. The source of stimulus for this response is a _____. [p.838]
   a. generic mouse pheromone
   b. signaling pheromone
   c. priming pheromone
   d. cue from male mice

____ 9. Female insects often attract mates by releasing sex pheromones. This is an example of a(n) _____ signal. [p.838]
   a. chemical
   b. visual
   c. acoustical
   d. tactile

____10. Male songbirds sing to stake out territories, attract females, and discourage males. This is an example of a(n) _____ signal. [p.838]
   a. chemical
   b. visual
   c. acoustical
   d. tactile

___11. An example of dominance hierarchy and self-sacrificing behavior is _____. [p.845]
   a. cannibalistic behavior of a breeding pair of herring gulls in a huge nesting colony
   b. members of wolf packs helping others by sharing food or fending off predators even though they do not breed
   c. clumps of regurgitating Australian sawfly caterpillars
   d. a huge colony of prairie dogs being ravaged by a parasite

___12. When musk oxen form a "ring of horns" against predators, it is _____. [p.842]
   a. a selfish herd
   b. cooperative predator avoidance
   c. self-sacrificing behavior
   d. dominance hierarchy

___13. Caring for nondescendant relatives favors the genes associated with helpful behavior and is classified as _____. [p.843]
   a. dominance hierarchy
   b. indirect selection
   c. altruism
   d. both b and c

___14. "_____" means only that a given trait has proved valuable in transmitting an individual's genes. [p.849]
   a. Dominance hierarchy
   b. Indirect selection
   c. Adaptive
   d. Altruism

___15. _____ also favors adults who direct parenting behavior toward relatives and so indirectly perpetuate their shared genes. [p.845]
   a. Indirect selection
   b. Moral selection
   c. Redirected selection
   d. Perpetuated selection

Chapter Objectives/Review Questions

1. What explains the fact that coastal and inland garter snakes of the same species have different food preferences? [p.834]
2. Describe the intermediate response obtained in Arnold's experiment with coastal and inland garter snakes. [p.834]
3. Tongue-flicking, body orientation, and strikes at prey by newborn garter snakes are good examples of _____ behavior. [p.834]
4. Describe the origin and formation of a song system. [pp.834–835]
5. Define the term *sign stimuli*. [p.835]
6. Describe and cite an example of a fixed action pattern. [p.835]
7. When animals process and integrate information gained from specific experiences and then use the information to vary or change responses to stimuli, it is _____ behavior. [p.836]
8. Define each of the following categories of learned behavior and give one example of each: imprinting, classical conditioning, operant conditioning, habituation, spatial or latent learning, and insight learning. [p.836]
9. What is meant by reproductive success? [p.837]
10. _____ behavior is any behavior that promotes the propagation of an individual's genes and tends to occur at increased frequency in successive generations. [p.837]
11. _____ behavior refers to the cooperative, interdependent relationships among individuals of the species. [p.837]
12. Distinguish between selfish behavior and altruistic behavior. [p.837]
13. A(n) _____ is an area that one or more individuals defend against competitors. [p.837]
14. Examples of _____ signals are chemical, visual, acoustical, and tactile. [p.838]
15. Define the roles of signalers and signal receivers. [p.838]
16. Distinguish between signaling and priming pheromones, and cite an example of each. [p.838]

17. A(n) _____ signal is illustrated by a zebra with laid-back ears and a gaping mouth. [p.838]
18. Describe one example of a threat display. [pp.838–839]
19. Ritualization is often developed to an amazing degree in _____ displays between potential mates. [p.839]
20. An example of a(n) _____ signal is the physical contact of bees in a hive maintaining physical contact during the dance of a returning foraging bee. [p.839]
21. When soldier termites detect ant scents meant for other ants and kill ants on cue, the termites are said to be _____ _____ of a signal meant for individuals of a different species. [p.839]
22. Assassin bugs covered with termite scent are able to use deception to hunt termite victims more easily and as such are acting as _____ signalers. [p.839]
23. Natural selection tends to favor communication signals that promote _____ success. [p.839]
24. Competition among members of one sex for access to mates and selection of mates are the result of a microevolutionary process called _____ _____. [p.840]
25. Discuss mate selection processes by female hangingflies and the female sage grouse. [pp.840–841]
26. List the costs and benefits of parenting in the example of adult Caspian terns. [p.841]
27. Explain the "cost–benefit approach" that evolutionary biologists use to find answers to questions about social life. [p.842]
28. Studies of Australian sawfly caterpillars indicate _____ predator avoidance. [p.842]
29. Define the term *selfish herd*; cite an example. [pp.842–843]
30. Members of baboon troops recognize a _____ _____ in which some individuals have adopted a subordinate status with respect to the others. [p.843]
31. List disadvantages to sociality. [p.844]
32. Hamilton's theory of _____ selection relates to caring for nondescendant relatives and how this favors genes associated with helpful behavior. [p.845]
33. With _____ behavior, an individual behaves in a self-sacrificing way that helps others but decreases its own chance of reproductive success. [p.845]
34. Discuss the self-sacrificing behavior of honeybees and termites in a eucalyptus forest in Australia. [pp.846–847]
35. Explain how DNA fingerprinting was used to establish that self-sacrificing mole-rats help to perpetuate a very high proportion of genes (alleles) that they carry—even though they are not the reproducing mole-rats. [p.848]
36. It is possible to test evolutionary hypotheses about the _____ value of human behaviors; discuss human adoption practices in this context. [p.849]
37. "_____" means only that a specified trait has proved beneficial in the transmission of the genes of an individual that are responsible for that trait. [p.849]

Integrating and Applying Key Concepts

Think about communication signals that humans use and list them. Do you believe a dominance hierarchy exists in human society? Think of examples.

48

COMMUNITY INTERACTIONS

Interactive Exercises

No Pigeon Is an Island [pp.852–853]

48.1. WHICH FACTORS SHAPE COMMUNITY STRUCTURE? [p.854]

48.2. MUTUALISM [p.855]

Selected Words: *partition* the fruit supply [p.853], *fundamental* niche [p.854], *realized* niche [p.854], *neutral* relationship [p.854], *obligatory* mutualism [p.855]

Boldfaced, Page-Referenced Terms

[p.854] habitat _____

_____

[p.854] community _____

_____

[p.854] niche _____

_____

[p.854] commensalism _____

_____

[p.854] mutualism _____

_____

[p.854] interspecific competition _____

_____

[p.854] predation _____

_____

[p.854] parasitism _____

_____

[p.854] symbiosis _____

_____

Fill-in-the-Blanks

The type of place where you will normally find an organism is its (1) _____ [p.854]. The populations of all species in a habitat associate with one another as a (2) _____ [p.854]. The (3) _____ [p.854] of an organism is defined by its "profession" in a community, in the sum of activities and relationships in which it engages to secure and use the resources necessary for its survival and reproduction. The (4) _____ [p.854] niche is the one that could prevail in the absence of competition and other factors that could constrain its acquisition and use of resources. The (5) _____ [p.854] niche is one that is more constrained and that shifts in large and small ways over time, as individuals of the species respond to a mosaic of changes. Each species has a (6) _____ [p.854] relationship with most species in its habitat; that is, they do not affect each other (7) _____ [p.854]. Tree-roosting birds are (8) _____ [p.854] with trees; the trees get nothing but are not harmed. With (9) _____ [p.854], benefits flow both ways between the interacting species. With (10) _____ [p.854] competition, disadvantages flow both ways between species. (11) _____ [p.854] and (12) _____ [p.854] are interactions that directly benefit one species and directly hurt the other. Commensalism, mutualism, and parasitism are all forms of (13) _____ [p.854]; it means "living together."

Complete the Table

14. Complete the following table to describe how each of the organisms listed is intimately dependent on the other for survival and reproduction in an obligatory mutualistic symbiotic interaction.

Organism	Dependency
[p.855] a. Yucca moth	
[p.855] b. Yucca plant	

48.3. COMPETITIVE INTERACTIONS [pp.856–857]

48.4. PREDATION [pp.858–859]

48.5. *Focus on the Environment:* THE COEVOLUTIONARY ARMS RACE [pp.860–861]

Selected Words: *intraspecific* competition [p.856], *interspecific* competition [p.856], *Paramecium* [p.856], *Plethodon* [p.857], *three-level* interaction [p.859], *Lithops* [p.860], *Dendrobates* [p.860], *model* for deception [p.860], *mimic* [p.860], *Dismorphia* [p.861], *Ranunculus* [p.861]

Boldfaced, Page-Referenced Terms

[p.856] competitive exclusion _____

_____

[p.857] resource partitioning _____

_____

[p.858] predators _____

_____

[p.858] prey _____

_____

[p.858] parasites _____

_____

[p.858] hosts _____

_____

[p.858] coevolution _____

_____

[p.858] carrying capacity _____

_____

[p.860] camouflage _____

_____

[p.860] warning coloration _____

_____

[p.860] mimicry _____

_____

Fill-in-the-Blanks

(1) _____ [p.856] competition means individuals of the same species compete with one another.

(2) _____ [p.856] competition occurs between populations of different species. (3) _____ [p.856] competition is usually the least intense of the two types. According to the concept of (4) _____ _____ [p.856], two species that require identical resources cannot coexist indefinitely.

In (5) _____ species [p.857], competing species coexist by subdividing a category of similar resources. A good example is the nine species of fruit-eating (6) _____ [p.857] in the same New Guinea forest. They all require the same resource: fruit. The pigeons overlap only slightly in their use of the resource, because each species specializes in fruits of a particular (7) _____ [p.857]. Another example is of three species of annual plants, each adapted to exploiting a different portion of the habitat. Each has a(n) (8) _____ [p.857] system that grows to a different depth in the soil from the others.

(9) _____ [p.858] are animals that feed on other living organisms, which are called their (10) _____ [p.858], but do not take up residence on or in them. Their (11) _____ [p.858] may or may not die from the interaction. In this respect they are unlike (12) _____ [p.858], which live in or on other living organisms—their (13) _____ [p.858]—and feed on specific (13) tissues for part of their life cycle. Their (13) may or may not die as a result of the interaction.

In a predator–prey cycle, predators are usually (14) _____ [p.858] common than their prey during the cycle. This pattern arises because of (15) _____ [p.858] lags in predator responses to changes in prey abundance. When prey density is low, predators have a harder time getting (16) _____ [p.858] and their population is (17) _____ [p.858]. Prey respond to the decline in predators by (18) _____ [p.858]. Both predator and prey populations grow until the predator increase causes the prey population to (19) _____ [p.858]. Predators continue to (20) _____ [p.858] and take out more prey. The lower prey density leads to (21) _____ [p.858] predators, and the population growth rate for predators (22) _____ [p.858]. Then a new cycle begins. This cycle seems to turn on (23) _____, herbivores, and carnivores. [p.859]

Many adaptations of predators (or parasites) and their victims arose by (24) _____ [p.858], the joint evolution of two (or more) species that exert selection pressure on each other as a result of close ecological interaction. Organisms that exhibit (25) _____ [p.860] have adaptations in form, patterning, color, and behavior that help them blend in with their surroundings and escape detection. Toxic types of prey often have (26) _____ _____ [p.860], or conspicuous patterns and colors that predators learn to recognize as "avoid me" signals. (27) _____ [p.860] is a term applied to prey organisms that bear close resemblance to dangerous, unpalatable, or hard-to-catch species. (28) _____-of-_____ [p.860] defenses refers to last-ditch tricks used by prey species for their survival as they are cornered or under attack. Predators may counter prey defenses with their own marvelous (29) _____ [p.861].

Matching

Match each of the following with the most appropriate answer. The same letter may be used more than once. Use only one letter per blank.

30. ___Cornered earwigs, skunks, and stink beetles producing awful odors [p.860]

31. ___A young, inexperienced bird will spear a yellow-banded wasp or an orange-patterned monarch butterfly—once [p.860]

32. ___Different species of New Guinea pigeons coexisting in the same environment by eating fruits of different sizes [p.857]

33. ___Canadian lynx and snowshoe hare [p.859]

34. ___Grasshopper mice plunging the noxious chemical-spraying tail end of their beetle prey into the ground to feast on the head end [p.861]

35. ___Resemblance of *Lithops*, a desert plant, to a small rock [p.860]

36. ___An inedible butterfly is a model for the edible *Dismorphia* [p.861]

37. ___In the same area, drought-tolerant foxtail grasses have a shallow, fibrous root system, mallow plants have a deeper taproot system, and smartweed has a taproot system that branches in topsoil and in soil below the roots of other species [p.857]

38. ___Aggressively stinging yellow jackets are the likely model for nonstinging edible wasps [p.861]

39. ___Two species of *Paramecium* requiring identical resources cannot coexist indefinitely [pp.856–857]

40. ___Baboon on the run turns to give canine tooth display to a pursuing leopard [pp.858,860]

41. ___Least bittern with coloration similar to surrounding withered reeds [p.860]

A. Predator–prey interactions
B. Camouflage
C. Warning coloration
D. Mimicry
E. Moment-of-truth defenses
F. Competitive exclusion
G. Resource partitioning
H. Predator responses to prey

48.6. PARASITIC INTERACTIONS [pp.862–863]

48.7. FORCES CONTRIBUTING TO COMMUNITY STABILITY [pp.864–865]

48.8. COMMUNITY INSTABILITY [pp.866–867]

48.9. *Focus on the Environment:* EXOTIC AND ENDANGERED SPECIES [pp.868–869]

Selected Words: *ecto*parasites [p.862], *endo*parasites [p.862], *holo*parasitic plants [p.862], *hemi*parasitic plants [p.862], *biological controls* [p.863], *facilitate* their own replacement [p.864], *natural* and *active* restoration of the climax community [p.865], *Pisaster* [p.866], *Mytilus* [p.866], *Litorrina littorea* [p.866], *Chondrus* [p.866], *Cryphonectria parasitica* [p.867], *Solenopsis invicta* [p.867], *Oryctolagus coniculus* [p.868], *myxamatosis* [p.869], *Pueraria lobata* [p.869], kudzu [p.869]

Boldfaced, Page-Referenced Terms

[p.862] microparasites _____

_____

[p.862] macroparasites _____

_____

[p.863] social parasites _____

_____

[p.863] parasitoids _____

_____

[p.864] ecological succession _____

_____

[p.864] pioneer species _____

_____

[p.864] climax community _____

_____

[p.864] primary succession _____

_____

[p.864] secondary succession _____

_____

[p.864] climax-pattern model _____

_____

[p.866] keystone species _____

_____

[p.867] geographic dispersal _____

_____

[p.867] jump dispersal _____

_____

[p.868] endangered species _____

_____

Matching

Choose the most appropriate answer for each term.

1. ____ectoparasites [p.862]
2. ____holoparasitic plants [p.862]
3. ____endoparasites [p.862]
4. ____microparasites [p.862]
5. ____hemiparasitic plants [p.862]
6. ____macroparasites [p.862]
7. ____social parasites [p.863]
8. ____biological controls [p.863]
9. ____parasitoids [p.863]

A. Parasites and parasitoids often touted as an alternative to chemical pesticides
B. Bacteria, viruses, protozoans, and sporozoans; microscopically small, reproduce rapidly
C. Live on a host's surface
D. Insect larvae that always kill what they eat; consume larvae or pupae of other insect species
E. Complete their life cycle by drawing on social behaviors of another species; the cowbird is a good example
F. Photosynthetic; they withdraw nutrients and water from young roots of host plants
G. Large, includes many flatworms, roundworms (nematodes), and arthropods such as fleas and ticks
H. Live inside a host's body; some species live on or in one or more hosts for their entire life cycle
I. Retain the capacity for photosynthesis but still withdraw nutrients and water from host plants; mistletoe is an example

Fill-in-the-Blanks

When a community develops in a sequence from pioneers to a stable end array of species that remain in equilibrium over some region, this is known as ecological (10) _____ [p.864]. (11) _____ [p.864] species are opportunistic colonizers of vacant habitats that enjoy high dispersal rates and rapid growth. As time passes, the (12) _____ [p.864] are replaced by species that are more competitive; these species are themselves then replaced until the array of species stabilizes under the prevailing habitat conditions. This persistent array of species is the (13) _____ [p.864] community. When pioneer species begin to colonize a barren habitat such as a volcanic island, it is known as (14) _____ [p.864] succession. Once established, the pioneers improve conditions for other species and often set the stage for their own (15) _____ [p.864].

When an area within a community is disturbed and then recovers to move again toward the climax state, it is known as (16) _____ [p.864] succession. Some scientists hypothesize that the colonizers (17) _____ [p.864] their own replacement while other scientists subscribe to the idea that the sequence of (18) _____ [p.864] depends on what species arrive first to compete against the species that could replace them. According to the (19) _____-_____ [p.864] model, a community is adapted to a total pattern of environmental factors—topography, climate, soil, wind, species interactions, recurring disturbances, chance events, and so on—that vary in their influence over a region.

Small-scale and recurring changes contribute to the internal dynamics of the (20) _____ [p.864] as a whole. For example, a tree falling in a tropical forest opens a gap in the forest canopy that allows more light to reach that gap. Here the growth of previously suppressed small trees and the germination of (21) _____ [p.865] or shade-intolerant species is encouraged. Giant sequoia trees in climax communities of the California Sierra Nevada are best maintained by modest (22) _____ [p.865] that eliminate trees and shrubs that compete with young sequoias but do not damage the older sequoias. Thus recurring small-scale changes are built into the overall workings of many communities. The example of secondary succession regrowth following Mount St. Helen's violent eruption in 1980 is classified as (23) _____ [p.865] restoration of the climax community. Attempts to reestablish biodiversity in key areas adversely affected or lost by agriculture and other human activities are known as (24) _____ [p.865] restoration.

Choice

For questions 25–33, choose from the following terms.

a. primary succession   b. secondary succession

25. ___ Following a disturbance, a patch of habitat or a community moves once again toward the climax state [p.864]

26. ___ Successional changes begin when a pioneer population colonizes a barren habitat [p.864]

27. ___ A disturbed area within a community recovers and moves again toward the climax state [p.864]

28. ___ Many plants in this succession arise from seeds or seedlings that are already present when the process begins [p.864]

29. ___ Many types are mutualists with nitrogen-fixing bacteria and initially outcompete other plants in nitrogen-poor habitats [p.864]

30. ___ Involves typically small plants, with brief life cycles, adapted to grow in exposed areas with intense sunlight, swings in temperature, and nutrient-deficient soil [p.864]

31. ___ A successional pattern that occurs in abandoned fields, burned forests, and storm-battered intertidal zones [p.864]

32. ___ May occur on a new volcanic island or on land exposed by the retreat of a glacier [p.864]

33. ___ Once established, the pioneers improve conditions for other species and often set the stage for their own replacement [p.864]

Fill-in-the-Blanks

A(n) (34) _____ [p.866] species is a dominant species that dictates community structure. *Pisaster*, a sea star, is a (35) _____ [p.866] species that controls abundances of mussels, limpets, chitons, and barnacles. If sea stars are removed from experimental plots, (36) _____ [p.866], the main prey of sea stars, become the strongest competitors; they crowd out seven other species of invertebrates. When sea stars are removed, the community of (36) shrinks from fifteen species to eight.

Matching

Choose the most appropriate answer for each term.

37. ___community stability [p.866]
38. ___endangered species [p.868]
39. ___keystone species [p.866]
40. ___geographic dispersal [p.867]
41. ___jump dispersal [p.867]
42. ___exotic species [p.868]

A. A rapid geographic dispersal mechanism, as when an insect might travel in a ship's hold from an island to the mainland
B. An outcome of forces that have come into an uneasy balance
C. Resident of an established community that has moved from its home range and successfully taken up residence elsewhere; Nile perch, European rabbits, and kudzu are examples
D. Species that are now on rare or threatened species lists or have already become extinct
E. Residents of established communities move out from their home range and successfully take up residence elsewhere; the process may be slow or rapid, as by expansion of home range or jump dispersal, or extremely slow as in continental drift
F. A dominant species that can dictate community structure; a sea star is an example

48.10. PATTERNS OF BIODIVERSITY [pp.870–871]

Boldfaced, Page-Referenced Terms

[p.871] distance effect _____

_____

[p.871] area effect _____

_____

Short Answer

1. List three factors responsible for creating the higher species-diversity values correlated with the distance of land and sea from the equator. [p.870]

a._____

_____

_____

_____

b._____

_____

_____

_____

c._____

_____

_____

_____

Fill-in-the-Blanks

The number of coexisting species is highest in the (2) _____ [p.870], and it systematically declines toward the poles. In 1965, a volcanic eruption formed a new island southwest of (3) _____ [p.870], which was named Surtsey. The new island served as a laboratory for studying (4) _____ [p.870]. Within six months, bacteria, fungi, seeds, flies, and some seabirds became established on the island. After two years one vascular plant appeared, and two years after that a moss plant was observed. These organisms were all colonists from (5) _____ [p.870].

Islands far from a source of potential colonists receive few colonizing species; the few that arrive are adapted for long-distance (6) _____ [p.871]. This is called the (7) _____ [p.871] effect. Larger islands tend to support more species than do small islands at equivalent distances from source areas. This is the (8) _____ [p.871] effect. (9) _____ [p.871] islands tend to have more and varied habitats, more complex topography, and extend farther above sea level. Thus, they favor species (10) _____ [p.871]. Also, being bigger (11) _____ [p.871], they may intercept more colonists.

Most importantly, extinctions suppress (12) _____ [p.871] on (13) _____ [p.871] islands. There, the (14) _____ [p.871] populations are far more vulnerable to storms, volcanic eruptions, diseases, and random shifts in birth and death rates. As for any island, the number of species reflects a balance between (15) _____ [p.871] rates for new species and (16) _____ [p.871] rates for established ones. Small islands that are distant from a source of colonists have low (17) _____ [p.871] and high (18) _____ [p.871] rates, so they support few species once the balance has been struck for their populations.

Problem

19. After considering the distance effect, the area effect, and species diversity patterns as related to the equator, answer the following question. There are two islands (B and C) of the same size and topography that are equidistant from the African coast (A), as shown in the following illustration. Which will have the higher species-diversity values? [pp.870–871]

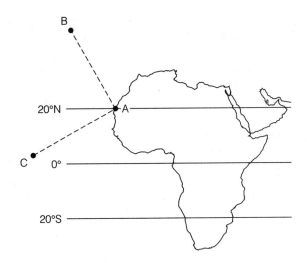

Self-Quiz

___ 1. All the populations of different species that occupy and are adapted to a given habitat are referred to as a(n) _____. [p.854]
   a. biosphere
   b. community
   c. ecosystem
   d. niche

___ 2. The range of all factors that influence whether a species can obtain resources essential for survival and reproduction is called the _____ of a species. [p.854]
   a. habitat
   b. niche
   c. carrying capacity
   d. ecosystem

___ 3. A one-way relationship in which one species benefits and the other is directly harmed is called _____. [p.854]
   a. commensalism
   b. competitive exclusion
   c. parasitism
   d. mutualism

___ 4. A lopsided interaction that directly benefits one species but does not harm or help the other much, if at all, is _____. [p.854]
   a. commensalism
   b. competitive exclusion
   c. predation
   d. mutualism

___ 5. An interaction in which both species benefit is best described as _____. [p.854]
   a. commensalism
   b. mutualism
   c. predation
   d. parasitism

___ 6. A striped skunk being pursued by a predator suddenly turns and releases its foul-smelling odor. This is an example of _____. [p.860]
   a. warning coloration
   b. mimicry
   c. camouflage
   d. moment-of-truth defense

___ 7. When an inexperienced predator attacks a yellow-banded wasp, the predator receives the pain of a stinger and will not attack again. This is an example of _____. [p.860]
   a. mimicry
   b. camouflage
   c. a prey defense
   d. warning coloration
   e. both c and d

___ 8. _____ is represented by foxtail grass, mallow plants, and smartweed because their root systems exploit different areas of the soil in a field. [p.857]
   a. Succession
   b. Resource partitioning
   c. A climax community
   d. A disturbance

___ 9. During the process of community succession, _____. [p.864]
   a. pioneer populations adapt to growing in habitats that cannot support most species
   b. pioneers set the stage for their own replacement
   c. species composition eventually is stable in the form of the climax community
   d. all of the above

___ 10. G. Gause used two species of *Paramecium* in a study that described _____. [p.856]
   a. interspecific competition and competitive exclusion
   b. resource partitioning
   c. the establishment of territories
   d. coevolved mutualism

___ 11. The most striking patterns of species diversity on land and in the seas relate to _____. [p.870]
   a. distance effect
   b. area effect
   c. immigration rate for new species
   d. distance from the equator

___12. The relationship between the yucca plant
and the yucca moth that pollinates it is best
described as _____. [p.855]
   a. camouflage
   b. commensalism
   c. competitive exclusion
   d. obligatory mutualism

Chapter Objectives/Review Questions

1. The type of place where you normally find a maple is its _____. [p.854]
2. List five factors that shape the structure of a biological community. [p.854]
3. An organism's _____ is the sum of activities and relationships in which it engages to secure and use the resources necessary for its survival and reproduction. [p.854]
4. Contrast fundamental niche and realized niche. [p.854]
5. The relation between a bird nesting in a tree, and the tree itself, can be characterized as _____. [p.854]
6. In forms of _____, each of the participating species reaps benefits from the interaction. [p.854]
7. In _____ competition, disadvantages flow both ways between species. [p.854]
8. _____ and _____ are interactions that directly benefit one species and directly hurt the other. [p.854]
9. Commensalism, mutualism, and parasitism are all forms of _____, which means "living together." [p.854]
10. Define and cite an example of obligatory mutualism. [p.855]
11. Define the terms *intraspecific competition* and *interspecific competition*. [p.856]
12. Describe a study that demonstrates laboratory evidence supporting the competitive exclusion concept. [pp.856–857]
13. Cite one example of resource partitioning. [p.857]
14. A predator gets food from other living organisms, its _____. [p.858]
15. _____ take up residence in or on other living organisms—their hosts—and feed on specific host tissues for part of the life cycle. [p.858]
16. Many of the adaptations of predators (or parasites) and their victims arose through _____. [p.858]
17. Generally describe an idealized cycling of predator and prey abundances. [pp.858–859]
18. In relation to predator–prey cycles, explain what is meant by a "three-level interaction." [p.859]
19. Completely define and give examples of the following prey defenses: warning coloration, mimicry, moment-of-truth defenses, and camouflage. [pp.860–861]
20. Describe two examples of how predators counter prey defenses with their own marvelous adaptations. [p.861]
21. What are the general body locations of ectoparasites and endoparasites? [p.862]
22. List the types of organisms classified as microparasites and those classified as macroparasites. [p.862]
23. Distinguish between holoparasitic plants and hemiparasitic plants. [p.862]
24. Discuss advantages and disadvantages of using parasites as biological controls. [p.863]
25. Define the term *ecological succession*. [p.864]
26. A _____ community is a stable, persistent array of species in equilibrium with one another and their habitat. [p.864]
27. Distinguish between primary and secondary succession. [p.864]
28. By the _____-_____ model, a community is adapted to a total pattern of environmental factors. [p.864]
29. Describe how fire disturbances positively affect a community of giant sequoias. [p.865]
30. Distinguish between natural and active restoration. [p.865]
31. Community _____ is an outcome of forces that have come into uneasy balance. [p.866]
32. Define what is meant by a *keystone species*. [p.866]
33. Describe the reasons why Robert Paine was able to identify a sea star (*Pisaster*) as a keystone species. [p.866]

34. Explain how the introduction of exotic species can be disastrous. List five specific examples of species introductions into the United States that have had adverse results. [pp.868–869]
35. Describe the distance effect and the area effect. [p.871]
36. Estimate qualitatively the differences in species diversity and abundance of organisms likely to exist on two islands with the following characteristics: Island A has an area of 6,000 square miles, and Island B has an area of 60 square miles; both islands lie at 10°N latitude and are equidistant from the same source area of colonizers. [pp.870–871]

Integrating and Applying Key Concepts

If you were Ruler of All People on Earth, how would you organize industry and human populations in an effort to solve our most pressing pollution problems?

Is there a *fundamental niche* that is occupied by humans? If you think so, describe the minimal abiotic and biotic conditions required by populations of humans in order to live and reproduce. (Note that *"thrive"* and *"be happy"* are not criteria.) If you do not think so, state why.

These minimal niche conditions can be viewed as resource categories that must be protected by populations if they are to survive. Do you believe that the cold war between the United States and the Soviet Union primarily involved protection of minimal niche conditions, or do you believe that the cold war was based on other, more (or less) important factors?

1. If the former, how do you think *minimal* niche conditions might have been guaranteed for all humans willing and able to accept certain responsibilities as their contribution toward enabling this guarantee to be met?
2. If the latter, identify what you think those factors are, and explain why you consider them more (or less) important than minimal niche conditions.

49

ECOSYSTEMS

Interactive Exercises

Crêpes for Breakfast, Pancake Ice for Dessert [pp.874–875]

49.1. THE NATURE OF ECOSYSTEMS [pp.876–877]

49.2. THE NATURE OF FOOD WEBS [pp.878–879]

49.3. *Focus on Science:* BIOLOGICAL MAGNIFICATION IN FOOD WEBS [p.880]

49.4. STUDYING ENERGY FLOW THROUGH ECOSYSTEMS [p.881]

49.5. *Focus on Science:* ENERGY FLOW AT SILVER SPRINGS [p.882]

Selected Words: *conservation biology* [p.875], *herbivores* [p.876], *carnivores* [p.876], *parasites* [p.876], *omnivores* [p.876], *scavengers* [p.876], *energy inputs* [p.876], *nutrient inputs* [p.876], *energy outputs* [p.876], *nutrient outputs* [p.876], *troph* [p.877], *cross-connecting* [p.877], *gross* primary productivity [p.881], *net* amount [p.881]

Boldfaced, Page-Referenced Terms

[p.876] primary producers _____

_____

[p.876] consumers _____

_____

[p.876] decomposers _____

_____

[p.876] detritivores _____

_____

[p.876] ecosystem _____

_____

[p.877] trophic levels _____

_____

[p.877] food chain _____

_____

[p.877] food webs _____

_____

[p.879] grazing food webs _____

_____

[p.879] detrital food webs _____

_____

[p.880] ecosystem modeling _____

_____

[p.880] biological magnification _____

_____

[p.881] primary productivity _____

_____

[p.881] biomass pyramid _____

_____

[p.881] energy pyramid _____

_____

Matching

Choose the most appropriate answer for each term.

1. ___primary producers [p.876]
2. ___consumers [p.876]
3. ___herbivores [p.876]
4. ___carnivores and parasites [p.876]
5. ___decomposers [p.876]
6. ___detritivores [p.876]
7. ___omnivores [p.876]
8. ___scavengers [p.876]
9. ___ecosystem [p.876]
10. ___energy inputs [p.876]
11. ___nutrient inputs [p.876]
12. ___energy outputs [p.876]
13. ___nutrient outputs [pp.876–877]
14. ___trophic levels [p.877]
15. ___food chain [p.877]
16. ___food webs [p.877]

A. Enters the ecosystem from the sun
B. Consumers that dine on animals, plants, fungi, even protistans and bacteria
C. Feed on the tissues of other organisms
D. Hierarchy of feeding relationships
E. A system of cross-connecting food chains
F. Feed on living animal tissues
G. Cannot be recycled, and over time most is lost to the environment
H. The autotrophs
I. Animals that ingest nonliving plant and animal tissues
J. Refers to a straight-line series of steps by which energy stored in autotroph tissues passes on through higher trophic levels
K. Consumers that eat only plants
L. Generally cycled, but some slips away
M. Fed by nonliving products and remains of producers and consumers
N. An array of organisms and their physical environment, interacting by a one-way flow of energy and a cycling of materials
O. Heterotrophs that ingest decomposing bits of organic matter, such as leaf litter
P. An example would be a creek that delivers dissolved materials into a lake

Choice

For questions 17–31, choose from the following trophic levels in a tallgrass prairie:

a. primary producers   b. first-level consumers   c. second-level consumers
d. third-level consumers   e. fourth-level consumers

17. ___ Garter snake, crow [pp.877–878]
18. ___ Larvae, earthworm [pp.877–878]
19. ___ Marsh hawk, mites [pp.877–878]
20. ___ Grasses, composites [pp.877–878]
21. ___ Saprobic fungi, bacteria [pp.877–878]
22. ___ Badger, coyote [pp.877–878]
23. ___ Prairie vole, pocket gopher [pp.877–878]
24. ___ Ticks, parasitic flies [pp.877–878]
25. ___ Gopher, ground squirrel [pp.877–878]
26. ___ Nitrifying bacteria [pp.877–878]
27. ___ Grasshoppers, moths, butterflies [pp.877–878]
28. ___ Weasel, badger, coyote [pp.877–878]
29. ___ The only category lacking heterotrophs [pp.877–878]
30. ___ Meadow frog, spiders [pp.877–878]
31. ___ Green plants [pp.877–878]

Fill-in-the-Blanks

When researchers compared the chains of different food webs, a pattern emerged: In most cases,
(32) _____ [p.878] that producers initially captured pass through no more than four or five trophic
levels. It simply makes no difference how much energy is available in the (33) _____ [p.878].
Remember, energy transfers never are (34) _____ [p.878] percent efficient; a bit of energy is lost at each
step. In time, the energy required to capture an organism at a higher (35) _____ [p.878] level would be
more than the amount of energy obtainable from it. Even rich ecosystems that are able to support many
species in complex food webs are not characterized by (36) _____ [p.878] food chains.

Field studies and computer simulations of the food webs of marine, freshwater, and terrestrial ecosystems
reveal more (37) _____ [p.879]. For example, the chains in food webs tend to be (38) _____ [p.879]
when environmental conditions vary. By contrast, the chains are longer in (39) _____ [p.879]
environments, such as zones of the deep ocean. The most complex webs have the greatest number of
(40) _____ [p.879] species, but their chains are (41) _____ [p.879]. Such webs are found in
(42) _____ [p.879]. The simplest food webs have more top (43) _____ [p.879].

Energy from a primary source flows in one direction through two categories of food webs. In
(44) _____ [p.879] food webs, the energy flows from photoautotrophs to herbivores, and then through
carnivores. By contrast, in (45) _____ [p.879] food webs, energy flows primarily from photoautotrophs
through detritivores and decomposers. In nearly all ecosystems, grazing and detrital food webs
(46) _____-_____ [p.879]. The amount of energy moving through food webs differs from one
ecosystem to the next and often varies with the (47) _____ [p.879]. In most cases, however, most of the
net primary production passes through the (48) _____ [p.879] food webs. Remains and wastes of
(49) _____ [p.879] and (50) _____ [p.879] are the basis of detrital food webs.

(51) _____ _____ [p.880] is a method of identifying and combining crucial bits of information
about an ecosystem through computer programs and models in order to predict the outcome of the next
disturbance. (52) _____ [p.880], a relatively stable hydrocarbon, is a synthetic organic pesticide.
Because it is insoluble in water, winds carry DDT in vapor form, transporting fine particles of it. DDT is also
highly soluble in (53) _____ [p.880], so it can accumulate in the tissues of organisms. Thus, DDT can
show (54) _____ _____ [p.880]. By this occurrence, a nondegradable or slowly degradable
substance becomes more and more (55) _____ [p.880] in the tissues of organisms at the higher trophic
levels of a food (56) _____ [p.880]. Most of the DDT from all organisms that a (57) _____ [p.880]
feeds on during its lifetime ends up in its own tissues. DDT and modified forms of it disrupt (58) _____
[p.880] activities and are often toxic to many aquatic and terrestrial animals.

The rate at which an ecosystem's primary producers capture and store a given amount of energy in a
specified time interval is the primary (59) _____ [p.881]. (60) _____ [p.881] primary productivity is
the total rate of photosynthesis for the ecosystem during the specified interval. The (61) _____ [p.881]
amount is the rate of energy storage in plant tissues in excess of the rate of aerobic respiration by the plants.

Other factors such as seasonal patterns and distribution through a given habitat also influence the amount of (62) _____ [p.881] primary production.

Ecologists often represent the trophic structure of an ecosystem in the form of an ecological (63) _____ [p.881]. In such forms, the primary (64) _____ [p.881] form a broad base for successive tiers of consumers above them. Some pyramids are based on (65) _____ [p.881] or the weight of all the members at each tier or trophic level. Sometimes, a pyramid of (66) _____ [p.881] can be "upside down," with the (67) [check one] ❏ smallest ❏ largest [p.881] tier on the bottom. Sometimes a more useful way to depict the diminishing flow of energy through ecosystems is with an (68) _____ [p.881] pyramid. (69) _____ [p.881] energy enters the pyramid's base, then diminishes through successive levels to its tip (the top carnivores). These pyramids have a (70) [check one] ❏ small ❏ large [p.881] energy base at the bottom and are always "right-side up."

The ecological study of energy flow at Silver Springs, Florida, found that about (71) [check one] ❏ 1 ❏ 5 [p.882] percent of all incoming solar energy was captured by producers prior to transfer to the next trophic level. It was also reported that about (72) [check one] ❏ 6 ❏ 8 [p.882] percent to (73) [check one] ❏ 10 ❏ 16 [p.882] percent of the energy entering one trophic level becomes available for organisms at the next level. In general, the study showed that the efficiency of the energy transfers is so (74) [check one] ❏ high ❏ low [p.882], ecosystems have usually no more than (75) [check one] ❏ 4 ❏ 6 [p.882] consumer trophic levels.

49.6. BIOGEOCHEMICAL CYCLES—AN OVERVIEW [p.883]
49.7. HYDROLOGIC CYCLE [pp.884–885]

Selected Words: *hydrologic* cycle [p.883], *atmospheric* cycles [p.883], *sedimentary* cycles [p.883]

Boldfaced, Page-Referenced Terms

[p.883] biogeochemical cycle _____

_____

[p.884] hydrologic cycle _____

_____

[p.884] watershed _____

_____

Short Answer

1. In what form(s) are elements used as nutrients usually available to producers? [p.883] _____

_____

2. How is the ecosystem's reserve of nutrients maintained? [p.883] _____

_____

_____

3. How does the amount of a nutrient being cycled through most major ecosystems compare with the amount entering or leaving in a given year? [p.883] _____

_____

_____

_____

4. What are the common environmental input sources for an ecosystem's nutrient reserves? [p.883] _____

_____

5. What are the output sources of nutrient loss for land ecosystems? [p.883] _____

_____

Complete the Table

6. Complete the following table to summarize the functions of the three types of biogeochemical cycles.

Biogeochemical Cycle	*General Function(s)*
[p.883] a. Hydrologic cycle	
[p.883] b. Atmospheric cycles	
[p.883] c. Sedimentary cycles	

Matching

Choose the most appropriate answer for questions 7–10; questions 11–14 may have more than one answer.

7. ___solar energy [p.884]

8. ___Earth's main water reservoirs [p.884]

9. ___watershed [p.884]

10. ___water and plants taking up water [p.885]

11. ___forms of precipitation falling to land [p.884]

12. ___deforestation [p.885]

13. ___have important roles in the global hydrologic cycle [p.884]

14. ___forms of atmospheric water [p.884]

A. Mostly rain and snow
B. Important in moving nutrients in biochemical cycles
C. Water vapor, clouds, and ice crystals
D. Where precipitation of a specified region becomes funneled into a single stream or river
E. May have long-term disruptive effects on nutrient availability for an entire ecosystem
F. Slowly drives water from the ocean into the atmosphere, to land, and back to the ocean—the main reservoir
G. Ocean currents and wind patterns
H. The oceans

Fill-in-the-Blanks

A(n) (15) _____ [p.884] is any region in which precipitation becomes funneled into a single stream or river. Most of the water that enters a watershed seeps into the (16) _____ [p.885] or becomes surface runoff that enters (17) _____ [p.885]. Plants withdraw water and its dissolved minerals from the soil, then they lose it by (18) _____ [p.885]. Watershed studies revealed the influence of (19) _____ [p.885] cover in the movements of (20) _____ [p.885] through the ecosystem phase of biogeochemical cycles.

Measurements of watershed inputs and outputs have many practical applications. In studies of young, undisturbed forests in the Hubbard Brook watersheds, each hectare lost only about 8 kilograms or so of (21) _____ [p.885]. Rainfall and the weathering of rocks brought in replacements of this element. Tree roots were also "mining" the soil, so (22) _____ [p.885] was being stored in a growing (23) _____ [p.885] of tree tissues. In experimental watersheds in the Hubbard Brook Valley, (24) _____ [p.885] caused a shift in nutrient outputs. Calcium and other nutrients (25) _____ [p.885] so slowly that deforestation may disrupt nutrient availability for entire (26) _____ [p.885]. This is especially the case for forests that cannot (27) _____ [p.885] themselves over the short term. (28) _____ [p.885] forests are like this.

49.8. CARBON CYCLE [pp.886–887]

49.9. *Focus on the Environment:* FROM GREENHOUSE GASES TO A WARMER PLANET? [pp.888–889]

49.10. NITROGEN CYCLE [pp.890–891]

49.11. SEDIMENTARY CYCLES [p.892]

Selected Words: Anabaena [p.890], Nostoc [p.890], Rhizobium [p.890], Azotobacter [p.890]

Boldfaced, Page-Referenced Terms

[p.886] carbon cycle _____

_____

[p.888] greenhouse effect _____

_____

[p.888] global warming _____

_____

[p.890] nitrogen cycle _____

_____

[p.890] nitrogen fixation _____

_____

[p.890] decomposition _____

_____

[p.890] ammonification _____

_____

[p.890] nitrification _____

_____

[p.891] denitrification _____

_____

[p.891] ion exchange _____

_____

[p.892] phosphorus cycle _____

_____

[p.893] eutrophication _____

_____

Matching

Choose the most appropriate answer for each term.

1. ___greenhouse gases [p.888]
2. ___carbon dioxide fixation [p.886]
3. ___carbon cycle [pp.886–887]
4. ___ways carbon enters the atmosphere [p.886]
5. ___greenhouse effect [p.888]
6. ___carbon dioxide (CO_2) [p.886]
7. ___oceans [p.886]
8. ___global warming [p.888]

A. Form of most of the atmospheric carbon
B. Aerobic respiration, fossil fuel burning, and volcanic eruptions
C. CO_2, CFCs, CH_4, and N_2O
D. Photosynthesizers incorporate carbon atoms into organic compounds
E. Holds most of the carbon in dissolved form
F. Carbon reservoirs → atmosphere and oceans → through organisms → carbon reservoirs
G. An effect that may be caused by greenhouse gases contributing to long-term higher temperatures at Earth's surface
H. Warming of Earth's lower atmosphere due to accumulation of certain gases

Matching

Choose the most appropriate answer for each term.

9. ___nitrogen cycle description [p.890]
10. ___nitrogen fixation [p.890]
11. ___decomposition and ammonification [p.890]
12. ___nitrification [p.890]
13. ___denitrification [p.891]

A. Ammonia or ammonium in soil is stripped of electrons, and nitrite (NO_2^-) is the result; other bacteria convert nitrite to nitrate (NO_3^-).
B. Bacteria convert nitrate or nitrite to N_2 and a bit of nitrous oxide (N_2O).
C. Bacteria and fungi break down nitrogen-containing wastes and plant and animal remains; released amino acids and proteins are used for growth with the excess given up as ammonia or ammonium ions that plants can use.
D. Occurs in the atmosphere (largest reservoir); only certain bacteria, volcanic action, and lightning can convert N_2 into forms that can enter food webs.
E. A few kinds of bacteria convert N_2 to ammonia (NH_3), which dissolves quickly in water to form ammonium (NH_4^+).

Short Answer

14. List reasons that an insufficient soil nitrogen supply is a problem for land plants. [p.891] _____

_____

_____

Fill-in-the-Blanks

The Earth's crust is the largest (15) _____ [p.892] for phosphorus just as it is for other minerals. In rock formations on land, phosphorus is typically in the form of (16) _____ [p.892]. By natural processes of weathering and soil erosion, phosphates enter rivers and streams that transport them to (17) _____ [p.892] sediments. Phosphorus slowly accumulates mainly on submerged (18) _____ [p.892] of continents. Millions of years go by. Where movements of (19) _____ [p.892] plates uplift part of the seafloor, phosphates become exposed on drained land surfaces. In time, weathering releases the phosphates from the exposed rocks, and the cycle's (20) _____ [p.892] phase begins again.

The (21) _____ [p.892] phase of the cycle is more rapid than the long-term geochemical phase. All (22) _____ [p.892] require phosphorus for synthesizing phospholipids, NADPH, ATP, nucleic acids, and other compounds. Plants take up dissolved, (23) _____ [p.892] forms of phosphate very rapidly. Herbivores obtain phosphorus by dining on (24) _____ [p.892]; carnivores obtain it by eating (25) _____ [p.892]. Both types of animals excrete phosphates in urine and feces.

Bacterial and fungal decomposers in the (26) _____ [p.892] release phosphates, then plants take up dissolved forms of this mineral. In this way, plants help (27) _____ [p.892] phosphorus rapidly through the ecosystem.

Sedimentary cycles, in combination with the (28) _____ [p.892] cycle, take most mineral elements through ecosystems on land and in water. (29) _____ [p.892] that evaporates from the ocean falls over land, and on its way back to the ocean, it transports silt and dissolved minerals that primary producers use as (30) _____ [p.892].

Of all minerals, (31) _____ [p.892] is the most prevalent limiting factor in natural (32) _____ [p.892] around the world. Soils hold little of it, and (33) _____ [p.892] tie up most of it in aquatic ecosystems. It helps that this element does not have a (34) _____ [p.892] phase, as nitrogen does, so ecosystems lose little of their scarce supply to the atmosphere. It also helps that this element resists (35) _____ [p.892].

Some human activities are lowering the (36) _____ [p.892] levels of natural ecosystems. This is especially the case for tropical and subtropical regions of the (37) _____ countries [p.892]. There, highly weathered soils have very little (38) _____ [p.892]. Enough is stored in (39) _____ [p.892] and is slowly released from decomposing organic matter to sustain undisturbed forests or grasslands. Harvesting trees or clearing grasslands for agriculture depletes the (40) _____ [p.892]. Crop yields start out low and soon become (41) _____ [p.892]. Fields are soon abandoned, but regrowth of natural (42) _____ [p.892] is sparse.

In developed countries, years of heavy phosphorus applications have raised the level of phosphorus in (43) _____ [p.893]. Without stringent monitoring, phosphorus becomes concentrated in eroded (44) _____ [p.893] and in runoff from agricultural fields. Phosphorus also is present in outflows from sewage treatment plants and from industries. It is in (45) _____ [p.893] from cleared land, even from lawns.

Dissolved phosphorus that enters streams, lakes, and estuaries can promote dense (46) _____ [p.893] blooms. (47) _____ [p.893] of the remains of all those algae depletes the water of (48) _____ [p.893], which kills off the fish and other organisms.

(49) _____ [p.893] is the name for nutrient enrichment of an ecosystem that is naturally low in nutrients. This is a natural process, but phosphorus inputs (50) _____ [p.893] it.

Self-Quiz

___ 1. An array of organisms and their physical environment, interacting by a one-way flow of energy and a cycling of materials, is a(n) _____. [p.876]
a. population
b. community
c. ecosystem
d. biosphere

___ 2. _____ ingest decomposing particles of organic matter. [p.876]
a. Herbivores
b. Parasites
c. Detritivores
d. Carnivores

___ 3. The members of feeding relationships are structured in a hierarchy, the steps of which are called _____. [p.877]
a. organism level
b. energy source level
c. eating level
d. trophic levels

___ 4. In grazing food webs, energy flows from _____. [p.879]
a. photoautotrophs through detritivores and decomposers
b. primary consumers through detritivores and decomposers
c. photoautotrophs to herbivores, then through carnivores
d. primary consumers to herbivores, then through carnivores

___ 5. Which of the following is a primary consumer? [p.877]
a. cow
b. dog
c. hawk
d. all of the above

___ 6. In a natural community, the primary consumers are _____. [p.876]
a. herbivores
b. carnivores
c. scavengers
d. decomposers

___ 7. A straight-line series of steps of who eats whom in an ecosystem is sometimes called a(n) _____. [p.877]
a. trophic level
b. food chain
c. ecological pyramid
d. food web

___ 8. Of the 1,700,000 kilocalories of solar energy that entered an aquatic ecosystem in Silver Springs, Florida, investigators determined that about _____ percent of incoming solar energy was trapped by photosynthetic autotrophs. [p.882]
a. 1
b. 10
c. 25
d. 74

___ 9. A biogeochemical cycle that deals with phosphorus and other nutrients that do not have gaseous forms is the _____ type. [p.883]
a. sedimentary
b. hydrologic
c. nutrient
d. atmospheric

____10. _____ is a process in which nitrogenous waste products or organic remains of organisms are decomposed by soil bacteria and fungi that use the amino acids being released for their own metabolism and release the excess as ammonia or ammonium in decay products, which plants take up. [p.890]
   a. Nitrification
   b. Ammonification
   c. Denitrification
   d. Nitrogen fixation

____11. In the carbon cycle, carbon enters the atmosphere through _____. [p.886]
   a. carbon dioxide fixation
   b. respiration, burning, and volcanic eruptions
   c. oceans and accumulation of plant biomass
   d. release of greenhouse gases

____12. _____ refers to an increase in concentration of a nondegradable (or slowly degradable) substance in organisms as it is passed along food chains. [p.880]
   a. Ecosystem modeling
   b. Nutrient input
   c. Biogeochemical cycle
   d. Biological magnification

Chapter Objectives/Review Questions

1. List the principal trophic levels in an ecosystem of your choice; state the source of energy for each trophic level and give one or two examples of organisms associated with each trophic level. [pp.876–877]
2. Distinguish among herbivores, carnivores, parasites, omnivores, and scavengers. [p.876]
3. A(n) _____ is an array of organisms and their physical environment, all interacting through a flow of energy and a cycling of materials. [p.876]
4. Explain why nutrients can be completely recycled but energy cannot. [pp.876–877]
5. In terms of an ecosystem, define energy inputs, nutrient inputs, energy outputs, and nutrient outputs. [p.876]
6. Members of an ecosystem fit somewhere in a hierarchy of energy transfers (feeding relationships) called _____ levels. [p.877]
7. Distinguish between food chains and food webs. [p.877]
8. Compare grazing food webs with detrital food webs. Present an example of each. [p.879]
9. Through _____ modeling, crucial bits of information about different ecosystem components are identified and used to build computer models for predicting outcomes of ecosystem disturbances. [p.880]
10. Describe how DDT damages ecosystems; discuss biological magnification. [p.880]
11. Explain how materials and energy enter, pass through, and exit an ecosystem. [p.881]
12. Ecological pyramids that are based on _____ are determined by the weight of all the members of each trophic level; _____ pyramids reflect the energy losses at each transfer to a different trophic level. [p.881]
13. Distinguish among primary productivity, gross primary productivity, and the net amount. [p.881]
14. Discuss the use of biomass pyramids and energy pyramids. [p.881]
15. In the study of energy flow at Silver Springs, the producers trapped _____ percent of the incoming solar energy, and only a little more than a _____ of the amount became fixed in new plant biomass. The producers used more than _____ percent of the fixed energy for their own metabolism. [p.882]
16. In a _____ cycle, ions or molecules of a nutrient are transferred from the environment into organisms, then back to the environment, part of which functions as a vast reservoir for them. [p.883]
17. Name and define the three categories of biogeochemical cycles. [p.883]
18. Discuss water movements through the hydrologic cycle. [pp.884–885]
19. Explain what studies in the Hubbard Brook watershed have taught us about the movement of substances (water, for example) through a forest ecosystem. [pp.884–885]
20. In a vital atmospheric cycle, carbon moves through the atmosphere and food _____ on its way to and from the ocean, sediments, and rocks. [p.886]

21. Certain gases cause heat to build up in the lower atmosphere, a warming action known as the _____ effect. [p.888]
22. As many researchers suspect, greenhouse gases may be contributing to long-term higher temperatures at the Earth's surface, an effect called global _____. [p.888]
23. The four greenhouse gases are carbon dioxide, CFCs, methane, and _____ oxide. [p.888]
24. A major element found in all proteins and nucleic acids moves in an atmospheric cycle called the _____ cycle. [p.890]
25. Define the chemical events that occur during nitrogen fixation, decomposition and ammonification, and nitrification. [p.890]
26. Through the process of _____, certain bacteria convert nitrate or nitrite to N_2 and N_2O. [p.891]
27. List the ways that humans impact the nitrogen cycle. [p.891]
28. Describe the geochemical and ecosystem phase of the phosphorus cycle. [p.892]
29. Define the term *eutrophication*, and discuss the causes of this process. [pp.892–893]

Integrating and Applying Key Concepts

In 1971, *Diet for a Small Planet* was published. Frances Moore Lappé, the author, felt that people in the United States of America wasted protein and ate too much meat. She said, "We have created a national consumption pattern in which the majority, who can pay, overconsume the most inefficient livestock products [cattle] well beyond their biological needs (even to the point of jeopardizing their health), while the minority, who cannot pay, are inadequately fed, even to the point of malnutrition." Cases of marasmus (a nutritional disease caused by prolonged lack of food calories) and kwashiorkor (caused by severe, long-term protein deficiency) have been found in Nashville, Tennessee, and on an Indian reservation in Arizona, respectively. Lappé's partial solution to the problem was to encourage people to get as much of their protein as possible directly from plants and to supplement that with less meat from the more efficient converters of grain to protein (chickens, turkeys, and hogs) and with seafood and dairy products. Most of us realize that feeding the hungry people of the world is not just a matter of distributing the abundance that exists—it is also a matter of political, economic, and cultural factors. Yet it is still valuable to consider applying Lappé's idea to our everyday living. Devise two full days of breakfasts, lunches, and dinners that would enable you to exploit the lowest acceptable trophic levels to sustain yourself healthfully.

50

THE BIOSPHERE

Interactive Exercises

Does a Cactus Grow in Brooklyn? [pp.896–897]

50.1. AIR CIRCULATION PATTERNS AND REGIONAL CLIMATES [pp.898–899]
50.2. OCEANS, LANDFORMS, AND REGIONAL CLIMATES [pp.900–901]

Selected Words: hydrosphere [p.897], lithosphere [p.897], "ozone layer" [p.898], *leeward* [p.901], *windward* [p.901], "mini-monsoons" [p.901]

Boldfaced, Page-Referenced Terms

[p.896] biogeography _____

_____

[p.897] biosphere _____

_____

[p.897] atmosphere _____

_____

[p.897] climate _____

_____

[p.898] temperature zones _____

_____

[p.900] oceans _____

_____

[p.901] rain shadow _____

_____

[p.901] monsoons _____

_____

Matching

Choose the most appropriate answer for each term.

1. ___biogeography [p.896]
2. ___biosphere [p.897]
3. ___hydrosphere [p.897]
4. ___lithosphere [p.897]
5. ___atmosphere [p.897]
6. ___climate [p.897]
7. ___ozone layer [p.898]
8. ___temperature zones [p.898]
9. ___oceans [p.900]
10. ___rain shadow [p.901]
11. ___leeward [p.901]
12. ___windward [p.901]
13. ___monsoons [p.901]
14. ___mini-monsoons [p.901]

A. One continuous body of water that covers 71 percent of Earth's surface
B. Made up of gases and airborne particles that envelop Earth
C. Of the world, defined by differences in solar heating at different latitudes and the modified air circulation patterns
D. The sum total of all places in which organisms live
E. Refers to the direction not facing a wind
F. An atmospheric region between 17 and 25 kilometers above sea level
G. Recurring sea breezes along coastlines; warmed air above land rises and cooler marine air moves in
H. The waters of Earth, including the ocean, polar ice caps, and other forms of liquid and frozen water
I. Refers to the direction from which the wind is blowing
J. Average weather conditions, such as temperature, humidity, wind speed, cloud cover, and rainfall, over time
K. The outer, rocky layer of Earth
L. An arid or semiarid region of sparse rainfall on the leeward side of high mountains
M. The study of the distribution of organisms, past and present, and of diverse processes that underlie the distribution patterns
N. Patterns of air circulation that affect conditions on the continents lying poleward of warm oceans; intensely heated land draws in moisture-laden air that forms above ocean water

Fill-in-the-Blanks

The numbered items in the following illustrations represent missing information. Complete the blanks in the narrative to supply that information.

The sun's rays are more concentrated in (15) _____ [p.898] regions than at the poles. The global pattern of air circulation starts as (16) _____ [p.898] equatorial air (17) _____ [p.898] and spreads northward and southward giving up much (18) _____ [p.898] as precipitation (warm air can hold more moisture than cold air) which supports luxuriant growth of tropical forests. The cooled, drier air then (19) _____ [p.899] at latitudes of about 30° N and S and becomes warmer and drier in areas where deserts form. Air even farther from the equator picks up some (20) _____ [p.899], and (21) _____ [p.899] to higher altitudes, cools, and then gives up (22) _____ [p.899] at latitudes of about 60° N and S to create another moist belt. The cooled, dry air then (23) _____ [p.899] at the polar regions, where the low temperatures and almost nonexistent precipitation give rise to the cold, dry, polar deserts. Earth's rotation then modifies the air circulation by deflecting it into worldwide belts of prevailing (24) _____ [p.899] and (25) _____ [p.899] winds.

The world's temperature zones, beginning at the equator and moving toward the poles, are the (26) _____ [p.899], the (27) _____ [p.899] temperate, the (28) _____ [p.899] temperate, and the (29) _____ [p.899]. Finally, the amount of the (30) _____ [p.899] radiation reaching the surface varies annually, owing to the Earth's (31) _____ [p.899] around the sun. This leads to seasonal changes in daylength, prevailing wind directions, and temperature. These factors influence the locations of different ecosystems.

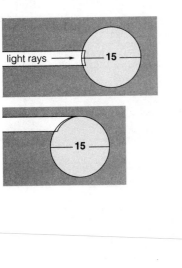

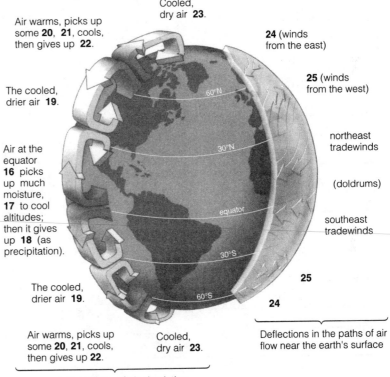

light rays → 15

15

Air warms, picks up some 20, 21, cools, then gives up 22.

Cooled, dry air 23.

24 (winds from the east)

25 (winds from the west)

The cooled, drier air 19.

60°N

northeast tradewinds

Air at the equator 16 picks up much moisture, 17 to cool altitudes; then it gives up 18 (as precipitation).

30°N

(doldrums)

equator

southeast tradewinds

30°S

The cooled, drier air 19.

60°S

25

24

Air warms, picks up some 20, 21, cools, then gives up 22.

Cooled, dry air 23.

Deflections in the paths of air flow near the earth's surface

Initial pattern of air circulation

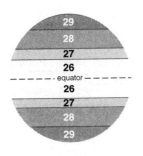

29
28
27
26
---- equator ----
26
27
28
29

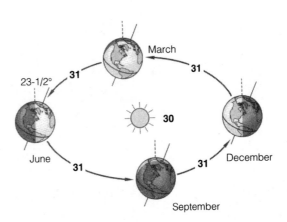

March

31

31

23-1/2°

31

30

June

31

September

December

31

50.3. REALMS OF BIODIVERSITY [pp.902–903]

50.4. SOILS OF MAJOR BIOMES [p.904]

50.5. DESERTS [p.905]

50.6. DRY SHRUBLANDS, DRY WOODLANDS, AND GRASSLANDS [pp.906–907]

Selected Words: soil *profile* [p.904], *shortgrass* prairie [p.906], *tallgrass* prairie [p.906], *monsoon* grasslands [p.907]

Boldfaced, Page-Referenced Terms

[p.903] biogeographic realms _____

_____

[p.903] biome _____

_____

[p.903] ecoregions _____

_____

[p.904] soils _____

_____

[p.905] deserts _____

_____

[p.905] desertification _____

_____

[p.906] dry shrublands _____

_____

[p.906] dry woodlands _____

_____

[p.906] grasslands _____

_____

[p.906] savannas _____

_____

Matching

Choose the most appropriate answer for each term.

1. ___biogeographic realms [p.903]
2. ___biome [p.903]
3. ___soils [p.904]
4. ___deserts [p.905]
5. ___desertification [p.905]
6. ___dry shrublands [p.906]
7. ___soil profile [p.904]
8. ___dry woodlands [p.906]
9. ___grasslands [p.906]
10. ___ecoregions [p.903]
11. ___savannas [p.906]
12. ___monsoon grasslands [p.907]

A. Mixtures of mineral particles and variable amounts of decomposing organic material (humus)
B. Areas receiving less than 25 to 60 centimeters of rain per year; local names include *fynbos* and *chaparral*
C. The conversion of grasslands and other productive biomes to dry wastelands
D. Sweep across much of the interior of continents, in the zones between deserts and temperate forests; warm temperatures prevail in summer, winters are extremely cold
E. Areas that dominate when annual rainfall is about 40 to 100 centimeters; dominant trees can be tall but do not form a dense, continuous canopy
F. Subdivision of biogeographic realms; a large region of land characterized by the climax vegetation of the ecosystems within its boundaries
G. Formed in land regions where the potential for evaporation exceeds sparse rainfall
H. Six vast land areas on the Earth, each with distinguishing plants and animals
I. Broad belts of grasslands with a smattering of shrubs or trees; rainfall averages 90 to 150 centimeters a year with prolonged seasonal droughts common
J. The layered structure of soils
K. Form in southern Asia where heavy rains alternate with a dry season; dense stands of tall, coarse grasses form, then die back and often burn in the dry season
L. Large areas representative of globally important biomes and water provinces that are vulnerable to extinction

Matching

Choose the most appropriate answer for each term.

13. ___tropical rain forest soil [p.904]
14. ___deciduous forest soil [p.904]
15. ___desert soil [p.904]
16. ___grassland soil [p.904]
17. ___coniferous soil [p.904]

A. A horizon: alkaline, deep, rich in humus
B. O horizon: scattered litter; A horizon: rich in organic matter above humus layer unmixed with minerals
C. O horizon: sparse litter; A–E horizons: continually leached
D. O horizon: pebbles, little organic matter; A horizon: shallow, poor soil
E. O horizon: well-defined, compacted mat of organic deposits resulting mainly from activity of fungal decomposers

Choice

For questions 17–27, choose from the following:

        a. deserts   b. dry shrublands   c. dry woodlands   d. grasslands   e. savannas

18. ___ Monsoon type that forms dense stands of tall, coarse plants in parts of southern Asia where heavy rains alternate with a dry season [p.907]

19. ___ Biome where the potential for evaporation greatly exceeds rainfall [p.905]

20. ___ Steinbeck's *Grapes of Wrath* speaks eloquently of the disruption of this biome [p.906]

21. ___ Local names for this biome include *fynbos* and *chaparral;* dominant plants often have hardened, tough, evergreen leaves [p.906]

22. ___ A biome in which dominant trees can be tall but do not form a dense canopy; includes eucalyptus woodlands of southwestern Australia and oak woodlands of California and Oregon biome [p.906]

23. ___ Home to deep-rooted evergreen shrubs, fleshy-stemmed, shallow-rooted cacti, saguaro, short prickly pear, and ocotillo biome [p.905]

24. ___ Broad belts of grasslands with a smattering of shrubs or trees; prolonged seasonal droughts are common biome [p.906]

25. ___ The dominant animals are grazing and burrowing species; grazing and periodic fires maintain the fringes of this biome [p.906]

26. ___ Within this biome, fast-growing grasses dominate where rainfall is low, but acacia and other shrubs grow where there is slightly more moisture [p.906]

27. ___ In summer, lightning-sparked, wind-driven firestorms can sweep through these biomes and swiftly burn shrubs with highly flammable leaves to the ground [p.906]

50.7. TROPICAL RAIN FORESTS AND OTHER BROADLEAF FORESTS [pp.908–909]

50.8. CONIFEROUS FORESTS [p.910]

50.9. ARCTIC AND ALPINE TUNDRA [p.911]

Selected Words: *tropical* deciduous forests [p.909], *monsoon* forests [p.909], *temperate* deciduous forests [p.909], *taiga* [p.910], *tuntura* [p.911], *arctic* tundra [p.911], *alpine* tundra [p.911]

Boldfaced, Page-Referenced Terms

[p.908] evergreen broadleaf forests _____

_____

[p.908] tropical rain forest _____

_____

[pp.908–909] deciduous broadleaf forests _____

_____

[p.910] coniferous forests _____

_____

[p.910] boreal forests _____

_____

[p.910] southern pine forests _____

_____

[p.911] tundra _____

_____

[p.911] permafrost _____

_____

Choice

For questions 1–16, choose from the following biomes:

a. deciduous broadleaf forests   b. coniferous forests   c. evergreen broadleaf forests   d. tundra

1. ___ Nearly continuous sunlight in summer; short plants grow and flower profusely with rapidly ripening seeds [p.911]

2. ___ Sweep across tropical zones of Africa, the East Indies and Malay Archipelago, Southeast Asia, South America, and Central America biome [p.908]

3. ___ Highly productive forest; decomposition and mineral cycling are rapid in the hot, humid climate biome; tropical rain forests [p.908]

4. ___ Within this temperate biome, complex forests of ash, beech, chestnut, elm, and deciduous oaks once stretched across northeastern North America. [p.909]

5. ___ Boreal forests or taiga; conifers are the primary producers [p.910]

6. ___ Spruce and balsam fir dominate the northern part [p.910]

7. ___ Biome of the temperate zone; cold winter temperatures; many trees drop all their leaves in winter [p.909]

8. ___ Great treeless plain between the polar ice cap and belts of boreal forests in Europe, Asia, and North America [p.911]

9. ___ Soils are highly weathered, humus-poor, and not good nutrient reservoirs [pp.908,911]

10. ___ Forests that dominate the coastal plains of the south Atlantic and Gulf states; dominant plants are adapted to the dry, sandy, nutrient-poor soil and to natural fires or controlled burns [p.910]

11. ___ Annual rainfall can exceed 200 centimeters and is never less than 130 centimeters. [p.908]

12. ___ Includes tropical deciduous forests, monsoon forests, and temperate deciduous forests [pp.908–909]

13. ___ Cone-bearing trees are the primary producers; most have needle-shaped leaves with a thick cuticle and recessed stomata—adaptations that help the trees conserve water through dry times [p.910]

14. ___ Pines, scrub oak, and wiregrass grow in New Jersey; palmettos grow below loblolly and other pines in the Deep South. [p.910]

15. ___ Just beneath the surface is a perpetually frozen layer, the permafrost [p.911]

16. ___ One type is known as alpine; it prevails at high elevations throughout the world, although there is no permafrost beneath the soil [p.911]

50.10. FRESHWATER PROVINCES [pp.912–913]

50.11. THE OCEAN PROVINCES [pp.914–915]

Selected Words: littoral zone [p.912], limnetic zone [p.912], profundal zone [p.912], *phyto*plankton [p.912], *zoo*plankton [p.912], *thermocline* [p.912], *oligotrophic* lakes [p.913], *eutrophic* lakes [p.913], *benthic* province [p.914], *pelagic* province [p.914]

Boldfaced, Page-Referenced Terms

[p.912] lake _____

_____

[p.912] spring overturn _____

_____

[p.912] fall overturn _____

_____

[p.913] eutrophication _____

_____

[p.913] streams _____

_____

[p.914] ultraplankton _____

_____

[p.914] marine snow _____

_____

[p.914] hydrothermal vents _____

_____

Fill-in-the-Blanks

A(n) (1) _____ [p.912] is a body of standing freshwater with three zones. The shallow, usually well-lit (2) _____ [p.912] zone extends around the shore to the depth at which rooted aquatic plants stop growing. The diversity of organisms is greatest here. The (3) _____ [p.912] zone is the open, sunlit water past the littoral and extends to a depth where photosynthesis is insignificant. Aquatic communities of (4) _____ [p.912] abound here. The (5) _____ [p.912] zone includes all open water below the depth at which wavelengths suitable for photosynthesis can penetrate. Bacterial decomposers in bottom sediments of this zone enrich the water with nutrients.

Water is densest at 4ºC; at this temperature, it sinks to the bottom of its basin, displacing the nutrient-rich bottom water upward and giving rise to spring and fall (6) _____ [p.912]. In spring, ice melts, daylength increases, and the surface waters of a lake slowly warm to (7) _____ ºC [p.912]. Surface winds cause a (8) _____ [p.912] overturn in which strong vertical water movements of water carry dissolved oxygen from a lake's surface layer to its depths, and nutrients released by decomposition are brought from the bottom sediments to the surface. The surface layer warms above 4ºC by midsummer, becomes less dense, and the lake has developed a middle layer, the (9) _____ [p.912] that prevents vertical mixing. When autumn comes, the upper layer (10) _____ [p.912] becomes denser, then sinks, and the thermocline vanishes. This is termed the (11) _____ [p.912] overturn. Water then mixes vertically, allowing dissolved oxygen to move (12) [check one] ❑ up ❑ down [p.912] and nutrients to move (13) [check one] ❑ up ❑ down [p.912]. Primary productivity of a lake corresponds with the seasons. Following a spring overturn, longer daylengths and cycled nutrients support (14) [check one] ❑ lower ❑ higher [p.912] rates of photosynthesis. By late summer, nutrient shortages are limiting photosynthesis. After the fall overturn, nutrient cycling drives a (15) [check one] ❑ short ❑ long [p.913] burst of primary activity. Not until spring will (16) _____ [p.913] productivity increase again.

Lakes have a trophic nature. (17) _____ [p.913] lakes are deep, nutrient-poor, and low in primary productivity. Lakes that are (18) _____ [p.913] are often shallow, nutrient-enriched, and high in primary productivity. Human activities can determine the trophic condition of lakes. The term (19) _____ [p.913] refers to nutrient enrichment of a lake resulting in reduced water transparency and a community rich in phytoplankton. (20) _____ [p.913] are flowing-water ecosystems that begin as freshwater springs or seeps. Three habitat types are found between headwaters and the river's end: riffles, pools, and (21) _____ [p.913].

Labeling and Matching

Label each numbered item in the accompanying illustration. Complete the exercise by matching and entering the letter of the proper description in the parentheses after each label.

22. _____ province ( ) [p.914]

23. _____ province ( ) [p.914]

24. _____ zone ( ) [p.914]

25. _____ zone ( ) [p.914]

A. The entire volume of ocean water
B. All the water above the continental shelves
C. Includes all sediments and rocks of the ocean bottom; begins with continental shelves and extends to deep-sea trenches
D. Water of the ocean basin

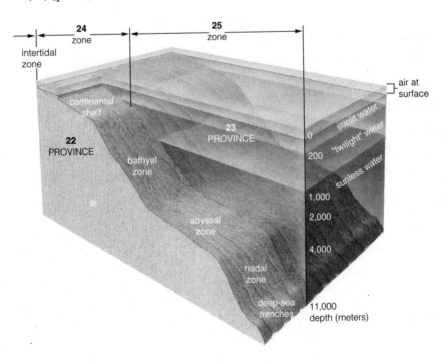

Choice

For questions 26–45, choose from the following:

a. stream ecosystems   b. lake ecosystems   c. ocean

26. ___ Riffles, pools, and runs [p.913]

27. ___ Hydrothermal vent communities of chemoautotrophic bacteria in the abyssal zone; possible sites of life's origin [pp.914–915]

28. ___ Ultraplankton contribute 70 percent of the primary productivity. [p.914]

29. ___ Can be oligotrophic or eutrophic [p.913]

30. ___ Submerged mountains, valleys, and plains [p.914]

31. ___ Begin as freshwater springs or seeps [p.913]

32. ___ Sewage discharges on adjacent lands can contribute to eutrophication of this water. [p.913]

33. ___ Average flow volume and temperature depend on rainfall, snowmelt, geography, altitude, and even shade cast by plants. [p.913]

34. ___ Spring and fall overturns [pp.912–913]

35. ___ Has benthic and pelagic provinces [p.914]

36. ___ Especially in forests, these waters import most of the organic matter that supports food webs. [p.913]

37. ___ Created by geologic processes such as retreating glaciers [p.912]

38. ___ They grow and merge as they flow downslope and then often combine. [p.913]

39. ___ Primary productivity there varies seasonally, just as it does on land. [p.914]

40. ___ The final successional stage is a filled-in basin. [p.913]

41. ___ Vast "pastures" of phytoplankton and zooplankton become the basis of detrital food webs. [p.914]

42. ___ Since cities formed, these waters have been sewers for industrial and municipal wastes. [p.913]

43. ___ Has a thermocline by midsummer [p.912]

44. ___ Littoral, limnetic, and profundal zones [p.912]

45. ___ Bathyal, abyssal, and hadal zones [p.914]

50.12. CORAL REEFS AND CORAL BANKS [pp.916–917]

50.13. MANGROVE WETLANDS [p.917]

50.14. ESTUARIES AND THE INTERTIDAL ZONE [pp.918–919]

50.15. *Focus on Science:* RITA IN THE TIME OF CHOLERA [pp.920–921]

Selected Words: *fringing reefs* [p.916], *atolls* [p.916], *barrier reefs* [p.916], *rocky, sandy,* and *muddy* shores [p.918], "downwelling" [p.919], *El Niño Southern Oscillation* (ENSO) [p.920], *Vibrio cholerae* [p.921]

Boldfaced, Page-Referenced Terms

[p.916] coral reefs _____

_____

[p.916] coral banks _____

_____

[p.917] mangrove wetland _____

_____

[p.918] estuary _____

_____

[p.918] intertidal zone _____

_____

[p.919] upwelling _____

_____

[p.919] El Niño _____

_____

Complete the Table

1. Complete the following table, which describes three types of coral reefs.

Reef Type	Description
[p.916] a. Atolls	
[p.916] b. Fringing reefs	
[p.916] c. Barrier reefs	

Choice

For questions 2–23, choose from the following:

    a. coral reefs and banks   b. estuary   c. intertidal zone   d. coastal upwelling
               e. ENSO   f. mangrove wetlands

2. ___ The fogbanks that form along the California coast are one outcome [p.919]

3. ___ *Lophelia* has constructed large ones in the cold waters of Norway's fjords [p.916]

4. ___ Waves batter its resident organisms [p.918]

5. ___ Massive dislocations in global rainfall patterns characterize this event, which corresponds to changes in sea surface temperatures and air circulation patterns [p.920]

6. ___ Organisms living there must constantly contend with the tides [p.918]

7. ___ Each sunlit, wave-resistant formation began with accumulated remains of marine organisms [p.916]

8. ___ Primary producers are phytoplankton, salt-tolerant plants that withstand submergence at high tide, and algae living in mud and on plants [p.918]

9. ___ In general, an upward movement of cold, deep, often nutrient-rich ocean waters occurring in equatorial currents as well as along the coasts of continents in both hemispheres [p.919]

10. ___ A partly enclosed coastal region where seawater swirls and slowly mixes with nutrient-rich freshwater from rivers, streams, and runoff from the surrounding land [p.918]

11. ___ Most of the substantial ones are formed in clear, warm waters between latitudes 25° north and south [p.917]

12. ___ Salt marshes are common [p.918]

13. ___ El Niño [p.920]

14. ___ Wind friction causes surface waters to begin moving and under the force of Earth's rotation, the slow-moving water is deflected west, away from a coast [p.919]

15. ___ Biodiversity is being threatened by bleaching, fast-growing seaweeds, and destruction of turtles and fishes that fed on seaweed populations [p.917]

16. ___ Commercial fishing industries of Peru and Chile depend on it [p.919]

17. ___ A recurring seesaw in atmospheric pressure in the western equatorial Pacific [p.920]

18. ___ Forests in sheltered regions along tropical coasts [p.917]

19. ___ Sandy and muddy shores [pp.918–919]

20. ___ In the Southern Hemisphere, commercial fishing industries of Peru depend on a wind-induced form of it [p.919]

21. ___ Ecosystems found along rocky and sandy coastlines [p.918]

22. ___ Forms are fringing, barrier, and atoll [p.916]

23. ___ Plants have shallow, spreading roots or branching prop roots that extend from the trunk; many have pneumatophores [p.917]

Self-Quiz

___ 1. The distribution of different types of ecosystems is influenced by _____. [pp.899, 901]
a. global air circulation patterns
b. variation in the amount of solar radiation reaching Earth through the year
c. surface ocean currents
d. all of the above

___ 2. In a(n) _____, nutrient-rich freshwater draining from the land mixes with seawater carried in on tides. [p.918]
a. pelagic province
b. rift zone
c. upwelling
d. estuary

___ 3. A biome with broad belts of grasslands and scattered trees adapted to prolonged dry spells is known as a _____. [p.906]
a. warm desert
b. savanna
c. tundra
d. taiga

___ 4. The _____ biome is located at latitudes of about 30° north and south, has limited vegetation, and has rapid surface cooling at night. [p.905]
a. shrublands
b. savanna
c. taiga
d. desert

___ 5. In tropical rain forests, _____. [p.908]
a. productivity is high
b. litter does not accumulate
c. soils are weathered and humus poor, and have poor nutrient reservoirs
d. decomposition and mineral cycling are extremely rapid
e. all of the above

___ 6. In a lake, the open sunlit water with its suspended phytoplankton is referred to as its _____ zone. [p.912]
a. epileptic
b. limnetic
c. littoral
d. profundal

___ 7. The lake's upper layer cools, the thermocline vanishes, lake water mixes vertically, and once again dissolved oxygen moves down and nutrients move up. This describes the _____. [p.912]
   a. spring overturn
   b. summer overturn
   c. fall overturn
   d. winter overturn

___ 8. The _____ is a permanently frozen, water-impermeable layer just beneath the surface of the _____ biome. [p.911]
   a. permafrost; alpine tundra
   b. hydrosphere; alpine tundra
   c. permafrost; arctic tundra
   d. taiga; arctic tundra

___ 9. _____ are air circulation patterns that influence the continents north or south of warm oceans; low pressure causes moisture-laden air above the neighboring ocean to move inland, resulting in heavy rains. [p.901]
   a. Geothermal ecosystems
   b. Upwelling
   c. Taigas
   d. Monsoons

___10. All the water above the continental shelves is in the _____. [p.914]
   a. neritic zone of the benthic province
   b. oceanic zone of the pelagic province
   c. neritic zone of the pelagic province
   d. oceanic zone of the benthic province

___11. Complex forests of ash, beech, birch, chestnut, elm, and oaks are found in the _____. [p.909]
   a. tropical deciduous forest
   b. monsoon forest
   c. temperate deciduous forest
   d. evergreen broadleaf forest

___12. Chemoautotrophic bacteria are the starting point for _____. [pp.914–915]
   a. hydrothermal vent communities
   b. desert communities
   c. lake communities
   d. coniferous forest communities
   e. coral reef communities

Chapter Objectives/Review Questions

1. _____ is the study of the distribution of organisms, past and present, and of diverse processes that underlie the distribution patterns. [p.896]
2. The _____ is the sum total of all places in which organisms live. [p.897]
3. _____ refers to average weather conditions, such as temperature, humidity, wind speed, cloud cover, and rainfall. [p.897]
4. State the reason that most forms of life depend on the ozone layer. [p.898]
5. _____ energy drives Earth's weather systems. [p.898]
6. Describe the causes of global air circulation patterns. [pp.898–899]
7. Describe how the tilt of Earth's axis affects annual variation in the amount of incoming solar radiation. [p.899]
8. Air _____ patterns, ocean _____, and landforms, in combination with global air circulation patterns, influence regional climates and help distribute nutrients in marine ecosystems. [p.901]
9. Mountains, valleys, and other aspects of topography influence _____ regional climates. [p.901]
10. Describe the cause of the rain shadow effect. [p.901]
11. Broadly, there are six distinct land realms, the _____ realms that were named by W. Sclater and then Alfred Wallace. [p.903]
12. Realms are further divided into _____. [p.903]
13. _____ are mixtures of mineral particles and variable amounts of decomposing organic material. [p.904]
14. List the major biomes and briefly characterize them in terms of climate, topography, and organisms. [pp.905–911]
15. The wholesale conversion of grasslands and other productive biomes to desertlike wastelands is known as _____. [p.905]
16. A _____ is a standing body of freshwater with littoral, limnetic, and profundal zones. [p.912]

17. Define the words *plankton, phytoplankton,* and *zooplankton.* [p.912]
18. Describe the spring and fall overturn in a lake in terms of causal conditions and physical outcomes. [pp.912–913]
19. _____ lakes are often deep, poor in nutrients, and low in primary productivity; _____ lakes are often shallow, rich in nutrients, and high in primary productivity. [p.913]
20. _____ refers to nutrient enrichment of a lake or some other body of water. [p.913]
21. Describe a stream ecosystem. [p.913]
22. Fully describe the benthic and pelagic provinces of the ocean. [p.914]
23. Within the pelagic province, all the water above the continental shelves is the _____ zone; the _____ zone is the water of the ocean basin. [p.914]
24. As much as 70 percent of the ocean's primary productivity may be the contribution of _____. [p.914]
25. Describe the unusual hydrothermal vent ecosystems. [pp.914–915]
26. List the three types of coral reefs, and describe their formation. [p.916]
27. State reasons why reef biodiversity is endangered. [pp.917–918]
28. At tropical latitudes, a different kind of nutrient-rich, saltwater ecosystem forms in tidal flats near the sea; it is called a _____ wetland. [p.917]
29. Descriptively distinguish between estuaries and intertidal zones. [p.918]
30. State the significance of ocean upwelling. [p.919]
31. Describe conditions of ENSO occurrence and how this phenomenon interrelates ocean surface temperatures, the atmosphere, and the land. [pp.920–921]
32. Describe how and why cholera outbreaks correlate with El Niño episodes in Bangladesh. [p.921]

Integrating and Applying Key Concepts

One species, *Homo sapiens,* uses about 40 percent of all of Earth's productivity, and its representatives have invaded every biome, either by living there or by dumping waste products there. Many of Earth's residents are being denied the minimal resources they need to survive, while human populations continue to increase exponentially. Can you suggest a better way of keeping Earth's biomes healthy while filling at least the minimal needs of all Earth's residents (not just humans)? If so, outline the requirements of such a system and devise a way in which it could be established.

51

HUMAN IMPACT ON THE BIOSPHERE

Interactive Exercises

An Indifference of Mythic Proportions [pp.924–925]

51.1. AIR POLLUTION—PRIME EXAMPLES [pp.926–927]
51.2. OZONE THINNING—GLOBAL LEGACY OF AIR POLLUTION [p.928]

Boldfaced, Page-Referenced Terms

[p.926] pollutants _____

_____

[p.926] thermal inversion _____

_____

[p.926] industrial smog _____

_____

[p.926] photochemical smog _____

_____

[p.926] peryoxacyl nitrates (PANs) _____

_____

[p.926] dry acid deposition _____

_____

[p.926] acid rain _____

_____

[p.928] ozone thinning _____

_____

[p.928] chlorofluorocarbons (CFCs) _____

_____

Fill-in-the-Blanks

(1) _____ [p.926] are substances with which ecosystems have had no prior evolutionary experience. Adaptive mechanisms are not in place to deal with them. When weather conditions trap a layer of dense, cool air beneath a layer of warm air, the situation is known as a(n) (2) _____ _____ [p.926]; this has been a key factor in some of the worst air pollution disasters. Where (3) _____ [p.926] are cold and wet, (4) _____ _____ [p.926] develops as a gray haze over industrialized cities that burn coal and other fossil fuels for manufacturing, heating, and generating electric power. In warm climates, (5) _____ _____ [p.926] develops as a brown haze over large cities located in natural basins. The key culprit is nitric oxide. After release from vehicles, it reacts with (6) _____ [p.926] in the air to form (7) _____ _____ [p.926]. When (7) is exposed to sunlight, it reacts with hydrocarbons, and (8) _____ [p.926] oxidants result. Most hydrocarbons come from spilled or partially burned (9) _____ [p.926]. The main oxidants are ozone and (10) _____ (peroxyacyl nitrates) [p.926]. Even traces can sting eyes, irritate lungs, and damage crops.

Oxides of (11) _____ [p.926] and (12) _____ [p.926] are among the worst pollutants. Coal-burning power plants, metal smelters, and factories emit most (13) _____ [p.926] dioxides. Motor vehicles, gas- and oil-burning power plants, and (14) _____-rich [p.926] fertilizers produce (15) _____ [p.926] oxides. During dry weather, fine particles of oxides may be briefly airborne and then fall to Earth as dry (16) _____ deposition [p.926]. When the oxides dissolve in atmospheric water, they form weak solutions of (17) _____ [p.926] and (18) _____ [p.926] acids. Strong winds may distribute them over great distances; when they fall to Earth in rain and snow, it is called wet acid deposition, or (19) _____ _____ [p.926]. (19) can be 10 to 100 times more acidic than normal rainwater, which has a pH of about (20) _____ [p.926]. The deposited acids eat away at marble buildings, metals, rubber, plastics, and even nylon stockings. They also have the potential to disrupt the physiology of organisms and the chemistry of ecosystems. (21) _____ (CFCs) [p.927] are compounds of chlorine, fluorine, and carbon that are odorless and invisible. They are major factors in reduction of the ozone layer in the atmosphere.

Choice

For questions 22–43, choose from the following aspects of atmospheric pollution.

    a. thermal inversion   b. industrial smog   c. photochemical smog   d. acid deposition
            e. chlorofluorocarbons   f. ozone layer

22. ___ Develops as a brown, smelly haze over large cities [p.926]

23. ___ Includes the "dry" and "wet" types [p.926]

24. ___ Contributes to ozone reduction more than any other factor [p.928]

25. ___ Where winters are cold and wet, this develops as a gray haze over industrialized cities that burn coal and other fossil fuels [p.926]

26. ___ Weather conditions trap a layer of cool, dense air under a layer of warm air [p.926]

27. ___ The cause of London's 1952 air pollution disaster, in which 4,000 died [p.926]

28. ___ Each year, from September through mid-October, it thins down at higher altitudes [p.928]

29. ___ Methyl bromide, a fungicide, will account for about 15 percent of its thinning if production does not stop [p.928]

30. ___ Intensifies a phenomenon called smog [p.926]

31. ___ Today most of this forms in cities of China, India, and other developing countries, as well as in coal-dependent countries of eastern Europe [p.926]

32. ___ Contains airborne pollutants, including dust, smoke, soot, ashes, asbestos, oil, bits of lead and other heavy metals, and sulfur oxides [p.926]

33. ___ Depending on soils and vegetation cover, some regions are more sensitive than others to this [p.927]

34. ___ Have been key factors in some of the worst local air pollution disasters [p.926]

35. ___ Chemically attack marble buildings, metals, mortar, rubber, plastic, and even nylon stockings [pp.926–927]

36. ___ Its reduction allows more ultraviolet radiation to reach Earth's surface [p.928]

37. ___ Reaches harmful levels where the surrounding land forms a natural basin, as it does around Los Angeles and Mexico City [p.926]

38. ___ Tall smokestacks were added to power plants and smelters in an unsuccessful attempt to solve this problem [p.927]

39. ___ As it thins, it lets more ultraviolet radiation reach Earth [p.928]

40. ___ A dramatic rise in skin cancers, eye cataracts, immune system weakening, and harm to photosynthesizers is related to its reduction [p.928]

41. ___ Found in refrigerators and air conditioners (as the coolants), solvents, and plastic foams [p.928]

42. ___ The key culprit is nitric oxide, produced mainly by vehicles [p.926]

43. ___ Oxides of sulfur and nitrogen dissolve in water to form weak solutions of sulfuric acid and nitric acid; winds may disperse them over great distances and they may fall to Earth with rain or snow [p.926]

51.3. WHERE TO PUT SOLID WASTES? WHERE TO PRODUCE FOOD? [p.929]

51.4. DEFORESTATION—AN ASSAULT ON FINITE RESOURCES [pp.930–931]

51.5. *Focus on Bioethics:* YOU AND THE TROPICAL RAIN FOREST [p.932]

Selected Words: *subsistence* agriculture [p.929], *animal-assisted* agriculture [p.929], *mechanized* agriculture [p.929]

Boldfaced, Page-Referenced Terms

[p.929] green revolution _____

_____

[p.929] deforestation _____

_____

[p.929] shifting cultivation _____

_____

Matching

Choose the most appropriate answer for each term.

1. ___throwaway mentality [p.929]
2. ___green revolution [p.929]
3. ___shifting cultivation [pp.930–931]
4. ___animal-assisted agriculture [p.929]
5. ___deforestation [p.930]
6. ___recycling [p.929]
7. ___watersheds of forested regions [p.930]
8. ___subsistence agriculture [p.929]
9. ___new genetic resources [p.932]
10. ___mechanized agriculture [p.929]

A. Agriculture in developing countries runs on energy inputs from sunlight and human labor
B. An affordable, technologically feasible alternative to "throwaway technology"
C. Act like giant sponges that absorb, hold, and gradually release water
D. Potential benefits to be obtained by genetic engineering and tissue culture methods in the rainforests
E. Runs on energy inputs from oxen and other draft animals
F. An attitude prevailing in the United States and other developed countries that greatly adds to solid waste accumulation
G. Requires massive inputs of fertilizers, pesticides, fossil fuel energy, and ample irrigation to sustain high-yield crops
H. Research directed toward improving crop plants for higher yields and exporting modern agricultural practices and equipment to developing countries
I. Trees are cut and burned, then ashes tilled into the soil; crops are grown for one to several seasons on quickly leached soils that become infertile
J. Removal of all trees from large land tracts; leads to loss of fragile soils and disrupts watersheds; greatest today in Brazil, Indonesia, Colombia, and Mexico

51.6. WHO TRADES GRASSLANDS FOR DESERTS? [p.933]

51.7. A GLOBAL WATER CRISIS [pp.934–935]

Selected Words: primary, secondary, and *tertiary* wastewater treatment [p.935]

Boldfaced, Page-Referenced Terms

[p.933] desertification _____

_____

[p.934] desalinization _____

_____

[p.934] salinization _____

_____

[p.934] water table _____

_____

[p.935] wastewater treatment _____

_____

Dichotomous Choice

Circle one of two possible answers given between parentheses in each statement.

1. Conversion of large tracts of grasslands, or rain-fed or irrigated croplands to a more barren state is known as (subsistence agriculture/desertification). [p.933]
2. Presently, (too many cattle in the wrong places/overgrazing on marginal lands) is the main cause of large-scale desertification. [p.933]
3. In Africa, (domestic cattle/native wild herbivores) trample grasses and compact the soil surfaces as they wander about looking for water. [p.933]
4. A 1978 study by biologist David Holpcraft demonstrated that African range conditions improved in land areas where (domestic cattle/native wild herbivores) were ranched. [p.933]
5. Without irrigation and conservation practices, grasslands that were converted for agriculture often end up as (deserts/forested watersheds). [p.933]

Fill-in-the-Blanks

Earth has a tremendous supply of water, but most is too (6) _____ [p.934] for human consumption or for agriculture. The removal of salt from seawater is called (7) _____ [p.934]. For most countries this process is not practical due to the costly fuel (8) _____ [p.934] necessary to drive it. Large-scale (9) _____ [p.934] accounts for nearly two-thirds of the human population's use of freshwater. Irrigation of otherwise useless soil can cause a salt buildup, or (10) _____ [p.934], caused by evaporation in areas of poor soil drainage. Land that drains poorly also becomes waterlogged and raises the (11) _____ _____ [p.934]. When the (11) is too close to the ground's surface, soil becomes saturated with (12) _____ [p.934] water, which can damage plant roots. A large problem is the fact that water tables are subsiding. For example, overdrafts have depleted half of the Ogallala aquifer that supplies irrigation water for (13) [check one] ❑ 40 ❑ 20 percent [p.934] of the croplands in the United States. Inputs of sewage, animal wastes, toxic chemicals, agricultural runoff, sediments, pesticides, and plant nutrients are all sources of water (14) _____ [p.934] that amplifies the problem of water scarcity. There are three levels of (15) _____ [p.935] treatment. They are primary, secondary, and (16) _____ [p.935] treatments. Most of the (14) is not treated adequately. If the current rates of population growth and water depletion hold, the amount of freshwater available for each person on the planet will be (17) [check one] ❐ 45–56 ❐ 55–66 ❐ 65–76 percent [p.935] less than it was in 1976. Water, not (18) _____ [p.935], may become the most important fluid of the twenty-first century. National, regional, and global (19) _____ [p.935] for water usage and water rights have yet to be developed.

Complete the Table

20. Complete the following table, which summarizes three levels of treatment methods for maintaining the water quality of polluted wastewater.

Treatment Method	Description
[p.935] a. Primary treatment	
[p.935] b. Secondary treatment	
[p.935] c. Tertiary treatment	

51.8. A QUESTION OF ENERGY INPUTS [pp.936–937]

51.9. ALTERNATIVE ENERGY SOURCES [p.938]

51.10. *Focus on Bioethics:* BIOLOGICAL PRINCIPLES AND THE HUMAN IMPERATIVE [p.939]

Selected Words: total energy [p.936], *net* energy [p.936], supertanker *Valdez* [p.936]

Boldfaced, Page-Referenced Terms

[p.936] fossil fuels _____

_____

[p.936] meltdown _____

_____

[p.938] solar–hydrogen energy _____

_____

[p.938] wind farms _____

_____

[p.938] fusion power _____

_____

Dichotomous Choice

Circle one of two possible answers given between parentheses in each statement.

1. Paralleling the (S-shaped/J-shaped) curve of human population growth is a dramatic rise in total and per capita energy consumption. [p.936]
2. The increase in per capita energy consumption is due to (increased numbers of energy users and to extravagant consumption and waste/energy used to locate, extract, transport, store, and deliver energy to consumers). [p.936]
3. (Total energy/Net energy) is that left over after subtracting the energy used to locate, extract, transport, store, and deliver energy to consumers. [p.936]
4. Fossil fuels are the carbon-containing remains of (plants/plants and animals) that lived hundreds of millions of years ago. [p.936]
5. Even with strict conservation efforts, known petroleum and natural gas reserves may be used up during the (current/next) century. [p.936]
6. The net energy (decreases/increases) as costs of extraction and transportation to and from remote areas increase. [p.936]
7. World coal reserves can meet human energy needs for several centuries, but burning releases sulfur dioxides into the atmosphere and adds to the global problem of (global photochemical smog/global acid deposition). [p.936]
8. By 1990 in the United States, it cost slightly more to generate electricity by nuclear energy than by using coal, but today it costs (more/less). [p.936]
9. The danger in the use of radioactivity as an energy supply during normal operation is with potential (radioactivity escape/meltdown). [p.936]
10. After nearly fifty years of research, scientists (have/have not) agreed on the best way to store high-level radioactive wastes. [p.937]
11. When electrodes in photovoltaic cells exposed to sunlight produce an electric current to split water molecules into oxygen and hydrogen gas (potential fuels), it is known as (fusion power/solar–hydrogen energy). [p.937]
12. California gets 1 percent of its electricity from (fusion power/wind farms). [p.938]
13. (Solar–hydrogen energy/Fusion power) involves a mimic of a process that takes place on the sun. [p.938]

Sequence-Classify

Arrange the consumption of world resources in correct hierarchical order. Enter the letter of the energy source of highest consumption next to 14, the letter of the next highest next to 15, and so on. Enter an (N) in the parentheses following the letter of the resource if it is nonrenewable and an (R) if the resource is renewable.

14. ___( ) A. Hydropower, geothermal, solar [p.936]

15. ___( ) B. Natural gas [p.936]

16. ___( ) C. Oil [p.936]

17. ___( ) D. Nuclear power [p.936]

18. ___( ) E. Biomass [p.936]

19. ___( ) F. Coal [p.936]

Self-Quiz

___ 1. Which of the following processes is not generally considered a component of secondary wastewater treatment? [p.935]
a. Screens and settling tanks remove sludge.
b. Microbial populations are used to break down organic matter.
c. All nitrogen, phosphorus, and toxic substances are removed.
d. Chlorine is often used to kill pathogens in the water.

___ 2. When fossil-fuel burning gives off dust, smoke, soot, ashes, asbestos, oil, bits of lead, other heavy metals, and sulfur oxides, it forms _____. [p.926]
a. photochemical smog
b. industrial smog
c. a thermal inversion
d. both a and c

___ 3. _____ result(s) when nitrogen dioxide and hydrocarbons react in the presence of sunlight. [p.926]
a. Photochemical smog
b. Industrial smog
c. A thermal inversion
d. Both a and c

___ 4. When weather conditions trap a layer of cool, dense air under a layer of warm air, _____ occurs. [p.926]
a. photochemical smog
b. a thermal inversion
c. industrial smog
d. acid deposition

___ 5. A major concern about the use of nuclear energy is _____. [pp.936–937]
a. excessive temperature
b. meltdown
c. waste disposal
d. all of the above

___ 6. Nitrogen oxides dissolve in atmospheric water to form a weak solution of sulfuric acid and nitric acid; this describes _____. [p.926]
a. photochemical smog
b. industrial smog
c. ozone and PANs
d. acid rain

___ 7. Which of the following statements is *false*? [p.934]
a. Ozone reduction allows more ultraviolet radiation to reach the Earth's surface.
b. CFCs enter the atmosphere and resist breakdown.
c. Salinization of soils aids plant growth and increases yields.
d. CFCs already in the air will be there for over a century.

___ 8. "Adequately reduces pollution but is largely experimental and expensive" describes _____ wastewater treatment. [p.935]
a. quaternary
b. secondary
c. tertiary
d. primary

___ 9. The statement "Photovoltaic cells exposed to sunlight produce an electric current that splits water molecules into oxygen and hydrogen gas" refers to _____. [p.938]
a. fusion power
b. wind energy
c. water power
d. solar–hydrogen energy

___10. Energy inputs from sunlight and human labor is the basis of _____. [p.929]
a. animal-assisted agriculture
b. subsistence agriculture
c. the green revolution
d. mechanized agriculture

Chapter Objectives/Review Questions

1. Identify the principal air pollutants, their sources, their effects, and the possible methods for controlling each pollutant. [pp.926–927]
2. During a _____ _____, weather conditions trap a layer of cool, dense air under a layer of warm air; trapped pollutants may reach dangerous levels. [p.926]
3. Distinguish photochemical smog from industrial smog. [p.926]
4. Explain what acid rain does to an ecosystem. Contrast those effects with the action of CFCs. [pp.926–928]
5. Discuss the significance of the effects of the ozone layer's thinning to life on Earth. [p.928]
6. List the key sources of air pollutants as related to ozone layer thinning. [p.928]
7. Under the banner of the _____ _____, research has been directed toward improving the genetics of crop plants for higher yields and exporting modern agricultural practices and equipment to the developing countries. [p.929]
8. _____ agriculture runs on energy inputs from sunlight and human labor; _____-_____ agriculture runs on energy inputs from oxen and other draft animals; _____ agriculture requires massive inputs of fertilizers, pesticides, and ample irrigation to sustain high-yield crops. [p.929]
9. Explain the repercussions of deforestation that are evident in soils, water quality, and genetic diversity in general. [pp.930–931]
10. _____ _____ involves cutting and burning trees, tilling ashes with soil, planting crops from one to several seasons, and then abandoning the clear plots. [pp.930–931]
11. Examine the effects that modern agriculture has wrought on desert and grassland ecosystems. [p.933]
12. _____ is the name for the conversion of large tracts of natural grasslands to a more desertlike condition. [p.933]
13. Explain why desalination is not a practical solution for the shortage of freshwater. [p.934]
14. Define primary, secondary, and tertiary wastewater treatment, and list some of the methods used in each. [p.935]
15. Explain the meaning of "the coming water wars." [p.935]
16. _____ energy refers to the amount left over after subtracting the energy that is used to locate, extract, transport, store, and deliver energy to consumers. [p.936]
17. Describe the dangers accompanying a meltdown. [pp.936–937]
18. Describe how our use of fossil fuels, solar energy, and nuclear energy affects ecosystems. [pp.936–937]
19. Briefly characterize solar–hydrogen energy, wind energy, and fusion power as alternative sources of energy. [p.938]
20. List five ways in which you could become personally involved in ensuring that institutions serve the public interest in a long-term, ecologically sound way. [p.939 and the whole chapter]

Integrating and Applying Key Concepts

1. If you were Ruler of All People on Earth, how would you encourage people to depopulate the cities and adopt a way of life by which they could supply their own resources from the land and dispose of their own waste products safely on their own land?
2. Explain why some biologists believe that the endangered species list now includes all species.

ANSWERS

Chapter 1 Concepts and Methods in Biology

Biology Revisited [pp.2–3]

1.1. DNA, ENERGY, AND LIFE [pp.4–5]
1. Cells; 2. DNA; 3. proteins; 4. amino; 5. Enzymes; 6. RNAs; 7. RNAs; 8. proteins; 9. reproduction; 10. development; 11. Energy; 12. transfer; 13. Metabolism; 14. photosynthesis; 15. ATP; 16. aerobic respiration; 17. responses; 18. receptors; 19. stimuli; 20. stimulus; 21. internal; 22. homeostasis.

1.2. ENERGY AND LIFE'S ORGANIZATION [pp.6–7]
1. F; 2. E; 3. H; 4. J; 5. G; 6. C; 7. L; 8. B; 9. M; 10. D; 11. K; 12. A; 13. I; 14. D; 15. G; 16. B; 17. J; 18. L; 19. E; 20. M; 21. A; 22. I; 23. F; 24. C; 25. H; 26. K; 27. producers; 28. consumers; 29. Decomposers; 30. energy; 31. cycling; 32. sun.

1.3. IF SO MUCH UNITY, WHY SO MANY SPECIES? [pp.8–9]
1. species; 2. genus; 3. species; 4. species; 5. genus; 6. species; 7. family; 8. order; 9. class; 10. phylum; 11. prokaryotic; 12. eukaryotic; 11. a. Plantae; b. Eubacteria; c. Fungi; d. Archaebacteria; e. Protista; f. Animalia; 12. D; 13. F; 14. A; 15. E; 16. B; 17. C; 18. G.

1.4. AN EVOLUTIONARY VIEW OF DIVERSITY [pp.10–11]
1. a; 2. b; 3. b; 4. a; 5. b; 6. b; 7. a; 8. a; 9. b; 10. b.

1.5. THE NATURE OF BIOLOGICAL INQUIRY [pp.12–13]

1.6. *Focus on Science:* **THE POWER OF EXPERIMENTAL TESTS** [pp.14–15]
1.7. THE LIMITS OF SCIENCE [p.15]
1. G; 2. A; 3. D; 4. B; 5. E; 6. C; 7. F; 8. O; 9. O; 10. C; 11. O; 12. C; 13. a. Hypotheses; b. Prediction; c. Scientific theory; d. Scientific experiments; e. Control group; f. Variables; g. Inductive logic; h. Deductive logic; 14. experiment; 15. control; 16. hypothesis; 17. prediction; 18. belief; 19. inductive logic; 20. deductive logic; 21. variable; 22. test predictions; 23. cause and effect; 24. sampling error; 25. quantitative; 26. subjective; 27. supernatural; 28. conviction.

Self-Quiz
1. d; 2. a; 3. b; 4. c; 5. d; 6. b; 7. d; 8. a; 9. c; 10. e.

Chapter 2 Chemical Foundations for Cells

Checking Out Leafy Clean-Up Crews [pp.20–21]

2.1. REGARDING THE ATOMS [p.22]
2.2. *Focus on Science:* **USING RADIOISOTOPES TO TRACK CHEMICALS AND SAVE LIVES** [p.23]
1. M; 2. C; 3. H; 4. D; 5. N; 6. B; 7. F; 8. K; 9. J; 10. L; 11. E; 12. G; 13. A; 14. I.

2.3. WHAT HAPPENS WHEN ATOM BONDS WITH ATOM? [pp.24–25]
1. C; 2. E; 3. I; 4. H; 5. A; 6. D; 7. F; 8. B; 9. G; 10. a. Calcium, Ca; b. Carbon, C; c. Chlorine, Cl; d. Hydrogen, H; e. Sodium, Na; f. Nitrogen, N; g. Oxygen, O; 11. equation; 12. formula; 13. yields; 14. reactants; 15. products; 16. 12; 17. 98 grams.

18.

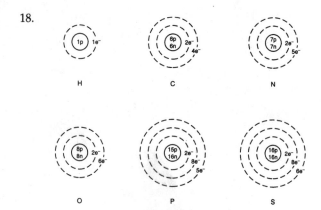

H   C   N

O   P   S

2.4. IMPORTANT BONDS IN BIOLOGICAL MOLECULES [pp.26–27]

1.

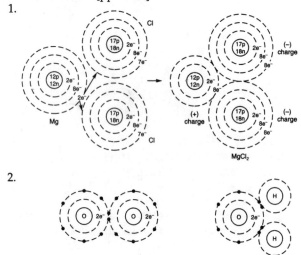

MgCl$_2$

2.

3. In a covalent bond, atoms share electrons to fill their outermost shells. In a nonpolar covalent bond, atoms attract shared electrons equally. An example is the H$_2$ molecule. In a polar covalent bond, atoms do not share electrons equally, and the bond is positive at one end, negative at the other (for example, the water molecule).
4. In the linear DNA molecule, the two nucleotide chains are held together by hydrogen bonds.

2.5. PROPERTIES OF WATER [pp.28–29]

1. polarity; 2. hydrophilic; 3. hydrophobic; 4. Temperature; 5. hydrogen; 6. evaporation; 7. ice; 8. cohesion; 9. solvent; 10. solutes; 11. dissolved; 12. hydration.

2.6. ACIDS, BASES, AND BUFFERS [pp.30–31]

1. N; 2. K; 3. F; 4. J; 5. H; 6. D; 7. I; 8. A; 9. B; 10. C; 11. O; 12. L; 13. E; 14. G; 15. M; 16. a. 7.3–7.5, slightly basic; b. 6.2–7.4, variable or slightly basic or slightly acid; c. 5.0–7.0, slightly acid to neutral; d. 1.0–3.0, acid; e. 7.8–8.3, basic; f. 3.0, acid.

Self-Quiz

1. c; 2. a; 3. d; 4. d; 5. c; 6. c; 7. a; 8. c; 9. a; 10. d; 11. e.

Chapter 3   Carbon Compounds in Cells

Carbon, Carbon, in the Sky—Are You Swinging Low and High? [pp.34–35]

3.1. PROPERTIES OF ORGANIC COMPOUNDS [pp.36–37]

1. methyl; 2. hydroxyl; 3. ketone; 4. amino; 5. phosphate; 6. carboxyl; 7. aldehyde; 8. Enzymes represent a special class of proteins that speed up specific metabolic reactions. Enzymes mediate five categories of reactions by which most of the biological molecules are assembled, rearranged, and broken apart; 9. a. A jug-

gling of internal bonds converts one type of organic compound into another; b. A molecule splits into two smaller ones; c. Through covalent bonding, two molecules combine to form a larger molecule; d. One or more electrons stripped from one molecule are donated to another molecule; e. One molecule gives up a functional group, which another molecule accepts;

10.

amino acid   amino acid   dipeptide

11. Hydrolysis reactions reverse the chemistry of condensation reactions; in the presence of water, large molecules are split into their component smaller molecules. Both condensation and hydrolysis require the presence of enzymes specific to the particular molecules involved.

3.2. CARBOHYDRATES [pp.38–39]

1.

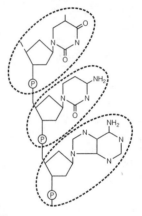

2. a. Sucrose, Oligosaccharide (disaccharide);
b. Deoxyribose, Monosaccharide; c. Glucose, Monosaccharide; d. Cellulose, Polysaccharide; e. Ribose, Monosaccharide; f. Lactose, Oligosaccharide (disaccharide); g. Glycogen, Polysaccharide; h. Chitin, Polysaccharide; i. Glycogen, Polysaccharide; j. Starch, Polysaccharide.

3.3. LIPIDS [pp.40–41]

1. a. unsaturated; b. saturated
2.

glycerol   three fatty acids   triglyceride (a complete fat molecule)

3. Phospholipid molecules have two fatty acid tails attached to a glycerol backbone; they have hydrophilic heads that dissolve in water. Phospholipids are the main structural materials of cell membranes; 4. b; 5. d; 6. e; 7. a; 8. b; 9. e; 10. e; 11. e; 12. d; 13. e; 14. a; 15. b; 16. b; 17. d; 18. b.

3.4. AMINO ACIDS AND THE PRIMARY
   ## STRUCTURE OF PROTEINS [pp.42–43]
3.5. HOW DOES A PROTEIN'S THREE-
   ## DIMENSIONAL STRUCTURE EMERGE? [pp.44–45]

3.6. *Focus on the Environment:* FOOD PRODUCTION AND A CHEMICAL ARMS RACE [p.46]
1. A; 2. C; 3. B
4.

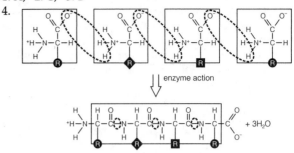

5. K; 6. F; 7. B; 8. J; 9. L; 10. A; 11. D; 12. I; 13. G; 14. E; 15. C; 16. H.

3.7. NUCLEOTIDES AND THE NUCLEIC ACIDS [p.47]
1. B; 2. A; 3. C; 4. Three as shown:

5. B; 6. C; 7. A; 8. D.

Self-Quiz
1. a. Lipids; b. Nucleic acids; c. Proteins; d. Proteins; e. Nucleic acids; f. Carbohydrates; g. Nucleic acids; h. Lipids; i. Nucleic acids; j. Lipids; k. Lipids; l. Carbohydrates; 2. d; 3. a; 4. c; 5. c; 6. c; 7. b; 8. d; 9. c; 10. b and e; 11. b.

Chapter 4   Cell Structure and Function

Animalcules and Cells Fill'd With Juices [pp.52–53]

4.1. BASIC ASPECTS OF CELL STRUCTURE AND
   ## FUNCTION [pp.54–55]
4.2. *Focus on Science:* MICROSCOPES—GATEWAYS
   ## TO CELLS [pp.56–57]
1. C; 2. H; 3. J; 4. F; 5. I; 6. K; 7. D; 8. B; 9. A; 10. G; 11. E; 12. Cell size is constrained by the surface-to-volume ratio. Past a certain point of growth, the surface area will not be sufficient to admit enough nutrients and to allow enough wastes to exit the cell. As a cell grows, the surface area increases with the square and the volume increases with the cube. 13. F; 14. E; 15. G; 16. C; 17. B; 18. A; 19. H; 20. D.

4.3. THE DEFINING FEATURES OF EUKARYOTIC CELLS [pp.58–61]

1. a. Nucleus; b. Ribosomes; c. Endoplasmic reticulum; d. Golgi body; e. Various vesicles; f. Mitochondria; g. Cytoskeleton; 2. A cell wall surrounding the cytoplasmic members of plant cells, a large central vacuole in a living plant cell, chloroplasts in the cytoplasm of plant cells, a pair of centrioles in the cytoplasm and close to the nucleus of animals cells.

4.4. THE NUCLEUS [pp.62–63]

1. First, the nucleus physically separates the cell's DNA from the metabolic machinery of the cytoplasm; this permits easier copying of DNA and the distribution of new DNA molecules into new cells following division. Second, the nuclear envelope helps control the exchange of substances and signals between the nucleus and the cytoplasm. 2. a. Nucleolus; b. Nuclear envelope; c. Chromatin; d. Nucleoplasm; e. Chromosome; 3. ribosomes; 4. cytoplasm; 5. cytomembrane; 6. endoplasmic; 7. Golgi; 8. DNA's; 9. proteins; 10. Lipids; 11. enzymes; 12. Vesicles.

4.5. THE CYTOMEMBRANE SYSTEM [pp.64–65]

1. I; 2. A; 3. B; 4. H; 5. C; 6. C; 7. F; 8. L; 9. K; 10. E; 11. D; 12. G; 13. J.

4.6. MITOCHONDRIA [p.66]
4.7. SPECIALIZED PLANT ORGANELLES [p.67]

1. b; 2. a; 3. c; 4. a; 5. e; 6. a; 7. d; 8. b; 9. e; 10. a; 11. d; 12. b; 13. b; 14. a; 15. c; 16. a; 17. b; 18. d; 19. a; 20. b.

4.8. COMPONENTS OF THE CYTOSKELETON [pp.68–69]
4.9. THE STRUCTURAL BASIS OF CELL MOBILITY [pp.70–71]

4.10. CELL SURFACE SPECIALIZATIONS [pp.72–73]

1. Protein; 2. movement; 3. tubulin; 4. plus; 5. MTOCs; 6. taxol; 7. actin; 8. motor; 9. cortex; 10. Intermediate filaments; 11. actin; 12. shortens; 13. cytoplasmic streaming; 14. flagella; 15. cilia; 16. centriole; 17. cilia; 18. dynein; 19. C; 20. E; 21. B; 22. A; 23. F; 24. D; 25. The presence of the complex carbohydrate molecules of lignin.

4.11. PROKARYOTIC CELLS—THE BACTERIA [pp.74–75]

1. bacteria; 2. nucleus; 3. wall; 4. permeable; 5. polysaccharides; 6. control; 7. proteins; 8. ribosomes; 9. cytoskeleton; 10. DNA; 11. nucleoid; 12. chromosome; 13. flagella; 14. pili; 15. protein; 16. Archaebacteria (Eubacteria); 17. Eubacteria (Archaebacteria); 18. prokaryotic.

Self-Quiz

1. Golgi body (J); 2. vesicle (G); 3. microfilaments (L); 4. mitochondrion (O); 5. chloroplast (M); 6. microtubules (H); 7. central vacuole (C); 8. rough endoplasmic reticulum (P); 9. ribosomes (R); 10. smooth endoplasmic reticulum (D); 11. DNA + nucleoplasm (Q); 12. nucleolus (K); 13. nuclear envelope (A); 14. nucleus (I); 15. plasma membrane (N); 16. cell wall (B); 17. microfilaments (L); 18. microtubules (H); 19. plasma membrane (N); 20. mitochondrion (O); 21. nuclear envelope (A); 22. nucleolus (K); 23. DNA + nucleoplasm (Q); 24. nucleus (I); 25. vesicle (G); 26. lysosome (E); 27. rough endoplasmic reticulum (P); 28. ribosomes (R); 29. smooth endoplasmic reticulum (D); 30. vesicle (G); 31. Golgi body (J); 32. centrioles (F); 33. d; 34. c; 35. d; 36. d; 37. b; 38. a; 39. c; 40. a; 41. b; 42. c; 43. b; 44. d; 45. c; 46. b, c, d, e; 47. a, b, c, d; 48. b, c, d, e; 49. b, c, d, e; 50. c, d; 51. a; 52. a, b, d; 53. b, d; 54. b, c, d, e; 55. b, c, d, e; 56. b, c, d, e; 57. b, c, d, e.

Chapter 5   A Closer Look at Cell Membranes

It Isn't Easy Being Single [pp.78–79]

5.1. MEMBRANE STRUCTURE AND FUNCTION [pp.80–81]
5.2. *Focus on Science:* TESTING IDEAS ABOUT CELL MEMBRANES [pp.82–83]

1. adhesion; 2. transport; 3. transport; 4. transport; 5. transport; 6. receptor; 7. recognition; 8. K; 9. J; 10. D; 11. A; 12. B; 13. C; 14. I; 15. H; 16. E; 17. F; 18. L; 19. G.

5.3. CROSSING SELECTIVELY PERMEABLE MEMBRANES [pp.84–85]
5.4. PROTEIN-MEDIATED TRANSPORT [pp.86–87]

1. permeability; 2. Concentration; 3. Gradient; 4. Diffusion; 5. steep; 6. equilibrium; 7. higher; 8. size;

9. electric; 10. pressure; 11. b; 12. c; 13. a; 14. c; 15. a; 16. b; 17. c; 18. b; 19. c; 20. b; 21. b; 22. a; 23. a; 24. c; 25. a. diffusion of water molecules (osmosis); b. diffusion; c. active transport; d. passive and active transport; e. diffusion; f. active transport; g. active transport; h. passive transport (facilitated diffusion).

5.5. MOVEMENT OF WATER ACROSS MEMBRANES [pp.88–89]

1. C; 2. F; 3. A; 4. H; 5. E; 6. B; 7. I; 8. D; 9. G; 10. T; 11. T; 12. tonicity; 13. hypotonic; 14. T; 15. isotonic; 16. less; 17. bulk flow; 18. T.

5.6. BULK TRANSPORT ACROSS MEMBRANES
[pp.90–91]
1. D; 2. E; 3. C; 4. A; 5. B.

Self-Quiz
1. d; 2. c; 3. a; 4. d; 5. c; 6. d; 7. d; 8. c; 9. d; 10. b.

Chapter 6   Ground Rules of Metabolism

You Light Up My Life [pp.94–95]

6.1. ENERGY AND THE UNDERLYING ORGANIZATION OF LIFE [pp.96–97]
1. energy; 2. work; 3. metabolism; 4. reactions;
5. kinetic energy; 6. heat; 7. I; 8. I; 9. II; 10. I; 11. II;
12. I; 13. I; 14. I; 15. II; 16. b; 17. e; 18. d; 19. a;
20. c; 21. second; 22. T; 23. T; 24. T; 25. increasing;
26. T; 27. bioluminescence; 28. luciferin; 29. ATP;
30. oxygen; 31. luciferase; 32. electrons.

6.2. ENERGY CHANGES AND CELLULAR WORK [pp.98–99]
6.3. A LOOK AT TYPICAL PHOSPHATE-GROUP TRANSFERS [p.100]
6.4. A LOOK AT TYPICAL ELECTRON TRANSFERS [p.101]
6.5. THE DIRECTIONAL NATURE OF METABOLISM [pp.102–103]
1. exergonic; 2. endergonic; 3. exergonic; 4. exergonic;
5. exergonic; 6. ATP; 7. adenosine triphosphate;
8. adenine; 9. ribose; 10. phosphate; 11. covalent;
12. hydrolysis; 13. energy; 14. energy; 15. ATP;
16. phosphorylation; 17. triphosphate; 18. ribose;
19. adenine; 20. adenosine triphosphate; 21. electron;
22. oxidized; 23. reduced; 24. energy; 25. most;
26. energy; 27. work; 28. ATP; 29. E; 30. B; 31. C;
32. A; 33. D; 34. products; 35. reactants; 36. reversible;
37. equilibrium; 38. rate (pace); 39. C; 40. B; 41. D;

42. F; 43. E; 44. A; 45. B; 46. A; 47. A; 48. B; 49. F;
50. G(A); 51. I; 52. D; 53. J; 54. A; 55. E; 56. C;
57. B(A); 58. H; 59. intermediate; 60. enzyme;
61. electron transport; 62. coenzymes; 63. linear;
64. cyclic; 65. branched.

6.6. ENZYME STRUCTURE AND FUNCTION [pp.104–105]
6.7. FACTORS INFLUENCING ENZYME ACTIVITY [pp.106–107]
6.8. *Focus on Health:* GROWING OLD WITH MOLECULAR MAYHEM [pp.108–109]
1. enzyme; 2. Enzymes; 3. equilibrium; 4. substrate;
5. active site; 6. induced-fit; 7. activation energy;
8. collision; 9. hydrogen; 10. hydrolysis; 11. rate
(pace); 12. activation energy without enzyme;
13. activation energy with enzyme; 14. catalyzed;
15. uncatalyzed; 16. energy released by the reaction;
17. D; 18. F; 19. C; 20. B; 21. A; 22. E; 23. 40°C;
24. 60°C; 25. C; 26. B; 27. A; 28. Temperature (pH,
saltiness); 29. pH or saltiness; 30. metabolism;
31. bonds; 32. 7; 33. tryptophan; 34. amino acids;
35. falls; 36. feedback inhibition; 37. enzyme;
38. allosteric; 39. allosteric binding; 40. rises;
41. hormones; 42. D; 43. E; 44. A; 45. B; 46. C.

Self-Quiz
1. d; 2. c; 3. e; 4. d; 5. a; 6. c; 7. d; 8. c; 9. d; 10. a.

Chapter 7   How Cells Acquire Energy

Sunlight and Survival [pp.112–113]

7.1. PHOTOSYNTHESIS—AN OVERVIEW [pp.114–115]
1. Autotrophs; 2. carbon dioxide; 3. Photosynthetic;
4. Englemann; 5. violet; 6. Heterotrophs; 7. animals;
8. aerobic respiration; 9. $12 H_2O + 6CO_2 \rightarrow 6O_2 + C_6H_{12}O_6 + 6H_2O$; 10. (a) Twelve; (b) carbon dioxide;
(c) oxygen; (d) glucose; (e) six; 11. light-dependent
(light-independent); 12. light-independent (light-dependent); 13. Carbon dioxide; 14. water; 15. glucose; 16.
thylakoid; 17. grana; 18. hydrogen; 19. stroma; 20. b;
21. a; 22. a; 23. b; 24. a; 25. a; 26. b; 27. a; 28. a;
29. a; 30. b; 31. a; 32. O_2; 33. ATP; 34. NADPH;
35. CO_2; 36. chloroplasts; 37. thylakoid membrane
system; 38. stroma.

7.2. SUNLIGHT AS AN ENERGY SOURCE [pp.116–117]
7.3. THE RAINBOW CATCHERS [pp.118–119]
1. thylakoid; 2. photon; 3. pigments; 4. chlorophylls;
5. red (blue); 6. blue (red); 7. Carotenoids; 8. photosystem; 9. light (photon); 10. electron; 11. acceptor;
12. H; 13. G; 14. E; 15. F; 16. B; 17. A; 18. C; 19. I;
20. J; 21. D.

7.4. THE LIGHT-DEPENDENT REACTIONS [pp.120–122]
7.5. A CLOSER LOOK AT ATP FORMATION IN CHLOROPLASTS [p.122]
1. a. A pigment cluster dominated by P700; b. Electrons
representing energy are ejected from P700 to an electron
acceptor but move over the electron transport system,

where some of the energy is used to produce ATP; c. A special chlorophyll molecule that absorbs wavelengths of 700 nanometers and then ejects electrons; d. A molecule that accepts electrons ejected from chlorophyll P700 and then passes electrons down the electron transport system; e. Electrons flow through this system, which is composed of a series of molecules bound in the thylakoid membrane that drive photophosphorylation; f. ADP undergoes photophosphorylation in cyclic photophosphorylation to become ATP; 2. electron acceptor molecule; 3. electron transport system; 4. photosystem II; 5. photosystem I; 6. photolysis; 7. NADPH; 8. ATP; 9. electrons; 10. thylakoid; 11. gradients; 12. water (H_2O); 13. ATP synthases; 14. ATP; 15. chemiosmotic; 16. NADPH; 17. electrons; (18–53). The following numbers should have a check mark (√): 20, 22, 25, 28, 30, 32, 33, 34, 35, 37, 40, 41, 42, 45, 50, and 51. All others should be blank.

7.6. THE LIGHT-INDEPENDENT REACTIONS [p.123]
7.7. FIXING CARBON—SO NEAR, YET SO FAR [pp.124–125]
7.8. *Focus on Science:* **LIGHT IN THE DEEP DARK SEA?** [p.125]

7.9. *Focus on the Environment:* **AUTOTROPHS, HUMANS, AND THE BIOSPHERE** [p.126]
1. carbon dioxide (D); 2. carbon fixation (E); 3. phosphoglycerate (F); 4. adenosine triphosphate (H); 5. NADPH (G); 6. phosphoglyceraldehyde (A); 7. phosphorylated glucose (B); 8. Calvin–Benson cycle (I); 9. ribulose bisphosphate (C); 10. ATP (NADPH); 11. NADPH (ATP); 12. carbon dioxide; 13. ribulose bisphosphate; 14. PGA; 15. fixation; 16. PGA; 17. PGAL; 18. six; 19. RuBP; 20. carbon dioxide; 21. PGALs; 22. phosphorylated glucose; 23. fixation; 24. light-dependent; 25. ATP (NADPH); 26. NADPH (ATP); 27. phosphorylated glucose; 28. photorespiration; 29. C3; 30. food (sugar); 31. oxygen; 32. oxaloacetate; 33. succulent; 34. night; 35. hydrothermal vent; 36. hydrogen sulfide; 37. carbon dioxide; 38. chlorophyll; 39. chemo-; 40. food (organic); 41. photo-; (42–85). The following numbers should have a check mark (√): 42, 47, 52, 54, 55, 56, 57, 59, 61, 65, 66, 71, 72, 73, 82, and 83. All others should be blank.

Self-Quiz
1. a; 2. b; 3. c; 4. a; 5. c; 6. d; 7. a; 8. b; 9. c; 10. d.

Chapter 8   How Cells Release Stored Energy

The Killers Are Coming! [pp.130–131]

8.1. HOW CELLS MAKE ATP [pp.132–133]
1. Adenosine triphosphate (ATP); 2. Oxygen withdraws electrons from the electron transport system and joins with H^+ to form water; 3. Glycolysis followed by some end reactions (called *fermentation*), and anaerobic electron transport. Some organisms (including humans) use fermentation pathways when oxygen supplies are low; many microbes rely exclusively on anaerobic pathways; 4. ATP; 5. photosynthesis; 6. aerobic respiration; 7. glycolysis; 8. pyruvate; 9. Krebs; 10. water; 11. ATP; 12. electrons; 13. transport; 14. phosphorylation; 15. ATP; 16. Oxygen; 17. anaerobic; 18. Fermentation; 19. electron transport; 20. $C_6H_{12}O_6 + 6O_2 \rightarrow 6CO_2 + 6H_2O$; 21. One molecule of glucose plus six molecules of oxygen (in the presence of appropriate enzymes) yield six molecules of carbon dioxide plus six molecules of water; 22. G; 23. I; 24. H; 25. D; 26. F; 27. J; 28. A; 29. B; 30. C; 31. E.

8.2. GLYCOLYSIS: FIRST STAGE OF ENERGY-RELEASING PATHWAYS [pp.134–135]
1. Autotrophic; 2. Glucose; 3. pyruvate; 4. ATP (NADH); 5. NADH (ATP); 6. glucose; 7. ATP; 8. PGAL; 9. phosphate; 10. hydrogen; 11. ATP; 12. water; 13. phosphate; 14. ATP; 15. substrate-level; 16. glucose; 17. pyruvate; 18. three; 19. D; 20. F; 21. B; 22. H; 23. G; 24. A; 25. E; 26. C.

8.3. SECOND STAGE OF THE AEROBIC PATHWAY [pp.136–137]
8.4. THIRD STAGE OF THE AEROBIC PATHWAY [pp.138–139]
1. acetyl-CoA; 2. Krebs; 3. electron transport; 4. thirty-six; 5. ATP; 6. carbon dioxide; 7. electrons; 8. NAD^+ (FAD); 9. FAD (NAD^+); 10. inner compartment; 11. inner membrane; 12. outer compartment; 13. outer membrane; 14. cytoplasm; 15. ATP; 16. oxygen (O_2); 17. $FADH_2$; 18. NADH; 19. electron transport system; 20. three; 21. two; 22. transport; 23. chemiosmotic; 24. synthases; 25. ATP; 26. oxygen; 27. mitochondria; 28. thirty-six (thirty-eight); 29. thirty-eight (thirty-six); 30. c; 31. a, b; 32. c; 33. a; 34. b; 35. a; 36. b; 37. b; 38. a, b, c; 39. b; 40. c; 41. a, b; 42. b; 43. c; 44. a; 45. c; 46. c; 47. a; 48. b; 49. a; 50. c; 51. c.

8.5. ANAEROBIC ROUTES OF ATP FORMATION [pp.140–141]
1. oxygen (O_2); 2. fermentation; 3. lactate; 4. ethanol; 5. carbon dioxide; 6. Anaerobic; 7. electron; 8. Glycolysis; 9. pyruvate; 10. NADH; 11. pyruvate; 12. lactate; 13. acetaldehyde; 14. carbon dioxide; 15. ethanol; 16. glycolysis; 17. NAD^+; 18. bacteria; 19. ATP; 20. sulfate; (21–72). With a checkmark (√): 21, 23, 28, 29, 31, 32, 34, 38, 43, 46, 49, 51, 54, 56, 58, 60, 65; all others lack a check.

8.6. ALTERNATIVE ENERGY SOURCES IN THE HUMAN BODY [pp.142–143]
8.7. *Commentary:* **PERSPECTIVE ON LIFE** [p.144]
1. Figure 8.12 shows how any complex carbohydrate or fat can be broken down and at least part of those molecules can be fed into the glycolytic pathway; 2. T; 3. T; 4. Page 97 (last paragraph) tells us that energy flows through time in one direction—from organized to less organized forms; thus energy cannot be completely recycled; 5. Page 132 tells us that Earth's first organisms were anaerobic fermenters; 6. fatty acids; 7. glycerol; 8. glycolysis; 9. amino acids; 10. Krebs cycle;

11. acetyl-CoA formation; 12. pyruvate; (13–102). With a check (√): 13, 15, 16, 17, 19, 22, 23, 25, 29, 31, 32, 39, 46, 47, 48, 49, 53, 54, 65, 66, 70, 71, 72, 80, 84, 87, 95, 102; all others lack a check; 103. e; 104. b; 105. d; 106. a; 107. c; 108. h; 109. e; 110. b; 111. b; 112. h; 113. i; 114. e; 115. c; 116. h; 117. g.

Self-Quiz
1. c; 2. c; 3. d; 4. b; 5. d; 6. d; 7. a; 8. d; 9. d; 10. d; 11. C; 12. A, B, D; 13. B, (C), D; 14. C, E; 15. B, C; 16. A, D; 17. A; 18. B, D; 19. E; 20. A, E.

Chapter 9   Cell Division and Mitosis

Silver in the Stream of Time [pp.148–149]

9.1. DIVIDING CELLS: THE BRIDGE BETWEEN GENERATIONS [p.150]
9.2. THE CELL CYCLE [p.151]
1. G; 2. J; 3. C; 4. E; 5. B; 6. A; 7. I; 8. H; 9. F; 10. D; 11. interphase; 12. mitosis; 13. G1; 14. S; 15. G2; 16. prophase; 17. metaphase; 18. anaphase; 19. telophase; 20. cytoplasm divided or cytokinesis; 21. 15; 22. 12; 23. 14; 24. 13; 25. 11; 26. 20; 27. 11.

9.3. THE STAGES OF MITOSIS—AN OVERVIEW [pp.152–153]
1. interphase-daughter cells (F); 2. anaphase (A); 3. late prophase (G); 4. metaphase (D); 5. cells at interphase (E); 6. early prophase (C); 7. transition to metaphase (B); 8. telophase (H).

9.4. DIVISION OF THE CYTOPLASM [pp.154–155]
1. a; 2. b; 3. a; 4. a; 5. b; 6. a; 7. b; 8. b; 9. a; 10. a.

9.5. A CLOSER LOOK AT THE CELL CYCLE [pp.156–157]
9.6. *Focus on Science:* **HENRIETTA'S IMMORTAL CELLS** [p.158]
1. F; 2. I; 3. H; 4. A; 5. D; 6. J; 7. B; 8. G; 9. E; 10. C; 11. HeLa cells are tumor cells taken from and named for a cancer patient, Henrietta Lacks, in 1951. Henrietta Lacks died (at age thirty-one) two months following her diagnosis of cancer. HeLa cells have continued to divide in culture and are used for cancer research in laboratories all over the world. The legacy of Henrietta Lacks continues to benefit humans everywhere.

Self-Quiz
1. a; 2. c; 3. d; 4. a; 5. d; 6. c; 7. e; 8. c; 9. c; 10. d; 11. d.

Chapter 10   Meiosis

Octopus Sex and Other Stories [pp.160–161]

10.1. COMPARING SEXUAL WITH ASEXUAL REPRODUCTION [p.162]
10.2. HOW MEIOSIS HALVES THE CHROMOSOME NUMBER [pp.162–163]
1. b; 2. a; 3. a; 4. b; 5. a; 6. b; 7. a; 8. b; 9. a; 10. b; 11. Meiosis; 12. gamete; 13. Diploid; 14. Diploid; 15. Meiosis; 16. sister chromatids; 17. sister chromatids; 18. one; 19. four; 20. two; 21. interphase preceding meiosis I; 22. chromosome; 23. haploid; 24. meiosis II; 25. twenty-three; 26. E ($2n = 2$); 27. D ($2n = 2$); 28. B ($2n = 2$); 29. A ($n = 1$); 30. C ($n = 1$).

10.3. A VISUAL TOUR OF THE STAGES OF MEIOSIS [pp.164–165]

10.4. A CLOSER LOOK AT KEY EVENTS OF MEIOSIS I [pp.166–167]
1. anaphase II (H); 2. metaphase II (F); 3. metaphase I (A); 4. prophase II (B); 5. telophase II (C); 6. telophase I (G); 7. prophase I (E); 8. anaphase I (D); 9. F; 10. I; 11. G; 12. D; 13. B; 14. J; 15. A; 16. H; 17. C; 18. E.

10.5. FROM GAMETES TO OFFSPRING [pp.168–169]
1. b; 2. c; 3. a; 4. c; 5. b; 6. a; 7. b; 8. a; 9. b; 10. a; 11. b; 12. c; 13. b; 14. a; 15. 2 ($2n$); 16. 5 (n); 17. 4 (n); 18. 1 ($2n$); 19. 3 (n); 20. A; 21. B; 22. E; 23. D; 24. C; 25. During prophase I of meiosis, crossing over and genetic recombination occur. During metaphase I of meiosis, the two members of each homologous chromosome assort independently of the other pairs. Fertilization is a chance mix of different combinations of alleles from two different gametes.

10.6. MEIOSIS AND MITOSIS COMPARED
[pp.170–171]
1. a. Mitosis; b. Mitosis; c. Meiosis; d. Meiosis; e. Meiosis; f. Meiosis; g. Mitosis; h. Mitosis; i. Meiosis; 2. C; 3. F; 4. D; 5. A; 6. B; 7. E; 8. 4; 9. 8; 10. 4; 11. 8; 12. 2.

Self-Quiz
1. a; 2. a; 3. d; 4. b; 5. c; 6. b; 7. d; 8. a; 9. b; 10. c.

Chapter 11   Observable Patterns of Inheritance

A Smorgasbord of Ears and Other Traits
[pp.174–175]

11.1. MENDEL'S INSIGHTS INTO PATTERNS OF INHERITANCE [pp.176–177]
11.2. MENDEL'S THEORY OF SEGREGATION [pp.178–179]
11.3. INDEPENDENT ASSORTMENT [pp.180–181]
1. I; 2. B; 3. D; 4. J; 5. H; 6. M; 7. O; 8. E; 9. F; 10. K; 11. A; 12. G; 13. N; 14. C; 15. L; 16. monohybrid; 17. Probability; 18. fertilization; 19. segregation; 20. testcross; 21. 1:1; 22. dihybrid; 23. independent assortment; 24. genotype: 1/2 *Tt*; 1/2 *tt*, phenotype: 1/2 tall; 1/2 short; 25. a. 1 tall : 1 short; 1 heterozygous tall : 1 homozygous short; b. all tall; 1 homozygous tall : 1 heterozygous tall; c. all short; all homozygous short; d. 3 tall : 1 short; 1 homozygous tall : 2 heterozygous tall : 1 homozygous short; e. 1 tall : 1 short; 1 homozygous short : 1 heterozygous tall; f. all tall; all heterozygous tall; g. all tall; all homozygous tall; h. all tall; 1 heterozygous tall : 1 homozygous tall; 26. a. 9/16; pigmented eyes, right-handed; b. 3/16 pigmented eyes, left-handed; c. 3/16 blue-eyed, right-handed; d. 1/16 blue-eyed, left-handed (note Punnett square below).

	BR	Br	bR	br
BR	BBRR	BBRr	BbRR	BbRr
Br	BBRr	BBrr	BbRr	Bbrr
bR	BbRR	BbRr	bbRR	bbRr
br	BbRr	Bbrr	bbRr	bbrr

27. Albino = *aa*, normal pigmentation = *AA* or *Aa*. The woman of normal pigmentation with an albino mother is genotype *Aa*; the woman received her recessive gene (*a*) from her mother and her dominant gene (*A*) from her father. It is likely that half of the couple's children will be albinos (*aa*) and half will have normal pigmentation but be heterozygous (*Aa*); 28. a. F_1: black trotter; F_2: nine black trotters, three black pacers, three chestnut trotters, one chestnut pacer; b. black pacer; c. *BbTt*; d. *bbtt*, chestnut pacers and *BBTT*, black trotters.

11.4. DOMINANCE RELATIONS [p.182]
11.5. MULTIPLE EFFECTS OF SINGLE GENES [p.183]
11.6. INTERACTIONS BETWEEN GENE PAIRS [pp.184–185]
1. a. Incomplete dominance; b. Codominance; c. Multiple alleles; d. Epistasis; e. Pleiotropy.
2. a. phenotype: all pink, genotype: all *RR'*; b. phenotype: all white, genotype: all *R'R'*; c. phenotype: 1/2 red; 1/2 pink, genotype: 1/2 *RR*; 1/2 *RR'*; d. phenotype: all red, genotype: all *RR*; 3. The man must have sickle-cell trait with the genotype $Hb^A Hb^S$, and the woman he married would have a normal genotype, $Hb^A Hb^A$. The couple could be told that the probability is 1/2 that any child would have sickle-cell trait and 1/2 that any child would have the normal genotype; 4. Both the man and the woman have the genotype $Hb^A Hb^S$. The probability of children from this marriage is: 1/4 normal, $Hb^A Hb^A$; 1/2 sickle-cell trait, $Hb^A Hb^S$; 1/4 sickle-cell anemia, $Hb^S Hb^S$; 5. Genotypes: 1/4 $I^A I^A$; 1/4 $I^A I^B$; 1/4 $I^A i$; 1/4 $I^B i$, phenotypes: 1/2 A; 1/4 AB; 1/4 B; 6. Genotypes: 1/4 $I^A I^B$; 1/4 $I^B i$; 1/4 $I^A i$; 1/4 *ii*, phenotypes: 1/4 AB; 1/4 B; 1/4 A; 1/4 O; 7. Genotypes: all $I^A i$, phenotypes: all A; 8. Genotypes: all *ii*, phenotypes: all O; 9. Genotypes: 1/4 $I^A I^A$; 1/2 $I^A I^B$; 1/4 $I^B I^B$, phenotypes: 1/4 A; 1/2 AB; 1/4 B; 10. 1/4 color; 3/4 white; 11. 3/4 color; 1/4 white; 12. 1/4 color; 3/4 white; 13. 3/8 black; 1/2 yellow; 1/8 brown; 14. The genotype of the male parent is *RrPp*, and the genotype of the female parent is *rrpp*. The offspring are 1/4 walnut comb, *RrPp*; 1/4 rose comb, *Rrpp*; 1/4 pea comb, *rrPp*; 1/4 single comb, *rrpp*; 15. The genotype of the walnut-combed male is *RRpp*, and the genotype of the single-combed female is *rrpp*. All offspring are rose comb with the genotype *Rrpp*.

11.7. HOW CAN WE EXPLAIN LESS PREDICTABLE VARIATIONS? [pp.186–187]
11.8. EXAMPLES OF ENVIRONMENTAL EFFECTS ON PHENOTYPE [p.188]
1. b; 2. b; 3. a; 4. b; 5. a.

Self-Quiz
1. d; 2. b; 3. a; 4. c; 5. d; 6. b; 7. e; 8. a; 9. a; 10. c; 11. d.

Chapter 12   Chromosomes and Human Genetics

The Philadelphia Story [pp.192–193]

12.1. THE CHROMOSOMAL BASIS OF INHERITANCE—AN OVERVIEW [p.194]
12.2. *Focus on Science:* KARYOTYPING MADE EASY [p.195]

1. Philadelphia; 2. Leukemias; 3. karyotype; 4. spectral; 5. genes; 6. homologous; 7. Alleles; 8. wild; 9. crossing over; 10. recombination; 11. sex; 12. autosomes; 13. karyotype; 14. in vitro; 15. colchicine; 16. metaphase; 17. centrifugation; 18. centromeres.

12.3. SEX DETERMINATION IN HUMANS [pp.196–197]
12.4. EARLY QUESTIONS ABOUT GENE LOCATIONS [pp.198–199]

1. E; 2. C; 3. A; 4. D; 5. B; 6. Two blocks of the Punnett square should be XX, and two blocks should be XY; 7. sons; 8. mothers; 9. daughters; 10. a. F_1 flies all have red eyes: 1/2 heterozygous red females : 1/2 red-eyed males; b. F_2 Phenotypes: females all have red eyes; males 1/2 red eyes, 1/2 white eyes; Genotypes: females are $X^W X^W$; 11. d.

12.5. RECOMBINATION PATTERNS AND CHROMOSOME MAPPING [pp.200–201]

1. Crossing over would be expected to occur twice as often between genes A and B as it would between genes C and D; 2. a; 3. Linked genes of the children: daughter 1 is *On/An*, daughter 2 is *ON/On*, and the son is *On/On*; 4. The son has genotype *On/On* because a crossover must have occurred during meiosis in his mother, resulting in the recombination *On*.

12.6. HUMAN GENETIC ANALYSIS [pp.202–203]

1. A; 2. E; 3. F; 4. B; 5. C; 6. G; 7. D; 8. A genetic abnormality is nothing more than a rare, uncommon version of a trait that society may judge as abnormal or merely interesting; a genetic disorder is an inherited condition that sooner or later causes mild to severe medical problems; a syndrome is a recognized set of symptoms that characterize a given disorder; a genetic disease is illness caused by a person's genes increasing susceptibility to infection or weakening the response to it; 9. a. Autosomal recessive; b. Autosomal dominant; c. X-linked recessive; d. X-linked recessive; e. Autosomal dominant; f. Changes in chromosome number; g. Autosomal dominant; h. Changes in chromosome structure; i. Changes in chromosome number; j. X-linked recessive; k. X-linked recessive; l. Autosomal dominant; m. Changes in chromosome number; n. X-linked recessive; o. Autosomal recessive.

12.7. INHERITANCE PATTERNS [pp.204–205]
12.8. *Focus on Health:* TOO YOUNG TO BE OLD [p.206]

1. b; 2. a and c; 3. b; 4. c; 5. a; 6. b; 7. b; 8. b; 9. a; 10. c; 11. c; 12. b; 13. b; 14. a; 15. b; 16. c; 17. a; 18. The woman's mother is heterozygous normal, *Gg*; the woman is also heterozygous normal, *Gg*. The man with galactosemia, *gg*, has two heterozygous normal parents, *Gg*. The two normal children are heterozygous normal, *Gg*; the child with galactosemia is *gg*; 19. Assuming the father is heterozygous with Huntington's disorder and the mother normal, the chances are 1/2 that the son will develop the disease; 20. If only male offspring are considered, the probability is 1/2 that the couple will have a color-blind son; 21. The probability is that 1/2 of the sons will have hemophilia; the probability is 0 that a daughter will express hemophilia; the probability is that 1/2 of the daughters will be carriers; 22. If the woman marries a normal male, the chance that her son would be color blind is 1/2. If she marries a color-blind male, the chance that her son would be color blind is also 1/2.

12.9. CHANGES IN CHROMOSOME STRUCTURE [pp.206–207]
12.10. CHANGES IN CHROMOSOME NUMBER [pp.208–209]

1. duplication (B); 2. inversion (C); 3. deletion (A); 4. translocation (D); 5. a. With aneuploidy, individuals have one extra or one less chromosome; a major cause of human reproductive failure; b. With polyploidy, individuals have three or more of each type of chromosome; common in flowering plants and some animals but lethal in humans; c. Nondisjunction is caused by a failure of one or more pairs of chromosomes to separate in mitosis or meiosis; some or all forthcoming cells will have too many or too few chromosomes; 6. All gametes will be abnormal; 7. One-half of the gametes will be abnormal; 8. About half of all flowering plant species are polyploids, but this condition is lethal for humans; 9. A tetraploid cell has four of each type of chromosome; a trisomic ($2n + 1$) individual will have three of one type of chromosome and two of every other type; a monosomic ($2n - 1$) individual will have only one of one type of chromosome but two of every other type; 10. c; 11. b; 12. c; 13. d; 14. a; 15. b; 16. d; 17. c; 18. a; 19. d.

12.11. *Focus on Bioethics:* PROSPECTS IN HUMAN GENETICS [pp.210–211]

1. a. Prenatal diagnosis; b. Abortion; c. Genetic counseling; d. Preimplantation diagnosis; e. Phenotypic treatments; f. Genetic screening.

Self-Quiz

1. d; 2. d; 3. b; 4. c; 5. b; 6. a; 7. c; 8. b; 9. b; 10. c.

Chapter 13   DNA Structure and Function

Cardboard Atoms and Bent-Wire Bonds
[pp.214–215]

13.1. DISCOVERY OF DNA FUNCTION [pp.216–217]
1. a. Meischer: identified "nuclein" from nuclei of pus cells and fish sperm; discovered DNA;   b. Griffith: discovered the transforming principle in *Streptococcus pneumoniae*; live, harmless R cells were mixed with dead S cells; R cells became S cells;   c. Avery: reported that the transforming substance in Griffith's bacteria experiments was probably DNA, the substance of heredity;   d. Hershey and Chase: worked with radioactive sulfur (protein) and phosphorus (DNA) labels; T4 bacteriophage and *E. coli* demonstrated that labeled phosphorus was in bacteriophage DNA and contained hereditary instructions for new bacteriophages;   2. virus;   3. bacterial;   4. viruses (bacteriophages);   5. proteins;   6. ^{35}S;   7. bacteriophage (viral);   8. ^{35}S;   9. ^{32}P;   10. genetic material;   11. DNA;   12. proteins.

13.2. DNA STRUCTURE [pp.218–219]
13.3. *Focus on Bioethics:* **ROSALIND'S STORY** [p.220]
1. A five-carbon sugar called *deoxyribose*, a phosphate group, and one of the four nitrogen-containing bases;   2. guanine (pu);   3. cytosine (py);   4. adenine (pu);   5. thymine (py);   6. deoxyribose (B);   7. phosphate group (G);   8. purine (C);   9. pyrimidine (A);   10. purine (E);   11. pyrimidine (D);   12. nucleotide (F);   13. T;   14. T;   15. F, sugar;   16. T;   17. T;   18. pairing;   19. constant;   20. sequence;   21. different;   22. Living organisms have so many diverse body structures and behave in different ways because the many different habitats of Earth have selected those genotypes most able to survive in those habitats. The remaining genotypes have perished. The directions that code for the building of those body structures and that enable the specific successful behaviors reside in DNA or in a few cases, RNA. All living organisms follow the same rules for base-pairing between the two nucleotide strands in DNA: adenine always pairs with thymine in undamaged DNA, and cytosine always pairs with guanine. All living organisms must extract energy from food molecules, and the reactions of glycolysis occur in virtually all of Earth's species. That means that similar enzyme sequences enable similar metabolic pathways to occur. Although virtually all living organisms on Earth use the same code and the same enzymes during replication, transcription, and translation, the particular array of proteins being formed differs from individual to individual even of the same species, according to the sequence of nitrogenous bases that make up the individual's chromosome(s). Therein lies the key to the enormous diversity of life on Earth: no two individuals have the exact same array of proteins in their phenotypes;   23. If you observe the sugar–phosphate sides of the DNA "ladder," you see that one strand runs from 5' to 3' and the other runs from 3' to 5'. The numerals 3' and 5' are used to identify specific carbon atoms in each deoxyribose molecule.

13.4. DNA REPLICATION AND REPAIR [pp.220–221]
13.5. *Focus on Science:* **DOLLY, DAISIES, AND DNA** [p.222]
1.

T – A		T – A
G – C		G – C
A – T		A – T
C – G		C – G
C – G		C – G
C – G		C – G

2. F, adenine bonds to thymine (during replication) or uracil (during transcription);   3. F, semiconservative;   4. T;   5. F, polymerases;   6. F, repair.

Self-Quiz
1. d;   2. d;   3. a;   4. d;   5. b;   6. d;   7. c;   8. a;   9. d;   10. d.

Chapter 14   From DNA to Proteins

Beyond Byssus [pp.224–225]

14.1. *Focus on Science:* **CONNECTING WITH PROTEINS** [pp.226–227]
14.2. HOW IS DNA TRANSCRIBED INTO RNA? [pp.228–229]
1. sequence;   2. gene;   3. transcription (translation);   4. translation (transcription);   5. transcription;   6. translation;   7. protein;   8. folded;   9. structural (functional);   10. functional (structural);   11. metabolic;   12. defective;   13. mutation;   14. enzyme;   15. genes;   16. a. ribosomal RNA; rRNA; RNA molecule that associates with certain proteins to form the ribosome, the "workbench" on which polypeptide chains are assembled;   b. messenger RNA; mRNA; RNA molecule that moves to the cytoplasm, complexes with the tRNA;   c. transfer RNA; tRNA; RNA molecule that moves into the cytoplasm, picks up a specific amino acid, and moves it to the ribosome where tRNA pairs with a specific mRNA code word for that amino acid;   17. RNA molecules are single-stranded, whereas DNA has two strands; uracil substitutes in RNA molecules for thymine in DNA molecules; ribose sugar is found in RNA, while DNA has deoxyribose sugar;   18. Both DNA replication and transcription

follow base-pairing rules; nucleotides are added to a growing RNA strand one at a time as in DNA replication; 19. Only one region of a DNA strand serves as a template for transcription; transcription requires different enzymes (three types of RNA polymerase); the results of transcription are single-stranded RNA molecules, but replication results in DNA, a double-stranded molecule; 20. C; 21. B; 22. E; 23. A; 24. D; 25. A-U-G-U-U-C-U-A-U-U-G-U-A-A-U-A-A-A-G-G-A-U-G-G-C-A-G-U-A-G; 26. DNA (E); 27. introns (B); 28. cap (F); 29. exons (A); 30. tail (D); 31. mature mRNA transcript (C).

14.3. DECIPERING THE mRNA TRANSCRIPTS
[pp.230–231]
14.4. HOW IS mRNA TRANSLATED? [pp.232–233]
1. F; 2. B; 3. G; 4. H; 5. C; 6. A; 7. E; 8. D; 9. a. initiation; b. chain elongation; c. chain termination; 10. mRNA transcript; AUG UUC UAU UGU AAU AAA GGA UGG CAG UAG; 11. tRNA anticodons: UAC AAG AUA ACA UUA UUU CCU ACC GUC AUC; 12. amino acids: (start) met phe tyr cys asn lys gly try gln stop; 13. amino acids; 14. three; 15. one; 16. mRNA;

17. codon; 18. mRNA; 19. assembly (synthesis); 20. Transfer; 21. amino acid; 22. protein (polypeptide); 23. codon; 24. anticodon; 25. initiation; 26. elongation; 27. termination; 28. release; 29. enzyme.

14.5. DO MUTATIONS AFFECT PROTEIN SYNTHESIS? [pp.234–235]
14.6. *Focus on Health:* MUTAGENIC RADIATION
[p.236]
1. mutagens; 2. substitution; 3. amino acid; 4. one; 5. frameshift; 6. transposable; 7. mutation; 8. mutagens; 9. ultraviolet; 10. pyrimidine; 11. Skin; 12. free; 13. alkylating; 14. mutation; 15. carcinogens; 16. protein; 17. evolutionary; 18. DNA (H); 19. transcription (J); 20. intron (E); 21. exon (A); 22. mature mRNA transcript (L); 23. tRNAs (C); 24. rRNA subunits (G); 25. mRNA (B); 26. anticodon (K); 27. amino acids (D); 28. tRNA (F); 29. ribosome–mRNA complex (I); 30. polypeptide (M).

Self-Quiz
1. c; 2. c; 3. b; 4. c; 5. a; 6. b; 7. a; 8. a; 9. d; 10. d.

Chapter 15   Controls Over Genes

When DNA Can't Be Fixed [pp.240–241]

15.1. OVERVIEW OF GENE CONTROL [p.242]
15.2. CONTROL IN BACTERIAL CELLS [pp.242–243]
1. a. Blocks transcription by preventing (RNA polymerases) from binding to DNA; this is negative transcription control; b. Activator proteins; c. Specific base sequences on DNA that serve as binding sites for control agents; before RNA assembly can occur on DNA, RNA polymerases must bind with the promoter site; d. Operators; 2. In any organism, gene controls operate in response to chemical changes within the cell or its surroundings; 3. transcription controls; 4. regulator; 5. promoter; 6. negative control; 7. low; 8. RNA polymerase (mRNA transcription); 9. blocks; 10. repressor protein; 11. operator; 12. needed (required); 13. regulator gene (K); 14. enyzme genes (G); 15. repressor protein (E); 16. promoter (J); 17. operator (B); 18. lactose operon (A); 19. repressor–operator complex (I); 20. RNA polymerase (D); 21. repressor–lactose complex (F); 22. lactose (C); 23. mRNA transcript (L); 24. lactose enzymes (H).

15.3. CONTROL IN EUKARYOTIC CELLS
[pp.244–245]
15.4. EVIDENCE OF GENE CONTROL [pp.246–247]
1. All cells in the body descend from the same zygote; as cells divide to form the body, they become specialized in composition, structure, and function—they differentiate through selective gene expression; 2. pretranscriptional control (D); 3. transcriptional control (E); 4. transcript processing control (B); 5. translational control (A); 6. post-translational control (C); 7. DNA (genes); 8. cell differentiation; 9. selective; 10. controls; 11. regulatory; 12. activators; 13. Transcription; 14. Barr; 15. mosaic; 16. anhidrotic ectodermal dysplasia; 17. selective; 18. F; 19. T.

15.5. EXAMPLES OF SIGNALING MECHANISMS
[pp.248–249]
15.6. *Focus on Science:* LOST CONTROLS AND CANCER [pp.250–251]
1. signals (molecules); 2. hormones; 3. receptors; 4. enhancer (hormone); 5. ecdysone; 6. amplification; 7. polytene; 8. prolactin; 9. receptors; 10. phytochrome; 11. cancer; 12. tumor; 13. metastasis; 14. T; 15. increase, high, start; 16. T; 17. T; 18. F, bring about; 19. T; 20. F, abnormal.

Self-Quiz
1. d; 2. b; 3. d; 4. b; 5. b; 6. c; 7. c; 8. a; 9. b; 10. c.

Chapter 16 Recombinant DNA and Genetic Engineering

Mom, Dad, and Clogged Arteries [pp.254–255]

16.1. A TOOLKIT FOR MAKING RECOMBINANT DNA [pp.256–257]

1. The bacterial chromosome, a circular DNA molecule, contains all the genes necessary for normal growth and development. Plasmids, small, circular molecules of "extra" DNA, carry only a few genes and are self-replicating; 2. F, small circles of DNA in bacteria; 3. T; 4. mutations; 5. recombinant DNA; 6. species; 7. amplify; 8. protein; 9. research; 10. Genetic engineering; 11. T; 12. F; 13. F; 14. T; 15. F; 16. B; 17. F; 18. E; 19. C; 20. A; 21. D; 22. protein; 23. engineered (changed, altered); 24. introns; 25. transcriptase; 26. mRNA; 27. cDNA; 28. Enzyme; 29. DNA; 30. cDNA; 31. transcript.

16.2. PCR—A FASTER WAY TO AMPLIFY DNA [p.258]
16.3. *Focus on Science:* DNA FINGERPRINTS [p.259]
16.4. HOW IS DNA SEQUENCED? [p.260]
16.5. FROM HAYSTACKS TO NEEDLES—ISOLATING GENES OF INTEREST [p.261]

1. a. A process that determines the order of nucleotides in a cloned or amplified DNA fragment; b. Short, synthesized sequences of nucleotides that base-pair with any complementary sequences in DNA; recognized as START tags by DNA polymerases; c. A collection of DNA fragments produced by restriction enzymes and incorporated into plasmids; d. An electrical field forces molecules (generally DNA or proteins) to move through a viscous medium and separate from each other according to their different physical and chemical properties; e. A method for amplifying DNA fragments in a test tube (see Fig. 16.5); f. Several procedures allow researchers to determine the nucleotide sequence of a DNA fragment (see Fig. 16.6); g. A very short length of DNA labeled with a radioisotope so that it is distinguishable from other DNA molecules in a sample; h. A unique array of RFLPs; 2. B; 3. D; 4. E; 5. C; 6. A; 7. F.

16.6. USING THE GENETIC SCRIPTS [p.262]
16.7. DESIGNER PLANTS [pp.262–263]
16.8. GENE TRANSFERS IN ANIMALS [pp.264–265]
16.9. *Focus on Bioethics:* WHO GETS ENHANCED? [p.265]
16.10. SAFETY ISSUES [p.266]

1. a. *E. coli* that produce human insulin, hemoglobin, interferon, and blood-clotting factors; b. We now have bacterial factories that produce various substances—even plastic. Also oils and textile fibers; c. Herbicide-resistant cotton plants. Tobacco plants that produce hemoglobin; d. Bacteria that break down crude oil into less toxic compounds. Bacteria that absorb excess phosphates or heavy metals; 2. The more closely related two species are, the greater the extent of nucleic acid hybridization and the more similar are their metabolic pathways; 3. If we learn about the genes, we may understand better the nature of a pathogen's attack strategy and be able to have advance warning about its likely plan of attack; 4. In 1970, a new strain of the fungus that causes Southern corn leaf blight destroyed most of the corn crop in the United States. Because corn plants were genetically similar, most were killed. If different kinds had been planted, more of the crop would more likely have survived; 5. Geneticists splice the genes into the Ti plasmid from *Agrobacterium tumefaciens,* which infects many species of flowering plants. Sometimes electric shocks or chemicals can deliver modified genes into plant cells. Some researchers blast microscopic particles coated with DNA into the plant cells. 6. Human serum albumin helps control blood pressure. It would be much easier to obtain large quantities of the protein from abundant supplies of milk instead of having to separate it from large quantities of donated human blood; 7. Researchers are working to sequence the estimated 3 billion nucleotides present in human chromosomes; 8. If human collagen can be produced, it may be used to correct various skin, cartilage, and bone disorders; 9. The plasmid of *Agrobacterium* can be used as a vector to introduce desired genes into cultured plant cells; *Agrobacterium* was used to deliver a firefly gene into cultured tobacco plant cells; 10. Certain cotton plants have been genetically engineered for resistance to worm attacks; 11. In separate experiments the rat and human somatotropin genes became integrated into the mouse DNA. The mice grew much larger than their normal littermates; 12. Bacteria that have specific proteins on their cell surfaces facilitate ice crystal formation on whatever substrate the bacteria are located; bacteria without the ability to synthesize those proteins ("ice-minus") have been genetically engineered and were sprayed on strawberry plants. Nothing bad happened; 13. In 1996, researchers produced a genetic duplicate of an adult ewe by fusing a reprogrammed mammary gland cell with an egg from which the nucleus was removed. When the fused cell was implanted into a surrogate mother, Dolly was born. Her existence means that genetically engineered clones of domestic animals may supply reliable quantities of specific proteins for research and for medical uses; 14. ice-minus; 15. body cells; 16. gene therapy; 17. eugenic engineering.

Self-Quiz

1. a; 2. b; 3. c; 4. d; 5. b; 6. a; 7. c; 8. d; 9. d; 10. a.

Chapter 17   Emergence of Evolutionary Thought

Legends of the Flood [pp.270–271]

17.1. EARLY BELIEFS, CONFOUNDING DISCOVERIES [pp.272–273]
1. G; 2. H; 3. J; 4. I; 5. D; 6. E; 7. F; 8. C; 9. B;
10. A.

17.2. A FLURRY OF NEW THEORIES [pp.274–275]
17.3. DARWIN'S THEORY TAKES FORM [pp.276–277]
17.4. *Focus on Science:* **REGARDING THE FIRST OF THE "MISSING LINKS"** [p.278]
1. b; 2. a; 3. b; 4. a; 5. a; 6. b; 7. a; 8. b; 9. a; 10. b;
11. b; 12. b; 13. a. John Henslow; b. Cambridge University; c. H.M.S. *Beagle*; d. Charles Lyell; e. Thomas Malthus; f. Galápagos Islands; g. Natural selection; h. Alfred Wallace; i. *Archaeopteryx*; 14. F(D); 15. D(F); 16. B; 17. E; 18. A; 19. C; 20. G.

Self-Quiz
1. c; 2. d; 3. d; 4. b; 5. d; 6. c; 7. b; 8. a; 9. c; 10. e.

Chapter 18   Microevolution

Designer Dogs [pp.280–281]

18.1. INDIVIDUALS DON'T EVOLVE—POPULATIONS DO [pp.282–283]
18.2. *Focus on Science:* **WHEN IS A POPULATION NOT EVOLVING?** [pp.284–285]
1. M; 2. P; 3. M; 4. B; 5. P; 6. M/B; 7. B; 8. P; 9. B;
10. M; 11. D; 12. C; 13. E; 14. B; 15. A; 16. Phenotypic variation is caused by the effects of genes and the environment. The environment can affect how the genes governing traits are expressed in an individual. For example, ivy plants grown from cuttings of the same parent all have the same genes, but the amount of sunlight affects the genes governing leaf growth. Ivy plants growing in full sun have smaller leaves than those growing in full shade. Offspring inherit genes, not phenotype; 17. There are five conditions: no genes are undergoing mutation; a very, very large population; the population is isolated from other populations of the species; all members of the population survive and reproduce; and mating is random; 18. a. 0.64 *BB*, 0.16 *Bb*, 0.16 *Bb*, and 0.04 *bb*; b. genotypes: 0.64 *BB*, 0.32 *Bb*, and 0.04 *bb*; phenotypes: 96% black, 4% gray; c.

Parents	B sperm	b sperm
0.64 *BB*	0.64	0
0.32 *Bb*	0.16	0.16
0.04 *bb*	0	0.04
Totals =	0.80	0.20

19. Find (b) first, then (c), and finally (a). a. $2pq = 2 \times (0.9) \times (0.1) = 2 \times (0.09) = 0.18 = 18\%$, which is the percentage of heterozygotes; b. $p^2 = 0.81$, $p = \sqrt{0.81} = 0.9 =$ the frequency of the dominant allele; c. $p + q = 1$, $q = 1.00 - 0.9 = 0.1 =$ the frequency of the recessive allele; 20. a. homozygous dominant $= p^2 \times 200 = (0.8)^2 \times 200 = 0.64 \times 200 = 128$ individuals; b. $q = (1.00 - p) = 0.20$; homozygous recessive $= q^2 \times 200 = (0.2)^2 \times 200 = (0.04) \times (200) = 8$ individuals; c. heterozygotes $= 2pq \times 200 = 2 \times 0.8 \times 0.2 \times 200 = 0.32 \times 200 = 64$ individuals. Check: $128 + 8 + 64 = 200$; 21. If $p = 0.70$, since $p + q = 1$, $0.70 + q = 1$; then $q = 0.30$, or 30 percent; 22. If $p = 0.60$, since $p + q = 1$, $0.60 + q = 1$; then $q = 0.40$; thus, $2pq = 0.48$, or 48 percent; 23. D; 24. H; 25. I; 26. J; 27. G; 28. C; 29. A; 30. F; 31. B; 32. E; 33. population; 34. Hardy–Weinberg; 35. gene frequency; 36. genetic equilibrium; 37. mutations.

18.3. NATURAL SELECTION REVISITED [p.285]
18.4. DIRECTIONAL CHANGE IN THE RANGE OF VARIATION [pp.286–287]
18.5. SELECTION AGAINST OR IN FAVOR OF EXTREME PHENOTYPES [pp.288–289]
18.6. SPECIAL TYPES OF SELECTION [pp.290–291]
18.7. GENE FLOW [p.291]
18.8. GENETIC DRIFT [pp.292–293]
1. a. the intermediate phenotypes are favored; gall-making flies; b. allele frequencies shift in a steady, consistent direction in response to a new environment or a directional change in an old one; light to dark forms of peppered moths; c. forms at both ends of the phenotypic range are favored, and intermediate forms are selected against; finches on the Galápagos Islands; 2. a. directional; b. disruptive; c. stabilizing; 3. natural; 4. Stabilizing; 5. Directional; 6. Disruptive; 7. Gall-making flies; 8. Sickle-cell anemia; 9. balancing; 10. allele; 11. balanced polymorphism; 12. sexual dimorphism; 13. selection; 14. Sexual; 15. sexual; 16. fitness; 17. c, d; 18. e; 19. c; 20. b; 21. d; 22. b; 23. a; 24. c, d; 25. b; 26. b; 27. c; 28. e; 29. a; 30. e; 31. a; 32. b; 33. a; 34. b; 35. a; 36. a; 37. a; 38. b; 39. a; 40. b; 41. In the founder effect, a few individuals leave a population and establish a new one; by chance, allele frequencies will

differ from the original population. In bottlenecks, disease, starvation, or some other stressful situation nearly eliminates a population; relative allele frequencies are randomly changed; 42. genetic drift; 43. Gene; 44. Natural selection; 45. Natural selection.

1. b; 2. a; 3. d; 4. c; 5. a; 6. b; 7. c; 8. b; 9. d; 10. c; 11. a; 12. b.

Chapter 19   Speciation

The Case of the Road-Killed Snails [pp.296–297]

19.1. ON THE ROAD TO SPECIATION [pp.298–299]
19.2. REPRODUCTIVE ISOLATING MECHANISMS [pp.300–301]
1. c; 2. a; 3. b; 4. a; 5. b; 6. a; 7. c; 8. c; 9. a; 10. c; 11. C (pre); 12. A (pre); 13. B (post); 14. G (pre); 15. H (post); 16. E (pre); 17. D (pre); 18. F (post); 19. f; 20. c; 21. a; 22. e; 23. d; 24. b; 25. a. one; b. two; c. Considerable divergence begins between B and C; d. D; e. B and C.

19.3. SPECIATION IN GEOGRAPHICALLY ISOLATED POPULATIONS [pp.302–303]
19.4. MODELS FOR OTHER SPECIATION ROUTES [pp.304–305]
1. a. Sympatric; b. Allopatric; c. Parapatric; 2. sympatric; 3. parapatric; 4. allopatric; 5. allopatric; 6. allopatric; 7. a. Seven pairs; b. AA; c. Seven pairs; d. BB; e. The chromosomes fail to pair in meiosis; f. Fertilization of nonreduced gametes produced a fertile

tetraploid; g. Sterile hybrid, ABD; h. Fertilization of nonreduced gametes resulted in a fertile hexaploid; i. *Triticum monococcum*; the unknown wild wheat; *T. tauschii.*

19.5. PATTERNS OF SPECIATION [pp.306–307]
1. a. Horizontal branching; b. Many branchings of the same lineage at or near the same point in geologic time; c. A branch that ends before the present; d. Softly angled branching; e. Vertical continuation of a branch; f. A dashed line; 2. J; 3. D; 4. G; 5. I; 6. H; 7. E; 8. A; 9. F; 10. B; 11. C; 12. Speciation; 13. species; 14. Divergence; 15. reproductive isolating; 16. geographic; 17. polyploidy (hybridization); 18. hybridization (polyploidy).

1. c; 2. d; 3. d; 4. a; 5. c; 6. c; 7. b; 8. a; 9. c; 10. d; 11. e; 12. a; 13. c; 14. b.

Chapter 20   The Macroevolutionary Puzzle

He Sees Seashells . . . So Where's the Sea? [pp.310–311]

20.1. FOSSILS—EVIDENCE OF ANCIENT LIFE [pp.312–313]
20.2. EVIDENCE FROM COMPARATIVE MORPHOLOGY [pp.314–315]
20.3. EVIDENCE FROM PATTERNS OF DEVELOPMENT [pp.316–317]
20.4. EVIDENCE FROM COMPARATIVE BIOCHEMISTRY [pp.318–319]
1. F; 2. J; 3. I; 4. A; 5. E; 6. H; 7. C; 8. D; 9. B; 10. G; 11. D; 12. C; 13. E; 14. A; 15. B; 16. D; 17. E; 18. B; 19. F; 20. A; 21. G; 22. C; 23. Cenozoic; 24. Mesozoic; 25. Paleozoic; 26. Proterozoic; 27. Archean; 28. Cretaceous; 29. Jurassic; 30. Permian; 31. Cambrian; 32. 65; 33. 240; 34. 570; 35. 2,500; 36. 4,600; 37. T; 38. F, low; 39. F, convergence; 40. T; 41. F, differences; 42. T; 43. c; 44. a; 45. e; 46. b; 47. D; 48. D; 49. D; 50. B; 51. B; 52. B; 53. D; 54. B; 55. M; 56. M.

20.5. IDENTIFYING SPECIES, PAST AND PRESENT [pp.320–321]
20.6. FINDING EVOLUTIONARY RELATIONSHIPS AMONG SPECIES [pp.322–323]
20.7. *Focus on Science:* **CONSTRUCTING A CLADOGRAM** [pp.324–325]
20.8. HOW MANY KINGDOMS? [pp.326–327]
20.9. *Focus on Science:* **ALONG CAME DEEP GREEN** [p.327]
1. G; 2. C; 3. J; 4. H; 5. B; 6. A; 7. E; 8. D; 9. I; 10. F; 11. kingdom; phylum (or division); class; order; family; genus; species; 12. kingdom: Plantae; division: Anthophyta; class: Dicotyledonae; order: Asterales; family: Asteraceae; genus: *Archibaccharis*; species: *lineariloba*; 13. Systematics relies on (a) taxonomy, the naming and identification of organisms; (b) phylogenetic reconstruction, the identification of evolutionary patterns that unite different organisms; and (c) classification, the grouping of organisms using systems that consist of many hierarchical levels, or ranks that reflect evolutionary relationships; 14. a. Differences and similarities between

organisms are compared, in a relatively imprecise, subjective manner. Groups of organisms are defined by traditionally accepted morphological features, such as pollen size and morphology, leaf size, leaf shape, and flower structure. Such subjective methods may result in several different interpretations of the same evidence; b. Using cladistics, organisms are linked or grouped together that have shared characteristics (homologies); the similarities are derived from a common ancestor; 15. Monophyletic lineages are lineages that share a common evolutionary heritage; 16. The cladogram is a branching diagram (a type of family tree) that represents the patterns of relationships of organisms, based on their derived traits; 17. An outgroup is the cladogram taxon exhibiting the fewest derived characteristics; 18. A derived trait is one that is shared by members of a lineage but not present in their common ancestor; 19. Cladograms portray relative relationships among organisms. Distribution patterns of many homologous structures are examined. Taxa closer together on a cladogram share a more recent common ancestor than those that are farther apart. Cladograms do not convey direct information about ancestors and descendants; 20. a. The numbered traits could be discrete morphological, physiological, and behavioral traits; b. trait 8; c. trait 4; d. trait 5; e. Taxa that are closer together share a more recent ancestor than those that are farther apart; f. taxon A; g. A monophyletic taxon is one in which all members have the same derived traits and share a single node on the cladogram; h. The monophyletic taxa and their derived traits are BCDEFG (derived trait 2), CDEFG (derived trait 4), CD (derived trait 5), EFG (derived trait 8), EF (derived traits 9 and 10); 21. B; 22. E; 23. A; 24. D; 25. C.

20.10. EVIDENCE OF A CHANGING EARTH [pp.328–329]
20.11. *Focus on Science:* DATING PIECES OF THE MACROEVOLUTIONARY PUZZLE [p.330]

1. uniformity; 2. erosion; 3. sedimentary; 4. magnetic; 5. Pangea; 6. seafloor; 7. plate tectonics; 8. Volcanic; 9. evolution; 10. radioactive decay; 11. relative; 12. radiometric dating; 13. volcanic; 14. half-life; 15. 40; 16. 11,460.

Self-Quiz

1. d; 2. b; 3. a; 4. c; 5. b; 6. d; 7. e; 8. a; 9. a; 10. b; 11. c; 12. c; 13. d.

Chapter 21   The Origin and Evolution of Life

In the Beginning . . . [pp.334–335]

21.1. CONDITIONS ON THE EARLY EARTH [pp.336–337]
21.2. EMERGENCE OF THE FIRST LIVING CELLS [pp.338–339]

1. f; 2. e; 3. a; 4. c; 5. d; 6. b; 7. f; 8. c; 9. b; 10. a; 11. d; 12. f; 13. e; 14. c; 15. a; 16. b; 17. e; 18. c; 19. a; 20. b.

21.3. ORIGIN OF PROKARYOTIC AND EUKARYOTIC CELLS [pp.340–341]
21.4. *Focus on Science:* WHERE DID ORGANELLES COME FROM? [pp.342–343]

1. b; 2. b; 3. a; 4. b; 5. a; 6. a; 7. b; 8. a; 9. a; 10. b; 11. a; 12. b; 13. a; 14. a; 15. b; 16. organelles; 17. gene; 18. plasma; 19. channel; 20. ER; 21. genes; 22. plasmids; 23. prokaryotic; 24. endosymbiosis; 25. eukaryotes; 26. Electron; 27. oxygen; 28. bacteria; 29. ATP; 30. host; 31. incapable; 32. mitochondria; 33. ATP; 34. DNA; 35. chloroplasts; 36. host; 37. eubacteria; 38. protistans; 39. Archaebacterial; 40. Eukaryotes; 41. Eubacterial; 42. Eukaryote; 43. Animals; 44. Thermophiles; 45. Animals; 46. Fungi; 47. Mitochondria; 48. Chloroplasts; 49. Plants; 50. Paleozoic; 51. Mesozoic; 52. Fungi; 53. Plants.

21.5. LIFE IN THE PALEOZOIC ERA [pp.344–345]
21.6. LIFE IN THE MESOZOIC ERA [pp.346–347]
21.7. *Focus on Science:* HORRENDOUS END TO DOMINANCE [p.348]
21.8. LIFE IN THE CENOZOIC ERA [p.349]
21.9. SUMMARY OF EARTH AND LIFE HISTORY [pp.350–351]

1. G; 2. H; 3. E; 4. K; 5. D; 6. L; 7. F; 8. M; 9. A; 10. N; 11. C; 12. J; 13. B; 14. I; 15. G, H; 16. E; 17. D, F, K, L; 18. M, N; 19. A, B, C, J, N; 20. a. Mesozoic, 240–205; b. Archean, 4,600–3,800; c. Paleozoic, 435–360; d. Paleozoic, 550–500; e. Archean, 4,600–3,800; f. Mesozoic, 135–65; g. Paleozoic, 360–280; h. Proterozoic, 2,500–570; i. Archean, 3,800–2,500; j. Paleozoic, 290–240; k. Proterozoic, 700–500; l. Mesozoic, 181–65; m. Proterozoic, 2,500–570; n. Cenozoic, 65–1.65; o. Cenozoic, 1.65–present; p. Paleozoic, 440–435; q. Paleozoic, 435–360.

Self-Quiz

1. b; 2. c; 3. c; 4. b; 5. e; 6. d; 7. a; 8. d; 9. c; 10. b.

Crossword

```
1E    2G  Y  3M  N  O  S  P  4E  R  M      5P  E  R  6M  I  A  N      7G  L  O  8B  A  9L
N     O     O              R          M              R     R     A
10D I N  O  S   A  U  R      A      11P  R  O  T  O  C  E  L  L           O     U
O     D     S              R          C              I     R     A
S     W     12S  T  R  O  M  A  T  O  L  I  T  E  13C  Y  C  A  D     L     A
Y     A  14F              T          N        R              I     S
15M A N  T  L   E      16C  R  E  17A  C  E  O  U  S           I     I
B     A     O              R     R           S     18P 19A N  G  E  A
I        W  20P  A  L  E  O  Z  O  I  C  22M     T        N
O           E  21L        Z        H     E     23E     G        24B
S           R     U        O        E     S  25P  R  O  T  I  S  T  A  N
I        26E D  I  A  C  A  R  A  N     I     A     O     N     O     N
S           N        E     27C  E  N  O  Z  O  I  C        S     G
            G        N              O              P
28R        29A S  T  E  R  O  I  D     I     30T  H  E  O  R  Y
N           I                       C              R
31A R C  H  A  E  B  A  C  T  E  R  I  U  M        32I  M  P  A  C  T
```

Chapter 22   Bacteria and Viruses

The Unseen Multitudes [pp.354–355]

22.1. CHARACTERISTICS OF BACTERIA [pp.356–357]
22.2. BACTERIAL GROWTH AND REPRODUCTION [pp.358–359]
1. b; 2. d; 3. c; 4. e; 5. c; 6. prokaryotic; 7. plasmids; 8. wall; 9. capsule; 10. micrometers; 11. prokaryotic fission; 12. cocci; 13. bacilli; 14. spiralla; 15. positive.

22.3. BACTERIAL CLASSIFICATION [p.359]
22.4. MAJOR BACTERIAL GROUPS [p.360]
22.5. ARCHAEBACTERIA [pp.360–361]
22.6. EUBACTERIA [pp.362–363]
1. Archaebacteria; 2. peptidoglycan; 3. eubacteria; 4. cyanobacteria; 5. nitrogen-fixation; 6. sulfur; 7. nitrogen; 8. *Rhizobium*; 9. endospores; 10. temperature; 11. *Clostridium botulinum* (*C. tetani*); 12. *Clostridium tetani* (*C. botulinum*); 13. Lyme disease or Rocky Mountain spotted fever; 14. *Borrelia* (*Rickettsia*); 15. membrane receptors; 16. sunlight; 17. decomposers; 18. Actinomycetes; 19. *Streptomyces*; 20. K; 21. d, L; 22. c, C; 23. c, J; 24. a, G; 25. c, E; 26. a, A; 27. b, I; 28. c, H; 29. e, B; 30. c, K; 31. c, N; 32. c, D; 33. c, M; 34. a, F.

22.7. THE VIRUSES [pp.364–365]
22.8. VIRAL MULTIPLICATION CYCLES [pp.366–367]
22.9. *Focus on Health:* **INFECTIOUS PARTICLES TINIER THAN VIRUSES** [p.367]
22.10. *Focus on Health:* **THE NATURE OF INFECTIOUS DISEASES** [pp.368–369]
1. a. Nonliving, infectious agents, smaller than the smallest cells; require living cells to act as hosts for their replication; not acted on by antibiotics; b. The core can be DNA or RNA; the capsid can be protein and/or lipid; c. Bacteriophage viruses may use the lytic pathway, in which the virus quickly subdues the host cells and replicates itself and descendants are released as the cell undergoes lysis; or they may use a temperate pathway, in which viral genes remain inactive inside the host cell during a period of latency, which may be a long time, before activation and lysis; 2. There can be multiple answers (see Table 22.3 in text); a. Possible answers include Herpes simplex (a herpesvirus), rhinovirus (a picornavirus), and HIV (a retrovirus); b. Herpes simplex: DNA virus. Initial infection is a lytic cycle that causes Herpes (sores) on mucous membranes on mouth or genitals. Recurrent infections are lysogenic. Most cells are in nerves and skin. No immunity. No cure.

Rhinovirus: RNA virus. Causes the *common cold*. Host cells are generally mucus-producing cells of respiratory tract. Ebola virus: RNA virus. Kills 70–90 percent of its victims. No vaccine or treatment. The disease starts with high fever and flulike aches. Within a few days nausea, vomiting and diarrhea begin. Blood vessels are destroyed as viruses damage circulatory tissues. Blood seeps into the tissues surrounding blood vessels and out through all of the body's openings. Patients die of circulatory shock. HIV: RNA virus. Host cells are specific white blood cells. Lysogenic cycle has a latency period that may last longer than a year before host tests positive for HIV. As white blood cells are destroyed, the host's immune system is progressively destroyed (AIDS). No cure exists; 3. virus; 4. nucleic acid; 5. protein coat (viral capsid); 6. Viruses; 7. *Herpesvirus* (or *Varicella*); 8. viroids; 9. Retroviruses (HIV); 10. lysogenic; 11. Nanometers; 12. micrometers; 13. 86,000; 14. latency; 15. prions; 16. Rhinoviruses, RNA; 17. Retroviruses, RNA; 18. Herpesviruses, DNA; 19. C; 20. H; 21. E; 22. D; 23. K; 24. I; 25. G; 26. L; 27. F; 28. B; 29. A; 30. J.

Self-Quiz

1. d; 2. a; 3. A, C; 4. A; 5. A; 6. B, E; 7. B, F; 8. A, D; 9. A, D; 10. E; 11. F; 12. D; 13. B; 14. C; 15. A.

Crossword Puzzle

```
[1 B]     [2 P] H  O  T  O  A  U  T  O  G  R  A  P  H  S         [3 F] I  [4 S] S  I  O  N
  A        E                                                           P           [5 G]
  C       [6 P] A  T  H  O  G  E  N  S    [7 C] O  N  J  U  G  A  T  I  O  N              R
  I        T                                                 R                           A
  L        I            [8 A] N  T  I  B  I  O  T  I [9 C] [10 F] L  A  G  E  L  [11 L] U  M
[12 L] A   D                                          H     L                     Y     [15 C]
  U        O  [13 A]  [14 C] Y  A  N  O  B  A  C  T  E  R  I  A           T              H
  S        G   R                                      M                   I              E
       [16 G] L  Y  C  O  C  A  L  Y  X       [17 P] R  O  K  A  R  Y  O  T  I  C          M
          Y    H                                      H                                   O
[18 T]    C    A      [19 P] H  O  T  O  H  E  T  E  R  O  T  R  O  P  H  S                A
  H       A    E                                      T                                   A
  E       N   [20 B] A  C  T  E  R  I  O  P  H  A  G  E              [21 L]                U
  R       A    A                                      R     [22 P]        Y               T
  M   [23 M] I  C  R  O  O  R  G  A  N  I  S  M        O     L  [24 P] I  L  U  S          O
  O        T                                          T     A              O             T
[25 P] A  N  D  E  M  I  C     [26 E] N  D  O  S  P  O  R  E  S              G             R
  H       R                           O        M                           E             O
  I  [27 E] P  I  D  E  M  I  C  [28 H] A  L  O  P  H  I  L  E              N             P
  L       A                            H     D                             I             H
  E  [29 M] E  T  H  A  N  O  G  E  N  S              [30 C] O  C  C  U  S
```

Chapter 23   Protistans

Kingdom at the Crossroads [pp.372–373]

23.1. PARASITIC AND PREDATORY MOLDS [pp.374–375]

1. a. E; b. E; c. P; d. E; e. P; f. E; g. E; 2. a. A C, D, E; b. F, G, J, K; c. none represented among the choices; d. B; e. H, I, (K), L; f. O; g. B; h. H, I, L; i. P; j. A, E; k. F, G, K; l. M, N, Q; 3. Chytrids; 4. water molds (Oomycota); 5. brown; 6. slime molds; 7. fruiting bodies; 8. phagocytic predators; 9. fungi; 10. slugs (amoebas); 11. chitin; 12. chytrids; 13. parasitic; 14. mycelium; 15. decomposers; 16. enzymes; 17. saprobes; 18. parasites; 19. *Saprolegnia*; 20. *Plasmopara viticola* (downy mildew).

23.2. THE ANIMAL-LIKE PROTISTANS [p.376]
23.3. AMOEBOID PROTOZOANS [pp.376–377]
23.4. CILIATED PROTOZOANS [pp.378–379]
23.5. ANIMAL-LIKE FLAGELLATES [p.380]
23.6. SPOROZOANS [p.381]
23.7. *Focus on Health:* MALARIA AND THE NIGHT-FEEDING MOSQUITOES [p.382]

1. pseudopods; 2. Foraminiferans; 3. microtubules; 4. radiolarians; 5. freshwater; 6. contractile vacuoles; 7. gullet; 8. enzyme-filled vesicles (food vacuoles); 9. trypanosomes; 10. *Plasmodium*; 11. mosquito; 12. gametes; 13. C, E; 14. C, E, I; 15. C, E, K; 16. A, L; 17. D, H; 18. B, F; 19. B, F, G; 20. A; 21. D; 22. C; 23. B; 24. E.

23.8. REGARDING "THE ALGAE" [p.383]
23.9. EUGLENOIDS [p.383]
23.10. CHRYSOPHYTES AND DINOFLAGELLATES [pp.384–385]

23.11. RED ALGAE [p.386]
23.12. BROWN ALGAE [p.387]
23.13. GREEN ALGAE [pp.388–389]

1. chloroplasts; 2. eyespot; 3. heterotrophic; 4. algae; 5. Chrysophytes; 6. diatoms; 7. fucoxanthin; 8. silica (glass); 9. diatom shells; 10. filtering; 11. phytoplankton; 12. Dinoflagellates (*Gymnodinium*); 13. agar; 14. marine; 15. walls; 16. brown; 17. algin; 18. photosynthetic; 19. cellulose (pectins, polysaccharides); 20. starch; 21. a. chlorophyll *a*, phycobilins; b. cyanobacteria; c. agar, used as a moisture-preserving agent and culture medium; carrageenan is a stabilizer of emulsions; d. *Bonnemaisonia, Eucheuma, Porphyra*; e. chlorophyll *a*, c_1, and c_2, fucoxanthin and other carotenoids; f. chrysophytes; g. algin, used as a thickener, emulsifier and stabilizer of foods, cosmetics, medicines, paper and floor polish; also are sources of mineral salts and fertilizer; h. *Postelsia* (sea palm), *Sargassum, Laminaria, Macrocystis*; i. chlorophylls *a* and *b*; j. plants; k. chlorophytes form much of the phytoplankton base of many food webs that support humans; l. *Volvox, Ulva, Spirogyra, Chlamydomonas*; 22. zygote, F; 23. resistant zygote, B; 24. meiosis and germination, H; 25. asexual reproduction, C; 26. gamete production, E; 27. gametes meet, G; 28. cytoplasmic fusion, A; 29. fertilization, D; 30. (+): a, c, d, e, g; (–): b, f, h, i, j.

Self-Quiz

1. b; 2. c; 3. b; 4. b; 5. a; 6. a; 7. d; 8. c; 9. a; 10. b; 11. d; 12. A, E; 13. A, I; 14. A, B; 15. A, E, J; 16. A, D; 17. A, H; 18. A, F, G; 19. A, C; 20. B; 21. E; 22. D; 23. A; 24. C.

Chapter 24   Fungi

Ode to the Fungus Among Us [pp.392–393]

24.1. CHARACTERISTICS OF FUNGI [p.394]

1. Symbiosis; 2. mutualism; 3. lichen; 4. mycorrhizae; 5. decomposers; 6. D; 7. E; 8. C; 9. H; 10. G; 11. B; 12. A; 13. F.

24.2. CONSIDER THE CLUB FUNGI [pp.394–395]

1. C; 2. D; 3. B; 4. A; 5. E; 6. reproductive; 7. mycelium; 8. club; 9. Nuclear; 10. diploid; 11. Meiosis; 12. spores; 13. cytoplasmic; 14. dikaryotic.

24.3. SPORES AND MORE SPORES [pp.396–397]

1. hyphae; 2. gametangia; 3. zygospore; 4. spores; 5. mycelium; 6. E; 7. C; 8. D; 9. A; 10. F; 11. B.

24.4. THE SYMBIONTS REVISITED [pp.398–399]
24.5. *Focus on Science:* A LOOK AT THE UNLOVED FEW [p.400]

1. Symbiosis; 2. mutualism; 3. lichen; 4. mycobiont; 5. photobiont; 6. sac; 7. hypha; 8. cytoplasm; 9. mycobiont (photobiont); 10. photobiont (mycobiont); 11. layers; 12. hostile; 13. fungus; 14. photobiont; 15. photobiont's; 16. shelter; 17. parasite; 18. mycorrhizae; 19. efficiently; 20. ectomycorrhiza; 21. temperate; 22. club; 23. endomycorrhizae; 24. penetrate; 25. zygomycetes; 26. absorptive; 27. soil; 28. a. *Cryphonectria parasitica*; b. *Allomyces capsulatus*; c. *Epidermophyton floccosum*; d. *Venturia inequalis*; e. *Claviceps purpurea*.

Self-Quiz
1. *Polyporus* (C); 2. *Pilobolus* (B); 3. fly agaric mushroom (C); 4. cup fungus (A); 5. algae and fungi—lichen (E); 6. algae and fungi—lichen (E); 7. big laughing mush-

room (C); 8. *Penicillium* (A); 9. b; 10. d; 11. d; 12. c; 13. a; 14. a; 15. c; 16. c; 17. d; 18. a; 19. a and c; 20. b.

Chapter 25 Plants

Pioneers in a New World [pp.402–403]

25.1. EVOLUTIONARY TRENDS AMONG PLANTS
   [pp.404–405]
1. C; 2. D; 3. E; 4. A; 5. B; 6. a. Well-developed root systems; b. Well-developed shoot systems; c. Lignin production; d. Xylem and phloem; e. Cuticle production; f. Stomata; g. Gametophytes; h. Sporophytes; i. Spores; j. Heterospory; k. Pollen grains; l. Seeds.

25.2. BRYOPHYTES [pp.406–407]
1. air; 2. T; 3. rhizoids; 4. T; 5. mosses; 6. sporophytes; 7. T; 8. Peat; 9. water; 10. nonvascular; 11. gametophytes; 12. gametophytes; 13. sperms; 14. Fertilization; 15. zygote; 16. sporophyte; 17. Meiosis; 18. spores; 19. gametophytes.

25.3. EXISTING SEEDLESS VASCULAR PLANTS
   [pp.408–409]
25.4. *Focus on the Environment:* **ANCIENT CARBON**
   TREASURES [p.410]
1. c; 2. b; 3. e; 4. a; 5. d; 6. d; 7. e; 8. c; 9. a; 10. d; 11. e; 12. d; 13. c; 14. c; 15. e; 16. b; 17. d; 18. d; 19. b; 20. e; 21. rhizome; 22. sorus; 23. Meiosis; 24. spores; 25. spores; 26. gametophyte; 27. gametophyte; 28. sperm; 29. egg; 30. fertilization; 31. zygote; 32. embryo.

25.5. THE RISE OF THE SEED-BEARING PLANTS
   [p.411]
1. C; 2. E; 3. A; 4. B; 5. D.

25.6. GYMNOSPERMS—PLANTS WITH "NAKED"
   SEEDS [pp.412–413]
25.7. A CLOSER LOOK AT THE CONIFERS
   [pp.414–415]
1. b; 2. d; 3. a; 4. b; 5. c; 6. a; 7. e; 8. b; 9. e; 10. c; 11. d; 12. e; 13. sporophyte; 14. cones; 15. cones; 16. ovule; 17. meiosis; 18. gametophyte; 19. eggs; 20. meiosis; 21. pollen; 22. Pollination; 23. tube; 24. Sperm; 25. Fertilization; 26. embryo.

25.8. ANGIOSPERMS—THE FLOWERING, SEED-
   BEARING PLANTS [p.416]
25.9. VISUAL OVERVIEW OF FLOWERING PLANT
   LIFE CYCLES [p.417]
25.10. *Commentary:* **SEED PLANTS AND PEOPLE**
   [pp.418–419]
1. B; 2. C; 3. E; 4. F; 5. D; 6. A; 7. H; 8. G; 9. G; 10. F; 11. I; 12. B; 13. J; 14. E; 15. K; 16. A; 17. D; 18. H; 19. C.

Self-Quiz
1. a. gametophyte, none or simple vascular tissue, no; b. sporophyte, yes, no; c. sporophyte, yes, no; d. sporophyte, yes, no; e. sporophyte, yes, yes; f. sporophyte, yes, yes; 2. b; 3. c; 4. a; 5. c; 6. d; 7. b; 8. a; 9. d; 10. c; 11. c.

Chapter 26 Animals: The Invertebrates

Madeleine's Limbs [pp.422–423]

26.1. OVERVIEW OF THE ANIMAL KINGDOM
   [pp.424–425]
26.2. PUZZLES ABOUT ORIGINS [p.426]
26.3. SPONGES—SUCCESS IN SIMPLICITY
   [pp.426–427]
1. O; 2. I; 3. H; 4. E; 5. P; 6. A; 7. F; 8. K; 9. C; 10. J; 11. D; 12. G; 13. N; 14. M; 15. L; 16. B; 17. a. Placozoa; b. Sponges; c. Cnidaria; d. Turbellarians, flukes, tapeworms; e. Nematoda; f. Rotifera; g. Snails, slugs, clams, squids, octopuses; h. Annelida; i. Crustaceans, spiders, insects; j. Echinodermata; 18. c; 19. a;

20. c; 21. d; 22. c; 23. b; 24. c; 25. a; 26. c; 27. b; 28. H; 29. K; 30. D; 31. J; 32. B; 33. I; 34. C; 35. E; 36. A; 37. G; 38. F.

26.4. CNIDARIANS—TISSUES EMERGE [pp.428–429]
26.5. VARIATIONS ON THE CNIDARIAN BODY
   PLAN [pp.430–431]
26.6. COMB JELLIES [p.431]
1. Cnidaria; 2. nematocysts; 3. medusa; 4. polyp; 5. gastrodermis; 6. epidermis; 7. epithelium; 8. nerve; 9. contractile; 10. nerve; 11. mesoglea; 12. jellyfishes; 13. hydrostatic; 14. mesoglea; 15. both (two); 16. gonads; 17. planulas; 18. reef; 19. 20°C (68°F);

20. calcium; 21. skeletons; 22. nutrient; 23. Portuguese; 24. toxin; 25. Atlantic; 26. planula; 27. teams; 28. feeding polyp; 29. reproductive polyp; 30. female medusa; 31. planula; 32. multiple; 33. Ctenophora; 34. mesoderm; 35. mirror; 36. radial; 37. eight; 38. forward; 39. do not; 40. lips; 41. two.

26.7. ACOELOMATE ANIMALS—AND THE SIMPLEST ORGAN SYSTEMS [pp.432–433]

1. b, c; 2. a; 3. c; 4. a; 5. c; 6. c; 7. c; 8. d; 9. a; 10. a; 11. b; 12. d; 13. a; 14. c; 15. d; 16. a; 17. c; 18. c; 19. d; 20. c; 21. b; 22. c; 23. d; 24. d; 25. b; 26. branching gut; 27. pharynx (protruding); 28. brain; 29. nerve cord; 30. ovary; 31. testis; 32. planarian (genus name = *Dugesia*); 33. no; 34. yes; 35. no coelom or acoelomate.

26.8. ROUNDWORMS [p.434]
26.9. *Focus on Health:* A ROGUE'S GALLERY OF WORMS [pp.434–435]
26.10. ROTIFERS [p.436]

1. roundworm; 2. no; 3. a false coelom (pseudocoelom); 4. rotifer; 5. bilateral; 6. yes; 7. false coelom; 8. Rotifers have a crown of cilia at the head end that assists in swimming and in wafting food toward the mouth; its rhythmic motions reminded early microscopists of a turning wheel; 9. b; 10. a; 11. a; 12. a; 13. b; 14. a; 15. a; 16. b; 17. a; 18. a; 19. b; 20. b; 21. a; 22. b; 23. a; 24. b; 25. a; 26. a; 27. sexually; 28. larvae; 29. human; 30. larvae; 31. human; 32. Larvae; 33. intermediate; 34. human; 35. small intestine; 36. proglottids; 37. organs; 38. proglottids; 39. feces; 40. larval; 41. intermediate.

26.11. TWO MAJOR DIVERGENCES [p.436]
26.12. A SAMPLING OF MOLLUSCAN DIVERSITY [p.437]
26.13. EVOLUTIONARY EXPERIMENTS WITH MOLLUSCAN BODY PLANS [pp.438–439]

1. b; 2. b; 3. a; 4. b; 5. a; 6. b; 7. a; 8. a; 9. b; 10. a; 11. III; 12. I; 13. II; 14. mouth; 15. anus; 16. gill; 17. heart; 18. radula; 19. foot; 20. shell; 21. stomach; 22. mouth; 23. gill; 24. mantle; 25. muscle; 26. foot; 27. stomach; 28. internal shell; 29. mantle; 30. reproductive organ; 31. gill; 32. ink sac; 33. tentacle; 34. mollusk; 35. shell; 36. mantle; 37. gills; 38. foot; 39. radula; 40. eyes (tentacles); 41. tentacles (eyes); 42. d; 43. b; 44. b; 45. a; 46. d; 47. b; 48. c; 49. d; 50. b; 51. d; 52. d; 53. b; 54. d; 55. c; 56. b; 57. d; 58. c; 59. d; 60. b; 61. d; 62. b; 63. d; 64. c; 65. b.

26.14. ANNELIDS—SEGMENTS GALORE [pp.440–441]

1. F; 2. L; 3. B; 4. I; 5. D; 6. G; 7. N; 8. C; 9. K; 10. J; 11. A; 12. H; 13. E; 14. M; 15. coelom;

16. cuticle; 17. nerve cord; 18. seta; 19. nephridium; 20. body wall; 21. hearts; 22. vessels; 23. mouth; 24. brain; 25. nerve cord; 26. earthworm; 27. Annelida; 28. segmentation and a closed circulatory system; 29. yes; 30. bilateral; 31. yes.

26.15. ARTHROPODS—THE MOST SUCCESSFUL ORGANISMS ON EARTH [p.442]

1. e; 2. f; 3. a; 4. b; 5. f; 6. a; 7. f; 8. d; 9. b; 10. c; 11. e; 12. d; 13. a; 14. Arthropods, as a group, have the highest number of species, occupy the most habitats, and have very efficient defenses against predators and competitors, and the capacity to exploit the greatest amounts and kinds of foods.

26.16. A LOOK AT SPIDERS AND THEIR KIN [p.443]
26.17. A LOOK AT THE CRUSTACEANS [pp.444–445]
26.18. HOW MANY LEGS? [p.445]

1. E; 2. G; 3. F; 4. D; 5. C; 6. B; 7. A; 8. an exoskeleton; 9. lobsters and crabs; 10. similar; 11. Lobsters and crabs; 12. Barnacles; 13. Copepods; 14. barnacles; 15. barnacles; 16. molts; 17. millipedes; 18. centipedes; 19. Millipedes; 20. Centipedes; 21. cephalothorax; 22. abdomen; 23. swimmerets; 24. legs; 25. cheliped; 26. antennae; 27. lobster; 28. crustaceans; 29. exoskeleton and jointed legs; 30. yes; 31. bilateral; 32. yes; 33. poison gland; 34. brain; 35. heart; 36. spinnerets; 37. book lung; 38. chelicerates.

26.19. A LOOK AT INSECT DIVERSITY [pp.446–447]
26.20. *Focus on Health:* UNWELCOME ARTHROPODS [pp.448–449]

1. J; 2. F; 3. E; 4. B; 5. I; 6. A; 7. D; 8. G; 9. H; 10. C; 11. d; 12. b; 13. a; 14. c; 15. a; 16. b; 17. c; 18. b; 19. d; 20. b.

26.21. THE PUZZLING ECHINODERMS [pp.450–451]

1. deuterostomes; 2. echinoderms; 3. calcium carbonate; 4. skeleton; 5. radial; 6. bilateral; 7. brain; 8. nervous; 9. arm; 10. tube; 11. water; 12. ampulla; 13. muscle; 14. whole; 15. digesting; 16. anus; 17. lower stomach; 18. upper stomach; 19. anus; 20. gonad; 21. coelom; 22. digestive gland; 23. eyespot; 24. tube feet; 25. starfish; 26. deuterostome; 27. water vascular; 28. a. brittle star; b. sea urchin; c. feather star (crinoid); d. sea cucumber; 29. Echinodermata; 30. A water vascular system and a body wall with spines, spicules, or plates; 31. radial.

Self-Quiz

1. a; 2. c; 3. a; 4. d; 5. b; 6. c; 7. d; 8. c; 9. d; 10. a; 11. a; 12. b; 13. g, G; 14. b, DHJ; 15. f, BE; 16. J, F; 17. d, K; 18. k, N; 19. i, A; 20. e, CM; 21. a, I; 22. c, O; 23. h, L.

Chapter 27   Animals: The Vertebrates

Making Do (Rather Well) With What You've Got
[pp.454–455]

27.1. THE CHORDATE HERITAGE [p.456]
27.2. INVERTEBRATE CHORDATES [pp.456–457]
27.3. EVOLUTIONARY TRENDS AMONG THE VERTEBRATES [pp.458–459]
27.4. EXISTING JAWLESS FISHES [p.459]
27.5. EXISTING JAWED FISHES [pp.460–461]
1. traits (characteristics); 2. nerve cord; 3. pharynx;
4. notochord; 5. invertebrate; 6. vertebrates;
7. lancelets; 8. filter-feeding; 9. gill slits; 10. Tunicates (sea squirts); 11. tadpoles; 12. notochord; 13. metamorphosis; 14. larva; 15. mutation; 16. sex organs;
17. cephalochordates (lancelets); 18. vertebrae;
19. predators; 20. jaws; 21. brain; 22. paired;
23. fleshy; 24. lungs; 25. circulatory; 26. tunicate adult (sea squirt); 27. lancelet; 28. notochord; 29. pharynx with gill slits; 30. urochordates; 31. cephalochordates;
32. ancestral vertebrates; 33. pharynx with gill slits;
34. anus; 35. atrial opening; 36. oral opening;
37. pharyngeal gill slits; 38. anus; 39. notochord;
40. dorsal, tubular nerve cord; 41. early jawless fish (agnathan); 42. supporting structure; 43. gill slit;
44. placoderm; 45. jaws; 46. acorn worms; 47. tunicates; 48. lancelets; 49. lampreys, hagfishes; 50. early jawed fishes; 51. cartilaginous fishes 52. bony fishes;
53. Amphibia; 54. Reptilia; 55. Aves; 56. Mammalia;
57. sharks; 58. gill slits; 59. bony; 60. ray-finned;
61. C; 62. G; 63. J; 64. F; 65. M; 66. H; 67. N; 68. I;
69. B; 70. O; 71. D; 72. E; 73. E; 74. K; 75. L; 76. A;
77. all jawless; 78. filter feeders; 79. jaws; 80. retain the cartilage skeleton; 81. bone; 82. ray-finned fishes;
83. lobe-finned fishes; 84. branch ⑤; 85. branch ④;
86. Silurian; 87. about 375 million years ago.

27.6. AMPHIBIANS [pp.462–463]
27.7. THE RISE OF REPTILES [pp.464–465]
27.8. MAJOR GROUPS OF EXISTING REPTILES
[pp.466–467]
1. lungs; 2. fins; 3. brains; 4. balance; 5. two;
6. three; 7. atrium; 8. insects; 9. salamanders;
10. water; 11. reproduce; 12. toxins; 13. limb;
14. shelled (amniote); 15. turtles (tuataras); 16. tuataras (turtles); 17. internal; 18. Carboniferous; 19. turtles;
20. Carboniferous; 21. Triassic; 22. Cretaceous;
23. Birds; 24. four; 25. synapsid; 26. Carboniferous;
27. G; 28. E; 29. C; 30. B; 31. D; 32. A; 33. F; 34. A;
35. A; 36. C; 37. C; 38. D; 39. D; 40. E; 41. a. dry, scaly skin; b. four-chambered heart; c. feather development; d. hair development; e. loss of limbs.

27.9. BIRDS [pp.468–469]
27.10. THE RISE OF MAMMALS [pp.470–471]
27.11. A PORTFOLIO OF EXISTING MAMMALS
[pp.472–473]
1. embryo (notochord); 2. albumin; 3. yolk sac;
4. reptiles; 5. feathers; 6. sternum (breastbone); 7. air cavities; 8. amniote; 9. temperature; 10. keeled;
11. high; 12. oxygen, O_2; 13. 4; 14. hummingbird;
15. ostrich; 16. mammals; 17. hair; 18. platypus;
19. pouched (marsupials); 20. placental; 21. four;
22. hair; 23. dentition; 24. placenta; 25. synapsids;
26. therians; 27. dinosaurs; 28. convergent evolution.

27.12. EVOLUTIONARY TRENDS AMONG THE PRIMATES [pp.474–475]
27.13. FROM PRIMATES TO HOMINIDS [pp.476–477]
27.14. EMERGENCE OF EARLY HUMANS
[pp.478–479]
27.15. *Focus on Science:* **OUT OF AFRICA—ONCE, TWICE, OR . . .** [p.480]
1. mammals; 2. hair (mammary glands); 3. Orders;
4. Primates; 5. depth; 6. prehensile; 7. gripping;
8. smell; 9. brains; 10. culture; 11. handbones;
12. teeth; 13. 60; 14. rodents; 15. insects; 16. Miocene;
17. drier; 18. grasslands; 19. dryopiths; 20. Asia;
21. 6; 22. 4; 23. L; 24. N; 25. A; 26. F; 27. E; 28. H;
29. M; 30. I; 31. J; 32. C; 33. K; 34. D; 35. B; 36. G;
37. O; 38. b; 39. c; 40. c; 41. c; 42. c; 43. b; 44. d;
45. a; 46. a; 47. c; 48. Hominids; 49. five; 50. Australopiths; 51. brain; 52. manual dexterity; 53. *habilis;*
54. 2.4; 55. *Homo erectus;* 56. fire; 57. tool; 58. Neandertals; 59. *Homo erectus;* 60. biochemical; 61. immunological; 62. Africa; 63. 100,000; 64. 40,000; 65. *Homo sapiens;* 66. anthropoids; 67. hominoids; 68. hominids;
69. Gorilla; 70. Chimpanzee; 71. *Australopithecus afarensis* (Lucy); 72. *Homo erectus.*

Self-Quiz
1. d; 2. a; 3. d; 4. c; 5. c; 6. a; 7. b; 8. a; 9. a; 10. b;
11. d; 12. B; 13. H; 14. C; 15. D; 16. A; 17. E; 18. G;
19. F; 20. d, H; 21. b, B; 22. i, F; 23. g, D; 24. h, A;
25. c, E; 26. a, I; 27. e, G; 28. f, C; 29. 5. early amphibian, Amphibia; 30. Arctic fox, Mammalia; 31. soldier fish, Osteichthyes; 32. Ostracoderm, Agnatha; 33. owl, Aves; 34. tortoise, Reptilia; 35. shark, Chondrichthyes;
36. coelacanth, Osteichthyes (lobe-finned fish); 37. reef ray, Chondrichthyes; 38. tunicate, Urochordata;
39. lancelet, Cephalochordata.

Chapter 28   Biodiversity in Perspective

The Human Touch [pp.484–485]

28.1. ON MASS EXTINCTIONS AND SLOW RECOVERIES [pp.486–487]
28.2. *Focus on the Environment:* **CASE STUDY: THE ONCE AND FUTURE REEFS** [pp.488–489]
28.3. *Focus on Science:* **RACHEL'S WARNING** [pp.490–491]

1. biodiversity; 2. mass extinction; 3. adaptive radiations; 4. million; 5. dinosaurs (mosasaurs); 6. mosasaurs (dinosaurs); 7. K–T; 8. Cenozoic; 9. dodo; 10. 6th; 11. human; 12. 300; 13. overharvesting; 14. habitat loss; 15. deforestation; 16. 50 percent; 17. 85–90 percent; 18. 98 percent; 19. predators; 20. disease; 21. breeding; 22. islands; 23. lakes; 24. biogeography; 25. 10; 26. 69; 27. Exotic; 28. whales; 29. coral reefs; 30. bleaching; 31. Rachel Carson; 32. environmental.

28.4. CONSERVATION BIOLOGY [pp.492–493]
28.5. RECONCILING BIODIVERSITY WITH HUMAN DEMANDS [pp.494–495]

1. humanity; 2. 9; 3. conservation biology; 4. survey; 5. methods; 6. Hot spots; 7. pharmaceutical; 8. tropical rain forests; 9. ecoregion; 10. strip logging; 11. contours; 12. road; 13. saplings; 14. nutrients; 15. growth; 16. riparian; 17. floods; 18. runoffs; 19. endemic; 20. ungulates; 21. watering; 22. population growth.

Self-Quiz
1. a; 2. b; 3. c; 4. c; 5. e; 6. a; 7. b; 8. c; 9. d; 10. d.

Crossword Puzzle

```
¹P  R   I   M  ²A  T   E       ³H  O  ⁴M   I   N   O  ⁵I   D
                U                       A               N
⁶H      ⁷A   S  T   E   R   ⁸S      M       ⁹B       C
 O              T               A       M       ¹⁰I   R   I   S
¹¹M O   L   A   R               P       A       P       S
 I              R               I       L       E       O
 N      ¹²C  U   L   T   U   R   E       ¹³A  D   O   R   E
 I              O               N           A
¹⁴D R   Y   O   P   I   T   H   S   ¹⁵E      L   ¹⁶T
                I           ¹⁷C  A   N   I   N   E   S
¹⁸E R   E   C   T   U   S               R       S       E
                H               L       M       T
    ¹⁹P  L   A   S   T   I   C   I   T   Y               H
```

Chapter 29   Plant Tissues

Plants Versus the Volcano [pp.498–499]

29.1. OVERVIEW OF THE PLANT BODY [pp.500–501]
1. ground tissue (C); 2. vascular tissues (E); 3. dermal tissues (D); 4. shoot system (A); 5. root system (B); 6. E; 7. D; 8. B; 9. F; 10. A; 11. C; 12. radial; 13. tangential; 14. transverse or cross; 15. shoot apical meristem (D); 16. shoot transitional meristems (B); 17. root transitional meristems (E); 18. root apical meristem (A); 19. lateral meristems (C); 20. a. Epidermis; primary; b. Ground tissues; primary; c. Vascular tissues; primary; d. Vascular tissues; secondary; e. Periderm; secondary.

29.2. TYPES OF PLANT TISSUES [pp.502–503]
1. sclerenchyma; 2. sclerenchyma; 3. collenchyma; 4. sclerenchyma; 5. parenchyma; 6. sclerenchyma; 7. b; 8. c; 9. a; 10. a; 11. b; 12. b; 13. a; 14. c; 15. a;

16. a; 17. c; 18. pits (xylem); 19. cytoplasm (xylem);
20. tracheids (xylem); 21. vessel (xylem); 22. vessel
(xylem); 23. sieve (phloem); 24. companion (phloem);
25. sieve (phloem); 26. sieve (phloem); 27. a. Vessel
members and tracheids; no; conduct water and dissolved
minerals absorbed from soil, mechanical support.
b. Sieve-tube members and companion cells; yes; transport sugar and other solutes; 28. a. Primary plant body;
cutin in the cuticle layer over epidermal cells restricts
water loss and resists microbial attack; openings (stomata) permit water vapor and gases to enter and leave the
plant; b. Secondary plant body; replaces epidermis to
cover roots and stems; 29. a. One; in threes or multiples
thereof; usually parallel; one pore or furrow; distributed
throughout ground stem tissue; b. Two, in fours or fives
or multiples thereof; usually netlike; three pores or pores
with furrows; positioned in a ring in the stem.

29.3. PRIMARY STRUCTURE OF SHOOTS
[pp.504–505]
1. primordium; 2. apical meristem; 3. bud; 4. shoot;
5. leaf; 6. procambium; 7. protoderm; 8. procambium;
9. ground meristem; 10. epidermis; 11. cortex;
12. procambium; 13. pith; 14. primary xylem;
15. primary phloem; 16. epidermis; 17. ground tissue;
18. vascular bundle; 19. sclerenchyma cells; 20. air
space; 21. xylem vessel; 22. sieve-tube; 23. companion
cell; 24. epidermis; 25. cortex; 26. vascular bundle;
27. pith; 28. vessels; 29. meristematic cells; 30. sieve
tube; 31. fibers.

29.4. A CLOSER LOOK AT LEAVES [pp.506–507]
1. blade; 2. petiole (leaf stalk); 3. axillary bud; 4. node;
5. stem; 6. sheath; 7. dicot; 8. monocot; 9. palisade
mesophyll (D); 10. spongy mesophyll (B); 11. lower
epidermis (A); 12. stoma (C); 13. leaf vein (E).

29.5. PRIMARY STRUCTURE OF ROOTS [pp.508–509]
1. root hair (H); 2. endodermis (G); 3. pericycle (B);
4. epidermis (E); 5. cortex (D); 6. root apical meristem
(F); 7. root cap (A); 8. endodermis (G); 9. pericycle (B);
10. primary phloem (C); 11. primary; 12. lateral;
13. taproot; 14. lateral; 15. fibrous; 16. hairs; 17. vascular; 18. pericycle; 19. cortex; 20. pith; 21. air;
22. oxygen; 23. endodermis; 24. control; 25. pericycle;
26. lateral; 27. cortex.

29.6. ACCUMULATED SECONDARY GROWTH—
THE WOODY PLANTS [pp.510–511]
29.7. A LOOK AT WOOD AND BARK [pp.512–513]
1. C; 2. K; 3. N; 4. F; 5. E; 6. H; 7. G; 8. L; 9. B;
10. A; 11. J; 12. M; 13. D; 14. I; 15. O; 16. vessel (E);
17. early wood (F); 18. late wood (B); 19. bark (D);
20. one (C); 21. and 22. two and three (A); 23. vascular
cambium (G).

Self-Quiz
1. b; 2. a; 3. b; 4. c; 5. a; 6. d; 7. d; 8. d; 9. c; 10. c;
11. b; 12. b.

Chapter 30   Plant Nutrition and Transport

Flies for Dinner [pp.516–517]

30.1. SOIL AND ITS NUTRIENTS [pp.518–519]
1. F; 2. C; 3. I; 4. B; 5. H; 6. A; 7. J; 8. E; 9. G;
10. D; 11. hydrogen; 12. thirteen; 13. ionic; 14. hydrogen; 15. macronutrients; 16. micronutrients; 17. a. iron,
micronutrient; b. potassium, macronutrient; c. magnesium, macronutrient; d. chlorine, micronutrient;
e. manganese, micronutrient; f. molybdenum, micronutrient; g. nitrogen, macronutrient; h. sulfur, macronutrient; i. boron, micronutrient; j. zinc, micronutrient;
k. calcium, macronutrient; l. copper, micronutrient;
m. phosphorus, macronutrient.

30.2. HOW DO ROOTS ABSORB WATER AND
MINERAL IONS? [pp.520–521]
1. exodermis (B); 2. root hair (E); 3. epidermis (H);
4. vascular cylinder (J); 5. cortex (C); 6. endodermis (I);
7. cytoplasm (A); 8. water movement (G); 9. Casparian
strip (D); 10. endodermal cell wall (F); 11. C; 12. D;
13. A; 14. B; 15. E; 16. mutualism; 17. Nitrogen
"fixed" by bacteria; 18. root nodules; 19. scarce minerals that the fungus is better able to absorb; 20. obtaining
sugars and nitrogen-containing compounds.

30.3. HOW IS WATER TRANSPORTED THROUGH
PLANTS? [pp.522–523]
1. water; 2. xylem; 3. tracheids (vessels); 4. vessels
(tracheids); 5. transpiration; 6. hydrogen; 7. cohesion;
8. tension; 9. roots; 10. xylem.

30.4. HOW DO STEMS AND LEAVES CONSERVE
WATER? [pp.524–525]
1. 90 percent; 2. T; 3. T; 4. stomata; 5. opens; 6. T;
7. T; 8. decrease; 9. opens; 10. T; 11. night; 12. day.

30.5. HOW ARE ORGANIC COMPOUNDS
DISTRIBUTED THROUGH PLANTS? [pp.526–527]
1. photosynthesis; 2. starch; 3. proteins; 4. seeds;
5. fats; 6. Protein; 7. starches; 8. solutes; 9. sucrose;
10. sucrose; 11. D; 12. F; 13. B; 14. G; 15. A; 16. C;
17. E.

Self-Quiz
1. c; 2. e; 3. c; 4. d; 5. b; 6. d; 7. b; 8. a; 9. b; 10. c.

Chapter 31   Plant Reproduction

A Coevolutionary Tale [pp.530–531]

31.1. REPRODUCTIVE STRUCTURES OF FLOWERING PLANTS [pp.532–533]
31.2. *Focus on the Environment:* **POLLEN SETS ME SNEEZING** [p.533]
1. sporophyte (C); 2. flower (B); 3. meiosis (D); 4. gametophyte (F); 5. gametophyte (E); 6. fertilization (A); 7. sepal; 8. petal; 9. stamen; 10. filament; 11. anther; 12. carpel; 13. stigma; 14. style; 15. ovary; 16. ovule; 17. receptacle; 18. I; 19. C; 20. G; 21. F; 22. B; 23. E; 24. H; 25. A; 26. D.

31.3. A NEW GENERATION BEGINS [pp.534–535]
1. anther; 2. pollen sac; 3. microspore mother; 4. meiosis; 5. microspores; 6. pollen tube; 7. sperm-producing; 8. pollen; 9. stigma; 10. male gametophyte; 11. ovule; 12. integuments; 13. meiosis; 14. megaspores; 15. megaspore; 16. megaspore; 17. mitosis; 18. eight; 19. embryo sac; 20. female gametophyte; 21. two; 22. endosperm; 23. egg; 24. double fertilization; 25. endosperm; 26. embryo; 27. seed; 28. coat; 29. Chemical and molecular cues guide a pollen tube's growth through tissues of the style and the ovary, toward the egg chamber and sexual destiny; 30. The embryo sac is the site of double fertilization; 31. a. Fusion of one egg nucleus (n) with one sperm nucleus (n); the plant embryo ($2n$); eventually develops into a new sporophyte plant; b. Fusion of one sperm nucleus (n) with the endosperm mother cell ($2n$); endosperm tissues ($3n$); nourishes the embryo within the seed.

31.4. FROM ZYGOTE TO SEEDS AND FRUITS [pp.536–537]
1. a. "Seed leaves" that develop from two lobes of meristematic tissue of the embryo; b. Seeds are mature ovules; c. Integuments of the ovule harden into the seed coat; d. A fruit is a mature ovary; 2. nucleus; 3. vacuole; 4. zygote; 5. embryo; 6. embryo; 7. seed coat; 8. shoot tip; 9. cotyledons; 10. sporophyte; 11. endosperm; 12. root tip; 13. mature; 14. fruit; 15. mitotic; 16. sporophyte; 17. ovule; 18. fruit; 19. cotyledons; 20. two; 21. one; 22. endosperm; 23. germinates; 24. thin; 25. enzymes; 26. seedling; 27. ovule; 28. endosperm; 29. ovary; 30. coat; 31. seed; 32. ovule; 33. fruits; 34. Simple; 35. aggregate; 36. multiple; 37. accessory; 38. endocarp; 39. Mesocarp; 40. Exocarp; 41. pericarp; 42. e; 43. b; 44. d; 45. c; 46. f; 47. b; 48. c; 49. c.

31.5. DISPERSAL OF FRUITS AND SEEDS [p.538]
31.6. *Focus on Science:* **WHY SO MANY FLOWERS AND SO FEW FRUITS?** [p.539]
1. c; 2. c; 3. b; 4. a; 5. a; 6. c; 7. b; 8. b; 9. b; 10. a; 11. Suppose the presumed "excess" flowers are formed strictly to produce pollen for export to other plants. Pollen grains are small, and in terms of energy, are inexpensive to produce, compared to large, calorie-rich, seed-containing fruits. Thus, for a fairly small investment, a plant might reap a large reward in offspring that carry its genes.

31.7. ASEXUAL REPRODUCTION OF FLOWERING PLANTS [pp.540–541]
1. E; 2. G; 3. I; 4. F; 5. H; 6. D; 7. B; 8. C; 9. A; 10. D; 11. E; 12. A; 13. B; 14. C.

Self-Quiz
1. b; 2. c; 3. d; 4. c; 5. c; 6. d; 7. b; 8. c; 9. c; 10. d.

Chapter 32   Plant Growth and Development

Foolish Seedlings and Gorgeous Grapes [pp.544–545]

32.1. PATTERNS OF EARLY GROWTH AND DEVELOPMENT—AN OVERVIEW [pp.546–547]
1. embryo; 2. germination; 3. environmental; 4. imbibition; 5. ruptures; 6. aerobic; 7. meristematic; 8. root; 9. primary root; 10. germination; 11. heritable (genetic); 12. genes; 13. zygote; 14. genes; 15. cytoplasmic; 16. metabolic; 17. selective; 18. hormones; 19. Interactions; 20. cotyledons; 21. hypocotyl; 22. primary; 23. cotyledons; 24. foliage; 25. primary; 26. cotyledons; 27. coleoptile; 28. branch; 29. primary; 30. foliage; 31. stem; 32. adventitious; 33. branch; 34. primary; 35. prop; 36. foliage; 37. coleoptile.

32.2. HORMONAL EFFECTS ON PLANT GROWTH AND DEVELOPMENT [pp.548–549]
1. e; 2. c; 3. a; 4. d; 5. a; 6. e; 7. a; 8. b; 9. a; 10. c; 11. e; 12. c; 13. e; 14. d; 15. F; 16. H; 17. E; 18. J; 19. G; 20. C; 21. I; 22. B; 23. A; 24. K.

32.3. ADJUSTMENTS IN THE RATE AND DIRECTION OF GROWTH [pp.550–551]
1. a; 2. c; 3. d; 4. b; 5. d; 6. b; 7. a; 8. a; 9. c; 10. b.

32.4. BIOLOGICAL CLOCKS AND THEIR EFFECTS [pp.552–553]
1. D; 2. I; 3. H; 4. J; 5. B; 6. A; 7. G; 8. C; 9. F; 10. E; 11. Pr; 12. Pfr; 13. Pfr; 14. Pr; 15. response; 16. Photoperiodism; 17. Pfr; 18. enzymes; 19. longer; 20. shorter; 21. mature; 22. long-day; 23. short-day.

32.5. LIFE CYCLES END, AND TURN AGAIN [pp.554–555]
32.6. *Commentary:* **THE RISE AND FALL OF A GIANT** [p.556]
1. b; 2. e; 3. d; 4. b; 5. a; 6. a; 7. d; 8. c; 9. a; 10. a;
11. d; 12. a; 13. a.

Self-Quiz
1. b; 2. d; 3. c; 4. a; 5. d; 6. a; 7. d; 8. d; 9. d; 10. c.

Chapter 33   Tissues, Organ Systems, and Homeostasis

Meerkats, Humans, It's All the Same [pp.560–561]

33.1. EPITHELIAL TISSUE [pp.562–563]
1. C; 2. F; 3. B; 4. H; 5. G; 6. D; 7. A; 8. E; 9. Epithelial; 10. Simple epithelium; 11. Stratified epithelium; 12. Tight; 13. Adhering; 14. Gap; 15. Tight; 16. peptic; 17. Exocrine; 18. epithelial; 19. Endocrine; 20. hormones; 21. bloodstream.

33.2. CONNECTIVE TISSUE [pp.564–565]
33.3. MUSCLE TISSUE [p.566]
33.4. NERVOUS TISSUE [p.567]
33.5. *Focus on Science:* **FRONTIERS IN TISSUE RESEARCH** [p.567]
1. b; 2. c; 3. a; 4. c; 5. b; 6. a; 7. c; 8. a; 9. c; 10. b; 11. a. Adipose; b. Cartilage; c. Blood; d. Bone; 12. smooth; 13. cardiac; 14. smooth; 15. skeletal; 16. Skeletal; 17. smooth; 18. striped; 19. Skeletal; 20. move internal organs; 21. cardiac; 22. Nervous; 23. neurons; 24. Neuroglia; 25. neuron; 26. neurons; 27. a designer organ; 28. connective, D, 8, 10; 29; epithelial, G, 1, 5, 11; 30. muscle, I, 7, (10); 31. muscle, J, 4, (10); 32. connective, E, 6, 10, (13);

33. connective, B, 9; 34. epithelial, H, 1, 5, 11; 35. connective, K, 1, 14; 36. nervous, L, 2; 37. muscle, C, 12; 38. epithelial, F, 1, 5, 11; 39. connective, A, 3, 13.

33.6. ORGAN SYSTEMS [pp.568–569]
1. cranial; 2. spinal; 3. thoracic; 4. abdominal; 5. pelvic; 6. a. Mesoderm; b. Endoderm; c. Ectoderm; 7. superior; 8. distal; 9. proximal; 10. posterior; 11. transverse; 12. inferior; 13. anterior; 14. frontal; 15. circulatory (D); 16. respiratory (E); 17. urinary (= excretory) (A); 18. skeletal (C); 19. endocrine (J); 20. reproductive (G); 21. digestive (I); 22. muscular (H); 23. nervous (B); 24. integumentary (F); 25. lymphatic (K).

33.7. HOMEOSTASIS AND SYSTEMS CONTROL [pp.570–571]
1. D; 2. G; 3. B; 4. I; 5. E; 6. C; 7. F; 8. A; 9. H.

Self-Quiz
1. d; 2. b; 3. c; 4. c; 5. d; 6. c; 7. c; 8. d; 9. a; 10. d;
11. d; 12. a; 13. A; 14. G; 15. E; 16. J; 17. H; 18. I;
19. F; 20. B; 21. D; 22. C.

Chapter 34   Information Flow and the Neurons

TORNADO! [pp.574–575]

34.1. NEURONS—THE COMMUNICATION SPECIALISTS [pp.576–577]
34.2. A CLOSER LOOK AT ACTION POTENTIALS [pp.578–579]
1. neurons; 2. Neuroglial; 3. Sensory; 4. interneurons; 5. motor; 6. cell body; 7. Dendrites; 8. information (signals, stimuli); 9. axon; 10. cell body; 11. axon; 12. endings; 13. voltage differential; 14. resting membrane potential; 15. negative; 16. millivolts; 17. action potential (nerve impulse); 18. potassium; 19. Na; 20. Channel; 21. Transport; 22. potassium; 23. sodium; 24. 30; 25. 10; 26. channel; 27. gates; 28. sodium–potassium pumps; 29. trigger zone; 30. channel proteins; 31. gates; 32. sodium–potassium pump; 33. lipid bilayer; 34. localized; 35. Graded; 36. duration; 37. trigger zone; 38. disturbance; 39. all-or-nothing; 40. threshold; 41. action potential; 42. threshold; 43. resting membrane potential; 44. milliseconds; 45. millivolts.

34.3. CHEMICAL SYNAPSES [pp.580–581]
34.4. PATHS OF INFORMATION FLOW [pp.582–583]
34.5. *Focus of Health:* **SKEWED INFORMATION FLOW** [p.584]
1. chemical synapse; 2. neurotransmitters; 3. Acetylcholine (ACh); 4. Serotonin; 5. inhibitory; 6. Endorphins; 7. excitatory; 8. inhibitory; 9. postsynaptic; 10. Synaptic integration; 11. summed; 12. excitatory; 13. hyperpolarizing; 14. myelin; 15. unsheathed node; 16. 120; 17. axon; 18. myelin sheath; 19. blood vessels; 20. axons; 21. unsheathed (exposed) node; 22. Schwann cell (myelin sheath); 23. sensory; 24. inter; 25. motor; 26. receptor endings or dendrites; 27. peripheral axon; 28. cell body; 29. axon; 30. axon endings; 31. reflex; 32. stretch reflex; 33. muscle spindles; 34. spinal cord; 35. neurotransmitters; 36. contract; 37. acetylcholine; 38. H, F; 39. E; 40. K; 41. I; 42. J; 43. B; 44. C; 45. G; 46. A, D.

Self-Quiz
1. a; 2. a; 3. b; 4. c; 5. d; 6. a; 7. b; 8. d; 9. c; 10. a.

Chapter 35   Integration and Control: Nervous Systems

Why Crack the System? [pp.586–587]

35.1. INVERTEBRATE NERVOUS SYSTEMS
[pp.588–589]
1. nervous; 2. neurons; 3. body; 4. seas; 5. nervous;
6. radial; 7. nerve; 8. sensory; 9. reflex; 10. body;
11. bilateral; 12. midsagittal; 13. two; 14. nerve;
15. ganglia; 16. bilateral; 17. plexuses; 18. sensory;
19. selection; 20. Cephalization.

**35.2. VERTEBRATE NERVOUS SYSTEMS—AN
OVERVIEW** [pp.590–591]
1. a. Midbrain; b. Forebrain; c. Hindbrain; 2. B; 3. D;
4. G; 5. H; 6. I; 7. F; 8. C; 9. E; 10. A.

35.3. THE MAJOR EXPRESSWAYS [pp.592–593]
1. autonomic; 2. sympathetic; 3. parasympathetic;
4. midbrain; 5. medulla oblongata; 6. cervical; 7. thoracic; 8. lumbar; 9. sacral; 10. spinal cord; 11. ganglion; 12. nerve; 13. vertebra; 14. intervertebral disk;
15. meninges; 16. gray matter; 17. white matter; 18. c;
19. a; 20. b; 21. d; 22. a; 23. c; 24. d; 25. d; 26. b;
27. d; 28. b; 29. b; 30. a; 31. b; 32. d; 33. c; 34. d.

**35.4. FUNCTIONAL DIVISIONS OF THE
VERTEBRATE BRAIN** [pp.594–595]
1. neural; 2. brain; 3. adult; 4. gray; 5. sensory;
6. cerebellum; 7. gray; 8. interneurons; 9. reticular;
10. cerebral; 11. neural; 12. cerebrospinal; 13. blood–
brain; 14. narrow; 15. plasma; 16. D; 17. B; 18. F;
19. A; 20. E; 21. H; 22. G; 23. C.

35.5. A CLOSER LOOK AT THE HUMAN CEREBRUM
[pp.596–597]
35.6. *Focus on Science:* **SPERRY'S SPLIT-BRAIN
EXPERIMENTS** [p.598]
1. cerebrum; 2. hypothalamus; 3. thalamus; 4. pineal;
5. medulla oblongata; 6. pons; 7. cerebellum; 8. mid-

brain; 9. optic; 10. corpus callosum; 11. a. Limbic
system; b. Cerebral cortex, temporal lobe; c. Right
cerebral hemisphere; d. Cerebral cortex, frontal lobe;
e. Corpus callosum; f. Cerebral cortex, occipital lobe;
g. Left cerebral hemisphere; h. Cerebral cortex, parietal
lobe; i. Cerebral cortex; 12. Sperry demonstrated that
signals across the corpus callosum coordinate the functioning of the two cerebral hemispheres, each of which
had responded to visual signals from the opposite side of
the body.

35.7. MEMORY [p.599]
1. touch; 2. hearing; 3. vision; 4. hippocampus;
5. amygdala; 6. smell; 7. prefrontal cortex; 8. basal
ganglia; 9. thalamus and hypothalamus; 10. motor
cortex; 11. corpus striatum; 12. Memory; 13. Learning;
14. Short-term; 15. long-term; 16. sensory; 17. irrele-
vant; 18. skills; 19. long-term; 20. Skills; 21. input;
22. amygdala (hippocampus); 23. hippocampus (amygdala); 24. amygdala; 25. hippocampus; 26. fact;
27. basal; 28. long-term; 29. Skill; 30. sensory;
31. corpus striatum; 32. cerebellum; 33. Amnesia;
34. skills; 35. Parkinson's; 36. Alzheimer's; 37. hippocampus; 38. information.

35.8. STATES OF CONSCIOUSNESS [p.600]
35.9. *Focus on Health:* **DRUGGING THE BRAIN**
[pp.600–601]
35.10. *Focus on Science:* **THE NOT-QUITE-COMPLETE
TEEN BRAIN** [pp.602–603]
1. C; 2. E; 3. B; 4. F; 5. A; 6. D; 7. a; 8. a; 9. a;
10. c; 11. b; 12. c; 13. d; 14. d; 15. a.

Self-Quiz
1. a; 2. d; 3. e; 4. b; 5. d; 6. c; 7. d; 8. e; 9. c; 10. e.

Chapter 36   Sensory Reception

Different Strokes for Different Folks [pp.606–607]

**36.1. SENSORY RECEPTORS AND PATHWAYS—AN
OVERVIEW** [pp.608–609]
1. nerve pathways; 2. brain regions; 3. sensation;
4. perception; 5. sensory receptors; 6. stimulus;
7. Chemoreceptors; 8. Mechanoreceptors; 9. photoreceptors; 10. thermoreceptors; 11. action potentials;
12. brain; 13. stimulus intensity; 14. frequency;
15. number; 16. frequency; 17. sensory adaptation;
18. change; 19. Stretch; 20. length; 21. somatic sensations; 22. D; 23. C; 24. A; 25. B; 26. A; 27. E; 28. E;
29. B; 30. D; 31. B; 32. B; 33. A; 34. C; 35. B.

36.2. SOMATIC SENSATIONS [pp.610–611]
36.3. SENSES OF TASTE AND SMELL [p.612]
1. somatosensory cortex; 2. head; 3. ear; 4. cerebral
hemisphere; 5. mouth; 6. hand; 7. touch; 8. cold;
9. skin; 10. skeletal; 11. mechanoreceptors; 12. Free;
13. Pain; 14. nociceptors; 15. action potentials;
16. referred pain; 17. Mechanoreceptors; 18. skin;
19. free nerve endings (C); 20. Ruffini endings (A);
21. Meissner's corpuscle (D); 22. dermis; 23. epidermis;
24. Pacinian corpuscle (B); 25. Chemo; 26. sensory;
27. taste buds; 28. chemo; 29. nose; 30. five; 31. olfactory; 32. Chemo; 33. smell.

36.4. SENSE OF BALANCE [p.613]
36.5. SENSE OF HEARING [pp.614–615]
1. amplitude; 2. frequency; 3. higher; 4. mechanoreceptors; 5. middle; 6. cochlea; 7. organ of Corti; 8. semicircular canals; 9. middle ear bones (hammer, anvil, stirrup); 10. cochlea; 11. auditory nerve; 12. tympanic membrane/eardrum; 13. oval window; 14. basilar membrane; 15. tectorial membrane.

36.6. SENSE OF VISION [pp.616–617]
36.7. STRUCTURE AND FUNCTION OF VERTEBRATE EYES [pp.618–619]
36.8. DISORDERS OF THE HUMAN EYE [pp.620–621]

36.9. Focus on Science: CASE STUDY: FROM SIGNALING TO VISUAL PERCEPTION [pp.622–623]
1. photons; 2. Photoreception; 3. Vision; 4. ocelli; 5. Eyes; 6. cornea; 7. retina; 8. ommatidia; 9. focal point; 10. Visual accommodation; 11. Farsighted; 12. Cone; 13. fovea; 14. vitreous body; 15. cornea; 16. iris; 17. lens; 18. aqueous humor; 19. ciliary muscles; 20. retina; 21. fovea; 22. optic nerve; 23. blind spot/optic disk; 24. sclera.

Self-Quiz
1. d; 2. c; 3. d; 4. e; 5. d; 6. a; 7. c; 8. b; 9. b; 10. a.

Chapter 37 Endocrine Control

Hormone Jamboree [pp.626–627]

37.1. THE ENDOCRINE SYSTEM [pp.628–629]
1. E; 2. A; 3. B; 4. F; 5. C; 6. D; 7. a. 1, six releasing and inhibiting hormones; synthesizes ADH, oxytocin; b. 2, ACTH, TSH, FSH, LH, GSH; c. 2, stores and secretes two hypothalamic hormones, ADH and oxytocin; d. 3, sex hormones of opposite sex, cortisol, aldosterone; e. 3, epinephrine, norepinephrine; f. 4, estrogens, progesterone; g. 5, testosterone; h. 6, melatonin; i. 7, thyroxine and triiodothyronine; j. 8, parathyroid hormone (PTH); k. 9, thymosins; l. 10, insulin, glucagon, somatostatin.

37.2. SIGNALING MECHANISMS [pp.630–631]
1. a; 2. b; 3. a; 4. b; 5. b; 6. a; 7. b; 8. b; 9. b; 10. b.

37.3. THE HYPOTHALAMUS AND PITUITARY GLAND [pp.632–633]
1. A (H); 2. P (G); 3. A (A); 4. A (I); 5. A (D); 6. I (B); 7. P (E); 8. A (C); 9. A (F); 10. hypothalamus; 11. posterior; 12. anterior; 13. intermediate; 14. releasers; 15. inhibitors.

37.4. EXAMPLES OF ABNORMAL PITUITARY OUTPUT [p.634]
1. a. Gigantism; b. Pituitary dwarfism; c. Diabetes insipidus; d. Acromegaly.

37.5. SOURCES AND EFFECTS OF OTHER HORMONES [p.635]
1. a. D (c); b. I (e); c. F (h); d. A (g); e. H (k); f. E (d); g. C (a); h. K (j); i. B (i); j. J (f); k. G (b); l. F (l); m. A (m); n. B (o); o. C (n).

37.6. FEEDBACK CONTROL OF HORMONAL SECRETIONS [pp.636–637]
1. E; 2. F; 3. D; 4. H; 5. I; 6. J; 7. A; 8. G; 9. B; 10. C; 11. thyroid; 12. metabolic (metabolism);

13. iodine; 14. iodide; 15. TSH; 16. goiter; 17. Hypothyroidism; 18. hypothyroid; 19. Hyperthyroidism; 20. gonads; 21. hormones; 22. testes; 23. ovaries; 24. testosterone; 25. secondary; 26. gametes.

37.7. RESPONSES TO LOCAL CHEMICAL CHANGES [pp.638–639]
1. parathyroid; 2. PTH; 3. calcium; 4. calcium; 5. reabsorption; 6. D_3; 7. intestinal; 8. rickets; 9. a. Glucagon, Causes glycogen (a storage polysaccharide) and amino acids to be converted to glucose in the liver (glucagon raises the glucose level); b. Insulin, Stimulates glucose uptake by liver, muscle, and adipose cells; promotes synthesis of proteins and fats, and inhibits protein conversion to glucose (lowers the glucose level); c. Somatostatin, Helps control digestion; can block secretion of insulin and glucagon; 10. rises; 11. excessive; 12. energy; 13. ketones; 14. insulin; 15. Glucagon; 16. type 1 diabetes; 17. type 1 diabetes; 18. type 2 diabetes; 19. Type 2 diabetes.

37.8. HORMONES AND THE ENVIRONMENT [pp.640–641]
1. b; 2. a; 3. a; 4. a; 5. b; 6. a; 7. a; 8. a; 9. a; 10. b; 11. The thyroid gland. Preliminary evidence suggests that frog embryos raised in hot-spot water (with as many as twenty kinds of dissolved pesticides—and with high deformity rates) had few or no deformity symptoms when supplied with extra thyroid hormones.

Self-Quiz
1. a; 2. e; 3. d; 4. b; 5. e; 6. d; 7. b; 8. a; 9. c; 10. a; 11. N; 12. E; 13. Q; 14. K; 15. D; 16. F; 17. J; 18. A; 19. I; 20. B; 21. R; 22. G; 23. L; 24. C; 25. P; 26. O; 27. M; 28. H.

Chapter 38   Protection, Support, and Movement

Of Men, Women, and Polar Huskies [pp.644–645]

38.1. INTEGUMENTARY SYSTEM [pp.646–647]
38.2. A LOOK AT HUMAN SKIN [p.648]
38.3. *Focus on Health:* SUNLIGHT AND SKIN [p.649]
1. epidermis; 2. dermis; 3. hypodermis; 4. adipose;
5. hair; 6. sensory neuron; 7. sebaceous (oil) gland;
8. smooth muscle; 9. hair follicle; 10. sweat gland;
11. blood vessel; 12. epidermis; 13. dermis; 14. hypodermis; 15. a; 16. c; 17. d; 18. b; 19. c; 20. c; 21. e;
22. melanin; 23. Oil; 24. acne; 25. Hirsutism;
26. vitamin D; 27. calcium; 28. sunlight; 29. cholesterol; 30. intestine; 31. immune; 32. Langerhans;
33. cold sores; 34. Granstein.

38.4. TYPES OF SKELETONS [pp.650–651]
38.5. CHARACTERISTICS OF BONE [pp.652–653]
38.6. HUMAN SKELETAL SYSTEM [pp.654–655]
1. contract (relax); 2. relax (contract); 3. contractile;
4. opposition; 5. hydrostatic; 6. exoskeleton; 7. Bones;
8. muscles; 9. Haversian; 10. Red marrow; 11. calcium
(phosphate); 12. phosphate (calcium); 13. osteoblasts;
14. hormonal; 15. osteoporosis; 16. axial; 17. appendicular; 18. Synovial; 19. cartilage; 20. synovial;
21. osteoarthritis; 22. rheumatoid arthritis; 23. nutrient
canal; 24. contains yellow marrow; 25. compact bone;
26. spongy bone; 27. connective tissue covering (perios-
teum); 28. Haversian system; 29. blood vessel;
30. mineral deposits (calcium phosphate); 31. osteocyte
(bone cell); 32. cranium; 33. clavicle; 34. sternum;
35. scapula; 36. radius; 37. carpal bones; 38. femur;
39. tibia; 40. tarsal bones; 41. metatarsals.

38.7. SKELETAL-MUSCULAR SYSTEMS [pp.656–657]
38.8. A CLOSER LOOK AT MUSCLES [pp.658–659]
38.9. ENERGY FOR CONTRACTION [p.660]
38.10. CONTROL OF CONTRACTION [pp.660–661]
38.11. PROPERTIES OF WHOLE MUSCLES [p.662]
**38.12. *Focus on Science:* PEBBLES, FEATHERS, AND
    MUSCLE MANIA** [p.663]
1. D; 2. C; 3. A; 4. E; 5. B; 6. F; 7. triceps brachii;
8. pectoralis major; 9. external oblique; 10. rectus abdominis; 11. quadriceps femoris; 12. tibialis anterior;
13. gastrocnemius; 14. trapezius; 15. deltoid; 16. biceps
brachii; 17. biceps contracts; 18. triceps relaxes;
19. biceps relaxes; 20. triceps contracts; 21. cardiac;
22. skeletal; 23. bones; 24. tendons; 25. Skeletal;
26. bones; 27. joints; 28. biceps brachii; 29. triceps
brachii; 30. excitability; 31. neuron; 32. action potential; 33. motor; 34. myofibrils; 35. sarcomeres;

36. myosin (thick); 37. actin (thin); 38. sliding-filament;
39. myosin; 40. sarcomere; 41. ATP; 42. creatine phosphate; 43. Contraction (Shortening); 44. bone;
45. calcium; 46. sarcoplasmic reticulum; 47. actin;
48. active transport; 49. action potentials; 50. sarcoplasmic reticulum; 51. Its oxygen-requiring reactions provide most of the ATP needed for muscle contraction
during prolonged, moderate exercise; 52. It provides
the energy to make myosin filaments slide along actin
filaments; 53. These are used to clear the actin binding
sites of any obstacles to cross-bridge formation with
myosin heads; 54. This supplies phosphate to change
ADP → ATP, which powers muscle contraction for a
short time because creatine phosphate supplies are limited; 55. Stored in muscles and in the liver, glucose is
stored by the animal body in this form of starch; 56. An
anaerobic pathway in which glucose is broken down to
yield a small amount of ATP; this pathway operates
during intense exercise; 57. This supplies commands
(signals) to muscle cells to contract and relax; 58. Projections from the thick filaments of myosin that bind to
actin sites and form temporary cross-bridges. Making
and breaking these cross-bridge attachments cause
myosin filaments to be pulled to the center of a sarcomere; 59. The repetitive unit of muscle contraction. Many
sarcomeres constitute a myofibril. Many myofibrils constitute a muscle cell; 60. This is the endoplasmic reticulum of a muscle cell. It stores calcium ions and releases
them in response to incoming signals from motor neurons. It uses active transport to bring the calcium ions
back inside; 61. dephosphorylation of creatine phosphate; 62. glycolysis alone; 63. aerobic respiration;
64. ATP; 65. diameter; 66. motor; 67. motor unit;
68. muscle twitch; 69. twitch; 70. Tetanus; 71. mitochondria; 72. myofibrils; 73. muscle; 74. muscle cell
(muscle fiber); 75. myofibril; 76. sarcomere;
77. myosin filament; 78. actin filament; 79. axon of
motor neuron serving one motor unit; 80. motor neuron
endings; 81. axon of another motor neuron serving
another motor unit; 82. individual muscle cells;
83. time that stimulus is applied; 84. contraction phase;
85. relaxation phase; 86. time (seconds); 87. force;
88. Anabolic steroids; 89. muscle; 90. acne; 91. testes;
92. facial; 93. menstrual cycles; 94. clitoris; 95. aggres-
sion (mania).

Self-Quiz
1. e; 2. c; 3. c; 4. d; 5. b; 6. d; 7. a; 8. c; 9. d; 10. b;
11. e; 12. d; 13. a; 14. e; 15. b; 16. c; 17. b; 18. d;
19. a; 20. e.

Chapter 39   Circulation

Heartworks [pp.666–667]

39.1. CIRCULATORY SYSTEMS—AN OVERVIEW [pp.668–669]

1. nutrients (food); 2. wastes; 3. closed; 4. Blood; 5. heart; 6. interstitial fluid; 7. heart; 8. rapidly; 9. capillary; 10. solutes; 11. lymphatic; 12. digestive (respiratory); 13. respiratory (digestive); 14. respiratory; 15. urinary; 16. heart(s); 17. blood vessels; 18. hearts; 19. open circulatory system; blood is pumped into short tubes that open into spaces in the body's tissues, mingles with tissue fluids, then is reclaimed by open-ended tubes that lead back to the heart; 20. closed circulatory system; blood flow is confined within blood vessels that have continuously connected walls and is pumped by 5 pairs of "hearts."

39.2. CHARACTERISTICS OF BLOOD [pp.670–671]
39.3. *Focus on Health:* BLOOD DISORDERS [p.672]
39.4. BLOOD TRANSFUSION AND TYPING [pp.672–673]

1. a. Plasma proteins; b. Red blood cells; c. Neutrophils; d. Lymphocytes; e. Platelets; 2. connective; 3. pH; 4. iron; 5. cell count; 6. 50 to 60; 7. red bone marrow; 8. Stem cells; 9. Neutrophils; 10. Platelets; 11. nucleus; 12. four; 13. nucleus; 14. nine; 15. AB; 16. Agglutination (clumping); 17. stem (F); 18. red blood (E); 19. platelets (C); 20. neutrophils (A); 21. B (D); 22. T (D); 23. monocytes (A); 24. macrophages (A); 25. A; 26. I; 27. D; 28. G; 29. F; 30. H; 31. B; 32. C; 33. E; 34. E.

39.5. HUMAN CARDIOVASCULAR SYSTEM [pp.674–675]
39.6. THE HEART IS A LONELY PUMPER [pp.676–677]

1. pulmonary; 2. systemic; 3. oxygen; 4. oxygen; 5. heart; 6. atria; 7. ventricles; 8. systole; 9. diastole; 10. ventricles; 11. atrial; 12. jugular; 13. superior vena cava; 14. pulmonary; 15. hepatic; 16. renal; 17. inferior vena cava; 18. iliac; 19. femoral; 20. femoral;

21. iliac; 22. abdominal; 23. renal; 24. brachial; 25. coronary; 26. pulmonary; 27. carotid; 28. aorta; 29. left pulmonary veins; 30. left semilunar valve; 31. left ventricle; 32. inferior vena cava; 33. right atrioventricular valve; 34. right pulmonary artery; 35. superior vena cava.

39.7. BLOOD PRESSURE IN THE CARDIOVASCULAR SYSTEM [pp.678–679]
39.8. FROM CAPILLARY BEDS BACK TO THE HEART [pp.680–681]
39.9. *Focus on Health:* CARDIOVASCULAR DISORDERS [pp.682–683]

1. aorta; 2. Arteries; 3. ventricles; 4. pressure; 5. resistance; 6. elastic; 7. little; 8. does not drop much; 9. arterioles; 10. nervous; 11. Arterioles; 12. pressure; 13. medulla oblongata; 14. beat more slowly; 15. contract less forcefully; 16. vasodilation; 17. capillary; 18. endothelial; 19. capillary bed (diffusion zone); 20. interstitial; 21. diffusion; 22. bulk flow (blood pressure); 23. veins; 24. venules; 25. Veins; 26. venules; 27. valves; 28. 50–60; 29. vein; 30. artery; 31. arteriole; 32. capillary; 33. smooth muscle, elastic fibers; 34. valve; 35. F, pressure; 36. F, blood pressure cannot remain constant because it passes through various kinds of vessels that have varied structures; 37. stroke; 38. coronary occlusion; 39. Plaque; 40. low; 41. thrombus.

39.10. HEMOSTASIS [p.684]
39.11. LYMPHATIC SYSTEM [pp.684–685]

1. hemostasis; 2. platelet plug formation; 3. coagulation; 4. collagen; 5. insoluble; 6. d; 7. c; 8. e; 9. b; 10. a; 11. tonsils; 12. thymus; 13. thoracic; 14. spleen; 15. lymph node(s); 16. lymphocytes; 17. bone marrow; 18. Lymph; 19. fats; 20. small intestine.

Self-Quiz

1. d; 2. e; 3. c; 4. e; 5. e; 6. a; 7. b; 8. a; 9. a; 10. a; 11. I; 12. G; 13. M; 14. H; 15. B; 16. O; 17. F; 18. N; 19. E; 20. L; 21. A; 22. J; 23. D; 24. C; 25. K.

Chapter 40   Immunity

Russian Roulette, Immunological Style [pp.688–689]

40.1. THREE LINES OF DEFENSE [p.690]
40.2. COMPLEMENT PROTEINS [p.691]
40.3. INFLAMMATION [pp.692–693]

1. mucous; 2. Lysozyme; 3. Gastric; 4. bacterial; 5. phagocytic; 6. clotting; 7. Phagocytic (Macrophage); 8. complement system; 9. histamine; 10. capillaries;

11. nonspecific; 12. specific; 13. lymphocytes; 14. immune; 15. secrete histamine and prostaglandins that change permeability of blood vessels in damaged or irritated tissues; 16. attack parasitic worms by secreting corrosive enzymes; 17. the most abundant white blood cells; they quickly phagocytize bacteria and reduce them to molecules that can be used for other purposes; 18. slow, "big eaters"; engulf and digest foreign agents, and clean up dead and damaged tissues.

40.4. THE IMMUNE SYSTEM [pp.694–695]
40.5. LYMPHOCYTE BATTLEGROUNDS [p.696]
40.6. CELL-MEDIATED RESPONSES [pp.696–697]
40.7. ANTIBODY-MEDIATED RESPONSES
   [pp.698–699]

1. nonspecific; 2. immune system; 3. MHC marker; 4. nonself; 5. lymphocytes; 6. B cell; 7. T cell; 8. thymus; 9. viruses; 10. tumor; 11. identity; 12. antigen; 13. antigen-presenting; 14. effector; 15. helper T; 16. cytotoxic T; 17. cell-mediated; 18. antibodies; 19. antibody-mediated; 20. memory; 21. antigen-presenting cells (macrophages); 22. virgin helper T cells; 23. memory T cells; 24. intracellular; 25. virgin B cells; 26. effector B cells; 27. antibodies; 28. extracellular; 29. a; 30. d; 31. e; 32. b; 33. c; 34. natural killer; 35. perforins; 36. T cell; 37. cytotoxic T; 38. helper T; 39. antigen–MHC; 40. virgin B; 41. antibody; 42. effector B; 43. B cells; 44. primary immune response; 45. memory cells; 46. antigens; 47. IgA; 48. IgE; 49. IgG; 50. IgM.

40.8. Focus on Health: CANCER AND
   IMMUNOTHERAPY [p.699]
40.9. IMMUNE SPECIFICITY AND MEMORY
   [pp.700–701]

40.10. DEFENSES ENHANCED, MISDIRECTED, OR
   COMPROMISED [pp.702–703]
40.11. Focus on Health: AIDS—THE IMMUNE
   SYSTEM COMPROMISED [pp.704–705]

1. immunotherapy; 2. cancer; 3. monoclonal antibodies; 4. hybrid; 5. lymphocytes; 6. lymphokine (interleukin); 7. immunoglobulin; 8. Antibodies; 9. recombination; 10. antibodies; 11. antigen; 12. virgin B; 13. memory cells; 14. effector cells; 15. clonal selection hypothesis; 16. antigen; 17. immunization; 18. active; 19. primary immune response; 20. memory cells; 21. b, d, e, h, i, j; 22. d, e, g; 23. a, b, c, f, j; 24. g, i, (j); 25. g, i, (j); 26. b, d, f, i; 27. g, i, (j); 28. b, d, e, h, i, j; 29. b, d, e, h, i; 30. Allergy; 31. Autoimmune disease; 32. Myasthenia gravis; 33. Rheumatoid arthritis; 34. human immunodeficiency virus; 35. male homosexuals; 36. retrovirus; 37. reverse transcriptase; 38. body fluids; 39. helper T (antigen-presenting); 40. macrophages; 41. 20,800,000; 42. 33.6 million.

Self Quiz

1. d; 2. b; 3. b; 4. a; 5. a; 6. e; 7. e; 8. e; 9. a; 10. d; 11. H; 12. D; 13. E; 14. J; 15. B; 16. G; 17. C; 18. I; 19. F; 20. A.

Age Vaccination Is Administered

Disease	(a) At birth	(b) At 2 mos.	(c) 1–4 mos.	(d) 4 mos.	(e) 6 mos.	(f) 6–18 mos.	(g) 12–15 mos.	(h) 12–18 mos.	(i) 4–6 yrs.	(j) 11–12 yrs.
21. Diphtheria		√		√	√			√	√	√
22. *Hemophilus influenzae*				√	√		√			
23. Hepatitis B	√	√	√			√				√
24. Measles							√		√	(√)
25. Mumps							√		√	(√)
26. Polio		√		√		√			√	
27. Rubella							√		√	(√)
28. Tetanus		√		√	√			√	√	√
29. Whooping cough (pertussis)		√		√	√			√	√	

Chapter 41   Respiration

Conquering Chomolungma [pp.708–709]

41.1. THE NATURE OF RESPIRATION [p.710]
41.2. INVERTEBRATE RESPIRATION [p.711]
41.3. VERTEBRATE RESPIRATION [pp.712–713]
1. gill; 2. countercurrent flow; 3. tracheas; 4. 21;
5. aerobic metabolism; 6. O_2 (oxygen); 7. carbon dioxide (CO_2); 8. respiration; 9. pressure gradient;
10. high; 11. lowest; 12. high; 13. lower; 14. surface area; 15. partial pressure; 16. Hemoglobin; 17. lung;
18. airways (blood); 19. blood (airways); 20. water;
21. blood vessel in gill filament; 22. oxygen-poor blood;
23. oxygen-rich blood; 24. water; 25. blood; 26. partial pressure; 27. Diffusion; 28. carbon dioxide;
29. Hypoxia.

41.4. HUMAN RESPIRATORY SYSTEM [pp.714–715]
41.5. BREATHING—CYCLIC REVERSALS IN AIR PRESSURE GRADIENTS [pp.716–717]
1. diaphragm; 2. rib cage; 3. increases; 4. drops;
5. ventilating; 6. pleural sac; 7. larynx; 8. glottis;
9. bronchi; 10. bronchioles; 11. alveoli; 12. intercostal muscles; 13. diaphragm; 14. pharynx; 15. epiglottis;
16. voice box; 17. trachea; 18. bronchus; 19. bronchioles; 20. thoracic cavity; 21. abdominal cavity;

22. smooth muscle; 23. bronchiole; 24. alveolus, alveoli;
25. capillary; 26. a. alveoli; b. bronchial tree;
c. diaphragm; d. larynx; e. pharynx.

41.6. GAS EXCHANGE AND TRANSPORT [pp.718–719]
41.7. *Focus on Health:* WHEN THE LUNGS BREAK DOWN [pp.720–721]
41.8. RESPIRATION IN UNUSUAL ENVIRONMENTS [pp.722–723]
41.9. *Focus on Science:* RESPIRATION IN LEATHERBACK SEA TURTLES [pp.724–725]
1. oxygen; 2. aerobic respiration (electron phosphorylation); 3. carbon dioxide; 4. hemoglobin; 5. bicarbonate; 6. oxyhemoglobin (hemoglobin); 7. systemic (low-pressure); 8. partial pressure; 9. medulla oblongata; 10. Emphysema; 11. lung cancer; 12. N_2 (nitrogen gas); 13. joints; 14. decompression.

Self-Quiz
1. I; 2. F; 3. K; 4. L; 5. E; 6. H; 7. O; 8. G; 9. M;
10. B; 11. D; 12. A; 13. C; 14. N; 15. J; 16. c; 17. e;
18. a; 19. c; 20. a; 21. a; 22. a; 23. a; 24. a; 25. d.

Chapter 42   Digestion and Human Nutrition

Lose It—And It Finds Its Way Back [pp.728–729]

42.1. THE NATURE OF DIGESTIVE SYSTEMS [pp.730–731]
42.2. OVERVIEW OF THE HUMAN DIGESTIVE SYSTEM [p.732]
42.3. INTO THE MOUTH, DOWN THE TUBE [p.733]
1. Nutrition; 2. carbohydrates; 3. particles; 4. molecules; 5. absorbed; 6. circulatory; 7. respiratory;
8. carbon dioxide, CO_2; 9. urinary; 10. incomplete;
11. circulatory; 12. complete; 13. opening; 14. Motility;
15. secretion; 16. ruminants; 17. crown; 18. stomach;
19. small intestine; 20. anus; 21. accessory; 22. pancreas; 23. salivary amylase; 24. epiglottis; 25. esophagus; 26. salivary glands; 27. liver; 28. gallbladder;
29. pancreas; 30. anus; 31. large intestine; 32. small intestine; 33. stomach; 34. esophagus; 35. pharynx;
36. mouth (oral cavity).

42.4. DIGESTION IN THE STOMACH AND SMALL INTESTINE [pp.734–735]
42.5. ABSORPTION IN THE SMALL INTESTINE [pp.736–737]

42.6. DISPOSITION OF ABSORBED ORGANIC COMPOUNDS [p.738]
42.7. THE LARGE INTESTINE [p.739]
1. a. Mouth; b. Salivary glands; c. Stomach; d. Small intestine; e. Pancreas; f. Liver; g. Gallbladder;
h. Large intestine; i. Rectum; 2. starches; 3. salivary amylase; 4. disaccharide; 5. stomach; 6. small intestine; 7. amylase; 8. small intestine; 9. disaccharidases;
10. stomach; 11. pepsins; 12. small intestine; 13. pancreas; 14. amino acids; 15. Lipase; 16. small intestine;
17. fatty acid; 18. Bile; 19. gallbladder; 20. lipase;
21. small intestine; 22. small intestine; 23. Carboxypeptidase; 24. Pancreatic nucleases; 25. segmentation (peristalsis); 26. T; 27. cellular respiration; 28. T; 29. T;
30. small intestine; 31. constructing hormones, nucleotides, proteins, and enzymes; 32. monosaccharides, free fatty acids, and glycerol; 33. The three uses are (a) to construct components of cells and storage forms (such as glycogen) and specialized derivatives such as steroids and acetylcholine; (b) to convert to amino acids as needed; and (c) to serve as a source of energy.

42.8. HUMAN NUTRITIONAL REQUIREMENTS [pp.740–741]
42.9. VITAMINS AND MINERALS [pp.742–743]
42.10. *Focus on Science:* WEIGHTY QUESTIONS, TANTALIZING ANSWERS [pp.744–745]

1. a. 2,070; b. 2,900; c. 1,230; 2. 700; 3. Complex carbohydrates; 4. 40 to 60; 5. Phospholipids; 6. energy reserves; 7. 30; 8. essential fatty acids; 9. Proteins; 10. essential; 11. milk (soybeans, eggs, meats, wheat germ); 12. soybeans (milk, eggs, meats, wheat germ); 13. Vitamins; 14. Minerals; 15. Consulting Fig. 42.15 (men's column, 6'0") yields 178 pounds as his ideal weight. 195–178 = 17 pounds overweight; 16. Multiply 178 times 10 (see p. 726) to obtain 1780 kilocalories (the daily number of calories that *maintains* weight in the correct size range). The excess 17 pounds should be lost gradually by adopting an everyday exercise program that over many months would gradually eliminate the excess kilocalories that are stored mostly in the form of fat; The smallest range of serving sizes shown in Figure 42.14 will help keep the total caloric intake to about 1600 kcal; 17. a. 6 servings; b. bread, cereal, rice, pasta; 18. a. 2 servings; b. fruits; 19. a. 3 servings; b. vegetables; 20. a. 2 servings; b. milk, yogurt or cheese; 21. a. 2 servings; b. legume, nut, poultry, fish, or meats; 22. a. Scarcely any; b. added fats and simple sugars; 23.–39. Choose from Figure 42.14; 23. food pyramid; 24. carbohydrates; 25. bread; 26. 6–11; 27. vegetable; 28. 3–5; 29. fruit; 30. 2–4; 31. apples; 32. berries; 33. meat; 34. proteins; 35. 2–3; 36. amino acids; 37. milk; 38. 2–3; 39. 0.

Self-Quiz

1. b; 2. b; 3. a; 4. e; 5. d; 6. c; 7. d; 8. b; 9. b; 10. c; 11. D; 12. C; 13. H; 14. J; 15. B; 16. A; 17. I; 18. F; 19. E; 20. G.

Chapter 43   The Internal Environment

Tale of the Desert Rat [pp.748–749]

43.1. URINARY SYSTEM OF MAMMALS [pp.750–751]
1. metabolism; 2. urine; 3. respiratory surfaces; 4. sweating; 5. interstitial fluid; 6. metabolism; 7. ammonia; 8. urea; 9. uric acid; 10. kidney; 11. ureter; 12. urinary bladder; 13. urethra; 14. cortex; 15. medulla; 16. ureter; 17. kidneys; 18. nephrons; 19. bloodstream; 20. arterioles; 21. Bowman's capsule; 22. glomerular; 23. solutes; 24. glomerulus; 25. proximal tubule; 26. peritubular; 27. loop of Henle; 28. distal tubule; 29. collecting duct; 30. ureter; 31. urinary; 32. solutes; 33. extracellular; 34. glomerular capillaries; 35. proximal tubule; 36. Bowman's capsule; 37. distal tubule; 38. peritubular capillaries; 39. collecting duct; 40. loop of Henle.

43.2. URINE FORMATION [pp.752–753]
43.3. *Focus on Health:* WHEN KIDNEYS BREAK DOWN [p.754]
43.4. THE ACID–BASE BALANCE [p.754]
43.5. ON FISH, FROGS, AND KANGAROO RATS [p.755]
1. filtration; 2. tubular reabsorption; 3. tubular secretion; 4. glomerular; 5. nephron; 6. peritubular; 7. proximal tubule; 8. osmosis; 9. sodium; 10. dilute; 11. aldosterone; 12. sodium; 13. less; 14. ADH; 15. inhibited; 16. Kidney stones; 17. renal failure; 18. Hemodialysis; 19. T; 20. T; 21. Kidneys; 22. H^+; 23. 7.43; 24. Acids; 25. bases; 26. lowered; 27. H^+; 28. bicarbonate (HCO_3^-); 29. urinary; 30. loops of Henle; 31. water; 32. water; 33. solutes; 34. very dilute.

43.6. MAINTAINING THE BODY'S CORE TEMPERATURE [pp.756–757]
43.7. TEMPERATURE REGULATION IN MAMMALS [pp.758–759]
1. hypothalamus; 2. central (core); 3. pilomotor response; 4. Peripheral vasoconstriction; 5. hypothermia; 6. T; 7. T; 8. T; 9. D; 10. B; 11. A; 12. C; 13. G; 14. F; 15. E.

Self-Quiz
1. c; 2. e; 3. d; 4. b; 5. d; 6. b; 7. d; 8. e; 9. a; 10. b; 11. d.

¹U	R	²E	T	³H	R	A		⁴A	L	D	O	⁵S	T	E	R	O	N	E		⁶F	L	U	I	D
R		X		E				D				S								I				
I		T		N	⁷U	R	I	C				⁸C	A	P	S	U	L	E		L		⁹B		
N		R		L		E				¹⁰R	E	N	I	N		E		¹¹G	L	O	¹²M	E	R	U	L	U	S
E		A		E				¹⁰R	E	N	I	N				¹¹G	E	A	A				
¹³U	R	E	T	E	R						L		¹⁴K	I	D	N	E	Y	D	T	D			
R		L									O					U		I						¹⁵P
I		L		¹⁶T	U	B	U	L	E		N		¹⁷R	E	N	A	L		O		E		R
N		U						¹⁸C			L				N			O				
A	¹⁹L	O	O	P		²⁰A	M	M	O	N	I	A		²¹U	R	E	A					X
R		A						R								I								
Y	²²R	E	²³A	B	S	O	R	P	T	I	O	N				²⁴S	Y	S	T	E	M		
			D					E													A		
	²⁵N	E	P	H	R	O	N		²⁶E	X	C	R	E	T	I	O	N		²⁷D	I	S	T	A	L

Chapter 44   Principles of Reproduction and Development

From Frog to Frog and Other Mysteries [pp.762–763]

44.1. THE BEGINNING: REPRODUCTIVE MODES [pp.764–765]

44.2. STAGES OF DEVELOPMENT—AN OVERVIEW [pp.766–767]

1. asexual; 2. environmental; 3. variation; 4. reproductive timing; 5. Viviparous; 6. ovoviviparous; 7. oviparous; 8. Gamete formation; 9. egg; 10. fertilization; 11. zygote; 12. Cleavage; 13. blastomeres; 14. Gastrulation; 15. nervous; 16. gut; 17. skeleton; 18. embryonic; 19. Development; 20. F, rich in lipids and proteins; 21. T; 22. T; 23. T; 24. T; 25. F; 26. B; 27. C; 28. A; 29. E; 30. D; 31. a. mesoderm; b. ectoderm; c. endoderm; d. mesoderm; e. ectoderm; f. mesoderm; g. endoderm; h. mesoderm; i. mesoderm.

44.3. EARLY MARCHING ORDERS [pp.768–769]

44.4. HOW DO SPECIALIZED TISSUES AND ORGANS FORM? [pp.770–771]

1. F, nuclear; 2. F, by "maternal messages"; 3. T; 4. T; 5. T; 6. sperm entry; 7. gray crescent; 8. body axis; 9. cleavage; 10. blastula; 11. maternal messages; 12. cytoplasmic localization; 13. gastrulation; 14. primary; 15. gastrula; 16. yolk; 17. three; 18. cell differentiation; 19. genes; 20. activated; 21. neural tube; 22. cell differentiation; 23. morphogenesis; 24. Morphogenesis; 25. active cell migration; 26. Adhesive; 27. microtubules; 28. microfilaments; 29. apoptosis; 30. ectodermal sheets.

44.5. EXPERIMENTAL EVIDENCE OF CELL INTERACTIONS [pp.772–773]

44.6. *Focus on Science:* **TO KNOW A FLY** [pp.774–775]

44.7. AN UNDERLYING UNITY IN ANIMAL DEVELOPMENT [pp.776–777]

44.8. FROM THE EMBRYO ONWARD [p.777]

44.9. WHY DO ANIMALS AGE? [p.778]

44.10. *Commentary:* **DEATH IN THE OPEN** [p.779]

1. pattern formation; 2. embryonic induction; 3. embryonic induction; 4. Morphogens; 5. larva; 6. metamorphosis; 7. complete; 8. telomeres; 9. C; 10. D; 11. E; 12. B; 13. A; 14. F, internal; 15. T.

Self-Quiz

1. c; 2. b; 3. c; 4. a; 5. c; 6. e; 7. d; 8. b; 9. e; 10. a.

Chapter 45 Human Reproduction and Development

The Journey Begins [pp.782–783]

45.1. REPRODUCTIVE SYSTEM OF HUMAN MALES [pp.784–785]

45.2. MALE REPRODUCTIVE FUNCTION [pp.786–787]

1. mitotic (mitosis); 2. seminiferous; 3. meiosis (spermatogenesis); 4. sperm; 5. epididymis; 6. vas deferens; 7. urethra; 8. Seminal vesicles; 9. Prostate gland; 10. Bulbourethral; 11. Leydig; 12. Testosterone; 13. testosterone; 14. anterior; 15. hypothalamus; 16. decrease; 17. LH; 18. Sertoli; 19. increase; 20. hypothalamus; 21. anterior pituitary; 22. Sertoli cells; 23. Leydig cells.

45.3. REPRODUCTIVE SYSTEM OF HUMAN FEMALES [pp.788–789]

45.4. FEMALE REPRODUCTIVE FUNCTION [pp.790–791]

45.5. VISUAL SUMMARY OF THE MENSTRUAL CYCLE [p.792]

1. ovary; 2. oviduct; 3. uterus; 4. cervix; 5. myometrium; 6. endometrium; 7. vagina; 8. labia majora; 9. labia minora; 10. clitoris; 11. urethra; 12. Meiosis; 13. I; 14. 300,000; 15. follicle; 16. hypothalamus; 17. anterior pituitary; 18. estrogens; 19. ovulation; 20. LH; 21. menstruation; 22. endometrial; 23. corpus luteum; 24. progesterone; 25. blastocyst; 26. endometrium; 27. menstrual flow.

45.6. PREGNANCY HAPPENS [p.793]

45.7. FORMATION OF THE EARLY EMBRYO [pp.794–795]

45.8. EMERGENCE OF THE VERTEBRATE BODY PLAN [p.796]

45.9. WHY IS THE PLACENTA SO IMPORTANT? [p.797]

45.10. EMERGENCE OF DISTINCTLY HUMAN FEATURES [pp.798–799]

45.11. *Focus on Health:* **MOTHER AS PROTECTOR, PROVIDER, POTENTIAL THREAT** [pp.800–801]

1. oviduct; 2. implantation; 3. placenta; 4. eighth; 5. fetus; 6. embryonic disk; 7. amniotic cavity; 8. embryonic disk; 9. amniotic cavity; 10. yolk sac; 11. future brain of embryo; 12. amnion; 13. chorion; 14. yolk sac; 15. umbilical cord; 16. four; 17. pharyngeal arches; 18. somites; 19. six; 20. forelimb; 21. D; 22. B; 23. A; 24. C; 25. human chorionic gonadotropin (HCG); 26. corpus luteum; 27. gastrulation; 28. primitive streak; 29. neural tube; 30. Somites; 31. pharyngeal arches; 32. placenta; 33. umbilical cord; 34. nutrients; 35. wastes; 36. arms; 37. head; 38. muscles; 39. 95.

45.12. FROM BIRTH ONWARD [pp.802–803]

45.13. CONTROL OF HUMAN FERTILITY [pp.804–805]

45.14. *Focus on Health:* **SEXUALLY TRANSMITTED DISEASES** [pp.806–807]

45.15. *Focus on Bioethics:* **TO SEEK OR END PREGNANCY** [p.808]

1. 364,409; 2. 850,000; 3. more than 1,000,000; 4. abstinence; 5. Condoms; 6. diaphragm; 7. estrogens (progesterones); 8. progesterones (estrogens); 9. anterior pituitary; 10. tubal ligation; 11. A, F; 12. A, C; 13. D, E; 14. F; 15. F; 16. D; 17. C; 18. F; 19. C; 20. C, F; 21. C; 22. F; 23. D, F; 24. B; 25. A, B, C, D, E, F.

Self-Quiz

1. d; 2. e; 3. a; 4. b; 5. c; 6. e; 7. d; 8. a; 9. c; 10. a; 11. b; 12. d.

Chapter 46 Population Ecology

Tales of Nightmare Numbers [pp.812–813]

46.1. CHARACTERISTICS OF POPULATIONS [p.814]

46.2. *Focus on Science:* **ELUSIVE HEADS TO COUNT** [p.815]

46.3. POPULATION SIZE AND EXPONENTIAL GROWTH [pp.816–817]

1. K; 2. H; 3. D; 4. I; 5. B; 6. F; 7. A; 8. G; 9. J; 10. L; 11. C; 12. E; 13. immigration; 14. emigration; 15. migrations; 16. zero population; 17. per capita; 18. heads; 19. 0.4; 20. 0.1; 21. reproduction; 22. time; 23. 0.3; 24. $G = rN$; 25. J; 26. exponential; 27. doubling;

28. biotic potential; 29. a. It increases; b. It decreases; c. It must increase; 30. population growth rate; 31. It must decrease; 32. 100,000; 33. a. 100,000; b. 300,000.

46.4. LIMITS ON THE GROWTH OF POPULATIONS [pp.818–819]

46.5. LIFE HISTORY PATTERNS [pp.820–821]

46.6. *Focus on Science:* **NATURAL SELECTION AND THE GUPPIES OF TRINIDAD** [pp.822–823]

1. limiting; 2. Carrying capacity; 3. logistic; 4. carrying capacity; 5. increases; 6. decreases; 7. density-dependent; 8. density-independent; 9. dependent; 10. carrying capacity; 11. independent; 12. life history;

13. insurance; 14. cohort; 15. life; 16. "survivorship"; 17. Survivorship; 18. III; 19. I; 20. II; 21. I; 22. F; 23. A, E, H; 24. B, C, D; 25. G.

46.7. HUMAN POPULATION GROWTH [pp.824–825]
46.8. CONTROL THROUGH FAMILY PLANNING [pp.826–827]
46.9. POPULATION GROWTH AND ECONOMIC DEVELOPMENT [pp.828–829]
46.10. SOCIAL IMPACT OF NO GROWTH [p.829]
1. a. 1962–1963; b. 2025 or sooner; c. Depends on the age and optimism of the reader; 2. 6 billion; 3. T; 4. short; 5. T; 6. sidestepped; 7. cannot; 8. 9; 9. natural resources; 10. pollution; 11. Family planning; 12. birth; 13. two; 14. female; 15. fertility; 16. 3; 17. 6.5;

18. baby-boomers; 19. one-third; 20. thirties; 21. China; 22. stabilize; 23. reproductive; 24. b; 25. a; 26. b; 27. a; 28. b; 29. a; 30. B; 31. D; 32. A; 33. C; 34. industrial; 35. decreasing; 36. smaller; 37. transitional; 38. transitional; 39. economic; 40. immigration; 41. 16; 42. 4.7; 43. 21; 44. 25; 45. fifty; 46. 25; 47. 1; 48. 3; 49. 3; 50. 12.9; 51. 258; 52. growth; 53. social; 54. older; 55. economic; 56. postponed; 57. cultural; 58. carrying capacity.

Self-Quiz
1. d; 2. a; 3. b; 4. a; 5. d; 6. d; 7. a; 8. b; 9. a; 10. a; 11. c; 12. c; 13. b; 14. d; 15. b; 16. a; 17. a.

Chapter 47   Social Interactions

Deck the Nest With Sprigs of Green Stuff [pp.832–833]

47.1. BEHAVIOR'S HERITABLE BASIS [pp.834–835]
47.2. LEARNED BEHAVIOR [p.836]
1. C; 2. F; 3. E; 4. B; 5. G; 6. A; 7. D; 8. the banana slug; 9. ate; 10. inland; 11. were not; 12. genetic; 13. pineal gland; 14. increase; 15. estrogen; 16. testosterone; 17. Hormones; 18. a. Cuckoo birds are social parasites in that adult females lay eggs in the nests of other bird species; young cuckoos instinctively eliminate the natural-born offspring (eggs are maneuvered onto their backs and pushed out of the nest) and then receive the undivided attention of their unsuspecting foster parents; b. Young toads instinctively capture edible insects with sticky tongues; if a bumblebee is captured and then stings the tongue, the toad learns to leave bumblebees alone; 19. E; 20. C; 21. F; 22. A; 23. D; 24. B.

47.3. THE ADAPTIVE VALUE OF BEHAVIOR [p.837]
1. Natural selection; 2. increase; 3. evolution; 4. adaptive; 5. reproductive; 6. adaptive; 7. individual's; 8. a. Consideration of individual survival and production of offspring; b. Any behavior that promotes propagation of an individual's genes and tends to occur at increased frequency in future generations; c. Cooperative, interdependent relationships among individuals of the same species; d. Within a population, any behavior that increases an individual's chances to produce or protect offspring of its own, regardless of the consequences for the population; e. Within a population, a self-sacrificing behavior. The individual behaves in a way that helps others but decreases its own chances to produce offspring.

47.4. COMMUNICATION SIGNALS [pp.838–839]
1. i; 2. c; 3. e; 4. a; 5. g; 6. j; 7. f; 8. b; 9. h; 10. c; 11. g; 12. d.

47.5. MATES, PARENTS, AND INDIVIDUAL REPRODUCTIVE SUCCESS [pp.840–841]
1. a. Hangingflies; b. Sage grouse; c. Lions, sheep, elk, elephant seals, and bison; d. Caspian terns.

47.6. BENEFITS OF LIVING IN SOCIAL GROUPS [pp.842–843]
47.7. COSTS OF LIVING IN SOCIAL GROUPS [p.844]
47.8. EVOLUTION OF ALTRUISM [p.845]
47.9. *Focus on Science:* WHY SACRIFICE YOURSELF? [pp.846–847]
47.10. *Focus on Science:* ABOUT THOSE NAKED MOLE-RATS [p.848]
1. C; 2. A; 3. E; 4. D; 5. B; 6. a. Royal penguins, herring gulls, cliff swallows, and prairie dogs live in huge colonies and must compete for a share of the same food resources; b. Under crowded living conditions, the individual and its offspring are more likely to be weakened by pathogens and parasites that are more readily transmitted from host to host in crowded groups. Plagues spread like wildfire through densely crowded human populations; this is especially the case in settlements and cities with chronic infestations of rats and fleas, and with inadequate or nonexistent sewage treatment and medical care; c. Breeding pairs of herring gulls will quickly cannibalize their neighbors' eggs or young chicks; long-lived lions compete for permanent hunting territories even though they can live for an extended time between kills; a lion pride of three or more actually eat less well than one or a pair of lions; male lions intent on taking over a pride will show infanticidal behavior and will kill the cubs; group living is costly for lionesses in terms of food intake and reproductive success; they stick together to defend territories against smaller groups of rivals; aggressive males almost always kill the cubs of a single lioness, but occasionally a group of two or more lionesses can save some of the cubs; 7. subordinate; 8. altruistic; 9. reproductive; 10. dominant; 11. nonbreeding; 12. altruistic; 13. insect;

14. worker; 15. genes; 16. indirect selection; 17. parenting; 18. indirect; 19. reproduce; 20. altruistic; 21. genes; 22. vertebrates; 23. nonbreeding; 24. perpetuate; 25. genes (alleles).

47.11. AN EVOLUTIONARY VIEW OF HUMAN SOCIAL BEHAVIOR [p.849]

1. human; 2. a trait valuable in gene transmission; 3. redirecting adaptive behaviors; 4. strangers; 5. can; 6. will; 7. nonrelated; 8. is; 9. are; 10. nonrelative.

1. c; 2. d; 3. b; 4. d; 5. a; 6. b; 7. b; 8. c; 9. a; 10. c; 11. b; 12. b; 13. d; 14. c; 15. a.

Chapter 48 Community Interactions

No Pigeon Is an Island [pp.852–853]

48.1. WHICH FACTORS SHAPE COMMUNITY STRUCTURE? [p.854]
48.2. MUTUALISM [p.855]

1. habitat; 2. community; 3. niche; 4. fundamental; 5. realized; 6. neutral; 7. directly; 8. commensalistic; 9. mutualism; 10. interspecific; 11. Predation (Parasitism); 12. parasitism (predation); 13. symbiosis; 14. a. It cannot complete its life cycle in any other plant, and its larvae eat only yucca seeds; b. The yucca moth is the plant's only pollinator.

48.3. COMPETITIVE INTERACTIONS [pp.856–857]
48.4. PREDATION [pp.858–859]
48.5. *Focus on the Environment:* THE COEVOLUTIONARY ARMS RACE [pp.860–861]

1. Intraspecific; 2. Interspecific; 3. Interspecific; 4. competitive exclusion; 5. keystone; 6. pigeons; 7. size; 8. root; 9. Predators; 10. prey; 11. prey; 12. parasites; 13. hosts; 14. less; 15. time; 16. food; 17. declining; 18. increasing; 19. decline; 20. increase; 21. starving; 22. slows; 23. plants; 24. coevolution; 25. camouflage; 26. warning coloration; 27. Mimicry; 28. Moment-of-truth; 29. adaptations; 30. E; 31. C; 32. G; 33. A; 34. H; 35. B; 36. D; 37. G; 38. D; 39. F; 40. E; 41. B.

48.6. PARASITIC INTERACTIONS [pp.862–863]
48.7. FORCES CONTRIBUTING TO COMMUNITY STABILITY [pp.864–865]
48.8. COMMUNITY INSTABILITY [pp.866–867]
48.9. *Focus on the Environment:* EXOTIC AND ENDANGERED SPECIES [pp.868–869]

1. C; 2. F; 3. H; 4. B; 5. I; 6. G; 7. E; 8. A; 9. D; 10. succession; 11. Pioneer; 12. pioneers; 13. climax; 14. primary; 15. replacement; 16. secondary; 17. facilitate; 18. succession; 19. climax-pattern; 20. community; 21. pioneers; 22. fires; 23. natural; 24. active; 25. b; 26. a; 27. b; 28. b; 29. a; 30. a; 31. b; 32. a; 33. a; 34. keystone; 35. dominant; 36. mussels; 37. B; 38. D; 39. F; 40. E; 41. A; 42. C.

48.10. PATTERNS OF BIODIVERSITY [pp.870–871]

1. a. Resource availability tends to be higher and more reliable. Tropical latitudes have more sunlight of greater intensity, rainfall amount is higher, and the growing season is longer. Vegetation grows all year long to support diverse herbivores, etc.; b. Species diversity might be self-reinforcing. When a greater number of plant species compete and coexist, a greater number of herbivore species evolve because no herbivore can overcome the chemical defenses of all kinds of plants. Then more predators and parasites evolve in response to the diversity of prey and hosts; c. The rates of speciation in the tropics have exceeded those of background extinction. At higher latitudes, biodiversity has been suppressed during times of mass extinction; 2. tropics; 3. Iceland; 4. biodiversity; 5. Iceland; 6. dispersal; 7. distance; 8. area; 9. Larger; 10. diversity; 11. targets; 12. biodiversity; 13. small; 14. small; 15. immigration; 16. extinction; 17. immigration; 18. extinction; 19. Island C.

Self-Quiz
1. b; 2. b; 3. c; 4. a; 5. b; 6. d; 7. e; 8. b; 9. d; 10. a; 11. d; 12. d.

Chapter 49   Ecosystems

Crêpes for Breakfast, Pancake Ice for Dessert
[pp.874–875]

49.1. THE NATURE OF ECOSYSTEMS [pp.876–877]
49.2. THE NATURE OF FOOD WEBS [pp.878–879]
49.3. *Focus on Science:* **BIOLOGICAL MAGNIFICATION ON FOOD WEBS** [p.880]
49.4. STUDYING ENERGY FLOW THROUGH ECOSYSTEMS [p.881]
49.5. *Focus on Science:* **ENERGY FLOW AT SILVER SPRINGS** [p.882]

1. H; 2. C; 3. K; 4. F; 5. M; 6. O; 7. B; 8. I; 9. N;
10. A; 11. P; 12. G; 13. L; 14. D; 15. J; 16. E; 17. c;
18. b; 19. e; 20. a; 21. b; 22. d; 23. c; 24. e; 25. c;
26. a; 27. b; 28. d; 29. a; 30. c; 31. a; 32. energy;
33. ecosystem; 34. 100; 35. trophic; 36. lengthy;
37. patterns; 38. shorter; 39. stable; 40. herbivorous;
41. shortest; 42. grasslands; 43. carnivores; 44. grazing;
45. detrital; 46. cross-connect; 47. seasons; 48. detrital;
49. photosynthesizers; 50. consumers; 51. Ecosystem
modeling; 52. DDT; 53. fats; 54. biological magnification; 55. concentrated; 56. web; 57. consumer;
58. metabolic; 59. productivity; 60. Gross; 61. net;
62. net; 63. pyramid; 64. producers; 65. biomass;
66. biomass; 67. smallest; 68. energy; 69. Sunlight;
70. large; 71. 1; 72. 6; 73. 16; 74. low; 75. 4.

49.6. BIOGEOCHEMICAL CYCLES—AN OVERVIEW [p.883]
49.7. HYDROLOGIC CYCLE [pp.884–885]

1. Usually as mineral ions such as ammonium (NH_4^+);
2. Inputs from the physical environment and the cycling
activities of decomposers and detritivores; 3. The
amount of a nutrient being cycled through the ecosystem
is greater; 4. Common sources are rainfall or snowfall,
metabolism (such as nitrogen fixation), and weathering
of rocks; 5. Losses of mineral ions occurs by runoff;
6. a. Oxygen and hydrogen move in the form of water
molecules; b. A large portion of the nutrient is in the
form of atmospheric gas such as carbon and nitrogen
(mainly CO_2); c. Nutrients are not in gaseous forms;
nutrients move from land to the seafloor and only "return" to land through geological uplifting of long duration; phosphorus is an example; 7. F; 8. H; 9. D;
10. B; 11. A (C); 12. E; 13. G (H); 14. C; 15. watershed; 16. soil; 17. streams; 18. transpiration; 19. vegetation; 20. nutrients; 21. calcium; 22. calcium;
23. biomass; 24. deforestation; 25. cycle; 26. ecosystems; 27. regenerate; 28. Coniferous.

49.8. CARBON CYCLE [pp.886–887]
49.9. *Focus on the Environment:* **FROM GREENHOUSE GASES TO A WARMER PLANET?** [pp.888–889]
49.10. NITROGEN CYCLE [pp.890–891]
49.11. SEDIMENTARY CYCLES [p.892]

1. C; 2. D; 3. F; 4. B; 5. G; 6. A; 7. E; 8. D; 9. D;
10. E; 11. C; 12. A; 13. B; 14. Soil nitrogen compounds
are vulnerable to being leached and lost from the soil;
some fixed nitrogen is lost to air by denitrification; nitrogen fixation comes at high metabolic cost to plants that are
symbionts of nitrogen-fixing bacteria; losses of nitrogen
are enormous in agricultural regions through the tissues
of harvested plants, soil erosion, and leaching processes;
15. reservoir; 16. phosphates; 17. ocean; 18. shelves;
19. crustal; 20. geochemical; 21. ecosystem; 22. organisms; 23. ionized; 24. plants; 25. herbivores; 26. soil;
27. cycle; 28. hydrologic; 29. Rainfall; 30. nutrients;
31. phosphorus; 32. ecosystems; 33. sediments;
34. gaseous; 35. leaching; 36. phosphorus; 37. developing; 38. phosphorus; 39. biomass; 40. phosphorus;
41. nonexistent; 42. vegetation; 43. soils; 44. sediments;
45. runoff; 46. algal; 47. Decomposition; 48. oxygen;
49. Eutrophication; 50. accelerate.

Self-Quiz
1. c; 2. c; 3. d; 4. c; 5. a; 6. a; 7. b; 8. a; 9. a; 10. b;
11. b; 12. d.

Chapter 50   The Biosphere

Does a Cactus Grow in Brooklyn? [pp.896–897]

50.1. AIR CIRCULATION PATTERNS AND REGIONAL CLIMATES [pp.898–899]
50.2. OCEANS, LANDFORMS, AND REGIONAL CLIMATES [pp.900–901]

1. M; 2. D; 3. H; 4. K; 5. B; 6. J; 7. F; 8. C; 9. A;
10. L; 11. E; 12. I; 13. N; 14. G; 15. equatorial;
16. warm; 17. rises or ascends; 18. moisture; 19. descends; 20. moisture; 21. ascends; 22. moisture;
23. descends; 24. east (easterlies); 25. west (westerlies);
26. tropical; 27. warm; 28. cool; 29. cold; 30. solar
(sun's); 31. rotation.

50.3. REALMS OF BIODIVERSITY [pp.902–903]
50.4. SOILS OF MAJOR BIOMES [p.904]
50.5. DESERTS [p.905]
50.6. DRY SHRUBLANDS, DRY WOODLANDS, AND GRASSLANDS [pp.906–907]

1. H; 2. F; 3. A; 4. G; 5. C; 6. B; 7. J; 8. E; 9. D;
10. L; 11. I; 12. K; 13. C; 14. B; 15. D; 16. A; 17. E;
18. d; 19. a; 20. d; 21. b; 22. c; 23. a; 24. e; 25. d;
26. e; 27. b.

50.7. TROPICAL RAIN FORESTS AND OTHER BROADLEAF FORESTS [pp.908–909]
50.8. CONIFEROUS FORESTS [p.910]
50.9. ARCTIC AND ALPINE TUNDRA [p.911]
1. d; 2. c; 3. c; 4. a; 5. b; 6. b; 7. a; 8. d; 9. c, d;
10. b; 11. c; 12. a; 13. b; 14. b; 15. d; 16. d.

50.10. FRESHWATER PROVINCES [pp.912–913]
50.11. THE OCEAN PROVINCES [pp.914–915]
1. lake; 2. littoral; 3. limnetic; 4. plankton; 5. profundal; 6. overturns; 7. 4; 8. spring; 9. thermocline;
10. cools; 11. fall; 12. down; 13. up; 14. higher;
15. short; 16. primary; 17. Oligotrophic; 18. eutrophic;
19. eutrophication; 20. Streams; 21. runs; 22. benthic
(C); 23. pelagic (A); 24. neritic (B); 25. oceanic (D);
26. a; 27. c; 28. c; 29. b; 30. c; 31. a; 32. b; 33. a;
34. b; 35. c; 36. a; 37. b; 38. a; 39. c; 40. b; 41. c;
42. a; 43. b; 44. b; 45. c.

50.12. CORAL REEFS AND CORAL BANKS
[pp.916–917]

50.13. MANGROVE WETLANDS [p.917]
50.14. ESTUARIES AND THE INTERTIDAL ZONE
[pp.918–919]
50.15. *Focus on Science:* RITA IN THE TIME OF CHOLERA [pp.920–921]
1. a. Atolls are ring-shaped coral reefs that enclose or almost enclose a shallow lagoon; b. Fringing reefs form next to the land's edge in regions of limited rainfall, as on the leeward side of tropical islands; c. Barrier reefs form around islands or parallel with the shore of a continent. A calm lagoon forms behind them; 2. d; 3. a;
4. c; 5. e; 6. c; 7. a; 8. b; 9. d; 10. b; 11. a; 12. b;
13. e; 14. d; 15. a; 16. d; 17. e; 18. f; 19. c; 20. e;
21. c; 22. a; 23. f.

Self-Quiz
1. d; 2. d; 3. b; 4. d; 5. e; 6. b; 7. c; 8. c; 9. d; 10. c;
11. c; 12. a.

Chapter 51   Human Impact on the Biosphere

An Indifference of Mythic Proportions [pp.924–925]

51.1. AIR POLLUTION—PRIME EXAMPLES
[pp.926–927]
51.2. OZONE THINNING—GLOBAL LEGACY OF AIR POLLUTION [p.928]
1. Pollutants; 2. thermal inversion; 3. winters; 4. industrial smog; 5. photochemical smog; 6. oxygen; 7. nitrogen dioxide; 8. photochemical; 9. gasoline; 10. PANs;
11. sulfur (nitrogen); 12. nitrogen (sulfur); 13. sulfur;
14. nitrogen; 15. nitrogen; 16. acid; 17. sulfuric (nitric);
18. nitric (sulfuric); 19. acid rain; 20. 5; 21. Chlorofluorocarbons; 22. c; 23. d; 24. e; 25. b; 26. a; 27. b;
28. f; 29. f; 30. a; 31. b; 32. b; 33. d; 34. a (b); 35. d;
36. f; 37. c; 38. d; 39. e; 40. f; 41. e; 42. c; 43. d.

51.3. WHERE TO PUT SOLID WASTES? WHERE TO PRODUCE FOOD? [p.929]
51.4. DEFORESTATION—AN ASSAULT ON FINITE RESOURCES [pp.930–931]
51.5. *Focus on Bioethics:* YOU AND THE TROPICAL RAIN FOREST [p.932]
1. F; 2. H (G); 3. I; 4. E; 5. J; 6. B; 7. C; 8. A; 9. D;
10. G.

51.6. WHO TRADES GRASSLANDS FOR DESERTS?
[p.933]
51.7. A GLOBAL WATER CRISIS [pp.934–935]
1. desertification; 2. overgrazing on marginal lands;
3. domestic cattle; 4. native wild herbivores; 5. deserts;

6. salty; 7. desalinization; 8. energy; 9. agriculture;
10. salinization; 11. water table; 12. saline; 13. 20;
14. pollution; 15. wastewater; 16. tertiary; 17. 55–66;
18. oil; 19. policies; 20. a. Screens and settling tanks remove sludge, which is dried, burned, dumped in landfills, or treated further; chlorine is often used to kill pathogens in water, but does not kill them all; b. Microbial populations are used to break down organic matter after primary treatment but before chlorination; c. It removes nitrogen, phosphorus, and toxic substances, including heavy metals, pesticides, and industrial chemicals; it is largely experimental and expensive.

51.8. A QUESTION OF ENERGY INPUTS [pp.936–937]
51.9. ALTERNATIVE ENERGY SOURCES [p.938]
51.10. *Focus on Bioethics:* BIOLOGICAL PRINCIPLES AND THE HUMAN IMPERATIVE [p.939]
1. J-shaped; 2. increased numbers of energy users and to extravagant consumption and waste. 3. Net energy;
4. plants; 5. next; 6. decreases; 7. global acid deposition; 8. less; 9. meltdown; 10. have not; 11. solar–hydrogen energy; 12. wind farms; 13. Fusion power;
14. C (N); 15. B (N); 16. F (N); 17. A (R); 18. D (N);
19. E (R).

Self-Test
1. c; 2. b; 3. a; 4. b; 5. d; 6. d; 7. c; 8. c; 9. d; 10. b.